SCHAUM'S OUTLINE OF

THEORY AND PROBLEMS

of

THERMODYNAMICS

•

BY

MICHAEL M. ABBOTT, Ph.D.
HENDRICK C. VAN NESS, D. Eng.
Fluid, Chemical, and Thermal Processes Division
Rensselaer Polytechnic Institute

SCHAUM'S OUTLINE SERIES
McGRAW-HILL BOOK COMPANY
New York, St. Louis, San Francisco, Düsseldorf, Johannesburg, Kuala Lumpur, London, Mexico, Montreal, New Delhi, Panama, Rio de Janeiro, Singapore, Sydney, and Toronto

07-000040-9

2 3 4 5 6 7 8 9 0 SH SH 7 9 8 7 6 5 4 3

Preface

This outline presents the fundamental principles of classical thermodynamics, and illustrates by numerous worked examples and problems many of their applications in science and engineering. As a supplement, or as a primary textbook, it should prove useful at the undergraduate and first-year graduate level.

Chapters 1 through 5 form the core of the outline, and are appropriate for students in all areas of science and technology. The first and second chapters deal with basic principles and present the two fundamental laws of thermodynamics. Chapter 3 develops the mathematical framework of the subject, and is included at this point largely for subsequent reference. Thus it need not be given detailed study in sequence. Chapters 4 and 5 treat the behavior of PVT systems.

The remaining chapters are somewhat more specialized, 6 and 8 being devoted to the treatment of flow processes, of particular interest to engineers. Chapter 7 constitutes an introduction to chemical thermodynamics, and should be especially useful to chemists, chemical engineers, biologists, and materials scientists and engineers.

We acknowledge with thanks the contribution of certain examples and problems by Professors Joseph W. Bursik and Howard E. Cyphers of Rensselaer Polytechnic Institute, and wish to express particular appreciation for the discerning editorial assistance of Mr. David Beckwith, whose efforts materially enhanced the quality of this outline.

MICHAEL M. ABBOTT
HENDRICK C. VAN NESS

Rensselaer Polytechnic Institute
April 1972

CONTENTS

CONTENTS

Chapter 1

Fundamental Concepts and First Principles

1.1 BASIC CONCEPTS

Energy.

Thermodynamics is concerned with energy and its transformations. The laws of thermodynamics are general restrictions which nature imposes on all such transformations. These laws cannot be derived from anything more basic; they are primitive. Moreover, the expression of these laws requires the use of words that are themselves primitive in that they have no precise definitions and no synonyms. *Energy* is such a word, and it was used in the very first sentence of this paragraph. Energy is a mathematical abstraction that has no existence apart from its functional relationship to other variables or coordinates that do have a physical interpretation and which can be measured. For example, the kinetic energy of a given mass of material is a function of its velocity, and it has no other reality.

The first law of thermodynamics is merely a formal statement asserting that energy is conserved. Thus it represents a primitive statement about a primitive concept. Moreover, energy and the first law are coupled: The first law depends on the concept of energy, but it is equally true that energy is an *essential* thermodynamic function precisely because it allows formulation of the first law.

System and surroundings.

Any application of the first law to a discrete portion of the universe requires the definition of a *system* and its *surroundings.* A system can be any object, any quantity of matter, any region of space, etc., selected for study and set apart (mentally) from everything else, which then becomes the surroundings. The systems of interest in thermodynamics are finite, and the point of view taken is *macroscopic* rather than *microscopic.* That is to say, no account is taken of the detailed structure of matter, and only the coarse characteristics of the system, such as its temperature and pressure, are regarded as thermodynamic coordinates. These are advantageously dealt with because they have a direct relation to our sense perceptions and are measurable.

The imaginary envelope which encloses a system and separates it from its surroundings is called the *boundary* of the system, and it may be imagined to have special properties which serve either (a) to *isolate* the system from its surroundings or (b) to provide for *interaction* in specific ways between system and surroundings. An *isolated* system can exchange neither matter nor energy with its surroundings. If a system is not isolated, its boundaries may permit either matter or energy or both to be exchanged with its surroundings. If the exchange of matter is allowed, the system is said to be *open*; if only energy and not matter may be exchanged, the system is *closed* (but not isolated), and its mass is constant. The exchange of energy can occur by two modes, *heat* and *work.*

Potential energy and kinetic energy are considered in both mechanics and thermodynamics. These forms of energy result from the position and motion of a system as a whole and are regarded as the *external* energy of the system. The special province of thermodynamics is the energy *interior* to matter, the energy associated with the internal state of a system, and called *internal energy.* When a sufficient number of thermodynamic coordinates, such as temperature and pressure, are specified, the internal state of a system is determined and its internal energy is fixed.

State.

When a system is isolated, it is not affected by its surroundings. Nevertheless, changes may occur within the system that can be detected with measuring devices such as thermometers and pressure gages. However, such changes are observed to cease after a period of time, and the system is said to have reached a condition of *internal equilibrium* such that it has no further tendency to change.

For a closed system which may exchange energy with its surroundings, a final static condition may also eventually be reached such that the system is not only internally at equilibrium but also in *external equilibrium* with its surroundings.

An *equilibrium state* represents a particularly simple condition of a system, and is subject to precise mathematical description because in such a state the system exhibits a set of identifiable, reproducible properties. Indeed, the word *state* represents the totality of macroscopic properties associated with a system. Certain properties are readily detected with instruments, such as thermometers and pressure gages. The existence of other properties, such as internal energy, is recognized only indirectly. The number of properties which may be arbitrarily set at given values in order to fix the state of a system (that is, to fix *all* properties of the system) depends on the nature of the system. This number is generally small, and is the number of properties which may be selected as independent variables for a particular system. These properties then represent one set of thermodynamic coordinates for the system.

To the extent that a system exhibits a set of identifiable properties it has a thermodynamic state, whether or not the system is at equilibrium. Moreover, the laws of thermodynamics have general validity, and their application is not limited to equilibrium states.

The importance of equilibrium states in thermodynamics derives from the fact that a system at equilibrium exhibits a set of *fixed* properties which are independent of time and which may therefore be measured with precision. Moreover, such states are readily reproduced from time to time and from place to place.

Process.

When a closed system is displaced from equilibrium, it undergoes a *process,* during which its properties change until a new equilibrium state is attained. During such a process the system may be caused to interact with its surroundings so as to exchange heat and work in a way that produces in the system or surroundings changes considered desirable for one reason or another. Where only the total heat and total work of such a process are of interest, one need consider only the properties of the equilibrium end states.

Dimensions and units.

The fundamental and primitive concepts which underlie all physical measurements and all properties are time θ, distance l, mass m, absolute temperature T, and current i. For these primary *dimensions* there must be set up arbitrary scales of measure, each divided into specific *units* of size. The internationally accepted basic units for the five quantities are the second (s), the meter (m), the kilogram (kg), the kelvin (K), and the ampere (A). Each of these has a precise definition according to international agreement.

They form the basis for the SI (Système International) or International System of Units.

The Kelvin scale of absolute temperature is related to the more common Celsius scale, which takes 0(°C) as the freezing point of water and 100(°C) as the normal boiling point:

$$T(\mathrm{K}) = t(°\mathrm{C}) + 273.15$$

[For many calculations the approximate relation $T(\mathrm{K}) = t(°\mathrm{C}) + 273$ is entirely adequate.] When $T = 0(\mathrm{K})$, $t = -273.15(°\mathrm{C})$; thus the absolute zero of temperature is 273.15(°C) below the freezing point of water. In thermodynamics it is often necessary to employ absolute temperatures.

The mass of a system is often given by stating the number of *moles* it contains. A mole is the mass of a chemical species equal numerically to its molecular weight. Thus a kilogram mole of oxygen (O_2) contains 32 kilograms. In addition, the number of molecules in a kilogram mole is the same for all substances. This is also true for a gram mole, and in this case the number of molecules is Avogadro's number, equal to 6.0225×10^{23} molecules.

Molar properties are widely used. Thus the molar volume is the volume occupied by a mole of material, and its molar density is the reciprocal of the molar volume.

There are a number of secondary quantities having derived units which are important in thermodynamics. Examples are force, pressure, and density. Force is determined through Newton's second law of motion, $F = ma$, and has the basic unit $(\mathrm{kg})(\mathrm{m})/(\mathrm{s})^2$. The SI unit for this composite set is the newton (N). Pressure is defined as force per unit area, $(\mathrm{N})/(\mathrm{m})^2$, and density is mass per unit volume, $(\mathrm{kg})/(\mathrm{m})^3$.

The English engineering system of units also recognizes the second as the basic unit of time, and the ampere as the unit of current. However, absolute temperature is measured in rankines, where $T(\mathrm{R}) = 1.8 \times T(\mathrm{K})$. Fahrenheit temperatures are given by:

$$t(°\mathrm{F}) = T(\mathrm{R}) - 459.67$$

[For many applications the approximation $t(°\mathrm{F}) = T(\mathrm{R}) - 460$ is acceptable.] The foot (ft) is the usual unit of length and the pound mass ($\mathrm{lb_m}$) is the unit of mass. The molar unit is the pound mole.

The unit of force, the pound force ($\mathrm{lb_f}$), is defined without reference to Newton's second law, and as a result this law must be written so as to include a dimensional proportionality constant:

$$F = \frac{1}{g_c} ma$$

where $1/g_c$ is the constant. In the English engineering system

$$g_c = 32.1740 \frac{(\mathrm{lb_m})(\mathrm{ft})}{(\mathrm{lb_f})(\mathrm{s})^2}$$

The unit of density is $(\mathrm{lb_m})/(\mathrm{ft})^3$, and the unit of pressure is $(\mathrm{lb_f})/(\mathrm{ft})^2$ or $(\mathrm{lb_f})/(\mathrm{in})^2$, often written (psi). Since pressure gages sometimes measure pressure relative to atmospheric pressure, the term *absolute pressure* is often used to distinguish thermodynamic pressure from *gage pressure*. Thus we have the abbreviations (psia) and (psig).

Relationships between units in common use are given in Appendix 1.

In the problems of this book we have in some cases used engineering units, and in others, SI units. When one works with SI units, the proportionality constant $1/g_c$ in Newton's law is unity, and

$$g_c = 1 \frac{(\mathrm{kg})(\mathrm{m})}{(\mathrm{N})(\mathrm{s})^2}$$

It will be our practice to carry the constant g_c along in all equations which derive from

Newton's law. This allows use of any self-consistent set of units, and requires only that the proper value of g_c be employed for the system used.

Example 1.1. The acceleration of gravity on the surface of Mars is $3.74(\mathrm{m})/(\mathrm{s})^2$ or $12.27(\mathrm{ft})/(\mathrm{s})^2$. The *mass* of a man as determined on earth is $168(\mathrm{lb_m})$. His *weight* on Mars is the *force* exerted on him by Martian gravity, and is determined by

$$F = \frac{ma}{g_c} = \frac{mg_{\mathrm{Mars}}}{g_c} = \frac{168(\mathrm{lb_m}) \times 12.27(\mathrm{ft})/(\mathrm{s})^2}{32.174(\mathrm{lb_m})(\mathrm{ft})/(\mathrm{lb_f})(\mathrm{s})^2} = 64.1(\mathrm{lb_f}) \quad \text{or} \quad 285(\mathrm{N})$$

Alternatively, the man's mass is 76.2(kg), and

$$F = \frac{mg_{\mathrm{Mars}}}{g_c} = \frac{76.2(\mathrm{kg}) \times 3.74(\mathrm{m})/(\mathrm{s})^2}{1(\mathrm{kg})(\mathrm{m})/(\mathrm{N})(\mathrm{s})^2} = 285(\mathrm{N})$$

Note that the man's mass on Mars is still $168(\mathrm{lb_m})$ or 76.2(kg).

PVT systems.

The simplest thermodynamic system consists of a fixed mass of an isotropic fluid uninfluenced by chemical reactions or external fields. Such systems are described in terms of the three measurable coordinates pressure P, volume V, and temperature T. In fact, they may be characterized as PVT systems. However, experiment shows that these three coordinates are not all independent, that fixing any two of them determines the third. Thus there must be an *equation of state* that interrelates these three coordinates for equilibrium states. This equation may be expressed in functional form as

$$f(P, V, T) = 0$$

When a specific equation is known to describe the PVT system considered, we may always solve for one of the coordinates in terms of the others. For example, $V = V(P, T)$.

The simplest example of an equation of state is that for an ideal gas:

$$PV = RT$$

where V is molar volume, R is the *universal gas constant*, and T is absolute temperature. Since R must have the units of PV/T, the units of R may clearly be (pressure) × (molar volume)/(temperature). For example, a common engineering value is

$$R = 10.73(\mathrm{psi})(\mathrm{ft})^3/(\mathrm{lb\ mole})(\mathrm{R})$$

Additional values for R are listed in Appendix 2.

1.2 MECHANICAL WORK

Work in thermodynamics always represents an exchange of energy between a system and its surroundings. Mechanical work occurs when a force acting on the system moves through a distance. As in mechanics this work is defined by the integral

$$W = \int F\,dl$$

where F is the component of the force acting in the direction of the displacement dl. In differential form this equation is written

$$\delta W = F\,dl \tag{1.1}$$

where δW represents a differential quantity of work.

It is not necessary that the force F actually cause the displacement dl, but it must be an external force. We adopt the usual sign convention that the value of δW is *negative* when work is done *on* the system and *positive* when work is done *by* the system.

In thermodynamics one often finds work done by a force distributed over an area, i.e. by a pressure P acting through a volume V, as in the case of a fluid pressure exerted on a piston. In this event, (*1.1*) is more conveniently expressed as

$$\delta W = P\,dV \tag{1.2}$$

where P is an external pressure exerted on the system.

The unit of work, and hence the unit of energy, comes from the product of force and distance or of pressure and volume. The SI unit of work and energy is therefore the newton-meter, which is called a joule (J). This is the one and only internationally recognized unit of energy of any kind. However, the calorie is still in common use. Power is the time rate of doing work, and the SI unit is the watt (W), which is defined as work done at the rate of 1(J)/(s).

In the English engineering system the unit of work and energy is often the foot-pound force. Another common unit is the Btu. The kilowatt and the horsepower are commonly used units of power. Conversion factors are listed in Appendix 1.

Example 1.2. A gas is confined initially to a volume V_i in a horizontal cylinder by a frictionless piston held in place by latches. When the piston is released, it is forced outward by the internal pressure of the gas acting on the interior piston face. A constant external pressure P acts on the external piston face and resists the motion of the piston. We will calculate the work of the system, taken as the piston, cylinder, and the gas, if the gas expands to a final volume V_f where the piston is in an equilibrium position with the gas pressure equal to the external pressure P.

From (*1.2*) we have for a constant external pressure P

$$W = P\,\Delta V = P(V_f - V_i)$$

If $P = 20\text{(psi)}$ and $V_f - V_i = 0.5\text{(ft)}^3$

$$W = 20(\text{lb}_f)/(\text{in})^2 \times 144(\text{in})^2/(\text{ft})^2 \times 0.5(\text{ft})^3 = 1440(\text{ft-lb}_f)$$

or

$$W = \frac{1440(\text{ft-lb}_f)}{778(\text{ft-lb}_f)/(\text{Btu})} = 1.85(\text{Btu})$$

or

$$W = 1.85(\text{Btu}) \times 1055(\text{J})/(\text{Btu}) = 1950(\text{J})$$

Note that we have made no use of the pressure of the gas in the cylinder. Even if we had been given the initial gas pressure, we could not have used it, because we must deal with forces external to the system. Had we taken the gas alone as our system, we would have needed the pressure exerted by the interior face of the piston on the gas and we would have needed to know this pressure as a function of V.

1.3 OTHER MODES OF THERMODYNAMIC WORK

In the preceding section we considered the work for systems described by the coordinates pressure, volume, and temperature, called PVT systems. Different kinds of systems, described by other coordinates, are also important, and they are subject to work done by forces other than pressure. Thus we have electrical work, work of magnetization, work of changing surface area, etc. The total work done on a system can be expressed as

$$\delta W = \sum_i Y_i\,dX_i$$

where Y_i represents a generalized force and dX_i represents a generalized displacement. The summation allows superposition of different work modes.

A major difficulty in application of thermodynamics to new types of systems is in proper identification of the forces and displacements. In general this can be determined only by experiment. Thermodynamics provides no *a priori* recipe for identification of the Y_i and X_i. However, as illustrated in the following example, *mechanical* work can be handled with little difficulty by appealing to (*1.1*).

Example 1.3. Consider the deformation of a bar of length l subjected to an axial tensile load F. As the force F is applied, the bar will elongate, and by (*1.1*) we have

$$\delta W = -F\,dl$$

The minus sign is required to satisfy our sign convention that work done *on* a system be negative. (Both F and dl are considered positive. For a compressive load both F and dl would be negative, and the minus sign would still be required.) Note that the volume of the bar may also change and that work may be done by the hydrostatic pressure exerted by the surrounding atmosphere. However, both the volume change and the pressure are small, and the associated work is taken to be negligible. Nevertheless, it could be included if desired.

It is common practice to express the force acting on a bar in terms of the stress, or force per unit area acting within the deformed material:

$$\text{stress} = \sigma = \frac{F}{A}$$

where A is the cross-sectional area of the bar. Also, elongations are referred to a characteristic length. Thus the *natural strain* ϵ is related to l by

$$d\epsilon = \frac{dl}{l}$$

The work may now be written in terms of stress and strain as

$$\delta W = -\sigma A l\,d\epsilon$$

Since Al is the volume of the bar

$$\delta W = -V\sigma\,d\epsilon$$

For a finite process

$$W = -\int_{\epsilon_1}^{\epsilon_2} V\sigma\,d\epsilon$$

and a relationship between σ and ϵ (such as Hooke's law) is required to evaluate the integral. In many applications V is nearly constant, and may be taken outside the integral.

1.4 HEAT

Heat, like work, is regarded in thermodynamics as energy in transit across the boundary separating a system from its surroundings. However, quite unlike work, heat transfer results from a temperature difference between system and surroundings, and simple contact is the only requirement for heat to be transferred by conduction. Heat is not regarded as being stored in a system. When energy in the form of heat is added to a system, it is stored as kinetic and potential energy of the microscopic particles that make up the system. The units of heat are those of work and energy.

The sign convention used for a quantity of heat Q is opposite to that used for work. Heat *added* to a system is given by a *positive* number, whereas heat *extracted* from a system is given by a *negative* number.

1.5 REVERSIBILITY

Because it is the system and not the surroundings that is of primary interest in any application of thermodynamics, it is essential that the equations of thermodynamics be expressed in terms of the properties of the system. The difficulty of doing this is suggested by the discussion of mechanical work in Sec. 1.2. Thus the expression for work as given by (*1.2*), $\delta W = P\,dV$, in general requires that P be the pressure external to the system. This was further illustrated by Example 1.2.

There is, however, a special kind of process for which it *is* always possible to base calculations on the properties of the system. This kind of process is termed *reversible,* and it plays a central role in the formulation of thermodynamics. A process is said to be reversible *if its direction can be reversed at any stage by an infinitesimal change in external conditions.* The implications of this definition can most easily be explained through specific examples.

Discussion of an adiabatic process.

Figure 1-1 represents a piston-and-cylinder apparatus which may be operated so as to bring about the compression or expansion of the gas trapped in the cylinder. We imagine the piston and cylinder to be perfect heat insulators so that no heat can be exchanged between the piston-cylinder apparatus (the system) and its surroundings. Any process carried out in the absence of such heat transfer is said to be *adiabatic,* and the label "adiabatic enclosure" on Fig. 1-1 indicates that no heat transfer can occur. Thus we consider here processes driven by mechanical forces only.

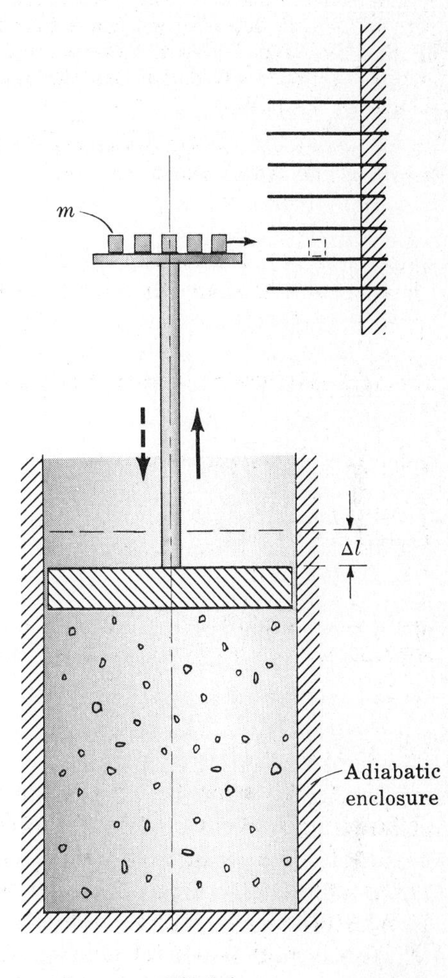

Fig. 1-1

The piston is shown initially held in place by a set of small weights, each of mass m, which may be removed as indicated. For a real apparatus the removal of a single small weight may or may not cause the piston to rise. It depends on the static friction between the piston and cylinder. The piston may well stick. If so, we presume that the removal of a number of the small weights will cause the piston to break free and to rise through a distance Δl, carrying with it the remaining weights and thus accomplishing useful work on the surroundings by virtue of the elevation of weights against the gravitational force. During this process the piston initially accelerates and eventually decelerates to reach a final position. In between, it may oscillate up and down while this motion is gradually damped out by friction between piston and cylinder and by "internal friction" resulting from the viscous nature of the gas itself. These are *dissipative effects,* and they prevent the piston from reaching an ultimate height as great as it would in their absence, thereby reducing the work obtained from the process. Clearly, such a process is *irreversible*; an *infinitesimal* change in external conditions is hardly adequate to reverse the direction of a piston moving with finite velocity. All real processes of any practical interest proceed at finite rates and are accompanied by dissipative effects, and all are therefore irreversible.

We can, however, *imagine* processes that are free of dissipative effects. For the process shown in Fig. 1-1, we must first imagine a frictionless piston. If there is no friction, then the initial state of the system is one of equilibrium, with the piston held in exact balance under the downward force of the weights and the upward pressure exerted by the gas. In this circumstance the internal pressure is equal to the force of gravity on the piston and all that it supports, divided by its cross-sectional area. Thus there is no distinction between an internal and an external pressure. The removal of a single weight from a system so perfectly in balance will certainly cause the piston to rise, but it will still accelerate to a finite velocity, and it will surely oscillate, only gradually settling down to a new equilibrium position as the result of the damping effect of the viscous gas. This process too is irreversible, because the piston attains finite velocities and the dissipative effects are by no means avoided. The only way we can imagine these effects to be eliminated is to suppose that the piston moves by infinitesimal steps. That is, we imagine the weights to be exceedingly small (grains of fine sand represent an approximation), so that when a mass dm is removed from the piston, it rises through a distance dl. The removal of so minute a mass from the piston perturbs the system but infinitesimally, the piston accelerates hardly at all, and there is only the slightest oscillation of the piston as it finds a new equilibrium position only a hair above its initial level. The continued removal of infinitesimal weights, one after the other, causes the gas to expand very slowly, but if we are prepared to wait long enough, we witness a process that produces a finite change in the state of the system and results in the gradual collection of infinitesimal weights at higher and higher elevations.

This imaginary process is unique, and is reversible, because at any stage its direction *can* be reversed if we take an infinitesimal weight from anywhere in the surroundings and place it on the piston. (This causes an infinitesimal change in external conditions.) The process can then be made to proceed in reverse through all the stages of the initial process merely by replacing the infinitesimal weights on the piston in exactly the reverse order of their earlier removal. In the end the initial state of the system is restored, as is also the initial state of the surroundings, except for the location of the infinitesimal weight used to initiate the reverse process.

It is this infinitesimal weight used to reverse the process that prevents an exact restoration of the surroundings. Recall that the *first* weight removed from the piston at the *start* of the forward expansion process is not raised at all. It remains at its initial level, and in that location is useless during the reverse compression process that brings the piston back to its initial level. Moreover, when the *last* weight is removed at the *end* of the forward expansion process, the piston rises a final increment dl above the level of this weight. Thus no infinitesimal weight taken from the piston is available at the maximum piston height to start the reverse compression process, and therefore a weight must be brought to this level. Any infinitesimal weight in the surroundings that is not otherwise needed will serve, and one obvious choice is the very first weight removed from the piston at the start of the forward process, since it cannot otherwise be used. The elevation of this weight requires an infinitesimal change in the surroundings, and the same conclusion is reached if an extra weight from the surroundings is considered.

Thus the result of the reverse process is the recompression of the gas in the cylinder to its initial state at the expense of just the work done during its initial expansion plus an infinitesimal amount of additional work required for the reversal of the process. It should be noted that the reverse compression process is just as reversible as was the initial expansion process. Thus we could equally well have considered the compression process first and then considered its reversal by an expansion process. In this case we would recover an amount of work in the expansion process just infinitesimally *less* than the work done on the system during compression. In either case if the initial process had been irreversible, the work of the process would differ by a finite amount from the work of a reversible restoration process.

Mechanical reversibility.

An important feature of the reversible process is that the system is never more than infinitesimally displaced from internal equilibrium. This implies that the system is always in an identifiable state of uniform temperature and pressure and that an equation of state is always applicable to the system. In addition the system is also never more than infinitesimally displaced from mechanical equilibrium with its surroundings, and this means that the internal pressure is always in virtual balance with the forces external to the system. Because of these circumstances, for a reversible process the expression of work (*1.2*), $\delta W = P\,dV$, may be evaluated through use of the *system* pressure for P. This is so, first, because the system pressure is well defined and is uniform, and, second, because the system pressure almost exactly balances the external forces. These two conditions are necessary but not sufficient for a process to be reversible. They are, however, both necessary and sufficient for the use of P as the system pressure for the calculation of work by (*1.2*). Processes that occur in such a way that these two conditions are fulfilled are said to be *mechanically reversible.*

Because the processes just considered were taken to be adiabatic, we have had no occasion to consider the role of heat transfer. The adiabatic enclosure of the system imposes a restraint on the system that makes irrelevant any question of external thermal equilibrium between the system and its surroundings. However, uniformity of temperature within the system implies internal thermal equilibrium.

Discussion of an isothermal process.

Figure 1-2 illustrates an apparatus for carrying out expansion and compression processes isothermally rather than adiabatically. It shows a piston-cylinder assembly which operates exactly as before, except that it is placed in contact with a *heat reservoir* at temperature T. A heat reservoir is a body capable of absorbing or giving off unlimited quantities of heat without any change of temperature. The atmosphere and the oceans approximate heat reservoirs, usually used as heat sink reservoirs. A continuously operating furnace and a nuclear reactor are equivalent to heat source reservoirs.

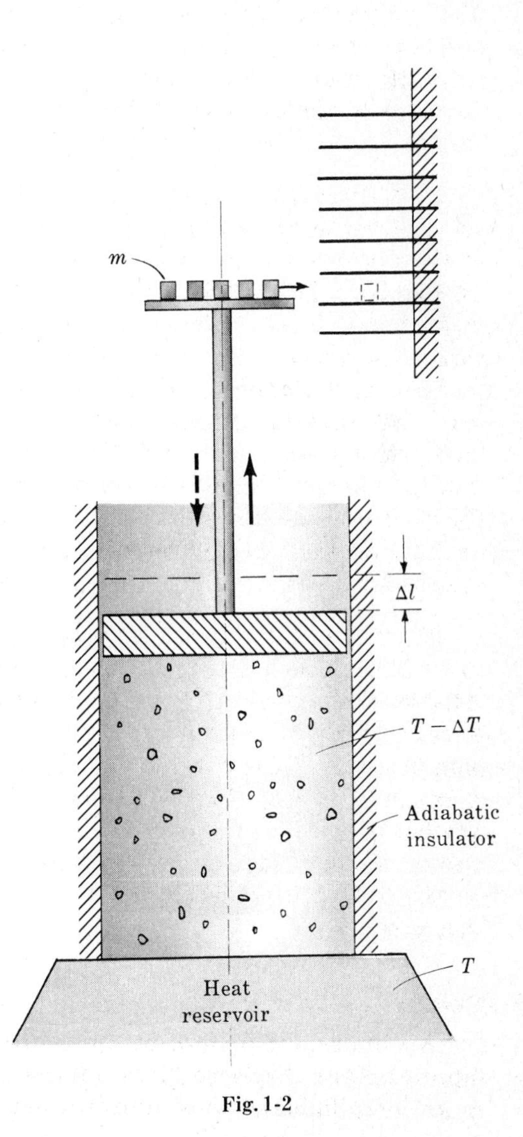

Fig. 1-2

The additional consideration which now enters with respect to Fig. 1-2 is that heat may flow from the heat reservoir (part of the surroundings) to the gas (part of the system) or the reverse. When a system comes to equilibrium with its surroundings by virtue of long contact, it is observed that both have the same temperature. This condition is called external thermal equilibrium. When their temperatures are different, we observe changes, and attribute the changes to a flow of heat from the warmer to the cooler region. Thus we regard a temperature difference as the driving force for heat transfer. When there is no temperature difference, there is no driving force, and no heat transfer. Irreversible processes occur as the result of finite driving forces; reversible processes as the result of infinitesimal driving forces. Thus heat transfer is irreversible when the temperature difference is finite, and becomes reversible only when the temperature difference is infinitesimal.

With respect to Fig. 1-2, the removal of a finite weight m from a frictionless piston causes finite changes in the system, and these include a temperature drop in the gas from its initial equilibrium value T to $T - \Delta T$. Thus heat flows at a finite rate from the heat reservoir to the gas. The process is irreversible in all respects, including the heat transfer. If the weight removed is infinitesimal, then the temperature drop is infinitesimal, and heat is transferred reversibly.

Summary.

Irreversibilities always lower the efficiencies of processes. Their effect in this respect is identical with that of friction, which is one cause of irreversibility. Conversely, no process more efficient than a reversible process can even be imagined. The reversible process is an abstraction, an idealization, which is never achieved in practice. It is, however, of enormous utility because it allows calculation of work from knowledge of the system properties alone. In addition it represents a standard of perfection that cannot be exceeded because:

(*a*) It places an upper limit on the work that may be *obtained* for a given work-producing process.

(*b*) It places a lower limit on the work *input* for a given work-requiring process.

Example 1.4. We wish to calculate the work done when an ideal gas expands isothermally and reversibly in a piston-and-cylinder assembly.

Since the process is mechanically reversible, P in (*1.2*) is the gas pressure and it may be eliminated by the ideal-gas law, $P = RT/V$, where V is the molar volume. Thus $\delta W = RT\,dV/V$, and integration from the initial to the final state at constant T gives the work per mole of gas:

$$W = RT \ln (V_f/V_i)$$

Since $P_i V_i = P_f V_f$, this may also be written

$$W = RT \ln (P_i/P_f)$$

For expansion of a gas at 300(K) from an initial pressure of 10(atm) to a final pressure of 1(atm), we have

$$W = 8314\text{(J)/(kg mole)(K)} \times 300\text{(K)} \times \ln 10 = 5{,}740{,}000\text{(J)/(kg mole)}$$

1.6 FIRST LAW OF THERMODYNAMICS

For a closed (constant-mass) system the first law of thermodynamics is expressed mathematically by

$$\Delta E = Q - W \tag{1.3}$$

where ΔE is the *total* energy change of the system, Q is heat added to the system, and W is work done by the system. The first law of thermodynamics merely gives quantitative expression to the principle of energy conservation. In words, it says that the total energy change of a closed system is equal to the heat transferred *to* the system minus the work done *by* the system.

The total energy change ΔE can be split up into several terms, each representing the change in energy of a particular form:

$$\Delta E = \Delta E_K + \Delta E_P + \Delta U$$

where ΔE_K is the change in kinetic energy, ΔE_P is the change in gravitational potential energy, and ΔU is the change in internal energy. By definition, the kinetic energy is

$$E_K = \frac{mu^2}{2g_c}$$

and the gravitational potential energy is

$$E_P = \frac{mzg}{g_c}$$

where u is the velocity, z is the elevation above a datum level, and g is the local acceleration of gravity. These energy functions are common to both mechanics and thermodynamics.

The internal energy function U, however, is peculiar to thermodynamics. It represents the kinetic and potential energies of the molecules, atoms, and subatomic particles that constitute the system on a microscopic scale. There is no known way to determine absolute values of U. Fortunately, only changes ΔU are needed and these can be found from experiment. It is also found by experiment that U is fixed whenever the state of the system is fixed. If ΔE is expanded in the first-law expression, we get the equation

$$\Delta E_K + \Delta E_P + \Delta U = Q - W$$

In the frequent case where the sum of the kinetic and potential energies of the system does not change, this equation becomes

$$\boxed{\Delta U = Q - W} \tag{1.4}$$

or in differential form

$$dU = \delta Q - \delta W \tag{1.5}$$

and all energy exchange with the surroundings serves to change just the internal energy. If in addition the process is *adiabatic*, then $Q = 0$ and (*1.4*) becomes

$$\Delta U = -W \quad \text{(adiabatic)}$$

This last equation shows that for a system changed adiabatically from one equilibrium state to another the work should be independent of *path*, for ΔU should depend only on the end states. Experiment shows it to be so, and this is the primary evidence that U is indeed a state function.

With respect to (*1.5*) note that the differential signs on Q and W are written δ, whereas we write dU. There is a fundamental difference between a property like U on the one hand and the quantities Q and W on the other. A property like U always has a value, dependent only on the state of the system. A process which changes the state changes U. Thus dU represents an infinitesimal *change* in U, and integration gives a difference between two values of the property:

$$\int_{U_1}^{U_2} dU = U_2 - U_1 = \Delta U$$

On the other hand Q and W are not properties of the system and depend on the *path* of the process. Thus δ is used to denote an infinitesimal *quantity*. Integration gives not a difference between two values but a finite quantity:

$$\int \delta Q = Q \quad \text{and} \quad \int \delta W = W$$

Thus integration of (*1.5*) yields (*1.4*).

Example 1.5.

(*a*) What velocity must be attained by a mass of $1(\text{lb}_\text{m})$ in order that it have a kinetic energy of 1(Btu)?

By definition of kinetic energy, $E_K = mu^2/2g_c$. Since the required energy is 1(Btu) or 778(ft-lb$_f$), we have

$$778(\text{ft-lb}_f) = \frac{1(\text{lb}_m) \times u^2(\text{ft})^2/(\text{s})^2}{2 \times 32.174(\text{lb}_m)(\text{ft})/(\text{lb}_f)(\text{s})^2}$$

and

$$u = \sqrt{778 \times 2 \times 32.174} = 223.75(\text{ft})/(\text{s})$$

(b) To what elevation must a mass of 1(kg) be raised in order that it have a potential energy of 1(J)?

By the definition of gravitational potential energy, $E_P = mzg/g_c$. For an energy of 1(J) or 1(N-m), we have

$$1(\text{N-m}) = \frac{1(\text{kg}) \times z(\text{m}) \times 9.81(\text{m})/(\text{s})^2}{1(\text{kg})(\text{m})/(\text{N})(\text{s})^2} \qquad \text{from which} \qquad z = 9.81(\text{m})$$

Example 1.6. Steel wool is contained in a cylinder in an atmosphere of pure oxygen. The cylinder is fitted with a frictionless piston which maintains the oxygen pressure constant at 1(atm). The iron in the steel wool reacts very slowly with the oxygen to form Fe_2O_3. Heat is removed from the apparatus during the process so as to keep the temperature constant at 25(°C). For the reaction of 2 gram moles of iron,

$$2Fe + \tfrac{3}{2}O_2 \longrightarrow Fe_2O_3$$

198,500(cal) of heat is removed. We wish to calculate Q and W for the process and ΔU of the system.

Both Fe and Fe_2O_3 are solids which occupy a negligible volume compared with the gaseous oxygen. We may therefore take the total volume V^t of the system to be that of the oxygen, which we assume is an ideal gas. Thus $V^t = nRT/P$, where n is the number of gram moles of gaseous oxygen present. During the process T and P are constant, and only n changes; so the total volume change is $\Delta V^t = RT\,\Delta n/P$.

Since pressure is constant, the work of the process is $W = P\,\Delta V^t = RT\,\Delta n$. According to the reaction equation $\frac{3}{2}$(g mole) of oxygen is consumed, and $\Delta n = -\frac{3}{2}$(g mole). Thus

$$W = -1.987(\text{cal})/(\text{g mole})(\text{K}) \times 298(\text{K}) \times \tfrac{3}{2}(\text{g mole}) = -890(\text{cal}) \quad \text{or} \quad -3720(\text{J})$$

The minus sign shows that work is done on the system by the piston.

We are given that $Q = -198{,}500$(cal) or $-830{,}500$(J). Thus

$$\Delta U = Q - W = -198{,}500 + 890(\text{cal}) = -197{,}610(\text{cal}) \quad \text{or} \quad -826{,}800(\text{J})$$

This decrease in internal energy reflects the changes in bond energies caused by the chemical reaction.

1.7 ENTHALPY

Special thermodynamic functions are defined as a matter of convenience. The simplest such function is the enthalpy H, explicitly defined for any system by the mathematical expression

$$\boxed{H = U + PV} \tag{1.6}$$

Since the internal energy U and the PV product both have units of energy, H also has energy units. Moreover, since U, P, and V are all properties of a system, H must be a property too.

Whenever a differential change occurs in a system its properties change. From (1.6) the change in H is related to other property changes by

$$dH = dU + d(PV) \tag{1.7}$$

Example 1.7. Show that the heat added to a closed PVT system undergoing a mechanically reversible, constant-pressure process is equal to ΔH.

By (1.7) for constant pressure

$$dH = dU + P\,dV$$

But for a mechanically reversible process, $\delta W = P\,dV$; and by (1.5), $dU = \delta Q - \delta W$. Therefore

$$dH = \delta Q$$

For a finite process

$$\Delta H = Q$$

Example 1.8. Liquid carbon dioxide at $-40(^\circ F)$ has a vapor pressure of 145.8(psia) and a specific volume of $0.0144(\text{ft})^3/(\text{lb}_m)$. At these conditions it is a *saturated liquid*, i.e. a liquid at its boiling point. We arbitrarily assign a value of 0.00 for its enthalpy, and this becomes the basis for all enthalpy and internal energy values.

The internal energy of the saturated liquid is calculated from (*1.6*):

$$U = H - PV = -PV = -145.8(\text{lb}_f)/(\text{in})^2 \times 144(\text{in})^2/(\text{ft})^2 \times 0.0144(\text{ft})^3/(\text{lb}_m)$$

$$= -302.33(\text{ft-lb}_f)/(\text{lb}_m)$$

or

$$U = \frac{-302.33(\text{ft-lb}_f)/(\text{lb}_m)}{778(\text{ft-lb}_f)/(\text{Btu})} = -0.389(\text{Btu})/(\text{lb}_m)$$

In SI units

$$U = -904.8(\text{J})/(\text{kg})$$

The latent heat of vaporization of carbon dioxide at $-40(^\circ F)$ and 145.8(psia) is $137.8(\text{Btu})/(\text{lb}_m)$. The specific volume of the *saturated vapor* produced by vaporization at these conditions is $0.6113(\text{ft})^3/(\text{lb}_m)$. We wish to determine both H and U for the vapor.

From Example 1.7 we have that $\Delta H = Q$ for a constant-pressure, reversible process in a closed system. Since the latent heat is measured by such a process

$$\Delta H = H_{\text{vap}} - H_{\text{liq}} = 137.8(\text{Btu})/(\text{lb}_m)$$

However

$$H_{\text{liq}} = 0 \quad \text{so that} \quad H_{\text{vap}} = 137.8(\text{Btu})/(\text{lb}_m)$$

The internal energy is given by

$$U_{\text{vap}} = H_{\text{vap}} - PV_{\text{vap}} = 137.8 - \frac{145.8 \times 144 \times 0.6113}{778} = 121.3(\text{Btu})/(\text{lb}_m)$$

In SI units

$$H_{\text{vap}} = 320{,}500(\text{J})/(\text{kg}) \qquad U_{\text{vap}} = 282{,}100(\text{J})/(\text{kg})$$

1.8 NOTATION

We have not yet adopted a policy that makes explicit the amount of material in a system to which our symbols refer. For example, the equation $H = U + PV$ can be written for any amount of material. The properties H, U, and V are *extensive*; that is, they are directly proportional to the mass of the system considered. Temperature T and pressure P are *intensive*, independent of the extent of the system. With respect to extensive properties, we will hereafter let the plain capital symbols such as U, H, and V refer to a unit amount of material, either to a unit mass or to a mole. With a unit mass as the basis the properties are often called *specific* properties, e.g. specific volume. With a mole as the basis they are called molar properties, e.g. molar enthalpy. To represent the total properties of a system we will usually multiply the specific or molar properties by the mass or number of moles of the system, e.g. mV, nH, etc.

1.9 HEAT CAPACITY

The amount of heat which must be added to a closed PVT system in order to accomplish a given change of state depends on how the process is carried out. Only for a reversible process where the path is fully specified is it possible to relate the heat to a property of the system. On this basis we define *heat capacity* in general by

$$C_X = \left(\frac{\delta Q}{dT}\right)_X$$

where X indicates that the process is reversible and the path is fully specified. We could define a number of heat capacities according to this prescription; however, for PVT systems only two are in common use. These are C_V, heat capacity at constant volume, and C_P, heat capacity at constant pressure. In both cases the system is presumed closed and to be of constant composition.

By definition

$$C_V = \left(\frac{\delta Q}{dT}\right)_V$$

and it represents the amount of heat required to increase the temperature by dT when the system is held at constant volume. That C_V is a property of the system follows from (*1.5*), which for a constant-volume, reversible process becomes $dU = \delta Q$, because no work can be done under the imposed restrictions. Thus

$$\boxed{C_V = \left(\frac{\partial U}{\partial T}\right)_V} \tag{1.8}$$

is an alternative definition of C_V and since U, T, and V are all properties of the system, C_V must also be a property.

Similarly,

$$C_P = \left(\frac{\delta Q}{dT}\right)_P$$

and it represents the amount of heat required to increase the temperature by dT when the system is heated in a reversible process at constant pressure. In Example 1.7 we showed that $\delta Q = dH$ for a reversible, constant-pressure process. Thus

$$\boxed{C_P = \left(\frac{\partial H}{\partial T}\right)_P} \tag{1.9}$$

is an alternative definition of C_P and shows C_P to be a property of the system.

We may also write (*1.8*) and (*1.9*) as

$$dU = C_V\,dT \qquad \text{(const. } V\text{)} \tag{1.10}$$

$$dH = C_P\,dT \qquad \text{(const. } P\text{)} \tag{1.11}$$

These equations relate properties only and do not depend on the process causing changes in the system. Thus for these equations no restriction with respect to reversibility is necessary. However, the change must be between equilibrium states.

Heat capacity of an ideal gas.

It is clear from (*1.10*) that in general the internal energy of a closed PVT system may be considered a function of T and V,

$$U = U(T, V)$$

and we can write

$$dU = \left(\frac{\partial U}{\partial T}\right)_V dT + \left(\frac{\partial U}{\partial V}\right)_T dV$$

In view of (*1.8*) this becomes

$$dU = C_V\,dT + \left(\frac{\partial U}{\partial V}\right)_T dV \tag{1.12}$$

Application of this equation requires values for $(\partial U/\partial V)_T$, and these must in general be determined by experiment. There is, however, one special and important case, that when $(\partial U/\partial V)_T = 0$, and this is part of the definition of an ideal gas. Thus the complete definition of an ideal gas requires that at all temperatures and pressures:

$$PV = RT \qquad \left(\frac{\partial U}{\partial V}\right)_T = 0$$

The ideal gas is, of course, an idealization, and no real gas exactly satisfies these equations over any finite range of temperature and pressure. However, all real gases approach ideal behavior at low pressures, and in the limit as $P \to 0$ do in fact meet the above requirements. Thus the equations for an ideal gas provide good approximations to real gas behavior at low pressures, and because of their simplicity are very useful.

For an ideal gas (*1.12*) becomes

$$dU = C_V\, dT \qquad \text{(ideal gas)} \tag{1.13}$$

an equation which is always valid for an ideal gas regardless of what kind of process is considered. Implicit in this statement is the fact that for an ideal gas both U and C_V are functions of temperature only, independent of V and P.

Example 1.9. We wish to show that for an ideal gas H is a function of temperature only and to find a relationship between C_P and C_V for an ideal gas.

By definition $H = U + PV$. But $PV = RT$, so that $H = U + RT$. Since U is a function of T only, this must also be the case for H.

Differentiating (*1.6*) with respect to T,

$$\frac{dH}{dT} = \frac{dU}{dT} + \frac{d(PV)}{dT}$$

where we have used total derivatives because H, U, and PV are functions of T only. This being the case we can also write

$$\left(\frac{\partial H}{\partial T}\right)_P = \frac{dH}{dT} = C_P \qquad \text{and} \qquad \left(\frac{\partial U}{\partial T}\right)_V = \frac{dU}{dT} = C_V$$

Also $d(PV)/dT = R$. The required relationship is then

$$C_P = C_V + R \tag{1.14}$$

This clearly implies that for an ideal gas C_P is also a function of T only.

From the equation $H = U + RT$ of the preceding example we get by differentiation $dH = dU + R\,dT$. But $dU = C_V\,dT$. Thus $dH = (C_V + R)\,dT$ and by (*1.14*) this becomes

$$dH = C_P\, dT \qquad \text{(ideal gas)} \tag{1.15}$$

an equation which is always valid for an ideal gas regardless of what kind of process is considered.

The ratio of heat capacities is often denoted by

$$\gamma = \frac{C_P}{C_V}$$

and is a useful quantity in calculations for ideal gases.

It is often convenient to imagine real gases to exist in a hypothetical *ideal-gas state.* Real gases in the zero-pressure limit obey the ideal-gas equations. If a real gas at a pressure approaching zero is imagined to remain an ideal gas as it is compressed to a finite pressure then the resulting state is known as an ideal-gas state. Real gases at zero pressure have properties which reflect their individuality, and this continues to be true for gases in the ideal-gas state. Thus we often use ideal-gas heat capacities. These are different for different gases and are functions of temperature only. The temperature dependence is often expressed as

$$C_P' = a + bT + cT^2$$

where a, b, and c are constants and the prime on C_P indicates an ideal-gas value. The effect of temperature is very small for monatomic gases such as helium and argon, and for these gases the molar heat capacities are very nearly given by

$$C'_V = 3\text{(cal)/(g mole)(K)} = 3\text{(Btu)/(lb mole)(R)}$$

$$C'_P = 5\text{(cal)/(g mole)(K)} = 5\text{(Btu)/(lb mole)(R)}$$

and

$$\gamma = 1.67$$

For diatomic gases such as O_2, N_2, and H_2 the heat capacities change rather slowly with temperature, and near room temperature have the approximate values

$$C'_V = 5\text{(cal)/(g mole)(K)} = 5\text{(Btu)/(lb mole)(R)}$$

$$C'_P = 7\text{(cal)/(g mole)(K)} = 7\text{(Btu)/(lb mole)(R)}$$

$$\gamma = 1.40$$

For polyatomic gases such as CO_2, NH_3, CH_4, etc., the heat capacities vary appreciably with temperature and differ from gas to gas. So no values can be given that are generally valid approximations. Values of γ usually are less than 1.3.

Example 1.10. We wish to show for an ideal gas with constant heat capacities undergoing a reversible, adiabatic compression or expansion that

$$\frac{T_2}{T_1} = \left(\frac{V_1}{V_2}\right)^{\gamma-1}$$

Since the process is adiabatic (*1.5*) becomes

$$dU = -\delta W = -P\,dV$$

But $dU = C_V\,dT$ and $P = RT/V$. Thus

$$C_V\,dT = -RT\frac{dV}{V} \quad \text{or} \quad \frac{dT}{T} = -\frac{R}{C_V}\frac{dV}{V}$$

By (*1.14*)

$$\frac{R}{C_V} = \frac{C_P - C_V}{C_V} = \gamma - 1$$

Therefore

$$\frac{dT}{T} = -(\gamma-1)\frac{dV}{V}$$

Integrating from (T_1, V_1) to (T_2, V_2) with γ constant,

$$\ln\frac{T_2}{T_1} = (\gamma-1)\ln\frac{V_1}{V_2} \quad \text{or} \quad \frac{T_2}{T_1} = \left(\frac{V_1}{V_2}\right)^{\gamma-1}$$

If an ideal gas for which $C_V = 3$ and $C_P = 5$(cal)/(g mole)(K) expands reversibly and adiabatically from an initial state $T_1 = 450$(K) and $V_1 = 3$(liter) to a final volume $V_2 = 5$(liter), the final temperature T_2 is given by

$$T_2 = T_1\left(\frac{V_1}{V_2}\right)^{\gamma-1} = 450\text{(K)} \times \left(\frac{3}{5}\right)^{0.67} = 320\text{(K)}$$

The work done during the process is

$$W = -\Delta U = -C_V\Delta T = -3\text{(cal)/(g mole)(K)} \times (320 - 450)\text{(K)}$$

$$= 390\text{(cal)/(g mole)} \quad \text{or} \quad 1630\text{(J)/(g mole)}$$

The positive value of W indicates correctly that in the expansion process work is done by the system.

The enthalpy change of the gas can also be calculated. By (*1.15*)

$$\Delta H = C_P\Delta T = 5\text{(cal)/(g mole)(K)} \times (-130)\text{(K)}$$

$$= -650\text{(cal)/(g mole)} \quad \text{or} \quad -2720\text{(J)/(g mole)}$$

Solved Problems

BASIC CONCEPTS (Sec. 1.1)

1.1. Find (*a*) the Kelvin temperature, (*b*) the Rankine temperature, (*c*) the Fahrenheit temperature, for the normal boiling point of water, 100(°C).

(*a*) Since $T(\text{K}) = t(°\text{C}) + 273.15$,

$$T(\text{K}) = 100 + 273.15 \quad \text{or} \quad T = 373.15(\text{K})$$

(*b*) By definition $T(\text{R}) = 1.8 \times T(\text{K})$. Thus

$$T(\text{R}) = 1.8 \times 373.15 \quad \text{or} \quad T = 671.67(\text{R})$$

(*c*) Since $T(°\text{F}) = T(\text{R}) - 459.67$,

$$T(°\text{F}) = 671.67 - 459.67 \quad \text{or} \quad T = 212(°\text{F})$$

1.2. (*a*) Does the rankine represent a larger or smaller temperature unit or interval than the kelvin? (*b*) What is the relationship between the temperature units, kelvin and Celsius degree? (*c*) What is the relationship between the rankine and Fahrenheit degree? (*d*) Is the Fahrenheit degree a larger or smaller unit than the Celsius degree?

(*a*) A temperature in rankines is always larger than the temperature in kelvins by a factor of 1.8. Thus the rankine must be a smaller unit of temperature or a smaller temperature interval than the kelvin, and is in fact smaller by a factor of 1.8.

(*b*) The Celsius temperature scale is obtained from the Kelvin scale merely by a shift of the zero point. Thus the Celsius degree and the kelvin represent temperature units or intervals of exactly the same size.

(*c*) The Fahrenheit temperature scale and the Rankine scale again differ only with respect to the zero point of the scale, and the Fahrenheit degree and the rankine are units of exactly the same size.

(*d*) The Fahrenheit degree is smaller than the Celsius degree by a factor of 1.8. They are related exactly as the rankine and the kelvin.

All these relationships are easily seen in Fig. 1-3.

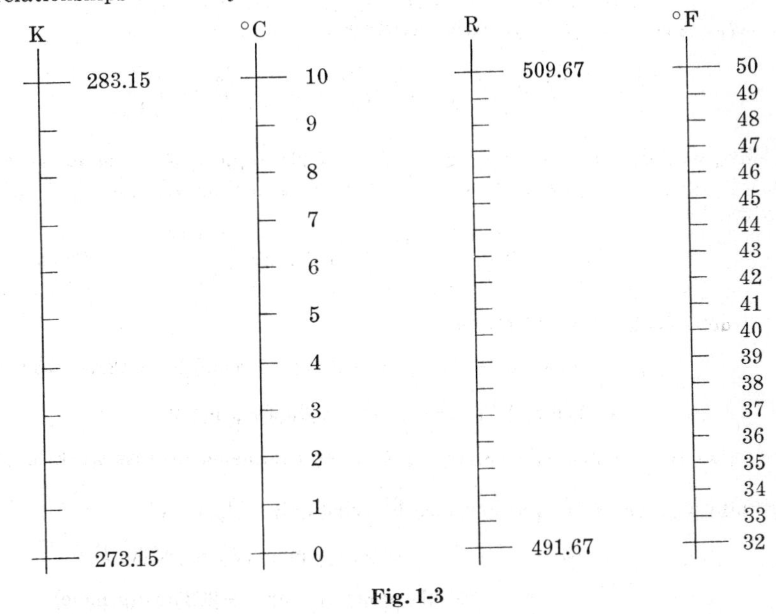

Fig. 1-3

1.3. Is the pound force, originally defined as the force necessary to accelerate a mass of $1(\text{lb}_\text{m})$ by $32.1740(\text{ft})/(\text{s})^2$, independent of the acceleration of gravity; for example, is it the same on the moon as on the earth?

The pound force is a fixed unit of force, no different in kind than the newton. Once defined, it is independent of any acceleration connected with its original operational definition. The present definition of the pound force establishes it in relation to the newton as *exactly* 4.4482216152605(N).

1.4. A deceleration of 25 g's, where $1\,\text{g} = 32.174(\text{ft})/(\text{s})^2$ or $9.8066(\text{m})/(\text{s})^2$, is sufficient to prove fatal to a human in most automobile accidents. What force acts on a man of $160(\text{lb}_\text{m})$, or 72.57(kg), when subjected to this deceleration?

We have

$$F = \frac{ma}{g_c} = -\frac{160(\text{lb}_\text{m}) \times 25 \times 32.174(\text{ft})/(\text{s})^2}{32.174(\text{lb}_\text{m})(\text{ft})/(\text{lb}_\text{f})(\text{s})^2} = -4000(\text{lb}_\text{f})$$

or

$$F = -4000(\text{lb}_\text{f}) \times 4.448(\text{N})/(\text{lb}_\text{f}) = -17{,}792(\text{N})$$

Alternatively

$$F = ma = -72.57(\text{kg}) \times 25 \times 9.8066(\text{m})/(\text{s})^2 = -17{,}792(\text{kg})(\text{m})/(\text{s})^2$$

However $1(\text{kg})(\text{m})/(\text{s})^2$ is defined as the newton. Thus $F = -17{,}792(\text{N})$. The minus sign originates with the negative acceleration or deceleration and ultimately indicates that the force applied to the man is in a direction opposite to his displacement.

WORK (Secs. 1.2 through 1.5)

1.5. Show that (*1.2*) is consistent with the basic defining equation for work, (*1.1*).

The work done by a fluid pressure P acting over a piston of area A and moving through a volume change dV is given by (*1.2*) as

$$\delta W = P\,dV$$

The physical situation is depicted in Fig. 1-4. A pressure P is exerted over the cross-hatched area A of the piston, which moves a distance dl into the cylinder. The cylinder volume is $V = Al$, and since A is constant

$$dV = A\,dl$$

Moreover, the pressure P is by definition the total force F on the piston face divided by the area A. Thus

$$P = \frac{F}{A}$$

Substitution into (*1.2*) gives

$$\delta W = P\,dV = \frac{F}{A} A\,dl = F\,dl$$

which is (*1.1*).

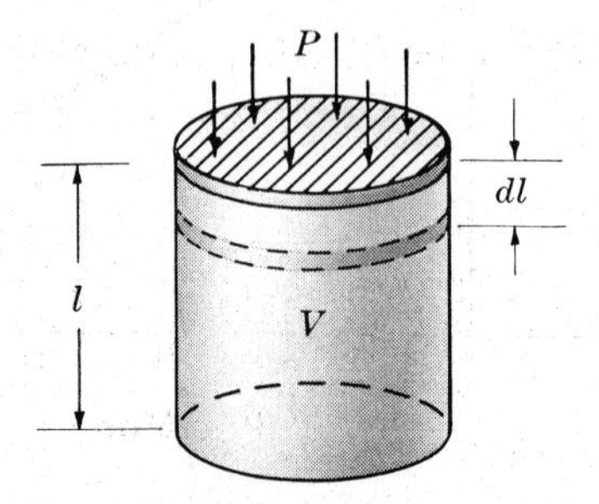

Fig. 1-4

1.6. A cylinder fitted with a sliding piston contains a volume of gas V which exerts a pressure P on the piston. The gas expands slowly, pushing the piston outward. Data taken during the process provide the following information on the relationship between P and V:

P, bars	V, cubic meters
15 (initial)	0.0300 (initial)
12	0.0361
9	0.0459
6	0.0644
4	0.0903
2 (final)	0.1608 (final)

Calculate the work done by the gas on the piston.

The work is given by (*1.2*) in integral form

$$W = \int_{V_i}^{V_f} P\,dV$$

The data given provide the relationship of P to V that is required for evaluation of the integral. This is conveniently accomplished graphically by means of a plot of P versus V (Fig. 1-5). The area below the curve between the initial and final values of V represents the work, and this is determined by some technique of graphical integration. The answer obtained will vary a bit with the technique used and with the care employed. The value determined here is $W = 0.641(\text{bar-m}^3)$, or

$$W = 0.641(\text{bar-m}^3) \times 10^5(\text{N})/(\text{m})^2(\text{bar}) = 0.641 \times 10^5(\text{N-m})$$

Since a newton-meter is a joule, $W = 64{,}100(\text{J})$.

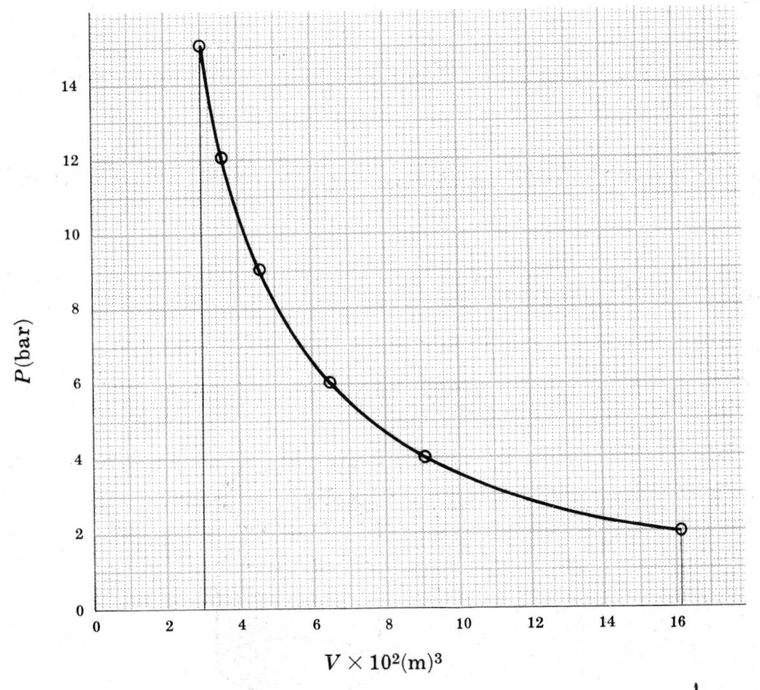

Fig. 1-5

1.7. The numerical *P-V* data of Problem 1.6 can be very closely fit by the equation $PV^{1.2} = 0.2232$. Rework Problem 1.6, making use of this expression for the evaluation of the work integral.

The *P-V* relation given is of the form $PV^\delta = k$. Substitution for *P* in the work integral gives

$$W = k\int_{V_i}^{V_f} \frac{dV}{V^\delta}$$

which upon integration becomes

$$W = \frac{k}{1-\delta}(V_f^{1-\delta} - V_i^{1-\delta}) = \frac{kV_f^{1-\delta} - kV_i^{1-\delta}}{1-\delta}$$

But

$$k = PV^\delta = P_iV_i^\delta = P_fV_f^\delta$$

and appropriate substitution for *k* provides

$$W = \frac{P_fV_f^\delta V_f^{1-\delta} - P_iV_i^\delta V_i^{1-\delta}}{1-\delta} = \frac{P_fV_f - P_iV_i}{1-\delta}$$

Substitution of numerical values gives

$$W = \frac{(2)(0.1608) - (15)(0.0300)}{1-1.2} = 0.6418(\text{bar-m}^3) \quad \text{or} \quad 64{,}180(\text{J})$$

Note that the expression $PV^\delta = k$ is purely empirical. Its use implies no assumption as to ideality of the gas in the cylinder. The use of an expression of this form to represent the *P-V* relation for a real gas in compression and expansion processes is fairly common.

1.8. Show that the reversible work of changing the area of a liquid surface is given by $\delta W = -\gamma\, dA$, where γ is surface tension and *A* is surface area.

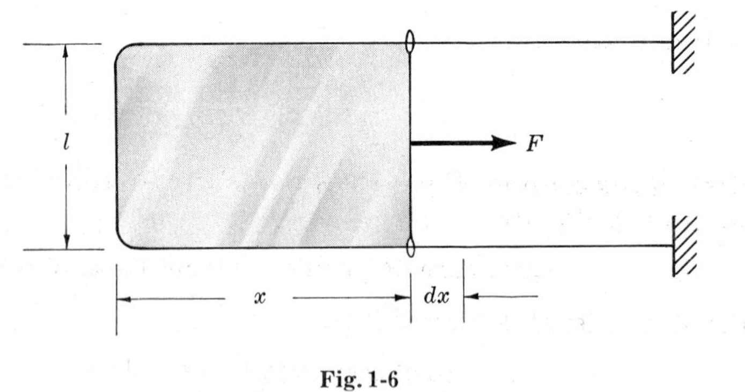

Fig. 1-6

Consider a film of liquid (such as a soap solution) held on a wire framework as represented in Fig. 1-6. The film is made up of two surfaces and a thin layer of liquid in between. The force *F* required to hold the movable wire in position is found by experiment to be proportional to *l*, *but independent of x*. This means that *F* is not influenced by the thickness of the film and must therefore result from stresses in the surfaces of the film. Thus we have *surface tension* defined as the force per unit length acting perpendicular to any line in the surface or at its boundary. When the force *F* is moved to the right, the film is extended, and liquid moves from the bulk region between the surfaces to the surfaces. Thus new surface is formed, not by stretching the original surface but by generation from the bulk of the film. Since there are two surfaces associated with the liquid film shown in the figure, the relation between the force *F* and the surface tension γ at equilibrium is

$$\gamma = \frac{F}{2l}$$

The work of moving F through the distance dx is

$$\delta W = -F\,dx$$

where the minus sign is inserted so as to conform with the sign convention. Since $F = 2l\gamma$, then for a reversible process

$$\delta W = -2l\gamma\,dx$$

However, $dA = 2l\,dx$ and therefore

$$\delta W = -\gamma\,dA$$

This is the work required to form the new surface dA. For pure materials γ is a function of temperature only, and the surface is considered a thermodynamic system for which the coordinates are γ, A, and T. Such a system is necessarily an *open* system, because it is not possible to have a surface without also having a bulk liquid phase.

1.9. What is the minimum work required to form the surface associated with a water mist having a particle radius of 5(micron)? Assume that 1(liter) of water at 50(°C) is to be turned into mist also at 50(°C). The surface tension of water at 50(°C) is 63(dyne)/(cm).

The surface-to-volume ratio of a spherical drop is

$$\frac{A}{V} = \frac{4\pi r^2}{\frac{4}{3}\pi r^3} = \frac{3}{r}$$

where r is the radius of the sphere. For a volume of 1(liter) or 1000(cm)^3 and a radius of 5(micron) or 5×10^{-4}(cm)

$$A = \frac{3V}{r} = \frac{3 \times 1000(\text{cm})^3}{5 \times 10^{-4}(\text{cm})} = 6 \times 10^6(\text{cm})^2$$

By Problem 1.8

$$W = -\int_0^A \gamma\,dA = -\gamma A = -63(\text{dyne})/(\text{cm}) \times 6 \times 10^6(\text{cm})^2 = -378 \times 10^6(\text{dyne-cm})$$

or since 1(J) is 10^7(dyne-cm)

$$W = -37.8(\text{J})$$

1.10. The universal gas constant R has units consistent with the ideal-gas law, $R = PV/T$. These units evidently are

(pressure) × (molar volume)/(temperature)

Show that the units of R may also be

(molar energy)/(temperature)

The product PV always has units of work or energy. By definition, pressure P is force per unit area, and has the units

(force)/(length)^2

The molar volume V has the units

(length)^3/(mole)

Thus for the PV product we have units of

[(force)/(length)^2] × [(length)^3/(mole)]

or

(force)(length)/(mole)

Since work or energy is the product of force and length or distance, the units of R may be

(molar energy)/(temperature)

The basic unit of pressure in the SI system is the (N)/(m)^2, and the basic unit of volume is the (m)^3. Thus the basic unit for $R = PV/T$ in the SI system is

$$[(\mathrm{N})/(\mathrm{m})^2] \times [(\mathrm{m})^3/(\mathrm{kg\ mole})]/(\mathrm{K})$$

This reduces to

$$[(\mathrm{N\text{-}m})/(\mathrm{kg\ mole})]/(\mathrm{K})$$

However, the newton-meter is defined as the joule. Thus the basic value for R in SI units is (see Appendix 2)

$$R = 8314(\mathrm{J})/(\mathrm{kg\ mole})(\mathrm{K})$$

With the gram mole as a basis we have

$$R = 8.314(\mathrm{J})/(\mathrm{g\ mole})(\mathrm{K})$$

FIRST LAW OF THERMODYNAMICS (Secs. 1.6 through 1.9)

1.11. By applying Newton's second law of motion to the special case of the translation of a rigid body, illustrate the origin of the kinetic and potential energy terms in the energy equation for a purely mechanical system.

Newton's second law for this system is

$$F = \frac{1}{g_c} ma = \frac{1}{g_c} m \frac{du}{dt} \tag{1}$$

where u is the velocity of the body and F is the total external force acting on the body parallel to its displacement dl. The total work W_t done by the body against the force is then

$$W_t = -\int F\, dl \tag{2}$$

Substituting (1) into (2), we obtain

$$W_t = -\int \frac{m}{g_c}\frac{du}{dt}\, dl = -\int \frac{m}{g_c}\frac{dl}{dt}\, du = -\int \frac{m}{g_c} u\, du = -\int \frac{m}{2g_c} d(u^2)$$

or

$$W_t = -\Delta\left(\frac{mu^2}{2g_c}\right) \tag{3}$$

The kinetic energy E_K is *defined* as

$$E_K = \frac{mu^2}{2g_c}$$

so (3) can be written

$$W_t = -\Delta E_K \tag{4}$$

Equation (4) is a perfectly general expression for the total mechanical work done by a rigid body in translation, and the equation is not based on any assumptions as to the nature of the force F. It proves convenient, however, to consider F the sum of two types of forces, *body* forces F_B and *surface* forces F_S:

$$F = F_B + F_S \tag{5}$$

Body forces are so called because they act throughout the *volume* of a system; surface forces act on an *area* of the bounding surface of a system. From (2) and (5), then, the total work can be considered the sum of two work terms:

$$W_t = W_B + W_S \tag{6}$$

where

$$W_B = -\int F_B\, dl \tag{7}$$

$$W_S = -\int F_S\, dl \tag{8}$$

Body forces are *conservative* forces. This means they can be derived from some function $\Phi(l)$, which depends only on the location of a system, by differentiation with respect to the position coordinate. Thus for the case at hand

$$F_B = -\frac{d\Phi(l)}{dl} \tag{9}$$

The function $\Phi(l)$ is called a *potential function,* and the utility of such a function will now become apparent. Substituting (*9*) in (*7*), we obtain

$$W_B = -\int\left[-\frac{d\Phi(l)}{dl}\right]dl = \int d\Phi$$

or

$$W_B = \Delta\Phi \qquad (10)$$

But the difference $\Delta\Phi$ depends only on the initial and final positions of the system, and not on the path followed between these positions. Thus the work done against body forces is *independent of path.* Defining the potential energy E_P as

$$E_P = \Phi$$

we can write (*10*) as

$$W_B = \Delta E_P \qquad (11)$$

The property expressed in (*9*) is not in general ascribable to surface forces (they are often *nonconservative* as in the case of a friction force) and we usually write expressions like (*8*) for the work done against such forces. Combination of (*6*) and (*11*) gives

$$W_t = \Delta E_P + W_S \qquad (12)$$

This equation is an alternative to (*4*), and the two equations may be applied to the same process. Doing this, and rearranging, we obtain the energy equation

$$-W_S = \Delta E_K + \Delta E_P \qquad (13)$$

The different origins of the two energy terms are apparent: ΔE_K arises from the general statement embodied in Newton's second law and is equal to the *total* work done by all the external forces on the system considered here, while ΔE_P follows from the decomposition of the total force into conservative and nonconservative terms, and is equal to that portion of the total work done by body forces. The justification for the seemingly arbitrary separation in (*5*), which led us to (*13*), is experience, for nature provides us with important examples of conservative forces, e.g. gravitational forces. When there are no surface forces (*13*) reduces to

$$\Delta E_K + \Delta E_P = 0 \qquad \text{or} \qquad E_K + E_P = \text{constant}$$

which is the familiar "principle of conservation of energy" of classical mechanics.

The work term in the first law of thermodynamics usually represents work done by surface forces, and the usual thermodynamic system is, of course, not a rigid body.

1.12. The gravitational force of attraction between homogeneous, spherical bodies A and B is given by

$$F_{\text{grav}} = -\frac{Gm_Am_B}{r^2} \qquad (1)$$

where G is a constant $[= 6.670 \times 10^{-11}(\text{N})(\text{m})^2/(\text{kg})^2]$, m_A and m_B are the masses of the two bodies, and r is the distance between their centers. Find an expression for the gravitational potential function Φ_{grav}, and show that the gravitational potential energy of a small body at a height z above the earth's surface is approximately

$$E_P = \frac{mg}{g_c}z \qquad (2)$$

where

$$g/g_c = \text{constant} = GM/R^2$$
$$m = \text{mass of the body}$$
$$M = \text{mass of the earth}$$
$$R = \text{radius of the earth}$$

The gravitational force is a body force and is conservative. From (*9*) of Problem 1.11, Φ_{grav} is related to F_{grav} by

$$F_{\text{grav}} = -\frac{d\Phi_{\text{grav}}}{dr} \qquad (3)$$

Combination of (*1*) and (*3*) gives

$$\frac{d\Phi_{\text{grav}}}{dr} = \frac{Gm_Am_B}{r^2}$$

which yields on integration an *exact* expression for Φ_{grav}:

$$\Phi_{grav} = -\frac{Gm_A m_B}{r} + C \tag{4}$$

where C is a constant of integration. The gravitational potential energy E_P is by definition equal to the gravitational potential function, so (4) can be written for an "earth-body" system as

$$E_P = -\frac{GMm}{(R+z)} + C$$

or

$$E_P = -\frac{GMm}{R\left(1+\frac{z}{R}\right)} + C \tag{5}$$

where $r \cong R + z$ from the statement of the problem. Now for small elevations $z/R \ll 1$, and the first two terms of the binomial expansion give

$$\left(1+\frac{z}{R}\right)^{-1} \cong 1 - \frac{z}{R}$$

Thus (5) becomes

$$E_P = -\frac{GMm}{R}\left(1-\frac{z}{R}\right) + C = \frac{GMm}{R^2}z + C' \tag{6}$$

where C' is a new constant for a given m. Since only *differences* in E_P are of importance, C' may be taken as zero: this is equivalent to choosing a datum of zero potential energy at the surface of the earth, and leads to the desired result (2).

1.13. An elevator with a mass of 5000(lb_m) or 2268(kg) rests at a level 25(ft) or 7.62(m) above the base of an elevator shaft. It is raised to 250(ft) or 76.2(m) above the base of the shaft, where the cable holding it breaks. The elevator falls freely to the base of the shaft where it is brought to rest by a strong spring. The spring assembly is designed to hold the elevator at the position of maximum spring compression by means of a ratchet mechanism. Assuming the entire process to be frictionless and taking g as the standard acceleration of gravity, calculate:

(*a*) The gravitational potential energy of the elevator in its initial position relative to the base of the shaft.

(*b*) The potential energy of the elevator in its highest position relative to the base of of the shaft.

(*c*) The work done in raising the elevator.

(*d*) The kinetic energy and velocity of the elevator just before it strikes the spring.

(*e*) The potential energy of the fully compressed spring.

(*f*) The energy of the system made up of the elevator and spring (1) at the start of the process, (2) when the elevator reaches its maximum height, (3) just before the elevator strikes the spring, and (4) after the elevator has come to rest. Use the numbers just given as subscripts on symbols to designate the stages of the process to which the symbols apply.

(*a*) The gravitational potential energy of the elevator at stage 1 is given by

$$E_{P_1} = \frac{mz_1 g}{g_c} = \frac{5000(\text{lb}_m) \times 25(\text{ft}) \times 32.174(\text{ft})/(\text{s})^2}{32.174(\text{ft})(\text{lb}_m)/(\text{s})^2(\text{lb}_f)} = 125{,}000(\text{ft-lb}_f)$$

Since 1(ft-lb_f) is equivalent to 1.3558(J), we also have

$$E_{P_1} = 169{,}480(\text{J})$$

Alternatively

$$E_{P_1} = \frac{mz_1 g}{g_c} = \frac{2268(\text{kg}) \times 7.62(\text{m}) \times 9.8066(\text{m})/(\text{s})^2}{1(\text{m})(\text{kg})/(\text{s})^2(\text{N})} = 169{,}480(\text{N-m}) = 169{,}480(\text{J})$$

(*b*) Similarly, the gravitational potential energy at stage 2 is calculated to be

$$E_{P_2} = 1{,}250{,}000(\text{ft-lb}_f) \quad \text{or} \quad 1{,}694{,}800(\text{J})$$

(*c*) The first law of thermodynamics applied between stage 1 and stage 2 reduces to

$$\Delta E_{P_{1\to 2}} = -W$$

because ΔE_K, ΔU, and Q are all presumed negligible during this process. Thus $-W = E_{P_2} - E_{P_1}$

or
$$W = E_{P_1} - E_{P_2} = -1{,}125{,}000(\text{ft-lb}_f) \quad \text{or} \quad -1{,}525{,}320(\text{J})$$

The minus sign merely means that work was done *on* the system rather than *by* the system.

(*d*) During the process from stage 2 to stage 3, the elevator falls freely, subject to no surface forces, and therefore $W = 0$. It is also assumed that $Q = 0$. Thus the first law becomes

$$\Delta E_{K_{2\to 3}} + \Delta E_{P_{2\to 3}} = 0 \quad \text{or} \quad E_{K_3} - E_{K_2} + E_{P_3} - E_{P_2} = 0$$

But E_{K_2} and E_{P_3} are zero. Therefore

$$E_{K_3} = E_{P_2} = 1{,}250{,}000(\text{ft-lb}_f) \text{ or } 1{,}694{,}800(\text{J})$$

Since $E_{K_3} = mu_3^2/2g_c$,

$$u_3 = \sqrt{2g_c E_{K_3}/m} = \sqrt{\frac{2 \times 32.174(\text{ft})(\text{lb}_m)/(\text{s})^2(\text{lb}_f) \times 1{,}250{,}000(\text{ft-lb}_f)}{5000(\text{lb}_m)}}$$

$$= 126.8(\text{ft})/(\text{s})$$

or
$$u_3 = \sqrt{\frac{2 \times 1(\text{m})(\text{kg})/(\text{s})^2(\text{N}) \times 1{,}694{,}800(\text{N-m})}{2268(\text{kg})}} = 38.66(\text{m})/(\text{s})$$

(*e*) Since the elevator comes to rest at the bottom of the shaft at stage 4, it ends up with zero potential and kinetic energy. All of its former energy is stored in the spring as elastic potential energy. The amount of this stored energy is $1{,}250{,}000(\text{ft-lb}_f)$ or $1{,}694{,}800(\text{J})$.

(*f*) If the elevator and the spring together are taken as the system, the initial energy of the system is the potential energy of the elevator, or $125{,}000(\text{ft-lb}_f)$. The total energy of the system can change only if energy is transferred between it and the surroundings. As the elevator is raised, work is done on the system by the surroundings in the amount of $1{,}125{,}000(\text{ft-lb}_f)$. Thus the energy of the system when the elevator reaches its maximum height is $125{,}000 + 1{,}125{,}000 = 1{,}250{,}000(\text{ft-lb}_f)$. Subsequent changes occur entirely within the system, with no energy transferred between the system and the surroundings. Hence the total energy of the system must remain constant at $1{,}250{,}000(\text{ft-lb}_f)$. It merely changes form from potential energy of position (elevation) of the elevator to kinetic energy of the elevator to potential energy of configuration of the spring.

This example serves to illustrate the application of the law of conservation of mechanical energy. It will be recalled that the entire process was assumed to occur without friction. The results obtained are exact only for such an idealized process, and would be approximations for the process as it actually occurs.

1.14. Water flows over a waterfall 100(m) in height. Consider 1(kg) of the water, and assume that no energy is exchanged between this 1(kg) and its surroundings. (*a*) What is the potential energy of the water at the top of the falls with respect to the base of the falls? (*b*) What is the kinetic energy of the water just before it strikes the bottom? (*c*) After the 1(kg) of water enters the river below the falls, what change has occurred in its state?

Taking the 1(kg) of water as the system, and noting that it exchanges no energy with its surroundings, we may set Q and W equal to zero and write the first-law energy equation as

$$\Delta E_K + \Delta E_P + \Delta U = 0$$

This equation applies to any part of the process.

(*a*)
$$E_P = \frac{mzg}{g_c} = \frac{1(\text{kg}) \times 100(\text{m}) \times 9.8066(\text{m})/(\text{s})^2}{1(\text{kg})(\text{m})/(\text{N})(\text{s})^2}$$

where g has been taken as the standard value. This gives

$$E_P = 980.66(\text{N-m}) \quad \text{or} \quad 980.66(\text{J})$$

(*b*) During the free fall of the water no mechanism exists for conversion of potential or kinetic energy into internal energy. Thus ΔU must be zero, and

$$\Delta E_K + \Delta E_P = E_{K_2} - E_{K_1} + E_{P_2} - E_{P_1} = 0$$

For practical purposes we may take $E_{K_1} = 0$ and $E_{P_2} = 0$. Then

$$E_{K_2} = E_{P_1} = 980.66(\text{J})$$

(*c*) As the 1(kg) of water strikes bottom and joins with other masses of water to reform a river, there is much turbulence, which has the effect of converting kinetic energy into internal energy. During this process ΔE_P is essentially zero, and therefore

$$\Delta E_K + \Delta U = 0 \quad \text{or} \quad \Delta U = E_{K_2} - E_{K_3}$$

However, the downstream river velocity is assumed to be small, and therefore E_{K_3} is negligible. Thus

$$\Delta U = E_{K_2} = 980.66(\text{J})$$

The overall result of the process is the conversion of potential energy of the water into internal energy of the water. This change in internal energy is manifested by a temperature rise of the water. Since energy in the amount of 4184(J)/(kg) is required for a temperature rise of 1(°C) in water, the temperature increase is 980.66/4184 = 0.234(°C).

1.15. The driver of a 3000(lb_m) car, coasting down a hill, sees a red light at the bottom, for which he must stop. His speed at the time the brakes are applied is 60 miles per hour (88 feet per second) and he is 100 feet vertically above the bottom of the hill. How much energy as heat must be dissipated by the brakes if wind and other frictional effects are neglected?

Equation (*1.3*) is applicable to the automobile as the system:

$$\Delta E = Q - W$$

The total energy change ΔE is made up of two parts, ΔE_K and ΔE_P, which account for kinetic- and potential-energy changes of the car:

$$\Delta E_K = \frac{m\,\Delta u^2}{2g_c} \quad \text{and} \quad \Delta E_P = \frac{mg\,\Delta z}{g_c}$$

Since there are no moving surface forces applied to the car, W is zero, and our energy equation becomes

$$Q = \frac{m\,\Delta u^2}{2g_c} + \frac{mg\,\Delta z}{g_c}$$

Taking g as the standard acceleration of gravity, we get

$$Q = \frac{3000(\text{lb}_\text{m}) \times [0 - 88^2](\text{ft})^2/(\text{s})^2}{2 \times 32.174(\text{lb}_\text{m})(\text{ft})/(\text{lb}_\text{f})(\text{s})^2} + \frac{3000(\text{lb}_\text{m}) \times 32.174(\text{ft})/(\text{s})^2 \times [0 - 100](\text{ft})}{32.174(\text{lb}_\text{m})(\text{ft})/(\text{lb}_\text{f})(\text{s})^2}$$

$$= -361{,}000 - 300{,}000 = -661{,}000(\text{ft-lb}_\text{f})$$

or

$$Q = \frac{-661{,}000(\text{ft-lb}_\text{f})}{778(\text{ft-lb}_\text{f})/(\text{Btu})} = -850(\text{Btu})$$

or $$Q = -661{,}000(\text{ft-lb}_f) \times 1.356(\text{J})/(\text{ft-lb}_f) = -896{,}200(\text{J})$$

The minus sign indicates that heat must pass from the system to the surroundings, as implied by the initial question.

1.16. A rigid tank which acts as a perfect heat insulator and which has a negligible heat capacity is divided into two unequal parts A and B by a partition. Different amounts of the same ideal gas are contained in the two parts of the tank. The initial conditions of temperature T, pressure P, and total volume V^t are known for both parts of the tank, as shown in Fig. 1-7.

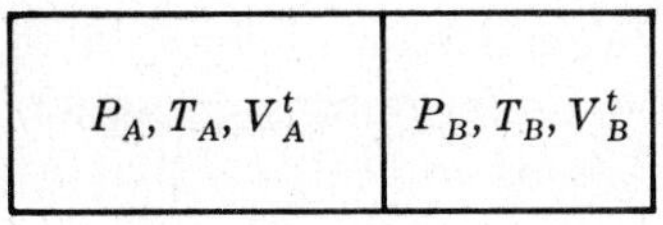

P_A, T_A, V_A^t	P_B, T_B, V_B^t

Fig. 1-7

Find expressions for the equilibrium temperature T and pressure P reached after removal of the partition. Assume that C_V, the molar heat capacity of the gas, is constant and that the process is adiabatic.

Since the tank is rigid, no external force acting on its contents (taken as the system) moves, and consequently $W = 0$. In addition, the process is adiabatic, and therefore $Q = 0$. Thus by the first law the total internal energy of the gas in the tank is constant. This may be expressed by the equation

$$\Delta U_A^t + \Delta U_B^t = 0$$

where the two terms refer to the quantities of gas initially in the two parts of the tank. For an ideal gas with constant heat capacities

$$\Delta U^t = nC_V \Delta T$$

Thus the energy equation may be written

$$n_A C_V(T - T_A) + n_B C_V(T - T_B) = 0 \tag{1}$$

or
$$n_A(T - T_A) + n_B(T - T_B) = 0 \tag{2}$$

where n_A and n_B are the numbers of moles of gas initially in parts A and B of the tank. By the ideal-gas law applied to the two initial compartments

$$n_A = \frac{P_A V_A^t}{RT_A} \qquad \text{and} \qquad n_B = \frac{P_B V_B^t}{RT_B} \tag{3}$$

Combination of (2) and (3) and solution for T leads directly to

$$T = T_A T_B \left(\frac{P_A V_A^t + P_B V_B^t}{P_A V_A^t T_B + P_B V_B^t T_A} \right) \tag{4}$$

Application of the ideal-gas law to the final state of the system gives the pressure as

$$P = \frac{nRT}{V^t} = \frac{(n_A + n_B)RT}{V_A^t + V_B^t}$$

Substitution for n_A and n_B by (3) and for T by (4) leads upon reduction to

$$P = \frac{P_A V_A^t + P_B V_B^t}{V_A^t + V_B^t}$$

This last result can be obtained in quite a different way. Multiplication of (1) by C_P/C_V gives

$$n_A C_P(T - T_A) + n_B C_P(T - T_B) = 0$$

For an ideal gas with constant heat capacities, $\Delta H^t = nC_P \Delta T$. Therefore the preceding equation may be written

$$\Delta H_A^t + \Delta H_B^t = 0$$

Since our original energy equation was $\Delta U_A^t + \Delta U_B^t = 0$ we may subtract to get

$$\Delta H_A^t - \Delta U_A^t + \Delta H_B^t - \Delta U_B^t = 0$$

However, by the definition of enthalpy it is generally true that $\Delta H^t - \Delta U^t = \Delta(PV^t)$. Therefore

$$\Delta(PV^t)_A + \Delta(PV^t)_B = 0 \qquad \text{or} \qquad PV_A^t - P_A V_A^t + PV_B^t - P_B V_B^t = 0$$

which yields the value of P obtained before.

1.17. In Example 1.10 an equation was derived to relate T and V for an ideal gas with constant heat capacities undergoing a reversible, adiabatic compression or expansion. Develop the analogous expression that relates T and P.

We start again with (*1.5*) applied to an adiabatic process:

$$dU = -\delta W = -P\,dV$$

Since the gas is ideal $dU = C_V\,dT$ and we have

$$C_V\,dT = -P\,dV$$

In Example 1.10 we eliminated P by the ideal-gas law. Here we will instead eliminate dV. Since $V = RT/P$

$$dV = \frac{-RT}{P^2}\,dP + \frac{R}{P}\,dT$$

and therefore

$$C_V\,dT = -P\left[\frac{-RT}{P^2}\,dP + \frac{R}{P}\,dT\right] = RT\frac{dP}{P} - R\,dT$$

By (*1.14*) $C_V + R = C_P$ and as a result

$$C_P\,dT = RT\frac{dP}{P}$$

or

$$\frac{dT}{T} = \frac{R}{C_P}\frac{dP}{P} = \left(\frac{C_P - C_V}{C_P}\right)\frac{dP}{P}$$

If the term in parentheses is divided by C_V, both in the numerator and denominator, then

$$\frac{dT}{T} = \left(\frac{\gamma - 1}{\gamma}\right)\frac{dP}{P}$$

Integration gives

$$\ln\frac{T_2}{T_1} = \left(\frac{\gamma - 1}{\gamma}\right)\ln\frac{P_2}{P_1} \qquad \text{or} \qquad \frac{T_2}{T_1} = \left(\frac{P_2}{P_1}\right)^{(\gamma-1)/\gamma}$$

1.18. A rigid cylinder contains a "floating" piston, free to move within the cylinder without friction. Initially, it divides the cylinder in half, and on each side of the piston the cylinder holds 1(lb mole) of the same ideal gas at 40(°F) and 1(atm). An electrical resistance heater is installed on side A of the cylinder as shown in Fig. 1-8, and it is energized so as to cause the temperature in side A to rise slowly to 340(°F). If the tank and the piston are perfect heat insulators and are of negligible heat capacity, calculate the amount of heat added to the system by the resistor. The molar heat capacities of the gas are constant and have the values $C_V = 3$(Btu)/(lb mole)(°F) and $C_P = 5$(Btu)/(lb mole)(°F).

We may write the first law to apply to the entire contents of the cylinder taken as the system:

$$\Delta U^t = Q - W$$

Since no force external to the system moves, $W = 0$. Hence

$$Q = n_A \Delta U_A + n_B \Delta U_B$$

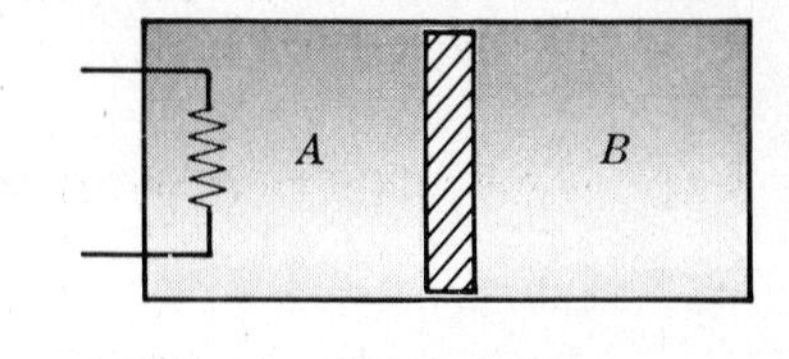

Fig. 1-8

However, $n_A = n_B = 1$, and for an ideal gas with constant heat capacity, $\Delta U = C_V \Delta T$. Thus

$$Q = C_V \Delta T_A + C_V \Delta T_B = C_V(\Delta T_A + \Delta T_B) \qquad (1)$$

In this equation ΔT_A is given by the data, but ΔT_B is unknown and must be found. As heat is added to A, the temperature and pressure of the gas in A rise, and the gas expands, pushing the piston to the right. This compresses the gas in B, and since the piston and cylinder are non-conducting and of negligible heat capacity, the process in B is adiabatic. The assumption of reversibility is also reasonable, because the piston is frictionless and the process is slow. For a reversible, adiabatic compression of an ideal gas with constant heat capacities, we have from Problem 1.17

$$\frac{T_{B_2}}{T_1} = \left(\frac{P_2}{P_1}\right)^{(\gamma-1)/\gamma} \qquad (2)$$

Note that no letter subscripts are needed on P, because the pressure is always uniform throughout the cylinder. The total volume of the cylinder V^t is constant and is given by the ideal-gas law applied both for the initial conditions and for the final conditions:

$$V^t = (n_A + n_B)RT_1/P_1 \qquad \text{(initial conditions)}$$

$$V^t = (n_A RT_{A_2} + n_B RT_{B_2})/P_2 \qquad \text{(final conditions)}$$

Equating these expressions and setting $n_A = n_B = 1$, we get

$$2\frac{P_2}{P_1} = \frac{T_{A_2}}{T_1} + \frac{T_{B_2}}{T_1} \qquad (3)$$

By combination of (2) and (3)

$$2\frac{P_2}{P_1} = \frac{T_{A_2}}{T_1} + \left(\frac{P_2}{P_1}\right)^{(\gamma-1)/\gamma}$$

The ratio P_2/P_1 can now be determined by trial.

$$T_{A_2} = 340 + 460 = 800(\text{R}), \quad T_1 = 40 + 460 = 500(\text{R}), \quad \frac{\gamma - 1}{\gamma} = \frac{5/3 - 1}{5/3} = 0.4$$

Thus $$2\left(\frac{P_2}{P_1}\right) = \frac{800}{500} + \left(\frac{P_2}{P_1}\right)^{0.4} \quad \text{and} \quad P_2/P_1 = 1.367$$

We now determine T_{B_2} by (2):

$$T_{B_2} = T_1\left(\frac{P_2}{P_1}\right)^{(\gamma-1)/\gamma} = 500(\text{R}) \times 1.367^{0.4} = 566.5(\text{R}) \text{ or about } 107(^\circ\text{F})$$

Thus by (1)

$$Q = 3[(340 - 40) + (107 - 40)] = 1100(\text{Btu}) \text{ or } 1{,}160{,}500(\text{J})$$

1.19. The basic equation of hydrostatics relating pressure to fluid depth is

$$dP = -\frac{\rho g}{g_c} dz$$

where ρ is the local density in mass per unit volume, g is the local acceleration of gravity, and z is elevation above a datum level in the fluid. Apply this equation to develop an expression for atmospheric pressure as a function of elevation above the surface of the earth.

For the temperatures and pressures involved air behaves essentially as an ideal gas, for which

$PV = RT$. The relation between ρ, the mass density of air, and V, the molar volume of air, is $\rho = M/V$, where M is the molecular weight of air. Substitution in the hydrostatic formula gives

$$dP = -\frac{Mg}{Vg_c}dz$$

and elimination of V by the ideal-gas law provides

$$\frac{dP}{P} = -\frac{Mg}{RTg_c}dz \tag{1}$$

In order to integrate this equation we need to know how the atmospheric temperature T varies with elevation.

The simplest assumption is that T is constant. In this case (1) can be integrated immediately:

$$\int_{P_0}^{P} \frac{dP}{P} = -\frac{Mg}{RT_0 g_c}\int_0^z dz$$

whence

$$\ln\frac{P}{P_0} = -\frac{Mgz}{RT_0 g_c} \tag{2}$$

or

$$P = P_0 e^{-(Mgz)/(RT_0 g_c)}$$

where P_0 is the pressure at the datum level, for which z is zero. Note that g has been assumed independent of z.

Equation (2) is known as the barometric equation, and is strictly applicable where T is independent of z. This condition obtains in the region of the atmosphere known as the stratosphere, at elevations between 36,000 and 80,000 feet. Below 36,000 feet there is continual movement of great masses of air upward and downward. These air masses expand and cool as they rise, and compress and warm as they descend. If we assume this process to be reversible and adiabatic, then the relation between T and P is given by

$$\frac{dT}{T} = \left(\frac{\gamma - 1}{\gamma}\right)\frac{dP}{P}$$

as shown in Problem 1.17. If in this equation we eliminate dP/P by (1) and solve for dT/dz, we get

$$\frac{dT}{dz} = -\left(\frac{\gamma - 1}{\gamma}\right)\left(\frac{Mg}{Rg_c}\right) = K \tag{3}$$

where the quantity K is a constant, provided that γ and g are taken as independent of T and z. This equation shows that the rate of change of temperature with elevation is constant, and measurements show that this is essentially so below 36,000 feet. Integration of (3) provides the linear equation $T = T_0 + Kz$, where T_0 is the temperature at $z = 0$. Substitution of this equation into (1) gives

$$\frac{dP}{P} = -\frac{Mg}{Rg_c}\frac{dz}{(T_0 + Kz)}$$

From (3) we note that

$$-\frac{Mg}{Rg_c} = \left(\frac{\gamma}{\gamma - 1}\right)K$$

Making this substitution and integrating:

$$\int_{P_0}^{P} \frac{dP}{P} = \left(\frac{\gamma}{\gamma - 1}\right)K\int_0^z \frac{dz}{(T_0 + Kz)}$$

or

$$\ln\frac{P}{P_0} = \left(\frac{\gamma}{\gamma - 1}\right)\ln\left(1 + \frac{Kz}{T_0}\right) \tag{4}$$

We have found two expressions, (2) and (4), which are based on the two extremes of possible temperature variation in the atmosphere. If we write an empirical but more general equation relating T and P in the atmosphere,

$$\frac{T}{T_0} = \left(\frac{P}{P_0}\right)^{(\delta - 1)/\delta}$$

where δ is a number between 1 and γ, then when $\delta = \gamma$ we have the adiabatic case, which yields (4), and when $\delta = 1$ we have the isothermal case, which yields (2). The actual situation is in between. Measurements show that $\delta = 1.24$, and the equation for atmospheric pressure as a function of elevation is exactly like (4) but with γ replaced by δ:

$$\ln \frac{P}{P_0} = \left(\frac{\delta}{\delta - 1}\right) \ln\left(1 + \frac{Kz}{T_0}\right) \tag{5}$$

where

$$K = \frac{dT}{dz} = -\left(\frac{\delta - 1}{\delta}\right)\left(\frac{Mg}{Rg_c}\right)$$

The following table gives results for various elevations as calculated by the three equations (2), (4), and (5).

z, feet	P, atmospheres		
	(2)	(4), $\gamma = 1.40$	(5), $\delta = 1.24$
1,000	0.9658	0.9657	0.9657
3,000	0.9010	0.8996	0.9000
5,000	0.8405	0.8367	0.8380
10,000	0.7064	0.6935	0.6978
20,000	0.4990	0.4608	0.4740
30,000	0.3525	0.2900	0.3121

Supplementary Problems

BASIC CONCEPTS (Sec. 1.1)

1.20. At what absolute temperature do the Celsius and Fahrenheit temperature scales give the same numerical value? *Ans.* 233.15(K) $\equiv$ 419.67(R)

1.21. The *volt* (V) and the *ohm* (Ω) are derived SI units. One (V) is defined as the difference in electrical potential between two points of a conducting wire carrying a constant current of 1(A), when the power dissipated between these two points is 1(W). One (Ω) is defined as the electrical resistance between two points of a conductor when a constant electrical potential difference of 1(V), applied between the two points, produces in this conductor a current of 1(A), the conductor not being the source of any electromotive force. Express the volt and the ohm in terms of *basic* SI units.

Ans. $1(\text{V}) = 1\,\frac{(\text{kg})(\text{m})^2}{(\text{s})^3(\text{A})}, \quad 1(\Omega) = 1\,\frac{(\text{kg})(\text{m})^2}{(\text{s})^3(\text{A})^2}$

1.22. What are the *basic* SI units for each of the following physical quantities?

(*a*) torque (*d*) frequency (*g*) wavelength
(*b*) electrical charge (*e*) momentum (*h*) moment of inertia
(*c*) molecular weight (*f*) electrical capacitance (*i*) surface tension

Ans. (*a*) $(\text{kg})(\text{m})^2/(\text{s})^2$ (*d*) $(\text{s})^{-1}$ (*g*) (m)
(*b*) (A)(s) (*e*) (kg)(m)/(s) (*h*) $(\text{kg})(\text{m})^2$
(*c*) (kg) (*f*) $(\text{A})^2(\text{s})^4/(\text{kg})(\text{m})^2$ (*i*) $(\text{kg})/(\text{s})^2$

WORK (Secs. 1.2 through 1.5)

1.23. (*a*) A certain gas obeys the equation of state $P(V - b) = RT$, where R is the universal gas constant, and b is a constant such that $0 < b < V$ for all V. Derive an expression for the work obtained from a reversible, isothermal expansion of one mole of this gas from an initial volume V_i to a final volume V_f. (b) If the gas in (a) were an ideal gas, would the same process produce more or less work?

Ans. (a) $W = RT \ln\left(\frac{V_f - b}{V_i - b}\right)$ (b) less

1.24. Figure 1-9 shows the relation of pressure to volume for a closed PVT system during a reversible process. Calculate the work done by the system for each of the three steps 12, 23, and 31, and for the entire process 1231.

Ans. $W_{12} = 19.43$(Btu), $W_{23} = -11.11$(Btu), $W_{31} = 0$, $W_{1231} = 8.32$(Btu)

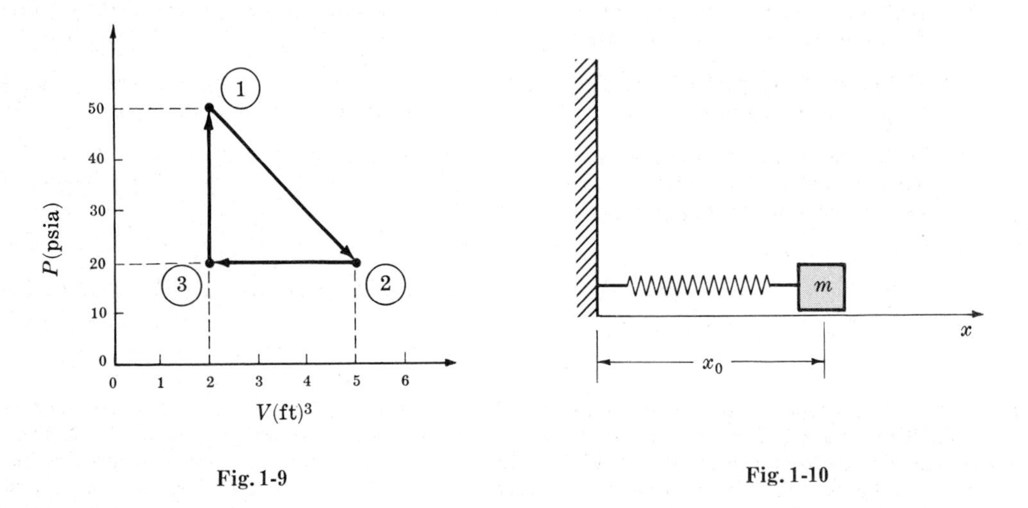

Fig. 1-9

Fig. 1-10

1.25. In Example 1.4 the *natural strain* ϵ was defined in terms of the length l of an axially stressed bar by the differential equation $d\epsilon = dl/l$. Similarly, the *engineering strain* ϵ' can be defined by the equation $d\epsilon' = dl/l_0$, where l_0 is the initial (unstrained) length of the bar. Show that ϵ and ϵ' are related by the equation $\epsilon = \ln(1 + \epsilon')$ and that the two measures of strain differ by less than 1% for ϵ' less than about 0.02.

1.26. (a) For a bar with no permanent strain, the stress σ is proportional to ϵ for small strains at constant temperature. Thus $\sigma = E\epsilon$ where E is a constant and a property of the material. Show that the work done when such a bar is strained reversibly by an axial tensile load is given by $W = -VE\epsilon^2/2$, where the volume V has been assumed constant. (b) A cylindrical brass rod with a diameter of 1(cm) and an initial length of 10(cm) is reversibly stretched by 0.01(cm). Calculate the work done if the volume remains constant. For brass, $E = 9 \times 10^{10}$(N)/(m)². *Ans.* (b) $W = -0.35$(J)

FIRST LAW OF THERMODYNAMICS (Secs. 1.6 through 1.9)

1.27. In some books the first law of thermodynamics for a closed system, equation (1.3), is written $\Delta E = Q + W$. Explain the sign convention on W.

Ans. W is work done *on* the system *by* the surroundings

1.28. (a) Show that the *exact* expression for g/g_c is

$$\frac{g}{g_c} = \frac{GM}{R^2\left(1 + \frac{z}{R}\right)}$$

where g/g_c is defined as the coefficient of mz in the expression for gravitational potential energy $E_P = (g/g_c)mz + \text{constant}$ [see (5) and (6) of Problem 1.12]. (b) At what distance from the surface of the earth does the acceleration g due to the earth's gravity become 0.01 of the value at the surface of the earth? The diameter of the earth is approximately 7900(mile). *Ans.* (b) 391,050(mile)

1.29. A mass m is attached to a spring as shown in Fig. 1-10. When left by itself, the system assumes the position shown, where x_0 is the equilibrium position of the center of mass. If m is displaced from x_0, an elastic restoring force F_e acts on the mass, and is given by

$$F_e = -k\xi \qquad (1)$$

where k is the spring constant and ξ is the displacement ($\xi = x - x_0$). The work done by m against F_e can be incorporated in the mechanical energy equation by treating F_e as a body force and proceeding as in Problems 1.11 and 1.12.

(*a*) What are the *basic* SI units for k?

(*b*) Show that the sign convention for force implied by (*1*) is consistent with the adopted coordinate system.

(*c*) Define the elastic potential function Φ_e *by* $F_e = -d\Phi_e/d\xi$. If the zero of elastic potential energy is taken at the equilibrium position, show that $E_P = (k/2)\xi^2$.

(*d*) It follows from part (*c*) and Problem 1.11 that the work W_B done by the system m against the body force F_e as m moves away from its equilibrium position is always positive. Why?

(*e*) Write the mechanical energy equation for the rectilinear translation of m if no forces other than F_e act on the mass. Assume that the *total* energy E_T of the system is known, and make use of the relationship $u = dx/dt = d\xi/dt \equiv \dot{\xi}$.

Ans. (*a*) (kg)/(s)^2

(*d*) Work is done *on* the spring by the mass both in compressing and extending it.

(*e*) $(m/g_c)\dot{\xi}^2 + k\xi^2 = 2E_T$

1.30. A scientist proposes to determine the heat capacities of liquids by use of a Joule calorimeter. In this device work is done by a paddle wheel on a liquid in an insulated container. The heat capacity is calculated from the measured temperature rise of the liquid and the measured work done by the paddle wheel. It is assumed that there is no heat exchanged between the liquid and its surroundings. To check this assumption, the scientist performs a preliminary experiment on 10(g mole) of benzene, for which C_P is 31.8(cal)/(g mole)(K). His data are as follows:

Work done by the paddle wheel = 1500(cal)

Temperature rise of the liquid = 4(°C)

If both C_P and the pressure on the liquid remain constant during the experiment, show that these results are not consistent with the stated assumptions, and offer an explanation for the inconsistency.

Ans. The data do not satisfy the first law. The disparity could be due to a heat leak to the surroundings of 228(cal) or to an error of measurement; for example, an error of 0.7(°C) in the temperature rise.

1.31. Figure 1-11 depicts two reversible processes undergone by one mole of an ideal gas. Curves T_a and T_b are isotherms, paths 23 and 56 are isobars, and paths 31 and 64 are isochores (paths of constant volume). Show that W and Q are the same for processes 1231 and 4564.

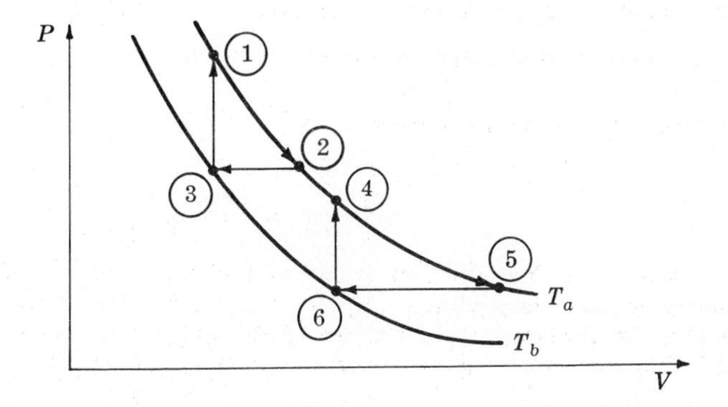

Fig. 1-11

1.32. An automobile tire is inflated to 25(psig) at the beginning of a trip. After three hours of high-speed driving pressure is 29(psig). What is the internal energy change of the air in the tire between pressure measurements? Assume that air is an ideal gas with a constant heat capacity $C_V = 5$(Btu)/(lb mole)(R) and that the internal volume of the tire remains constant at 2(ft)^3.

Ans. $\Delta U^t = 3.73$(Btu)

1.33. A perfectly insulated rigid container of total volume V^t is divided into two parts by a partition of negligible volume. One side of the partition contains n moles of an ideal gas with constant heat capacities at a temperature T_i, and the other side is evacuated. If the partition is broken, calculate Q, W, and ΔU for the ensuing process, and calculate the final temperature and pressure T_f and P_f of the gas.

Ans. $Q = 0$, $W = 0$, $\Delta U = 0$, $T_f = T_i$, $P_f = nRT_i/V^t$

Chapter 2

The Second Law of Thermodynamics

2.1 AXIOMATIC STATEMENTS OF THE FIRST AND SECOND LAWS. ENTROPY

In Chapter 1 it was pointed out that the concept of energy is primitive, and that the first law of thermodynamics, expressing the requirement that energy be conserved, is also primitive. The origin of the conservation principle was in mechanics, where it applies to rigid bodies in the absence of friction and is written in terms of the external forms of energy and mechanical work. This principle of conservation of mechanical energy is readily tested by experiment and found valid. It is also directly related to Newton's second law of motion. Although historically it took a very long time, the transition from a limited conservation principle in mechanics to an all-inclusive conservation law in thermodynamics does not today seem difficult. The key step is the recognition that heat is a form of energy and that the quantity called internal energy is an intrinsic property of matter. The test is experiment and experience, and now that all this is firmly established the simplest procedure is to formalize the basic principles of thermodynamics with a set of axioms, taken as valid from the beginning. The multitude of consequences which follow from these axioms by formal mathematical deduction have been amply compared with experiment already, and it is no longer necessary for each student to retrace the historical path that led to discovery of the fundamentals of thermodynamics. Application of the axioms presented below to familiar situations will lead to results easily recognized as valid, and no other justification of the axioms should be necessary.

The formal statements necessary to the first law of thermodynamics are provided by the two following axioms:

Axiom 1: There exists a form of energy, known as internal energy U, which is an intrinsic property of a system, functionally related to the measurable coordinates which characterize the system. For a closed system, not in motion, changes in this property are given by

$$\boxed{dU = \delta Q - \delta W} \tag{1.5}$$

Axiom 2 (First law of thermodynamics): The *total* energy of any system and its surroundings, considered together, is conserved.

The first axiom asserts the existence of a function called internal energy and provides a relationship connecting it with measurable quantities. This equation in no way explicitly defines internal energy: *there is no definition.* What is provided is a means for calculation of changes in this function. Absolute values are unknown. The second axiom depends upon the first, and is regarded as one of the fundamental laws of science.

Nature places a second restriction on all processes, and this is given formal statement by two additional axioms:

Axiom 3: There exists a property called *entropy S*, which is an intrinsic property of a system, functionally related to the measurable coordinates which characterize the system. For a reversible process changes in this property are given by

$$\boxed{dS = \frac{\delta Q}{T}} \tag{2.1}$$

Axiom 4 (Second law of thermodynamics): The entropy change of any system and its surroundings, considered together, is positive and approaches zero for any process which approaches reversibility.

The third axiom is the same in form and character as the first. It does for the entropy function what the first axiom does for the internal energy, asserting its existence and providing a relationship which connects it with measurable quantities. Again there is no explicit definition. Entropy is another primitive concept. Equation (*2.1*) allows calculation of changes in this function, but does not permit determination of absolute values. The fourth axiom clearly depends on the third and is, of course, its motivation. The second law of thermodynamics is a conservation law only for reversible processes, which are unknown in nature. All natural processes result in an increase in total entropy. The mathematical expression of the second law is simply:

$$\boxed{\Delta S_{\text{total}} \geqq 0} \tag{2.2}$$

where the label "total" indicates that both the system and its surroundings are included. The equality applies only to the limiting case of a reversible process.

Example 2.1. Show that any flow of heat between two heat reservoirs at temperatures T_H and T_C, where $T_H > T_C$, must be from the hotter to the cooler reservoir.

A heat reservoir is by definition a body with an infinite heat capacity, and its entropy change is given by (*2.1*) regardless of the source or sink of the heat. The reason is that no irreversibilities occur *within* the reservoir, and changes in the reservoir depend only on the quantity of heat transferred and not on where it comes from or goes to. When a finite quantity of heat is added to or extracted from a heat reservoir, it undergoes a finite entropy change *at constant temperature*, and (*2.1*) becomes

$$\Delta S = \frac{Q}{T}$$

Let the quantity of heat Q pass from one reservoir to the other. The magnitude of Q is the same for both reservoirs, but Q_H and Q_C are of opposite sign, for heat added to one reservoir (considered positive) is heat extracted from the other (considered negative). Therefore

$$Q_H = -Q_C$$

By (*2.1*)

$$\Delta S_H = \frac{Q_H}{T_H} = \frac{-Q_C}{T_H} \quad \text{and} \quad \Delta S_C = \frac{Q_C}{T_C}$$

Thus

$$\Delta S_{\text{total}} = \Delta S_H + \Delta S_C = \frac{-Q_C}{T_H} + \frac{Q_C}{T_C} = Q_C\left(\frac{T_H - T_C}{T_H T_C}\right)$$

According to the second law, (*2.2*), ΔS_{total} must be positive; therefore

$$Q_C(T_H - T_C) > 0$$

Since by the problem statement $T_H > T_C$, Q_C must clearly be positive, and must therefore represent heat *added* to the reservoir at T_C. Thus heat must flow from the higher-temperature reservoir at T_H to the lower-temperature reservoir at T_C, a result which is entirely in accord with experience.

Example 2.1 has dealt with pure heat transfer, for which the temperature difference $T_H - T_C$ is the driving force. For such a process our results show that ΔS_{total} becomes zero only for $T_H = T_C$. This is the condition of thermal equilibrium between the two heat reservoirs. Reversible heat transfer occurs when the two reservoirs have temperatures that differ only infinitesimally.

Example 2.2. What restrictions do the laws of thermodynamics impose on the production of work by a device that exchanges heat with the two reservoirs of Example 2.1, but which itself remains unchanged? Such a device is known as a *heat engine*.

Let us suppose that the device exchanges heat Q_H with the reservoir at T_H and Q_C with the reservoir at T_C. The symbols Q_H and Q_C refer to the heat reservoirs, and the signs on their numerical values are therefore determined by whether heat is added to or extracted from the heat reservoirs. Thus the entropy changes of the heat reservoirs are given by

$$\Delta S_H = \frac{Q_H}{T_H} \quad \text{and} \quad \Delta S_C = \frac{Q_C}{T_C}$$

These same heat quantities apply to the device or engine, but have opposite signs. Thus we let Q'_H and Q'_C be symbols for the heat exchanges taken with respect to the engine, where

$$Q_H = -Q'_H \quad \text{and} \quad Q_C = -Q'_C$$

The total entropy change resulting from any process involving just the engine and the heat reservoirs is

$$\Delta S_{\text{total}} = \Delta S_H + \Delta S_C + \Delta S_{\text{engine}}$$

Since the engine remains unchanged, the last term is zero, and we have

$$\Delta S_{\text{total}} = \frac{Q_H}{T_H} + \frac{Q_C}{T_C} \tag{1}$$

The first law as written for the *engine* has the form (*1.4*):

$$\Delta U = Q - W$$

where ΔU is the internal energy change of the engine, W is the work done by the engine and Q represents all heat transfer *with respect to the engine*. Thus

$$\Delta U_{\text{engine}} = Q'_H + Q'_C - W$$

Again, since the engine remains unchanged, $\Delta U_{\text{engine}} = 0$ and therefore

$$W = Q'_H + Q'_C = -Q_H - Q_C \tag{2}$$

Combination of (*1*) and (*2*) to eliminate Q_H gives

$$W = -T_H \Delta S_{\text{total}} + Q_C\left(\frac{T_H}{T_C} - 1\right) \tag{3}$$

This equation is valid within two limits. First, we are dealing with a work-producing device or engine, for which W must be positive, with a limiting value of zero. At this limit the engine is completely ineffective, and the process reduces to simple heat transfer between the reservoirs, and (*3*) reduces to the result of Example 2.1.

The second limit on (*3*) is represented by the reversible process, for which ΔS_{total} becomes zero by the second law, and W attains its maximum value for given values of T_H and T_C. In this case (*3*) reduces to

$$W = Q_C\left(\frac{T_H}{T_C} - 1\right) \tag{4}$$

From this result it is clear that for W to have a positive finite value Q_C must also be positive and finite. This means that even for the limiting case of reversible operation it is necessary that heat Q_C be *rejected from* the engine and *absorbed by* the cooler reservoir at T_C.

Combination of (*2*) and (*4*), first, to eliminate W and, second, to eliminate Q_C, provides the following two equations:

$$\frac{Q_C}{T_C} = \frac{-Q_H}{T_H} \tag{2.3}$$

and

$$\frac{W}{-Q_H} = 1 - \frac{T_C}{T_H} \tag{2.4}$$

These are known as *Carnot's equations*, and apply to all reversible heat engines operating between fixed temperature levels, i.e. to all *Carnot engines*.

The various quantities considered in this example are shown schematically in Fig. 2-1.

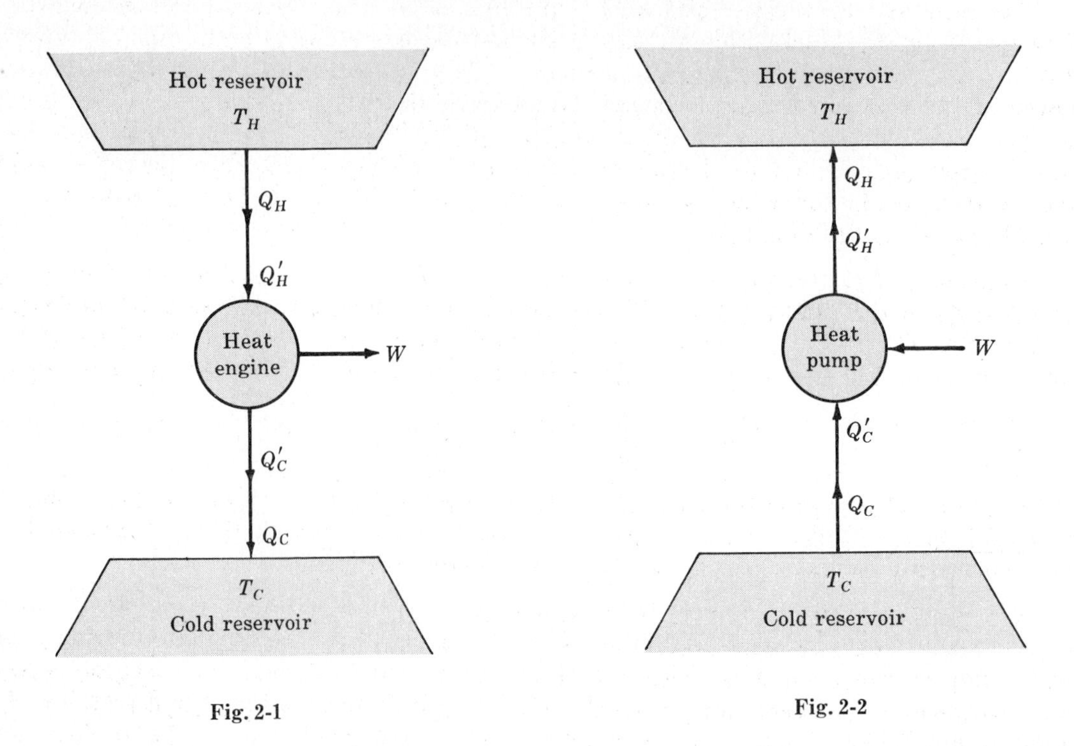

Fig. 2-1 Fig. 2-2

2.2 HEAT ENGINES AND HEAT PUMPS

In (*2.3*) and (*2.4*), applicable to Carnot engines, the Q's refer to the heat reservoirs, and since Q_H represents a negative number (heat out of the reservoir), the minus sign in front of Q_H makes the term positive. Equations (*2.3*) and (*2.4*) are commonly written without the minus signs, and in this event the sign convention for Q is disregarded, and all Q's are taken as positive numbers. This is perhaps better done with the aid of absolute-value signs:

$$\left|\frac{Q_C}{Q_H}\right| = \frac{T_C}{T_H} \tag{2.3a}$$

$$\left|\frac{W}{Q_H}\right| = 1 - \frac{T_C}{T_H} \tag{2.4a}$$

Equation (*2.3a*) shows that the ratio of heat discarded by a Carnot heat engine at T_C to the heat taken in at T_H is equal to the ratio of the two absolute temperatures. Thus the only way to make $|Q_C|$ zero is to discard heat to a reservoir at absolute zero of temperature. Since no reservoir at temperatures even approaching absolute zero is naturally available on earth, there is no practical means by which we can operate a heat engine so that it discards no heat.

This observation with respect to heat engines is so basic that its formal statement is often regarded as an alternative expression of the second law of thermodynamics: *It is impossible to construct an engine that, operating in a cycle, will produce no effect other than the extraction of heat from a reservoir and the performance of an equivalent amount of work.* This is the Kelvin-Planck statement of the second law. All heat engines must discard part of the heat they take in, and the natural heat reservoirs available to absorb this discarded heat are the atmosphere, lakes and rivers, and the oceans. The temperatures of these are of the order of 300(K).

The practical heat reservoirs at T_H are objects such as furnaces maintained at high temperature by the combustion of fossil fuels and nuclear reactors maintained at high temperature by fission of radioactive elements. The common components of all stationary power plants that generate our electricity are: a high-temperature source of heat; a heat engine, which may be very complex; and a sink for the discharge of waste heat, namely our environment. This discharge of waste heat to the environment cannot be avoided, and is responsible for what is known as thermal pollution. It is a consequence of the second law of thermodynamics.

Equation (*2.4a*) gives what is known as the *thermal efficiency* η of a Carnot heat engine, i.e. $|W|/|Q_H|$ is the fraction that $|W|$ represents of the heat $|Q_H|$ taken into the engine, or that part of $|Q_H|$ which can at best be obtained as work $|W|$. In practice, values of T_H are on the order of 600(K). If we use this figure and a value of $T_C = 300(K)$ in (*2.4a*) we get

$$\eta = \frac{|W|}{|Q_H|} = 1 - \frac{300}{600} = 0.5$$

as the practical limit for the efficiency of the conversion into work of the heat taken in by a Carnot heat engine. This figure is not attained in practice; thermal efficiencies run more commonly between 0.3 and 0.4, because of irreversibilities.

A reversible heat engine may be reversed, i.e. run as a heat pump or refrigerator, as shown schematically in Fig. 2-2. Equations (*2.3a*) and (*2.4a*) are equally valid for a Carnot heat pump, because all quantities are the same as for a Carnot heat engine operating between the same two temperature levels. The only difference is that the directions of heat transfer are reversed and work is required rather than performed. The work is used to "pump" heat from the heat reservoir at the cooler temperature T_C to the reservoir at the higher temperature T_H. Refrigerators work on this principle. The reservoir at T_C is the "cold box" or refrigerator and the reservoir at T_H is the environment to which heat is discarded.

The fact that work is required in order to produce a refrigeration effect is again a basic observation which may be formalized by another negative statement that gives expression to the second law of thermodynamics: *It is impossible to construct a device that, operating in a cycle, will produce no effect other than the transfer of heat from a cooler to a hotter body.* This is known as the Clausius statement of the second law.

The important quantity for a heat pump or refrigerator is the ratio of heat removed at the low temperature to the work required, $|Q_C|/|W|$. This quantity is called the *coefficient of performance* or *cooling energy ratio* ω, and is obtained for a Carnot heat pump by dividing (*2.3a*) by (*2.4a*):

$$\omega \equiv \frac{|Q_C|}{|W|} = \frac{T_C}{T_H - T_C} \tag{2.5}$$

This equation gives the maximum possible value of ω for given values of T_C and T_H.

Example 2.3. A freezer is to be maintained at a temperature of −40(°F) on a summer day when the ambient temperature is 80(°F). In order to maintain the freezer box at −40(°F) it is necessary to remove heat from it at the rate of 70(Btu)/(min). What is the maximum possible coefficient of performance of the freezer, and what is the minimum power that must be supplied to the freezer?

The maximum coefficient of performance is obtained when the freezer operates reversibly between the temperature levels of $T_C = -40(°F) \equiv 420(R)$ and $T_H = 80(°F) \equiv 540(R)$ and is given by (*2.5*) as

$$\omega = \frac{|Q_C|}{|W|} = \frac{420}{540 - 420} = 3.5$$

Since $|Q_C| = 70\text{(Btu)/(min)}$, the minimum power requirement is

$$|W| = \frac{70}{3.5} = 20.0\text{(Btu)/(min)}$$

or

$$|W| = \frac{20.0\text{(Btu)/(min)} \times 1055\text{(J)/(Btu)}}{60\text{(s)/(min)}} = 352\text{(J)/(s)} \quad \text{or} \quad 352\text{(W)}$$

The heat discarded to the environment at T_H is given by the first law, which in this case is written:

$$|Q_H| = |W| + |Q_C| = 20 + 70 = 90\text{(Btu)/(min)} \quad \text{or} \quad 1582\text{(J)/(s)}$$

2.3 ENTROPY OF AN IDEAL GAS

The first law for a closed PVT system is given in differential form by (*1.5*): $dU = \delta Q - \delta W$. Now for a reversible process we may write (*2.1*) as $\delta Q = T\,dS$, and in addition (*1.2*) gives $\delta W = P\,dV$. Combining these three equations, we have

$$\boxed{dU = T\,dS - P\,dV} \tag{2.6}$$

This is a perfectly general equation relating the properties of a closed PVT system, in spite of the fact that it was derived for a reversible process. We have simply taken an especially simple process for the purpose of derivation, and we find that the resulting equation contains properties only and must therefore be independent of any process. It is in fact the *fundamental property relation* for closed PVT systems and for PVT systems of fixed composition, and all other property relations are derived from it, as will be shown in the next chapter.

For the special case of an ideal gas, we have from (*1.13*) that $dU = C_V\,dT$, and (*2.6*) becomes

$$C_V\,dT = T\,dS - P\,dV$$

or

$$dS = C_V\frac{dT}{T} + \frac{P}{T}dV$$

However, by the ideal-gas law $P/T = R/V$, and we get

$$dS = C_V\frac{dT}{T} + R\frac{dV}{V} \quad \text{(ideal gas)} \tag{2.7}$$

This equation shows that the entropy of an ideal gas is a function of T and V, i.e., $S = S(T,V)$. If we integrate (*2.7*), the result is

$$S = \int C_V\frac{dT}{T} + R\ln V + S_0$$

where S_0 is a constant of integration, which we have no way to determine. However, we always deal with changes of state, for which only values of ΔS are required. The subtraction necessary to give ΔS automatically drops S_0 from the equation. The process is equivalent to integration of (*2.7*) between limits:

$$\Delta S = \int_{T_1}^{T_2} C_V\frac{dT}{T} + R\ln\frac{V_2}{V_1} \tag{2.8}$$

Equation (*2.7*) can be transformed into an equation which depends on the fact that the entropy of an ideal gas is a function of T and P, i.e. $S = S(T,P)$. Differentiating $PV = RT$ and then dividing by PV gives

$$\frac{dV}{V} + \frac{dP}{P} = \frac{dT}{T}$$

Eliminating dV/V between this and (*2.7*), and using $C_V = C_P - R$, we find:

$$dS = C_P \frac{dT}{T} - R\frac{dP}{P} \quad \text{(ideal gas)} \tag{2.9}$$

and

$$\Delta S = \int_{T_1}^{T_2} C_P \frac{dT}{T} - R \ln \frac{P_2}{P_1} \tag{2.10}$$

Example 2.4. For an ideal gas with constant heat capacities, (*2.7*) and (*2.9*) integrate to give

$$S = C_V \ln T + R \ln V + S_0 \qquad \text{and} \qquad S = C_P \ln T - R \ln P + S_0'$$

where S_0 and S_0' are constants of integration. Show that a third equation of this kind is

$$S = C_V \ln P + C_P \ln V + S_0''$$

Of the many ways to derive the required equation one of the simplest is to start with the first equation above and to eliminate T by the ideal-gas law. This gives

$$S = C_V \ln P + C_V \ln V - C_V \ln R + R \ln V + S_0$$

But

$$C_V \ln V + R \ln V = (C_V + R) \ln V = C_P \ln V$$

and

$$S_0 - C_V \ln R = S_0''$$

Combination of the last three equations yields the desired result. Note that V is *molar* volume.

Example 2.5. Derive equations relating T and V, T and P, and P and V for the reversible, adiabatic compression or expansion of an ideal gas with constant heat capacities.

For a reversible process $dS = \delta Q/T$. If in addition the process is adiabatic, then $\delta Q = 0$ and $dS = 0$. In other words a reversible, adiabatic process takes place with constant entropy i.e. is *isentropic.*

If dS is set equal to zero in (*2.7*) the result is

$$\frac{dT}{T} = -\frac{R}{C_V}\frac{dV}{V}$$

For an ideal gas $R = C_P - C_V$, and therefore

$$\frac{dT}{T} = -\left(\frac{C_P - C_V}{C_V}\right)\frac{dV}{V} = -(\gamma - 1)\frac{dV}{V}$$

where $\gamma = C_P/C_V$. Integration with γ constant yields

$$\ln \frac{T_2}{T_1} = \ln \left(\frac{V_1}{V_2}\right)^{\gamma-1} \qquad \text{or} \qquad \frac{T_2}{T_1} = \left(\frac{V_1}{V_2}\right)^{\gamma-1}$$

This same result was obtained in Example 1.10 by a different method. An equivalent statement is:

$$TV^{\gamma-1} = \text{constant}$$

If we treat (*2.9*) in an analogous way, we get

$$\frac{T_2}{T_1} = \left(\frac{P_2}{P_1}\right)^{(\gamma-1)/\gamma} \qquad \text{or} \qquad \frac{T}{P^{(\gamma-1)/\gamma}} = \text{constant}$$

If the two expressions just obtained for T_2/T_1 are equated, we find:

$$\left(\frac{V_1}{V_2}\right)^{\gamma} = \frac{P_2}{P_1} \qquad \text{or} \qquad PV^{\gamma} = \text{constant}$$

Example 2.6. Show that for an ideal gas with constant heat capacities the slope of a PV curve for a reversible, adiabatic process is negative and that it has a larger absolute value than the slope of a PV curve for an isotherm at the same values of P and V.

It was shown in Example 2.5 that the relation between P and V for a reversible, adiabatic (i.e. isentropic) process is

$$PV^{\gamma} = k \qquad \text{or} \qquad P = kV^{-\gamma}$$

where k is a constant. Differentiation of this equation gives

$$\frac{dP}{dV} = k(-\gamma)V^{-\gamma-1} = -\gamma(kV^{-\gamma})(V^{-1})$$

or
$$\frac{dP}{dV} = -\gamma\frac{P}{V} \qquad \text{(isentropic)}$$

Since γ, P, and V are all positive, dP/dV is negative.

For an isotherm:
$$PV = RT \qquad \text{or} \qquad P = RTV^{-1}$$

Differentiation gives
$$\frac{dP}{dV} = RT(-1)(V^{-2}) = -(RTV^{-1})V^{-1}$$

or
$$\frac{dP}{dV} = -\frac{P}{V} \qquad \text{(isothermal)}$$

Again the slope dP/dV is negative, but since $\gamma > 1$, the reversible, adiabatic (isentropic) curve has a steeper negative slope than the isothermal curve passing through the same point. This is illustrated in Fig. 2-3.

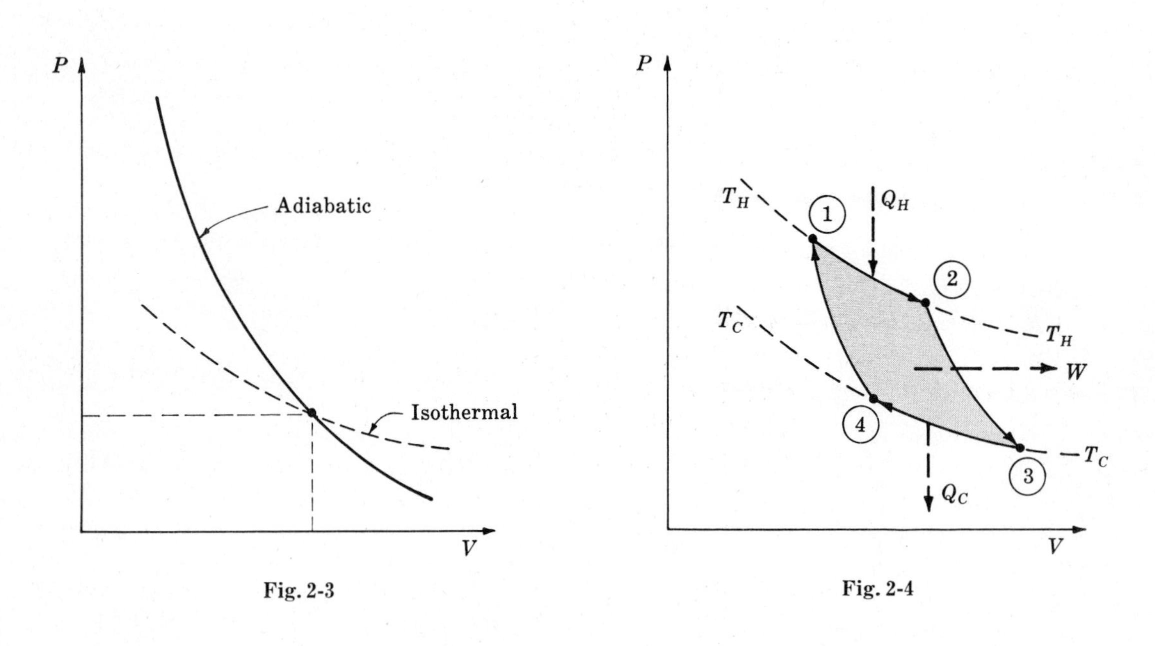

Fig. 2-3

Fig. 2-4

2.4 CARNOT CYCLE FOR AN IDEAL GAS

Our object now is to determine the shape of the cycle on a PV diagram that would be traced out by an ideal gas in a piston-and-cylinder apparatus acting as a Carnot heat engine operating between heat reservoirs at T_H and T_C.

The ideal gas in the piston-and-cylinder apparatus is the working medium of a heat engine. Since the engine must be capable of performing continuously and must itself not undergo any permanent change, it must return periodically to its initial state, i.e. it must operate in a cycle. As the ideal gas in the engine changes state, it goes through a series of values of P and V, and these represent a closed cycle on a PV diagram.

If the engine is to operate reversibly, it must absorb heat Q_H from the hot reservoir at temperature T_H, and *only* at T_H, and it must discard heat Q_C to the cold reservoir at temperature T_C, and *only* at T_C. This means that heat transfer may occur only at the temperatures T_H and T_C. Thus steps of the cycle during which heat is transferred must be isothermal, either at T_H or at T_C, and all other steps must represent adiabatic processes between these two temperatures. The only possible combination of such steps that represents a work-producing cycle is shown in Fig. 2-4, which displays a *Carnot cycle* for an ideal gas.

At point 1 the apparatus is in thermal equilibrium with the hot reservoir at T_H, and the piston is at the inner limit of its stroke. The gas is now allowed to expand reversibly and isothermally along path $1 \rightarrow 2$. During this step heat Q_H flows into the gas to compensate for the work done. At point 2 the apparatus is insulated from both heat reservoirs and the gas expands reversibly and adiabatically along path $2 \rightarrow 3$. As shown in Example 2.6, path $2 \rightarrow 3$ is steeper than path $1 \rightarrow 2$, and it leads away from the isotherm T_H, and terminates at isotherm T_C. At point 3 equilibrium is established with the cold reservoir at T_C and a reversible isothermal compression along path $3 \rightarrow 4$ carries the piston inward. The work done on the gas during this step is compensated by removal of heat Q_C. This step terminates at a point such that reversible, adiabatic compression will complete the cycle along path $4 \rightarrow 1$.

The cycle of steps just described encloses the shaded area of Fig. 2-4, and this area represents $\int P\,dV$ taken around the entire cycle, and is the net work of the engine for one complete cycle. This net work is the sum of the work terms for the various steps:

$$W = W_{12} + W_{23} + W_{34} + W_{41}$$

Processes $2 \rightarrow 3$ and $4 \rightarrow 1$ are adiabatic, and for them the first law and the relation (*1.13*) for an ideal gas give:

$$W_{23} = -\Delta U_{23} = -\int_{T_H}^{T_C} C_V\,dT$$

and
$$W_{41} = -\Delta U_{41} = -\int_{T_C}^{T_H} C_V\,dT = \int_{T_H}^{T_C} C_V\,dT$$

Evidently W_{23} and W_{41} are equal numerically but of opposite sign. Therefore they cancel, and

$$W = W_{12} + W_{34}$$

The terms on the right are for isothermal processes, and since the gas is ideal, there is no internal energy change. In this case the first law reduces to

$$Q_{12} = W_{12} \quad \text{and} \quad Q_{34} = W_{34}$$

Using the result of Example 1.3 for the work of a reversible, isothermal process, we get

$$Q_H = Q_{12} = W_{12} = RT_H \ln \frac{P_1}{P_2} \quad \text{and} \quad Q_C = Q_{34} = W_{34} = RT_C \ln \frac{P_3}{P_4}$$

Therefore
$$\frac{Q_C}{Q_H} = \frac{T_C \ln (P_3/P_4)}{T_H \ln (P_1/P_2)}$$

However, from Example 2.5 we have

$$\frac{T_H}{T_C} = \left(\frac{P_2}{P_3}\right)^{(\gamma-1)/\gamma} \quad \text{and} \quad \frac{T_H}{T_C} = \left(\frac{P_1}{P_4}\right)^{(\gamma-1)/\gamma}$$

Hence
$$P_2/P_3 = P_1/P_4 \quad \text{or} \quad P_1/P_2 = P_4/P_3$$

As a result
$$\frac{Q_C}{Q_H} = -\frac{T_C}{T_H}$$

which is (*2.3*). Written with absolute value signs it becomes (*2.3a*):

$$\frac{|Q_C|}{|Q_H|} = \frac{T_C}{T_H}$$

Combination with the first-law equation for the cycle,

$$|W| = |Q_H| - |Q_C|$$

leads immediately to (*2.4a*):

$$\frac{|W|}{|Q_H|} = 1 - \frac{T_C}{T_H}$$

The fact that we regenerate the basic expressions (*2.3a*) and (*2.4a*) by consideration of the properties of an ideal gas as it serves as the working medium of a Carnot heat engine means that the temperature used in the ideal-gas law is entirely consistent with the absolute temperature used in (*2.1*), the basic relation connecting entropy with measurable quantities. This has been implicitly assumed all along, but now we have established a formal connection.

Many cyclic processes are possible that do not operate on a Carnot cycle. Examples are the Otto cycle of gasoline engines and the Diesel cycle. Any reversible cyclic process is subject to the same kind of thermodynamic analysis as a Carnot cycle.

Example 2.7. Develop an expression for the thermal efficiency of a reversible heat engine operating on the Otto cycle with an ideal gas of constant heat capacity as the working medium.

The Otto cycle consists of two constant-volume steps during which heat is transferred, connected by two adiabatics, as shown in Fig. 2-5.

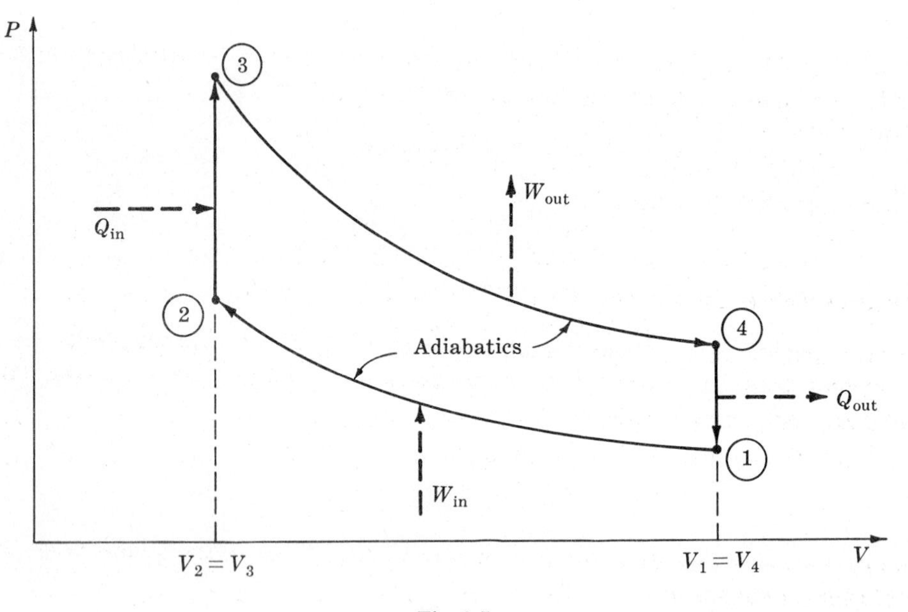

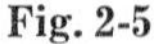

Fig. 2-5

For the various steps of the cycle, we have:

Step $1 \to 2$: $Q_{12} = 0$ $\quad W_{12} = -\Delta U_{12} = -C_V(T_2 - T_1)$

Step $2 \to 3$: $W_{23} = 0$ $\quad Q_{23} = \Delta U_{23} = C_V(T_3 - T_2)$

Step $3 \to 4$: $Q_{34} = 0$ $\quad W_{34} = -\Delta U_{34} = -C_V(T_4 - T_3)$

Step $4 \to 1$: $W_{41} = 0$ $\quad Q_{41} = \Delta U_{41} = C_V(T_1 - T_4)$

The thermal efficiency of the cycle is defined as

$$\eta = \frac{W_{\text{cycle}}}{Q_{\text{in}}} = \frac{W_{12} + W_{34}}{Q_{23}} = \frac{-C_V(T_2 - T_1) - C_V(T_4 - T_3)}{C_V(T_3 - T_2)} = 1 - \frac{T_4 - T_1}{T_3 - T_2}$$

Since steps $1 \rightarrow 2$ and $3 \rightarrow 4$ represent reversible, adiabatic processes, we may write (see Example 2.5):

$$\frac{T_1}{T_2} = \left(\frac{V_2}{V_1}\right)^{\gamma-1} \quad \text{and} \quad \frac{T_4}{T_3} = \left(\frac{V_3}{V_4}\right)^{\gamma-1}$$

However, $V_2 = V_3$ and $V_1 = V_4$, and therefore the right sides of the above equations are equal. As a result

$$\frac{T_1}{T_2} = \frac{T_4}{T_3} \quad \text{and} \quad T_4 = \frac{T_1 T_3}{T_2}$$

Elimination of T_4 in the equation for η gives

$$\eta = 1 - \frac{T_1}{T_2} \quad \text{or} \quad \eta = 1 - \left(\frac{V_2}{V_1}\right)^{\gamma-1}$$

The quantity V_1/V_2 is known as the *compression ratio* r, and this provides yet another equation for thermal efficiency:

$$\eta = 1 - \left(\frac{1}{r}\right)^{\gamma-1}$$

Clearly η gets larger as r increases, but η becomes unity only as $r \rightarrow \infty$.

2.5 ENTROPY AND EQUILIBRIUM

In the mathematical statement of the second law

$$\Delta S_{\text{total}} \geqq 0$$

the subscript "total" indicates that both the system and its surroundings must be taken into account. An alternative expression makes this explicit:

$$\Delta S_{\text{system}} + \Delta S_{\text{surroundings}} \geqq 0$$

The first law may be expressed similarly

$$\Delta E_{\text{system}} + \Delta E_{\text{surroundings}} = 0$$

where E represents energy in general.

An *isolated* system is a system completely cut off from its surroundings, and changes in such a system have no effect on the surroundings. In this case we need consider the system only, and our equations become

$$\Delta S_{\text{system}} \geqq 0$$

$$\Delta E_{\text{system}} = 0$$

For an isolated system the total energy is constant, and the entropy can only increase or in the limit remain constant.

If the system is in an equilibrium state, its properties, entropy included, do not change. Thus the equality $\Delta S_{\text{system}} = 0$ implies that equilibrium has been reached, whereas the inequality $\Delta S_{\text{system}} > 0$ implies a change toward equilibrium. Since the entropy of the system can only *increase*, this must mean that the equilibrium state of an isolated system is that state for which the entropy has its maximum value with respect to all possible variations. The mathematical condition for this maximum is

$$dS_{\text{system}} = 0 \quad \text{(isolated system)}$$

Example 2.8. Consider a cylinder closed at both ends and containing gas divided into two parts by a piston, as in Fig. 2-6 below. Let the cylinder be a perfect heat insulator so as to isolate the system it contains. The piston will be a heat conductor and will be imagined frictionless. How is P_1 related to P_2 and how is T_1 related to T_2 at equilibrium?

Since the system is isolated, we may apply the equilibrium criterion $dS_{\text{system}} = 0$. To do this we need expressions for dS_1^t and dS_2^t, and these come from the property relation (2.6) solved for dS. Thus

$$dS_1^t = \frac{dU_1^t}{T_1} + \frac{P_1}{T_1}dV_1^t \qquad dS_2^t = \frac{dU_2^t}{T_2} + \frac{P_2}{T_2}dV_2^t$$

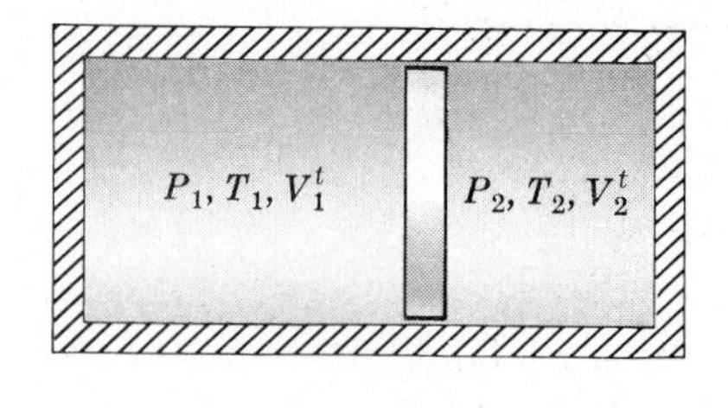

Fig. 2-6

The sum of these gives dS_{system}:

$$dS_{\text{system}} = \frac{dU_1^t}{T_1} + \frac{dU_2^t}{T_2} + \frac{P_1}{T_1}dV_1^t + \frac{P_2}{T_2}dV_2^t$$

Since the system is isolated and there are no kinetic- or potential-energy changes, its internal energy is constant. So is its volume. Thus

$$dU_2^t = -dU_1^t \quad \text{and} \quad dV_2^t = -dV_1^t$$

and we can write for equilibrium

$$dS_{\text{system}} = \left(\frac{1}{T_1} - \frac{1}{T_2}\right)dU_1^t + \left(\frac{P_1}{T_1} - \frac{P_2}{T_2}\right)dV_1^t = 0$$

The question now is how this equation can be valid for all imaginable variations in the independent variables U_1^t and V_1^t. Clearly the coefficients of dU_1^t and dV_1^t must be zero. Thus

$$\frac{1}{T_1} - \frac{1}{T_2} = 0 \quad \text{and} \quad \frac{P_1}{T_1} - \frac{P_2}{T_2} = 0$$

From these it is obvious that for equilibrium

$$T_1 = T_2 \quad \text{and} \quad P_1 = P_2$$

This answer was evident from the start. The important point is that our abstract criterion for equilibrium leads to results known to be correct. It can be applied to nontrivial problems as well.

Solved Problems

HEAT ENGINES AND HEAT PUMPS (Secs. 2.1 and 2.2)

2.1. (*a*) A reversible heat engine operates between a heat reservoir at $T_1 = 1200$(R) and another heat reservoir at $T_2 = 500$(R). If 100(Btu) of heat is transferred to the heat engine from the reservoir at T_1, how much work is done by the engine? (*b*) The engine would produce more work if the temperature of the cold reservoir could be lowered to 400(R). Someone suggests that a finite "cold box" be held at 400(R) by refrigeration, and proposes to remove heat from the "cold box" by a reversible heat pump which discharges heat to the original reservoir at $T_2 = 500$(R). The scheme is represented in Fig. 2-7. If $Q_1 = 100$(Btu), calculate the *net* work W of the process, and compare the result with the answer to part (*a*).

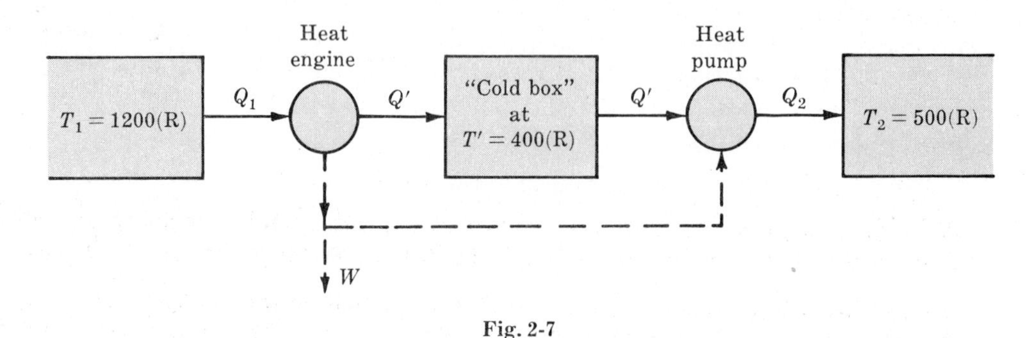

Fig. 2-7

(*a*) By (*2.4a*)

$$\frac{|W|}{|Q_1|} = 1 - \frac{T_2}{T_1} = 1 - \frac{500}{1200} = 0.5833$$

Thus

$$|W| = 0.5833|Q_1| = 58.33(\text{Btu}) \quad \text{or} \quad 61{,}540(\text{J})$$

(*b*) The result must be the same as in part (*a*), because the process represented is just a more complicated, but entirely equivalent, heat engine operating reversibly between the same two temperatures as the engine in part (*a*). However if one needs proof it is given as follows. The net work W is the difference between the total work of the heat engine and the work delivered to the heat pump:

$$W = |W|_{\text{engine}} - |W|_{\text{heat pump}} \tag{1}$$

For the engine (*2.4a*) becomes

$$|W|_{\text{engine}} = |Q_1|\left(1 - \frac{T'}{T_1}\right) = \frac{|Q_1|}{T_1}(T_1 - T') \tag{2}$$

Application of (*2.5*) to the heat pump gives

$$|W|_{\text{heat pump}} = \frac{|Q'|}{T'}(T_2 - T') \tag{3}$$

Hence

$$W = \frac{|Q_1|}{T_1}(T_1 - T') - \frac{|Q'|}{T'}(T_2 - T')$$

However, (*2.3a*) applied to the engine becomes

$$|Q'| = |Q_1|\frac{T'}{T_1}$$

from which

$$W = \frac{|Q_1|}{T_1}(T_1 - T') - \frac{|Q_1|}{T_1}(T_2 - T') = |Q_1|\left(\frac{T_1 - T_2}{T_1}\right)$$

the formula used in part (*a*).

2.2. A thermoelectric device cools a small refrigerator and discards heat to the surroundings at 25(°C). The maximum electric power for which the device is designed is 100(W). The heat load on the refrigerator (that is, the heat leaking through the walls which must be removed by the thermoelectric device) is 350(J)/(s). What is the minimum temperature that can be maintained in the refrigerator?

The thermoelectric device is a heat pump, subject to the same thermodynamic limitations as any other heat pump. The minimum temperature is attained when the device operates reversibly at maximum power. For a reversible heat pump we have from (*2.5*)

$$\left|\frac{W}{Q_C}\right| = \frac{T_H}{T_C} - 1 \quad \text{or} \quad T_C = \frac{T_H}{\left|\frac{W}{Q_C}\right| + 1}$$

We have the following values for substitution

$$T_H = 25 + 273 = 298(\text{K}) \qquad W = 100(\text{W}) = 100(\text{J})/(\text{s}) \qquad Q_C = 350(\text{J})/(\text{s})$$

As a result

$$T_C = \frac{298(\text{K})}{\frac{100}{350} + 1} = 232(\text{K}) \quad \text{or} \quad -41(°\text{C})$$

2.3. A kitchen is provided with a standard refrigerator and a window air conditioner. How is it that the refrigerator heats the kitchen whereas the air conditioner cools it?

Both the refrigerator and the air conditioner are heat pumps, and both most likely operate on the same principle and are mechanically very similar. The difference lies in the fact that heat discarded by the refrigerator enters the room (usually through coils on the back of the refrigerator) whereas heat is discarded by the air conditioner outside the window. The situation is depicted in Fig. 2-8.

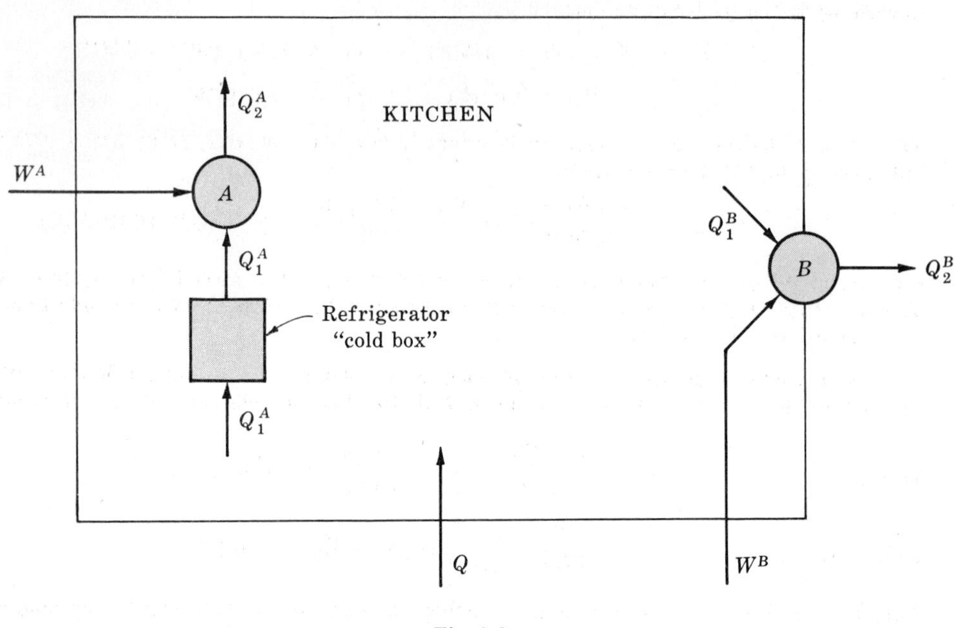

Fig. 2-8

The heat pumps are indicated by circles; the one designated A is in the refrigerator, and B is in the air conditioner. The work to drive both heat pumps is presumed to be supplied electrically, and this energy comes from outside the kitchen as shown. The refrigerator heat pump A removes heat $|Q_1^A|$ from the refrigerator "cold box" and discards heat $|Q_2^A|$ into the kitchen. An energy balance on heat pump A shows that $|Q_2^A| = |Q_1^A| + |W^A|$. Now if the "cold box" of the refrigerator stays at constant temperature, then the heat pump removes just the amount of heat that passes into the cold box through its walls from the kitchen. Thus the net effect of the refrigerator on the kitchen is to add energy as heat in an amount equal to the work $|W^A|$ of running the refrigerator.

The air conditioner removes heat $|Q_1^B|$ from the kitchen and discards heat $|Q_2^B|$ *outside* the kitchen. The air conditioner is merely a refrigerator that has the kitchen as its "cold box." The work $|W^B|$ enters heat pump B from outside the kitchen, and $|Q_2^B|$ is larger than $|Q_1^B|$ by just the amount of this work. If the kitchen is maintained at constant temperature then the heat $|Q_2^B|$ discarded by the air conditioner is equal to the heat $|Q|$ leaking through the walls of the kitchen from outside plus the work of running both the refrigerator and the air conditioner. That is

$$|Q_2^B| = |Q| + |W^A| + |W^B|$$

2.4. An inventor claims to have devised a cyclic engine which exchanges heat with reservoirs at 80(°F) and 510(°F), and which can produce 0.45(Btu) of work per Btu of heat extracted from the hot reservoir. Is his claim feasible?

The engine is cyclic and no net changes occur in its properties during its operation. Thus the results of Example 2.2 apply. Equation (*2*) of that example is

$$W = -Q_H - Q_C$$

where the Q's refer to the heat reservoirs. This equation expresses the first law for the engine, and no decision on the feasibility of the inventor's claims can be made on the basis of this equation, because no value is given for Q_C. However, substituting $Q_C = -W - Q_H$ into (*1*) of Example 2.2 gives as the total entropy change resulting from the operation of the engine

$$\Delta S_{\text{total}} = \frac{Q_H}{T_H} - \frac{W + Q_H}{T_C}$$

According to the problem statement

$$T_C = 80 + 460 = 540(\text{R}) \qquad T_H = 510 + 460 = 970(\text{R})$$

$$W = 0.45(\text{Btu}) \qquad Q_H = -1.00(\text{Btu})$$

The value of Q_H is negative because it refers to the hot reservoir, from which heat is extracted. Substitution of these values gives:

$$\Delta S_{\text{total}} = \frac{-1.00(\text{Btu})}{970(\text{R})} - \frac{[0.45 - 1.00](\text{Btu})}{540(\text{R})} = -0.000012(\text{Btu})/(\text{R})$$

for every Btu of heat taken from the reservoir at T_H. The second law requires that ΔS_{total} be positive or at least no less than zero for a process to be possible. We conclude that the inventor has claimed the impossible.

An alternative solution to the problem is to calculate the thermal efficiency $|W/Q_H|$ claimed for the engine and to compare this figure with the thermal efficiency of a reversible engine.

Thus

$$\frac{|W|}{|Q_H|}_{\text{claimed}} = \frac{0.45}{1.00} = 0.45$$

and, by (*2.4a*),

$$\frac{|W|}{|Q_H|}_{\text{reversible}} = 1 - \frac{540}{970} = 0.443$$

This second figure is the *maximum* possible efficiency for a heat engine exchanging heat with reservoirs at the stated temperatures, and since the claimed efficiency exceeds this, the claim is not possible.

2.5. (*a*) With reference to Example 2.7 and Fig. 2-5, take the working medium of an engine operating on the Otto cycle to be an ideal gas for which $C_V = 5$ and $C_P = 7(\text{Btu})/(\text{lb mole})(\text{R})$. The conditions at point 1 will be $P_1 = 1(\text{atm})$ and $T_1 = 140(°\text{F})$. The compression ratio of the engine is such as to make $P_2 = 15(\text{atm})$ and heat is added in step $2 \to 3$ sufficient to make $T_3 = 3940(°\text{F})$. Determine the values of Q and W for the four steps of the cycle, the net work of the cycle, and the thermal efficiency of the engine. (*b*) If the engine accomplishes the *same changes of state as in* (*a*), but if each step is now 75% efficient, recompute the thermal efficiency of the engine.

(*a*) In order to compute the required quantities by means of the equations of Example 2.7, we need first to determine the temperatures at the various points of the cycle. Given are

$$T_1 = 140 + 460 = 600(\text{R}) \qquad T_3 = 3940 + 460 = 4400(\text{R})$$

Since step $1 \to 2$ is reversible and adiabatic, we have

$$T_2 = T_1\left(\frac{P_2}{P_1}\right)^{(\gamma-1)/\gamma}$$

with

$$\gamma = \frac{C_P}{C_V} = \frac{7}{5} = 1.4 \qquad (\gamma - 1)/\gamma = 0.286$$

Thus

$$T_2 = (600)(15)^{0.286} = 1300(\text{R})$$

Example 2.7 provides the formula

$$T_4 = \frac{T_1 T_3}{T_2} = \frac{(600)(4400)}{1300} = 2030(\text{R})$$

Then the equations for Q and W of Example 2.7 become:

Step $1 \rightarrow 2$: $Q_{12} = 0$ $W_{12} = -\Delta U_{12} = -C_V(T_2 - T_1) = -3500(\text{Btu})/(\text{lb mole})$

Step $2 \rightarrow 3$: $W_{23} = 0$ $Q_{23} = \Delta U_{23} = C_V(T_3 - T_2) = 15{,}500(\text{Btu})/(\text{lb mole})$

Step $3 \rightarrow 4$: $Q_{34} = 0$ $W_{34} = -\Delta U_{34} = -C_V(T_4 - T_3) = 11{,}850(\text{Btu})/(\text{lb mole})$

Step $4 \rightarrow 1$: $W_{41} = 0$ $Q_{41} = \Delta U_{41} = C_V(T_1 - T_4) = -7150(\text{Btu})/(\text{lb mole})$

The thermal efficiency of the cycle can be calculated by the defining equation

$$\eta = \frac{W_{\text{cycle}}}{Q_{\text{in}}} = \frac{W_{12} + W_{34}}{Q_{23}} = \frac{-3500 + 11{,}850}{15{,}500} = 0.539$$

As shown in Example 2.7 it is also given by

$$\eta = 1 - \frac{T_1}{T_2} = 1 - \frac{600}{1300} = 0.539$$

The net work of the cycle is

$$W_{\text{cycle}} = W_{12} + W_{34} = -3500 + 11{,}850 = 8350(\text{Btu})/(\text{lb mole})$$

Thus when operating reversibly the engine delivers as work only about 54% of the heat supplied to the engine, and discards the remaining 46% to the surroundings as waste heat. This quantity is

$$Q_{41} = -7150(\text{Btu})/(\text{lb mole})$$

There is no way to avoid this discarding of heat from a heat engine, and it contributes substantially to the thermal pollution of our surroundings.

(*b*) For an engine that operates irreversibly, but goes through the *same changes of state* as the reversible engine of part (*a*), we may calculate the work of each step by assigning an efficiency to the process. For steps that produce work the efficiency is defined as the ratio of the actual work to the reversible work:

$$\text{efficiency} = W/W_{\text{rev}} \quad \text{(work produced)}$$

This quantity is always less than unity, because it reflects the extent to which irreversibilities reduce the work output in comparison with the reversible work. For steps that require work, the reversible work is the minimum work, for irreversibilities always increase the work requirement. So in this case the efficiency is defined by

$$\text{efficiency} = W_{\text{rev}}/W \quad \text{(work required)}$$

This efficiency too is always less than unity.

For the present problem, step $1 \rightarrow 2$ requires work, and for an efficiency of 75% we have

$$W_{12} = \frac{W_{\text{rev}}}{0.75} = \frac{-3500}{0.75} = -4670(\text{Btu})/(\text{lb mole})$$

From the first law we have

$$\Delta U_{12} = Q_{12} - W_{12}$$

but ΔU_{12} is the same as before, namely 3500(Btu)/(lb mole)

Thus $$Q_{12} = \Delta U_{12} + W_{12} = 3500 - 4670 = -1170(\text{Btu})/(\text{lb mole})$$

and we see that in the irreversible engine this step cannot be adiabatic. In fact 1170(Btu)/(lb mole) must be transferred to the surroundings.

Step $3 \rightarrow 4$ produces work, and for an efficiency of 75% the amount is given by

$$W_{34} = (0.75)W_{\text{rev}} = (0.75)(11{,}850) = 8900(\text{Btu})/(\text{lb mole})$$

The heat transfer is again calculated by the first law:

$$Q_{34} = \Delta U_{34} + W_{34} = -11{,}850 + 8900 = -2950(\text{Btu})/(\text{lb mole})$$

Steps $2 \rightarrow 3$ and $4 \rightarrow 1$ remain as before, because at constant volume no work is done. The net work of the cycle is

$$W_{\text{cycle}} = W_{12} + W_{34} = -4670 + 8900 = 4230(\text{Btu})/(\text{lb mole})$$

and
$$\eta = \frac{W_{cycle}}{Q_{23}} = \frac{4230}{15{,}500} = 0.273$$

Only the basic definition for thermal efficiency η can be used here, because the derived formulas of Example 2.7 depend upon the assumption of reversibility. It is worthy of note that use of an efficiency of 75% for each step of the cycle has reduced the thermal efficiency of the engine by a factor of about 2. Thus for the irreversible engine only 27.3% of the heat taken into the engine is delivered as work, and the remaining 72.7% is discarded as waste heat to the surroundings. This amounts to

$$Q_{out} = Q_{12} + Q_{34} + Q_{41} = -1170 - 2950 - 7150 = -11{,}270\text{(Btu)/(lb mole)}$$

These results are typical of heat engines in actual operation. A comparison of results for parts (*a*) and (*b*) follows.

		(*a*) Reversible	(*b*) Irreversible
Q_{12}	(Btu)/(lb mole)	0	−1170
W_{12}	(Btu)/(lb mole)	−3500	−4670
Q_{23}	(Btu)/(lb mole)	15,500	15,500
W_{23}	(Btu)/(lb mole)	0	0
Q_{34}	(Btu)/(lb mole)	0	−2950
W_{34}	(Btu)/(lb mole)	11,850	8900
Q_{41}	(Btu)/(lb mole)	−7150	−7150
W_{41}	(Btu)/(lb mole)	0	0
W_{cycle}	(Btu)/(lb mole)	8350	4230
Q_{out}	(Btu)/(lb mole)	−7150	−11,270
η		0.539	0.273

2.6. Derive an expression for the thermal efficiency of a reversible heat engine operating on the Diesel cycle with an ideal gas of constant heat capacity as the working medium.

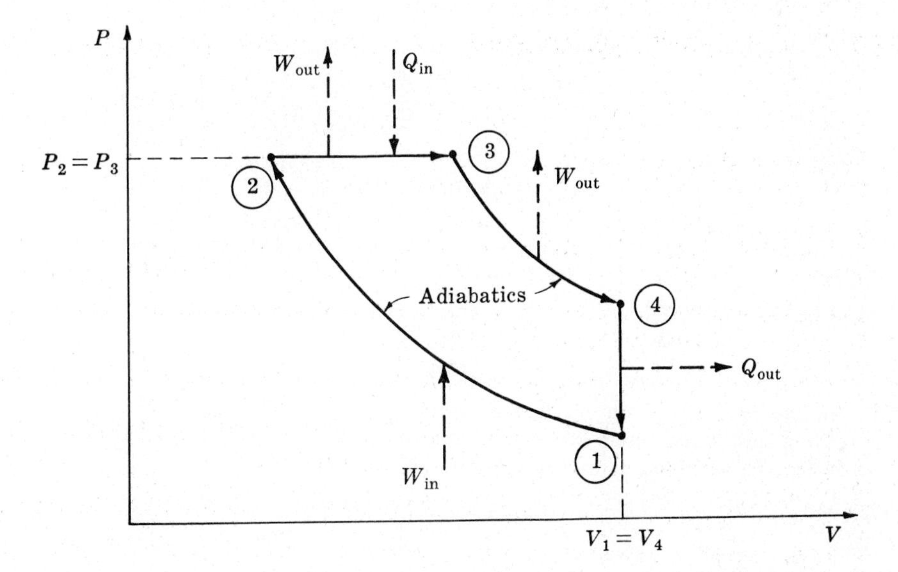

Fig. 2-9

As shown in Fig. 2-9, the Diesel cycle consists of a constant-pressure and a constant-volume step, during both of which heat is transferred, connected by two adiabatics.

For the four steps of the cycle, we have:

Step $1 \to 2$: $Q_{12} = 0 \quad W_{12} = -\Delta U_{12} = -C_V(T_2 - T_1)$

Step $2 \to 3$: $W_{23} = P_2(V_3 - V_2) = P_3(V_3 - V_2); \; \Delta U_{23} = C_V(T_3 - T_2)$

$$Q_{23} = W_{23} + \Delta U_{23} = P_2(V_3 - V_2) + C_V(T_3 - T_2)$$

Step $3 \to 4$: $Q_{34} = 0 \quad W_{34} = -\Delta U_{34} = -C_V(T_4 - T_3)$

Step $4 \to 1$: $W_{41} = 0 \quad Q_{41} = \Delta U_{41} = C_V(T_1 - T_4)$

The thermal efficiency of the cycle is

$$\eta = \frac{W_{\text{cycle}}}{Q_{\text{in}}} = \frac{W_{12} + W_{23} + W_{34}}{Q_{23}} = \frac{-C_V(T_2 - T_1) + P_2(V_3 - V_2) - C_V(T_4 - T_3)}{P_2(V_3 - V_2) + C_V(T_3 - T_2)}$$

But $P_2 = P_3$, $P_2V_2 = RT_2$, and $P_3V_3 = RT_3$, so that

$$\eta = \frac{-C_V(T_2 - T_1) + R(T_3 - T_2) - C_V(T_4 - T_3)}{R(T_3 - T_2) + C_V(T_3 - T_2)} = 1 - \frac{C_V(T_4 - T_1)}{(C_V + R)(T_3 - T_2)}$$

But

$$\frac{C_V}{C_V + R} = \frac{C_V}{C_P} = \frac{1}{\gamma}$$

so

$$\eta = 1 - \frac{1}{\gamma}\frac{T_4 - T_1}{T_3 - T_2} = 1 - \frac{1}{\gamma}\frac{T_1}{T_2}\frac{(T_4/T_1) - 1}{(T_3/T_2) - 1}$$

Now steps $1 \to 2$ and $3 \to 4$ are reversible, adiabatic processes, so (see Example 2.5)

$$\frac{T_1}{T_2} = \left(\frac{V_2}{V_1}\right)^{\gamma - 1} \qquad \frac{P_2}{P_1} = \left(\frac{V_1}{V_2}\right)^{\gamma}$$

$$\frac{P_4}{P_3} = \left(\frac{V_3}{V_4}\right)^{\gamma} = \left(\frac{V_2}{V_4}\frac{V_3}{V_2}\right)^{\gamma} = \left(\frac{V_2}{V_1}\frac{V_3}{V_2}\right)^{\gamma}$$

Thus

$$\frac{T_4}{T_1} = \frac{P_4V_4/R}{P_1V_1/R} = \frac{P_4}{P_1}\frac{V_4}{V_1} = \frac{P_4}{P_1} = \frac{P_4}{P_3}\frac{P_3}{P_2}\frac{P_2}{P_1} = \frac{P_4}{P_3}\frac{P_2}{P_1}$$

$$= \left(\frac{V_2}{V_1}\frac{V_3}{V_2}\right)^{\gamma}\left(\frac{V_1}{V_2}\right)^{\gamma} = \left(\frac{V_3}{V_2}\right)^{\gamma}$$

and

$$\frac{T_3}{T_2} = \frac{P_3V_3/R}{P_2V_2/R} = \frac{P_3}{P_2}\frac{V_3}{V_2} = \frac{V_3}{V_2}$$

from which

$$\eta = 1 - \frac{1}{\gamma}\left(\frac{V_2}{V_1}\right)^{\gamma - 1}\frac{(V_3/V_2)^{\gamma} - 1}{(V_3/V_2) - 1}$$

As it was for the Otto cycle (Example 2.7), the ratio V_1/V_2 is here called the *compression ratio* r; the ratio V_3/V_2 is known as the *cutoff ratio* and is designated r_c. The thermal efficiency η_D of the Diesel cycle can then be written

$$\eta_D = 1 - \left(\frac{1}{r}\right)^{\gamma - 1}\frac{r_c^{\gamma} - 1}{\gamma(r_c - 1)}$$

Heat engines operating both on the Otto and Diesel cycles (or rather the real counterparts of the idealizations presented here) are extensively employed as mechanical power sources, and it is of interest to compare their efficiencies. From Example 2.7, the ideal thermal efficiency of an Otto cycle is

$$\eta_O = 1 - \left(\frac{1}{r}\right)^{\gamma - 1}$$

and, for a fixed compression ratio r, η_D and η_O differ only in the term multiplying $(1/r)^{\gamma - 1}$, which is unity for the Otto cycle and

$$\frac{r_c^\gamma - 1}{\gamma(r_c - 1)}$$

for the Diesel cycle. Both the cutoff ratio and γ are greater than unity, and the above term can be shown to be always greater than unity for these conditions. Therefore, when compared at the same compression ratio, a reversible Otto cycle employing an ideal gas as working fluid is more efficient than a reversible Diesel cycle. However, preignition and knocking difficulties limit the compression ratios for real engines operating on the Otto cycle, so Diesel engines can be operated at greater compression ratios and resultant greater efficiencies than Otto engines.

2.7. A central power plant, whether the energy source is nuclear or fossil fuel, is a heat engine operating between the temperatures of the reactor or furnace and the surroundings, usually represented by a river or other body of water. Consider a modern nuclear power plant generating 750,000(kW) for which the reactor temperature is 600(°F) and a river is available with a water temperature of 70(°F). (*a*) What is the maximum possible thermal efficiency of the plant and what is the minimum amount of heat that must be discarded to the river? (*b*) If the actual thermal efficiency of the plant is 60% of the maximum, how much heat must be discarded into the river, and what will be the temperature rise of the river if it has a flow rate of 5800(ft)^3/(s)?

(*a*) The maximum thermal efficiency is given by (*2.4a*), which is applicable to a reversible heat engine operating between two fixed temperatures:

$$\eta = 1 - \frac{T_C}{T_H} = 1 - \frac{70 + 460}{600 + 460} = 0.5$$

Thus in this ideal situation 50% of the heat supplied by the nuclear reactor is converted into work and the other 50% is discarded to the river. Clearly, the amount of heat discarded is equal to the amount of work done, and the rate of heat transfer to the river is 750,000(kW) $\equiv$ 750,000,000(J)/(s) or

$$Q_C = \frac{750{,}000{,}000(\text{J})/(\text{s})}{1055(\text{J})/(\text{Btu})} = 710{,}900(\text{Btu})/(\text{s})$$

(*b*) If the actual value of η is 60% of that calculated in part (*a*) then

$$\eta = (0.6)(0.5) = 0.30$$

Since $\eta = |W|/|Q_H|$, we have

$$|Q_H| = \frac{|W|}{\eta} = \frac{750{,}000{,}000(\text{J})/(\text{s})}{0.30} = 2{,}500{,}000{,}000(\text{J})/(\text{s})$$

Of this, the 70% not converted into work is discarded to the river. Thus

$$Q_C = (0.7)(2.5 \times 10^9) = 1.75 \times 10^9(\text{J})/(\text{s}) \quad \text{or} \quad 1.659 \times 10^6(\text{Btu})/(\text{s})$$

The flow rate of the river is

$$5800(\text{ft})^3/(\text{s}) \times 62.4(\text{lb}_\text{m})/(\text{ft})^3 = 0.3619 \times 10^6(\text{lb}_\text{m})/(\text{s})$$

Since $Q_C = mC_P\,\Delta T$, we have

$$\Delta T = \frac{Q_C}{mC_P}$$

Taking the heat capacity of water as 1(Btu)/(lb$_\text{m}$)(°F), we get

$$\Delta T = \frac{1.659 \times 10^6(\text{Btu})/(\text{s})}{0.3619 \times 10^6(\text{lb}_\text{m})/(\text{s}) \times 1(\text{Btu})/(\text{lb}_\text{m})(°\text{F})} = 4.58(°\text{F})$$

This is the temperature rise of an entire river of moderate size occasioned by a power plant of typical size, and is the most commonly mentioned form of thermal pollution.

ENTROPY CALCULATIONS (Secs. 2.3 and 2.4)

2.8. Heat is transferred directly from a heat reservoir at 540(°F) to another heat reservoir at 40(°F). If the amount of heat transferred is 100(Btu), what is the total entropy change as a result of this process?

The two temperatures of the heat reservoirs are

$$T_H = 540 + 460 = 1000(\text{R}) \quad \text{and} \quad T_C = 40 + 460 = 500(\text{R})$$

The heat transferred is 100(Btu), and therefore

$$Q_H = -100(\text{Btu}) \quad (\text{heat out of reservoir } H)$$

$$Q_C = +100(\text{Btu}) \quad (\text{heat into reservoir } C)$$

Since for heat reservoirs $\Delta S = Q/T$, we have

$$\Delta S_H = \frac{Q_H}{T_H} = \frac{-100(\text{Btu})}{1000(\text{R})} = -0.10(\text{Btu})/(\text{R})$$

$$\Delta S_C = \frac{Q_C}{T_C} = \frac{100(\text{Btu})}{500(\text{R})} = 0.20(\text{Btu})/(\text{R})$$

Thus $$\Delta S_{\text{total}} = \Delta S_H + \Delta S_C = -0.10 + 0.20 = +0.10(\text{Btu})/(\text{R})$$

This same result is obtained from the general formula derived for this kind of process in Example 2.1:

$$\Delta S_{\text{total}} = Q_C\left(\frac{T_H - T_C}{T_H T_C}\right) = 100\left(\frac{1000 - 500}{1000 \times 500}\right) = +0.1(\text{Btu})/(\text{R})$$

2.9. For any closed PVT system show that

(*a*) $dS = \dfrac{C_V\,dT}{T}$ for a process at constant volume.

(*b*) $dS = \dfrac{C_P\,dT}{T}$ for a process at constant pressure.

(*a*) At constant volume the fundamental property relation (*2.6*) for a closed PVT system becomes $dU = T\,dS$. However, (*1.10*) shows that for such a system at constant volume $dU = C_V\,dT$. Combination of these two equations gives $dS = C_V\,dT/T$ (const. V).

(*b*) For a constant-pressure process (*1.7*) becomes $dH = dU + P\,dV$. Substitution for dU by (*2.6*) gives $dH = T\,dS - P\,dV + P\,dV = T\,dS$. However, (*1.11*) shows that for a closed PVT system at constant pressure $dH = C_P\,dT$. Combination of these last two equations provides $dS = C_P\,dT/T$ (const. P).

2.10. If 2(kg) of liquid water at 90(°C) is mixed adiabatically and at constant pressure with 3(kg) of liquid water at 10(°C), what is the total entropy change resulting from this process? For simplicity take the heat capacity of water to be constant at $C_P = 1(\text{cal})/(\text{g})(\text{K})$ or 4184(J)/(kg)(K).

The two masses of water in effect exchange heat under constant-pressure conditions until they reach the same final temperature T. Thus for constant C_P

$$Q_C = m_C C_P(T - T_C) = -Q_H = -m_H C_P(T - T_H)$$

where the subscripts C and H denote the initial conditions for the cooler and hotter masses of water. Thus

$$m_C(T - T_C) + m_H(T - T_H) = 0 \qquad \text{or} \qquad T = \frac{m_C T_C + m_H T_H}{m_C + m_H}$$

Substituting $T_C = 10 + 273 = 283(\text{K})$, $T_H = 90 + 273 = 363(\text{K})$, we find

$$T = \frac{(3)(283) + (2)(363)}{5} = 315(\text{K})$$

It was shown in Problem 2.9 that for a constant-pressure process

$$dS = \frac{C_P\,dT}{T}$$

Integration with C_P constant gives

$$\Delta S = C_P \ln \frac{T_f}{T_i}$$

Application of this equation for each of the two masses of water provides

$$\Delta S_C = C_P \ln \frac{T}{T_C} = 4184 \ln \frac{315}{283} = 448.2(\text{J})/(\text{kg})(\text{K})$$

$$\Delta S_H = C_P \ln \frac{T}{T_H} = 4184 \ln \frac{315}{363} = -593.4(\text{J})/(\text{kg})(\text{K})$$

Since the process is adiabatic, there is no entropy change of the surroundings, and

$$\Delta S_{\text{total}} = m_C\,\Delta S_C + m_H\,\Delta S_H = (3)(448.2) + (2)(-593.4) = +157.8(\text{J})/(\text{K})$$

2.11. In Example 2.4 it was shown that for an ideal gas with constant heat capacities the entropy can be expressed as a function of P and V by the equation

$$S = C_V \ln P + C_P \ln V + S_0''$$

(*a*) If the molar entropy of an ideal gas for which $C_V = 3$ and $C_P = 5(\text{cal})/(\text{g mole})(\text{K})$ is taken as zero at 1(atm) and 0(°C), evaluate the constant S_0''. What units must be used for P and V in the resulting equation for S? (*b*) Calculate S for $1000(\text{cm})^3$ of the ideal gas at 20(atm) and 50(°C).

(*a*) At the conditions $P = 1(\text{atm})$ and $T = 0(°\text{C})$ or 273.15(K), the ideal-gas law gives V as

$$V = \frac{RT}{P} = \frac{82.05(\text{cm}^3\text{-atm})/(\text{g mole})(\text{K}) \times 273.15(\text{K})}{1(\text{atm})} = 22{,}412(\text{cm})^3/(\text{g mole})$$

Substitution of values in the equation for S at the conditions for which S is zero gives

$$0 = [3(\text{cal})/(\text{g mole})(\text{K}) \times \ln 1] + [5(\text{cal})/(\text{g mole})(\text{K}) \times \ln 22{,}412] + S_0''$$

from which

$$S_0'' = -50.087(\text{cal})/(\text{g mole})(\text{K}) \qquad \text{and} \qquad S = C_V \ln P + C_P \ln V - 50.087$$

In this equation we have committed ourselves to units of atmospheres for P, cubic centimeters per gram mole for V and (cal)/(g mole)(K) for S, C_V, and C_P.

(*b*) For $1000(\text{cm})^3$ at 20(atm) and $50(°\text{C}) = 323.15(\text{K})$ we calculate the number of gram moles by the ideal-gas law:

$$n = \frac{PV^t}{RT} = \frac{20(\text{atm}) \times 1000(\text{cm})^3}{82.05(\text{cm}^3\text{-atm})/(\text{g mole})(\text{K}) \times 323.15(\text{K})} = 0.7543(\text{g mole})$$

The molar volume is

$$V = \frac{V^t}{n} = \frac{1000(\text{cm})^3}{0.7543(\text{g mole})} = 1325.73(\text{cm})^3/(\text{g mole})$$

The molar entropy as given by the equation of part (*a*) is

$$S = 3 \ln 20 + 5 \ln 1325.73 - 50.087 = -5.1512(\text{cal})/(\text{g mole})(\text{K})$$

Thus the entropy of $1000(\text{cm})^3$ or 0.7543(g mole) is

$$S^t = -5.1512(\text{cal})/(\text{g mole})(\text{K}) \times 0.7543(\text{g mole})$$

$$= -3.886(\text{cal})/(\text{K}) \quad \text{or} \quad -16.26(\text{J})/(\text{K})$$

2.12. One kg mole of an ideal gas is compressed isothermally at 127(°C) from 1(atm) to 10(atm) in a piston-and-cylinder arrangement. Calculate the entropy change of the gas, the entropy change of the surroundings, and the total entropy change resulting from the process, if (*a*) the process is mechanically reversible and the surroundings consist of a heat reservoir at 127(°C); (*b*) the process is mechanically reversible and the surroundings consist of a heat reservoir at 27(°C); (*c*) the process is mechanically irreversible, requiring 20% more work than the mechanically reversible compression, and the surroundings consist of a heat reservoir at 27(°C).

(*a*) The work of a mechanically reversible, isothermal *expansion* of an ideal gas was shown in Example 1.3 to be given by $W = RT \ln (P_i/P_f)$. Exactly the same expression applies to a mechanically reversible, isothermal *compression* of an ideal gas. Since T is $127 + 273$ or 400(K),

$$W = 8314(\text{J})/(\text{kg mole})(\text{K}) \times 400(\text{K}) \times \ln (1/10) = -7{,}657{,}500(\text{J})/(\text{kg mole})$$

The system contains just 1(kg mole), and therefore $W = -7{,}657{,}500(\text{J})$.

For an ideal gas the internal energy is a function of temperature only, and since T is constant, $\Delta U = 0$. Thus the first law requires that

$$Q = W = -7{,}657{,}500(\text{J})$$

This heat leaves the system (the gas) and flows to the surroundings (the heat reservoir). Thus

$$Q_{\text{surr}} = -Q = +7{,}657{,}500(\text{J})$$

The temperature of the system and of the surroundings is 127(°C) and is constant. Moreover the process is reversible, and therefore

$$\Delta S = \frac{Q}{T} = \frac{-7{,}657{,}500}{400} = -19{,}144(\text{J})/(\text{K})$$

$$\Delta S_{\text{surr}} = \frac{Q_{\text{surr}}}{T_{\text{surr}}} = \frac{7{,}657{,}500}{400} = +19{,}144(\text{J})/(\text{K})$$

and

$$\Delta S_{\text{total}} = \Delta S + \Delta S_{\text{surr}} = 0$$

Clearly, this result is required by the second law, because the process is reversible, completely reversible. Not only is the compression mechanically reversible, but also the heat transfer from system to surroundings, both at 127(°C), is reversible.

(*b*) The only difference between this process and that of part (*a*) is that the heat reservoir is at a temperature of 27(°C) rather than 127(°C), the system temperature. This means that the process is no longer completely reversible, because heat transfer is now across a finite temperature difference. However, the compression is still accomplished by a mechanically reversible process, and the system changes along exactly the same path as before. Thus ΔU, W, Q, Q_{surr}, and ΔS are unchanged from the values of part (*a*). However, ΔS_{surr} and ΔS_{total} are different:

$$\Delta S_{\text{surr}} = \frac{Q_{\text{surr}}}{T_{\text{surr}}} = \frac{7{,}657{,}500}{27 + 273} = \frac{7{,}657{,}500(\text{J})}{300(\text{K})} = 25{,}525(\text{J})/(\text{K})$$

$$\Delta S_{\text{total}} = \Delta S + \Delta S_{\text{surr}} = -19{,}144 + 25{,}525 = 6381(\text{J})/(\text{K})$$

Since the process is irreversible the second law requires that $\Delta S_{\text{total}} > 0$, and our result is clearly in accord with this requirement.

(*c*) The process described includes the same irreversibility as in part (*b*), but in addition the compression step is now mechanically irreversible, and because of this the work required is greater by 20%. Thus

$$W = -7{,}657{,}500 \times 1.20 = -9{,}189{,}000(\mathrm{J})$$

However, the process accomplishes exactly the same change of state in the gas as before, and as a result the property changes of the gas are the same. Thus $\Delta U = 0$, and the first law still requires that $Q = W$. Therefore

$$Q = -9{,}189{,}000(\mathrm{J}) \quad \text{and} \quad Q_{\text{surr}} = -Q = +9{,}189{,}000(\mathrm{J})$$

The entropy change of the heat reservoir (surroundings) is given by

$$\Delta S_{\text{surr}} = \frac{Q_{\text{surr}}}{T_{\text{surr}}} = \frac{9{,}189{,}000(\mathrm{J})}{300(\mathrm{K})} = 30{,}630(\mathrm{J})/(\mathrm{K})$$

The entropy change of the system is a property change like ΔU, and must have the same value as before, i.e. $\Delta S = -19{,}144(\mathrm{J})/(\mathrm{K})$. Thus

$$\Delta S_{\text{total}} = \Delta S + \Delta S_{\text{surr}} = -19{,}144 + 30{,}630 = 11{,}486(\mathrm{J})/(\mathrm{K})$$

Note that the entropy change of the system is *not* given as Q/T, because the process is mechanically irreversible, which means it is internally irreversible. The heat reservoir, however, suffers no internal irreversibilities. The irreversibility of heat transfer is in fact external to both the system and surroundings. The quantity Q/T for the system if calculated gives

$$\frac{Q}{T} = \frac{-9{,}189{,}000}{400} = -22{,}972(\mathrm{J})/(\mathrm{K})$$

and this *is* the entropy change in the system *caused by the heat transfer*. However, it is not the *entire* entropy change of the system, because the mechanical irreversibility within the system causes an increase in entropy that partially compensates the entropy decrease caused by heat transfer. Recognition that this is the case does not, however, provide a means for the calculation of entropy changes of systems subjected to mechanically irreversible processes. Fortunately, there is no need, because entropy is a state function, and its values are independent of the process causing a change of state.

The results for the three different parts of this problem are summarized as follows:

	$W = Q$ (J)	ΔU (J)	ΔS (J)/(K)	ΔS_{surr} (J)/(K)	ΔS_{total} (J)/(K)
(*a*)	−7,657,500	0	−19,144	+19,144	0
(*b*)	−7,657,500	0	−19,144	+25,525	+6,381
(*c*)	−9,189,000	0	−19,144	+30,630	+11,486

The degree of irreversibility of the three processes is reflected in the values of ΔS_{total}. For the completely reversible process this is, of course, zero. For parts (*b*) and (*c*) it is positive, with the value for (*c*) being larger than for (*b*), because (*c*) includes the irreversibility of (*b*) and a further irreversibility as well.

2.13. One mole of an ideal gas expands isothermally in a piston-and-cylinder device from an initial pressure of 2(atm) to a final pressure of 1(atm). The device is surrounded by the atmosphere which exerts a constant pressure of 1(atm) on the outer face of the piston. Moreover, the device is always in thermal equilibrium with the atmosphere, which constitutes a heat reservoir at a temperature of 540(R). During the expansion process a frictional force is exerted on the piston and this force varies in such a way that it always almost balances the net pressure forces on the piston. Thus the piston moves very slowly and experiences negligible acceleration. The piston and cylinder are good heat conductors. Determine the entropy change of the gas, the entropy change of the atmosphere, and the total entropy change brought about by the process.

If one takes as the system not only the gas but also the piston and cylinder, then clearly the only external force on the system is caused by P_0, the pressure of the surrounding atmosphere. The work done by the system is therefore $W = P_0 \Delta V$ where ΔV is given by the ideal-gas law:

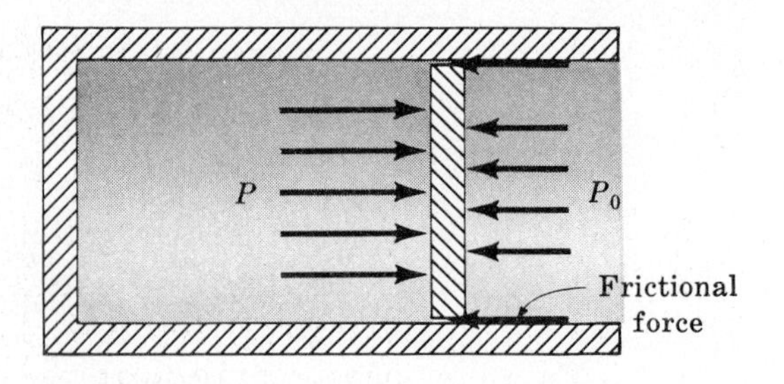

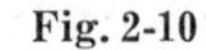

Fig. 2-10

$$\Delta V = V_2 - V_1 = \frac{RT}{P_2} - \frac{RT}{P_1}$$

$$= RT\left(\frac{P_1 - P_2}{P_1 P_2}\right)$$

Thus
$$W = \frac{P_0 RT}{P_1 P_2}(P_1 - P_2)$$

However, $P_0 = P_2$, so that
$$W = \frac{RT}{P_1}(P_1 - P_2)$$

This work is literally work done in pushing back the atmosphere. The first-law equation is $\Delta U = Q - W$. But $\Delta U = 0$, because the process is isothermal and for an ideal gas U is a function of T only. The properties of the piston and cylinder are also presumed to be constant at constant T. Hence

$$Q = W \quad \text{and} \quad Q_{\text{surr}} = -Q = -W$$

Moreover
$$\Delta S_{\text{surr}} = \frac{Q_{\text{surr}}}{T} = \frac{-W}{T} = \frac{-R}{P_1}(P_1 - P_2) = -\frac{1}{2}R$$

The entropy change of the gas is given by (*2.10*), which for constant T is

$$\Delta S = -R \ln \frac{P_2}{P_1} = -R \ln \frac{1}{2} = 0.6932\,R$$

The total entropy change is therefore

$$\Delta S_{\text{total}} = \Delta S + \Delta S_{\text{surr}} = 0.6932\,R - 0.5000\,R = 0.1932\,R$$

The process is irreversible on account of the frictional force acting on the piston, and this is clearly reflected in the positive value obtained for ΔS_{total}, even though the frictional force was never explicitly considered in the problem solution. The basic reason that it all comes out right is that entropy is a property, and the entropy change of the gas is computed by an equation that is independent of the process.

It is instructive to rework the problem with the gas alone taken as the system. Since the process occurs very slowly the pressure is uniform throughout the system, and is the pressure acting on the inner face of the piston. Processes such as this are sometimes said to be *quasi-static*, i.e. almost static. The notion of internal equilibrium is equivalent. Note that the frictional force does not act on the gas, which is the system, so that we may use the formula of Example 1.3 for the work of a reversible isothermal expansion of an ideal gas:

$$W = RT \ln \frac{P_1}{P_2} = RT \ln \frac{2}{1} = RT \ln 2$$

This is the work done by the gas on the piston, and by the first law $Q = W = RT \ln 2$. Thus heat must flow into the gas in an amount equal to the work done by it. The entropy change of the gas is then

$$\frac{Q}{T} = R \ln 2 = 0.6932\,R$$

just as before. Now the heat which flows to the gas comes from two sources. The frictional force acting between the piston and cylinder converts part of the work done by the gas into heat which in effect flows into the gas. This heat does not come from the surrounding atmosphere and does not affect the entropy of the surroundings. Only the *additional* heat required to maintain constant temperature comes from the surroundings and this is equal to the work done by the piston against the atmosphere, as previously calculated.

MISCELLANEOUS APPLICATIONS

2.14. Consider an ideal gas for which $C_V = 5$ and $C_P = 7$(Btu)/(lb mole)(°F). The initial state is 1(atm) and 70(°F). (*a*) One pound mole is heated at constant volume to 170(°F). Calculate ΔU, ΔH, ΔS, Q, and W. (*b*) One pound mole is heated at constant pressure to 170(°F). Calculate $\Delta U, \Delta H, \Delta S, Q$, and W if the process is reversible. (*c*) One pound mole is changed from its initial state to a final state of 10(atm) and 170(°F) by several different reversible processes. Which of the following quantities will be the same for all processes: $\Delta U, \Delta H, \Delta S, Q$, and W? (*d*) Repeat part (*c*) for irreversible processes.

For an ideal gas with constant heat capacities, regardless of the process

$$\Delta U = C_V \Delta T \qquad \text{from } (1.13)$$

$$\Delta H = C_P \Delta T \qquad \text{from } (1.15)$$

$$\Delta S = C_V \ln \frac{T_2}{T_1} + R \ln \frac{V_2}{V_1} \qquad \text{from } (2.8)$$

or

$$\Delta S = C_P \ln \frac{T_2}{T_1} - R \ln \frac{P_2}{P_1} \qquad \text{from } (2.10)$$

(*a*) Substitution into the first three of the above equations gives for one pound mole of gas:

$$\Delta U = 5 \times 100 = 500(\text{Btu})$$

$$\Delta H = 7 \times 100 = 700(\text{Btu})$$

$$\Delta S = 5 \ln \frac{170 + 460}{70 + 460} = 5 \ln 1.189 = 0.866(\text{Btu})/(\text{R})$$

Since the volume is constant $W = 0$ and by the first law

$$Q = \Delta U = 500(\text{Btu})$$

(*b*) As in part (*a*)

$$\Delta U = 500(\text{Btu})$$

$$\Delta H = 700(\text{Btu})$$

By the second equation for ΔS

$$\Delta S = 7 \ln 1.189 = 1.212(\text{Btu})/(\text{R})$$

From Example 1.7 we have that $Q = \Delta H = 700(\text{Btu})$ and by the first law

$$W = Q - \Delta U = 700 - 500 = 200(\text{Btu})$$

(*c*) The property changes ΔU, ΔH, and ΔS, will be the same for all processes.

(*d*) The answers here are identical to those of part (*c*). Property changes depend only on the initial and final states of the system, independent of the process, reversible or irreversible.

2.15. One gram mole of an ideal gas for which $C_V = 6$ and $C_P = 8(\text{cal})/(\text{g mole})(\text{K})$ expands adiabatically from an initial state at 340(K) and 5(atm) to a final state where its volume has doubled. Find the final temperature of the gas, the work done, and the entropy change of the gas, for (*a*) a reversible expansion and (*b*) a free expansion of the gas into an evacuated space (Joule expansion).

(*a*) For a reversible, adiabatic expansion of an ideal gas with constant heat capacities, we have from Example 2.5:

$$T_2 = T_1 \left(\frac{V_1}{V_2}\right)^{\gamma - 1}$$

Here $\gamma = C_P/C_V = 8/6 = 1.333$ and $V_1/V_2 = 0.5$. Thus

$$T_2 = 340 \times (0.5)^{0.333} = 269.9(\text{K})$$

By the first law for an adiabatic process, and for an ideal gas with constant heat capacities

$$W = -\Delta U = -C_V \Delta T = -6(269.9 - 340) = 420.6(\text{cal}) \quad \text{or} \quad 1760(\text{J})$$

By the second law, the entropy change for a reversible adiabatic process is zero.

(*b*) Expansion of the gas into an evacuated space produces no work. Since the process is also adiabatic both Q and W are zero. Thus the first law requires that $\Delta U = 0$, and for an ideal gas with constant heat capacities this means that $C_V \Delta T = 0$, which implies that $\Delta T = 0$. Hence $T_2 = T_1$.

The entropy change is given by (*2.8*), which becomes

$$\Delta S = R \ln \frac{V_2}{V_1} = 2 \ln 2 = 1.386(\text{cal})/(\text{K})$$

the value of $R = 2(\text{cal})/(\text{g mole})(\text{K})$ being used as an approximation so as to be consistent with the given values of C_V and C_P and with the equation $R = C_P - C_V$. The value of ΔS for the gas is also ΔS_{total}, because $\Delta S_{\text{surr}} = 0$ for an adiabatic process. The positive value of ΔS_{total} is required by the second law for this completely irreversible process.

2.16. In Example 2.2 an equation was derived for the work obtainable from a heat engine. Deduce a similar expression for a work-producing device which again exchanges heat with hot and cold reservoirs, but whose properties show a net change during the process.

As in Example 2.2, the entropy changes of the reservoirs are $\Delta S_H = Q_H/T_H$ and $\Delta S_C = Q_C/T_C$, where Q_H and Q_C refer to the reservoirs. There is in addition an entropy change for the device (engine), which we designate simply ΔS. The total entropy change accompanying the process is then

$$\Delta S_{\text{total}} = \frac{Q_H}{T_H} + \frac{Q_C}{T_C} + \Delta S \tag{1}$$

The first law written for the engine is

$$\Delta U = Q - W$$

where ΔU is the change in internal energy of the engine. Solving for W and noting that $Q = -Q_H - Q_C$, one then obtains as a consequence of the first law

$$W = -Q_H - Q_C - \Delta U \tag{2}$$

Combination of (*1*) and (*2*) to eliminate Q_H gives the desired expression:

$$W = -\Delta U - T_H(\Delta S_{\text{total}} - \Delta S) + Q_C\left(\frac{T_H}{T_C} - 1\right) \tag{3}$$

Although (*3*) is similar to (*3*) of Example 2.2, the two equations are for different kinds of processes. The processes described in Example 2.2 are cyclic, and represent methods for continuously converting heat to work, while those considered in this problem may not be cyclic or continuous. However, the two work expressions become identical if ΔU and ΔS are zero, as for a cyclic engine.

2.17. A mass of liquid water, $m = 10(\text{lb}_\text{m})$, initially in thermal equilibrium with the atmosphere at 70(°F) is cooled at constant pressure to 40(°F) by means of heat pumps operating between the mass of water and the atmosphere. What is the minimum work required? For water take $C_P = 1.0(\text{Btu})/(\text{lb}_\text{m})(\text{R})$.

The minimum work is required if the process is reversible, and we can imagine a series of reversible heat pumps operating so as to remove heat from the water at various temperature levels as the water cools from 70 to 40(°F) and discharging heat to the atmosphere at $T_0 = 70 + 460 = 530(\text{R})$. Each heat pump removes a differential amount of heat δQ and reduces the water temperature differentially. For a reversible heat pump (*2.5*) may be written

$$\frac{|W|}{|Q_C|} = \frac{T_H - T_C}{T_C} = \frac{T_H}{T_C} - 1$$

In the present notation T_H becomes T_0 and T_C becomes T, the water temperature; in addition the removal of a differential amount of heat δQ with the performance of a differential amount of work δW for a given temperature T, requires that the equation be cast in differential form. Thus

$$\frac{|\delta W|}{|\delta Q|} = \frac{T_0}{T} - 1$$

The process is depicted in Fig. 2-11.

We may remove the absolute value signs in the above equation by noting that the numerical value of δW is negative (work done on the process) and that the numerical value of δQ is also negative *provided that δQ is taken to refer to the water.* With this understanding, $\delta W/\delta Q$ is positive, consistent with the condition that $(T_0/T) > 1$. Thus we write

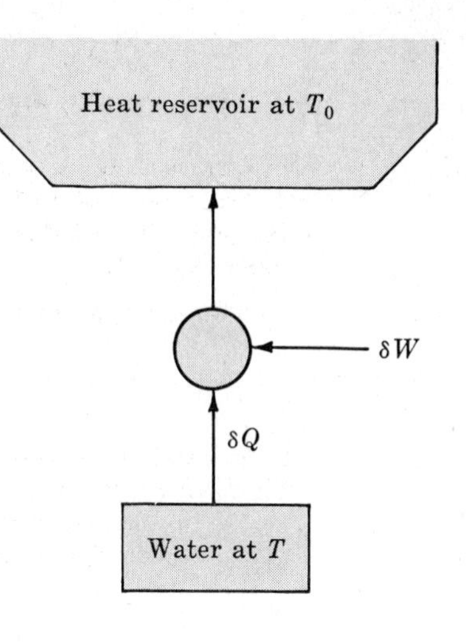

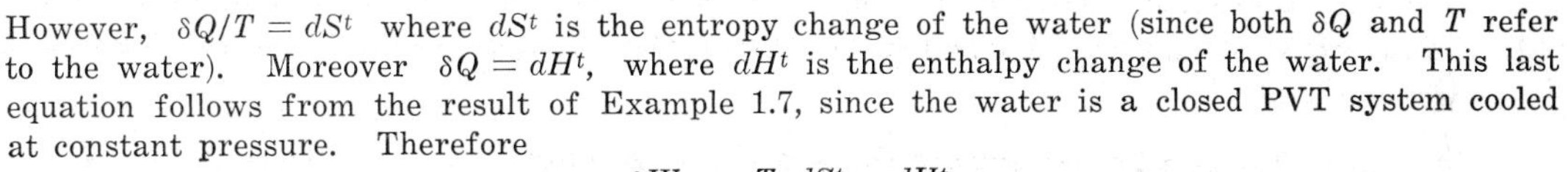
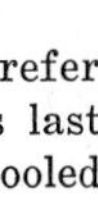

$$\frac{\delta W}{\delta Q} = \frac{T_0}{T} - 1 \qquad \text{or} \qquad \delta W = T_0 \frac{\delta Q}{T} - \delta Q$$

Fig. 2-11

However, $\delta Q/T = dS^t$ where dS^t is the entropy change of the water (since both δQ and T refer to the water). Moreover $\delta Q = dH^t$, where dH^t is the enthalpy change of the water. This last equation follows from the result of Example 1.7, since the water is a closed PVT system cooled at constant pressure. Therefore

$$\delta W = T_0\, dS^t - dH^t$$

and integration over the entire process, during which T_0 is constant, gives

$$W = T_0\, \Delta S^t - \Delta H^t \tag{1}$$

where ΔS^t and ΔH^t are property changes of the water. These are evaluated by

$$\Delta H^t = mC_P\, \Delta T$$

which follows from *(1.11)* when C_P is constant, and

$$\Delta S^t = mC_P \ln \frac{T_2}{T_1}$$

which follows similarly from the result of Problem 2.9(*b*). Finally, we have by substitution into *(1)*:

$$W = mC_P\left[T_0 \ln \frac{T_2}{T_1} - (T_2 - T_1)\right]$$

where subscripts 1 and 2 denote the initial and final water temperatures. Since

$$T_1 = 70 + 460 = 530(\text{R}) \qquad T_2 = 40 + 460 = 500(\text{R})$$

then $$W = 10(\text{lb}_\text{m}) \times 1.0(\text{Btu})/(\text{lb}_\text{m})(\text{R})\left[530(\text{R}) \times \ln \frac{500}{530} + 30(\text{R})\right] = -8.82(\text{Btu})$$

Supplementary Problems

HEAT ENGINES AND HEAT PUMPS (Secs. 2.1 and 2.2)

2.18. A reversible engine operates on the Carnot cycle between the temperature levels of 800(°F) and 80(°F). What is the percentage of the heat taken in that is converted into work? *Ans.* 57.1%

2.19. (*a*) Prove that the thermal efficiency of a reversible engine operating on the Carnot cycle is greater than that of a reversible Otto engine operating between the same temperature limits. Assume that the working fluid is an ideal gas with constant heat capacity. (*b*) At what compression ratio must a reversible Otto engine be run to obtain the same thermal efficiency as the Carnot engine of Problem 2.18? Assume that the working fluid is an ideal gas of constant heat capacity with $\gamma = 1.4$. *Ans.* (*b*) $r = 8.3$

2.20. Calculate the minimum work required to manufacture 5(lb$_m$) of ice cubes from water initially at 32(°F). Assume that the surroundings are at 80(°F). The latent heat of fusion of water at 32(°F) is 143.4(Btu)/(lb$_m$). *Ans.* 70.0(Btu)

2.21. The Joule cycle, shown in Fig. 2-12, consists of two constant-pressure steps connected by two adiabatics. Show that the thermal efficiency of a reversible heat engine operating on this cycle, with an ideal gas of constant heat capacity as the working medium, is

$$\eta = 1 - r_P^{(1-\gamma)/\gamma}$$

where $r_P = P_2/P_1 = P_3/P_4$ and is called the *pressure ratio*.

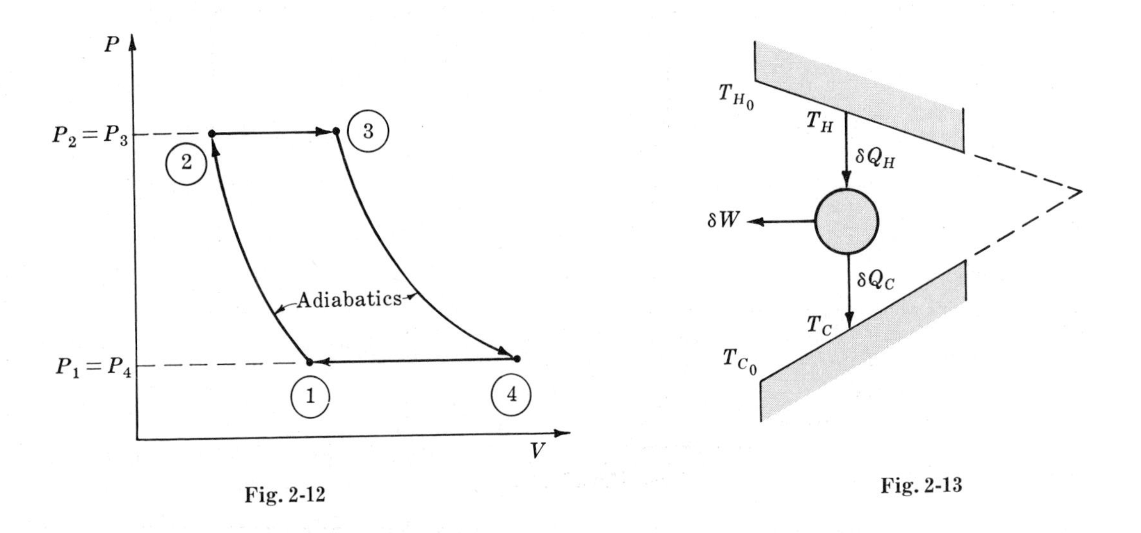

Fig. 2-12

Fig. 2-13

2.22. Figure 2-13 depicts a cyclic engine exchanging heat with two constant-pressure "reservoirs" with *finite* total heat capacities C_{PH} and C_{PC}. At the beginning of the process, the temperatures of the hot and cold "reservoirs" are T_{H_0} and T_{C_0}, respectively, and as the process progresses T_H decreases and T_C increases. (*a*) Find an equation for the maximum work obtainable by decreasing the hot "reservoir" temperature from T_{H_0} to any final temperature T_H. Express your result in terms of T_H, C_{PH}, C_{PC}, T_{H_0}, and T_{C_0}. (*b*) What is the lowest possible T_H consistent with the solution to part (*a*)?

Ans. (*a*) $W_{max} = C_{PH}(T_{H_0} - T_H) - C_{PC}T_{C_0}[(T_{H_0}/T_H)^{C_{PH}/C_{PC}} - 1]$

(*b*) $T_{H,min} = (T_{C_0}^{C_{PC}} T_{H_0}^{C_{PH}})^{1/(C_{PC}+C_{PH})}$

ENTROPY CALCULATIONS (Secs. 2.3 and 2.4)

2.23. (*a*) If the system in Problem 1.24 consists of 0.02(lb mole) of an ideal gas with constant heat capacity $C_V = 5$(Btu)/(lb mole)(R), calculate ΔS^t of the gas for steps 12, 23, and 31. (*b*) What would ΔS^t be for step 12 if this path were instead an isotherm?

Ans. (*a*) $\Delta S^t_{12} = +0.0367$(Btu)/(R), $\Delta S^t_{23} = -0.1283$(Btu)/(R), $\Delta S^t_{31} = +0.0916$(Btu)/(R)

(*b*) the same

2.24. (*a*) The temperature of an ideal gas with constant heat capacity is changed from T_1 to T_2. Show that ΔS of the gas is greater if the change in state occurs at constant pressure than if at constant volume. (*b*) The pressure of an ideal gas is changed from P_1 to P_2 by an isothermal process and by a constant-volume process. Show that ΔS for the gas is of opposite sign for the two processes.

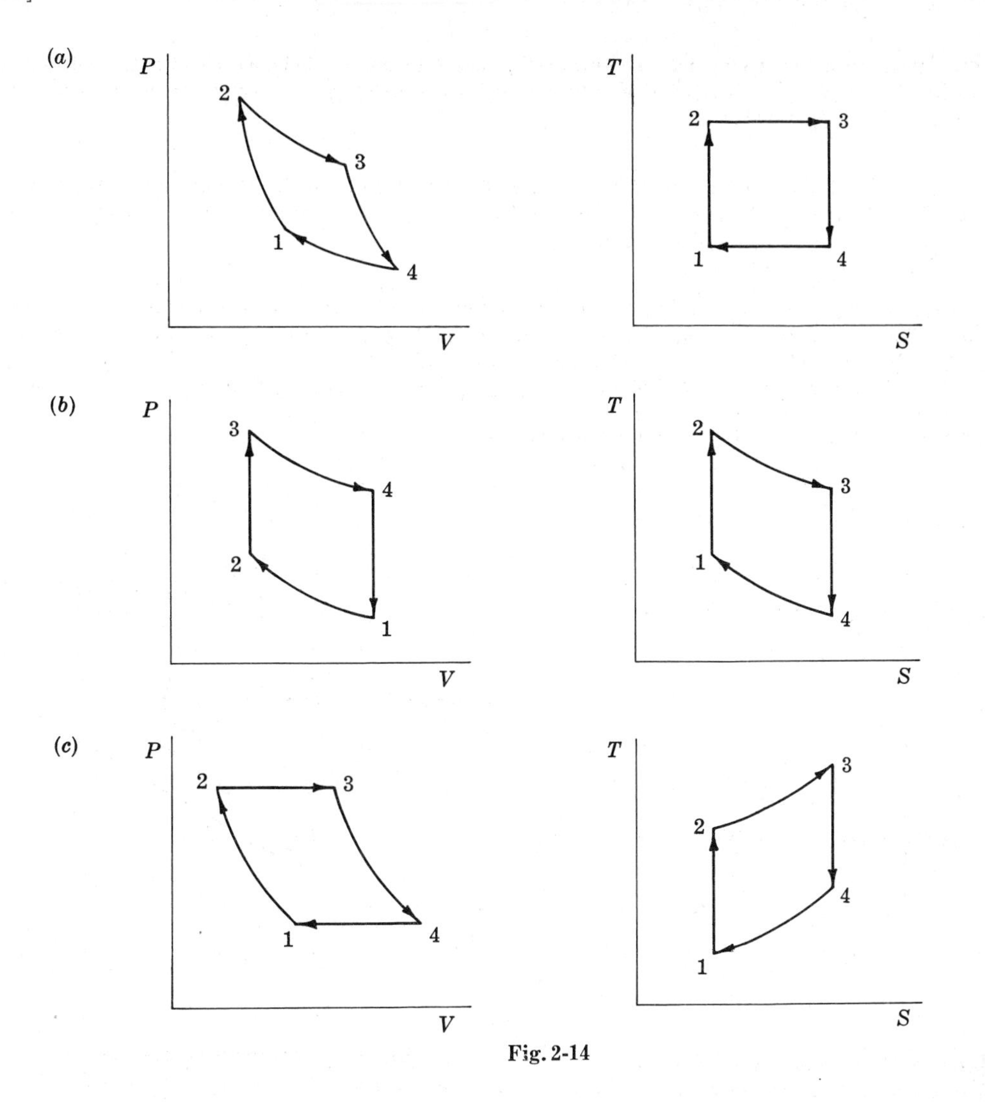

Fig. 2-14

2.25. Three pairs of diagrams are given in Fig. 2-14. The PV diagrams are for an ideal gas of constant heat capacity undergoing (*a*) a reversible Carnot cycle, (*b*) a reversible Otto cycle (Example 2.7), and (*c*) a reversible Joule cycle (Problem 2.21). Paths 12 and 34 are adiabatics in all three cases. Are the TS diagrams qualitatively consistent with the corresponding PV diagrams?

Ans. (*a*) yes; (*b*) no; (*c*) yes

2.26. Two bodies A and B with constant total heat capacities C_{PA} and C_{PB} and different initial temperatures T_{A_0} and T_{B_0} exchange heat with each other under isobaric conditions. No heat is transferred to or from the surroundings, and the temperatures of both bodies are assumed uniform at all times. (*a*) Derive an expression for the total entropy change ΔS of the two bodies as a function of T_A, the temperature of body A. (*b*) Demonstrate that the process is irreversible by showing that $\Delta S > 0$. (*c*) What is the equilibrium temperature of the system? (*Hint*: Problem 2.10 is an example of this type of problem.)

Ans. (*a*) $\Delta S = C_{PA} \ln (T_A/T_{A_0}) + C_{PB} \ln \left[\dfrac{T_{B_0} + (C_{PA}/C_{PB})(T_{A_0} - T_A)}{T_{B_0}} \right]$

(*c*) $T_A = T_B = (C_{PA}T_{A_0} + C_{PB}T_{B_0})/(C_{PA} + C_{PB})$

MISCELLANEOUS APPLICATIONS

2.27. An ideal gas for which $C_V = 3$ and $C_P = 5$(Btu)/(lb mole)(°F) expands adiabatically and reversibly from 3(atm) to 1(atm) in a piston-and-cylinder apparatus. If $T_1 = 600(°F)$, determine T_2, ΔU, ΔH, and W.

Ans. $T_2 = 223(°F)$, $\Delta U = -W = -1131(\text{Btu})/(\text{lb mole})$, $\Delta H = -1885(\text{Btu})/(\text{lb mole})$

2.28. In Example 2.1 a heat reservoir was defined as a body with an infinite heat capacity. Show why the temperature of such a body remains constant during the transfer of a finite amount of heat.

2.29. Show that the same result obtains for Example 2.1 if, instead of exchanging heat directly with the cold reservoir, the hot reservoir exchanges heat with an intermediate reservoir of temperature T_M, which in turn exchanges heat with the cold reservoir.

2.30. Figure 2-15 shows the relation of temperature to entropy for a closed PVT system during a reversible process. Calculate the heat added to the system for each of the three steps 12, 23, and 31, and for the entire process 1231.

Ans. $Q_{12} = 1600(\text{cal})$, $Q_{23} = -1000(\text{cal})$, $Q_{31} = 0$, $Q_{\text{total}} = 600(\text{cal})$

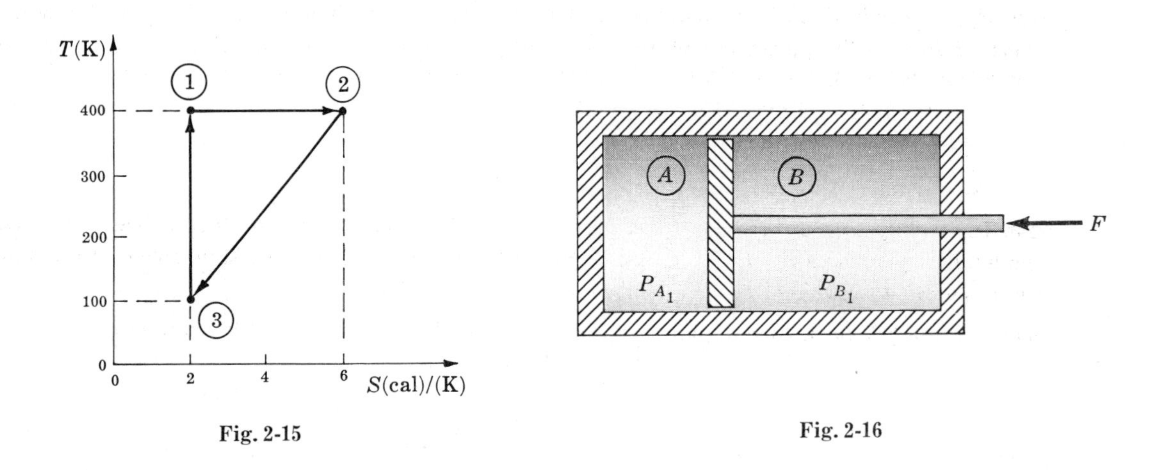

Fig. 2-15 Fig. 2-16

2.31. A horizontal cylinder, closed at both ends, is divided in half by a free piston. The cylinder neither absorbs nor conducts heat, and the piston is perfectly lubricated in the cylinder. The left half of the cylinder contains one lb mole of an ideal gas at 2(atm) and 60(°F). The gas has constant heat capacities, $C_V = 5$ and $C_P = 7(\text{Btu})/(\text{lb mole})(°\text{F})$. The right half of the cylinder is evacuated and the piston is temporarily restrained by latches. The following operations are carried out: (1) the latches are removed, allowing the gas on the left to expand as it pushes the piston to the right-hand end of the cylinder; (2) a small rod is inserted through the right-hand end of the cylinder and exerts a force on the piston, pushing it slowly back to its initial position.

(*a*) What is the gas temperature after process 1 once internal equilibrium is established?

(*b*) What is the gas temperature at the end of process 2, assuming this process to be reversible and adiabatic?

(*c*) What is the gas pressure at the end of process 2?

(*d*) What is W for process 2?

(*e*) What is ΔS for process 1 and for the combined processes 1 and 2?

Ans. (*a*) 60(°F), (*b*) 226(°F), (*c*) 2.64(atm), (*d*) −830(Btu), (*e*) 1.386(Btu)/(R) for both

2.32. Figure 2-16 shows a closed cylinder divided into two unequal chambers A and B by a piston that is free to move, except that it is initially kept from moving by a force F applied to a piston rod that extends through the right-hand end of the cylinder. Both the cylinder wall and the piston are perfect heat insulators, so that no heat is exchanged either between the chambers or with the surroundings. The two chambers contain fixed amounts of the same ideal gas, which may be assumed to have the constant heat capacities: $C_V = 5$ and $C_P = 7$(cal)/(g mole)(K). The piston has an area of $0.1(\text{m})^2$ and the cylinder has a free length of 1(m). The initial location of the piston is such that chamber A represents 1/3 of the total volume. The initial conditions are as follows:

$$P_{A_1} = 4(\text{atm}) \qquad P_{B_1} = 2(\text{atm})$$
$$T_{A_1} = 127(^\circ\text{C}) \qquad T_{B_1} = 27(^\circ\text{C})$$

The force F is slowly reduced so as to allow the piston to move to the right until the pressures in the two chambers are equal. Assuming the process to be reversible, determine $P_{A_2} = P_{B_2} = P_2$, the temperatures T_{A_2} and T_{B_2}, and the work done by the system against the external force F.

Ans. $P_2 = 2.62$(atm), $T_{A_2} = 81.4(^\circ\text{C})$, $T_{B_2} = 51.0(^\circ\text{C})$, $W = 276$(cal)

2.33. A closed cylinder is divided into two unequal chambers A and B by a piston that is free to move, except that it is initially kept from moving by a stop (see Fig. 2-17). Both the cylinder wall and the piston are perfect heat insulators, so that no heat is exchanged either between the chambers or with the surroundings. The two chambers contain fixed amounts of the same ideal gas, which may be assumed to have constant heat capacities. Chamber A initially contains n_A moles at T_{A_1} and P_{A_1}; chamber B initially contains n_B moles at T_{B_1} and P_{B_1}, where $P_{A_1} > P_{B_1}$. If the stop is removed so that the piston is free to move until the pressures in the two chambers equalize, will the change in entropy as a result of the process be less than, equal to, or greater than zero in:

(*a*) the entire system, consisting of A and B

(*b*) chamber A

(*c*) chamber B

Can the final pressure $P_{A_2} = P_{B_2} = P_2$ be determined by the methods of thermodynamics? Can the final temperatures T_{A_2} and T_{B_2} be determined by the methods of thermodynamics? How would the problem be changed if the piston were a heat conductor?

Ans. *Greater than* in each case, (*a*), (*b*), and (*c*). The final pressure can be determined:

$$P_2 = \frac{n_A T_{A_1} + n_B T_{B_1}}{\dfrac{n_A T_{A_1}}{P_{A_1}} + \dfrac{n_B T_{B_1}}{P_{B_1}}}$$

The final temperatures T_{A_2} and T_{B_2} can *not* be determined when the piston is nonconducting. When the piston is a conductor the problem becomes equivalent to Problem 1.16.

2.34. A system consisting of gas contained in a piston-cylinder device undergoes an *irreversible* process between initial and final equilibrium states that causes its internal energy to increase by 30(Btu). During the process the system receives heat in the amount of 100(Btu) from a heat reservoir at 1000(R).

The system is then restored to its initial state by a *reversible* process, during which the only heat transfer is between the system and the heat reservoir at 1000(R).

The entropy change of the heat reservoir as a result of *both* processes is +0.01(Btu)/(R). Calculate (*a*) the work done by the system during the first (irreversible) process, (*b*) the heat transfer with respect to the system during the second (reversible) process, (*c*) the work done by the system during the second process.

Ans. (*a*) 70(Btu), (*b*) −110(Btu), (*c*) −80(Btu)

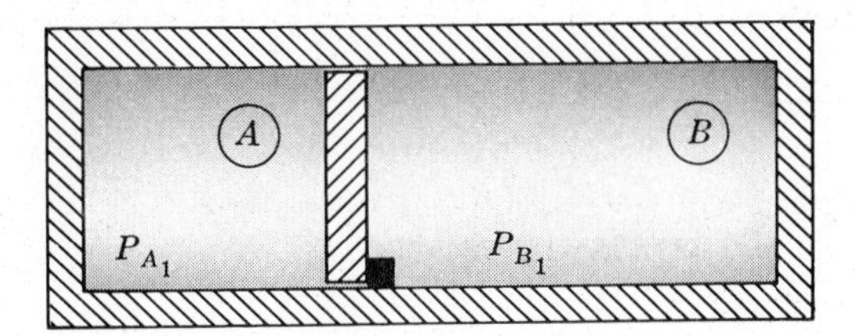

Fig. 2-17

Chapter 3

Mathematical Formulations of Thermodynamics

The quantitative relationships provided by the laws of thermodynamics find use in the solution of two quite different types of problems. The first is concerned with *processes,* and the equations employed deal with the relations between property changes of a system and the quantity of energy transferred between the system and its surroundings.

A second and equally important use of thermodynamics is in the elucidation of relationships among the equilibrium *properties* of a system. Derivation of these equations starts with a consideration of processes, because the laws of thermodynamics include the quantities Q and W, which are not properties but manifestations of processes. However, for reversible processes Q and W can be replaced by expressions involving properties only, and the resulting equations then become general relationships among equilibrium properties, no longer limited by the special kind of process initially chosen for the derivation. These properties are functions of the state variables, and are often called *state functions.* Purely mathematical considerations allow the derivation of a large number of equations interconnecting the state functions. This chapter is devoted to the development of such equations, and is a repository for numerous equations to be employed in later chapters.

3.1 EXACT DIFFERENTIALS AND STATE FUNCTIONS

Mathematical descriptions of the changes which occur in physical systems often lead to differential expressions of the form:

$$C_1\,dX_1 + C_2\,dX_2 + \cdots + C_n\,dX_n \equiv \sum (C_i\,dX_i) \tag{3.1}$$

where the X_i are independent variables, and the C_i are functions of the X_i. When it is possible to set the differential expression (3.1) equal to dY, the differential of a *function* Y, where

$$Y = Y(X_1, X_2, \ldots, X_n)$$

then the differential expression (3.1) is said to be *exact,* and we can write:

$$dY = C_1\,dX_1 + \cdots + C_n\,dX_n \equiv \sum (C_i\,dX_i) \tag{3.2}$$

Mathematics provides a *definition* for the differential of such a function:

$$dY = \left(\frac{\partial Y}{\partial X_1}\right)_{X_j} dX_1 + \left(\frac{\partial Y}{\partial X_2}\right)_{X_j} dX_2 + \cdots + \left(\frac{\partial Y}{\partial X_n}\right)_{X_j} dX_n \equiv \sum \left(\frac{\partial Y}{\partial X_i}\right)_{X_j} dX_i$$

where the subscript X_j on the partial derivatives indicates that all X_i are held constant except the one in the derivative considered. Since the X_i are independent, this last equation and (3.2) may be equated term by term to give:

$$C_1 = \left(\frac{\partial Y}{\partial X_1}\right)_{X_j}, \quad \ldots, \quad C_n = \left(\frac{\partial Y}{\partial X_n}\right)_{X_j} \quad \text{or} \quad C_i = \left(\frac{\partial Y}{\partial X_i}\right)_{X_j} \tag{3.3}$$

From this we see that when the differential expression (*3.1*) is exact, the C_i are interpreted as partial differential coefficients in the defining equation for dY, and each C_i and its corresponding X_i are said to be *conjugate* to each other.

If Y and its derivatives are continuous, then for any pair of independent variables X_k and X_l, we have the mathematical requirement that

$$\frac{\partial^2 Y}{\partial X_k\,\partial X_l} = \frac{\partial^2 Y}{\partial X_l\,\partial X_k}$$

From (*3.3*) we have

$$C_k = \left(\frac{\partial Y}{\partial X_k}\right)_{X_j} \quad \text{and} \quad C_l = \left(\frac{\partial Y}{\partial X_l}\right)_{X_j}$$

Therefore

$$\left(\frac{\partial C_k}{\partial X_l}\right)_{X_j} = \frac{\partial^2 Y}{\partial X_l\,\partial X_k} \quad \text{and} \quad \left(\frac{\partial C_l}{\partial X_k}\right)_{X_j} = \frac{\partial^2 Y}{\partial X_k\,\partial X_l}$$

and as a result we get the important equation:

$$\boxed{\left(\frac{\partial C_k}{\partial X_l}\right)_{X_j} = \left(\frac{\partial C_l}{\partial X_k}\right)_{X_j}} \tag{3.4}$$

This equation holds for any two pairs of conjugate variables (C_l, X_l) and (C_k, X_k) in an exact differential expression, and represents a condition that is both necessary and sufficient for the exactness of (*3.1*).

In property relation (*2.6*), $dU = T\,dS - P\,dV$, U is known to be a function of S and V, so that $T\,dS - P\,dV$ must be exact. For such property relations (*3.4*) is used not to test exactness but rather to provide additional thermodynamic relationships. Thus, from (*2.6*) we infer

$$\left(\frac{\partial T}{\partial V}\right)_S = -\left(\frac{\partial P}{\partial S}\right)_V$$

Example 3.1. Given the exact differential expression $dx = M\,dy + N\,dz - P\,dw$, write the result of application of (*3.4*).

The three equations which follow from (*3.4*) are

$$\left(\frac{\partial M}{\partial z}\right)_{y,w} = \left(\frac{\partial N}{\partial y}\right)_{z,w} \qquad \left(\frac{\partial M}{\partial w}\right)_{y,z} = -\left(\frac{\partial P}{\partial y}\right)_{w,z} \qquad \left(\frac{\partial N}{\partial w}\right)_{z,y} = -\left(\frac{\partial P}{\partial z}\right)_{w,y}$$

There are several other characteristics of exact differentials in addition to the exactness criterion (*3.4*) that are important in thermodynamics. If $dY = \Sigma\,(C_i\,dX_i)$ is an exact differential expression, then

I. The value of the integral $\Delta Y = \int_A^B \Sigma\,(C_i\,dX_i)$ is independent of the path followed from point A to point B.

II. The integral around any closed path $\oint dY = \oint \Sigma\,(C_i\,dX_i)$ is identically zero.

III. The function Y, defined only by $dY = \Sigma\,(C_i\,dX_i)$, can be determined only to within an additive constant.

Example 3.2. What applications have already been made in Chapters 1 and 2 of the characteristics of exact differentials just listed?

Axioms 1 and 3 near the beginning of Chapter 2 formally identify the internal energy and the entropy as intrinsic properties of a system. Thus U and S are functions of the state variables and their differentials must be exact. Earlier, in Chapter 1 (page 11) it was stated that ΔU should depend only on the end states of a system and should therefore be independent of path.

In Example 2.2 use was made of the fact that ΔS_{engine} and ΔU_{engine} are zero, because the engine operates in a cycle, returning periodically to its initial state.

It was also shown that integration of (*2.7*) or (*2.9*) leads to an equation for S that contains an integration constant S_0, for which no value can be determined.

The attributes of U and S characterized in items I and II are shared by all proper thermodynamic functions, and the differentials of all such functions are known from experiment and experience to be exact. These functions are variously called *state functions, state properties, variables of state,* or *point functions.*

Item III states the limit to which purely mathematical considerations can aid in the deduction of values for a thermodynamic function Y defined by means of (*3.2*). It does not rule out the existence of absolute values for Y, but implies that such values must be otherwise obtained.

Example 3.3. Differential expressions of the form (*3.1*) can be written which do *not* satisfy the exactness criterion (*3.4*). For example, consider $y\,dx - x\,dy$. Equation (*3.4*) is not satisfied because $1 \neq -1$. Hence there is *no* function of x and y whose differential is given by the original expression.

On the other hand, for the expression $y\,dx + x\,dy$, (*3.4*) *is* satisfied, and the expression is therefore exact. Indeed, $y\,dx + x\,dy$ is the differential of the function $z(x,y) = yx$.

Example 3.4. Certain differentials of thermodynamics are inexact. For example, the equation

$$\delta Q_{\text{rev}} = dU + P\,dV \tag{3.5}$$

expresses the first law for a PVT system undergoing a reversible process. Although the variables on the right are state functions, it will be demonstrated that Q_{rev} is not.

If U is considered a function of T and V, then

$$dU = \left(\frac{\partial U}{\partial V}\right)_T dV + \left(\frac{\partial U}{\partial T}\right)_V dT \tag{3.6}$$

Combination of (*3.5*) and (*3.6*) gives

$$\delta Q_{\text{rev}} = \left[\left(\frac{\partial U}{\partial V}\right)_T + P\right] dV + \left(\frac{\partial U}{\partial T}\right)_V dT$$

$$= M\,dV + N\,dT$$

Differentiation of M with respect to T and of N with respect to V results in

$$\left(\frac{\partial M}{\partial T}\right)_V = \frac{\partial^2 U}{\partial T\,\partial V} + \left(\frac{\partial P}{\partial T}\right)_V \qquad \left(\frac{\partial N}{\partial V}\right)_T = \frac{\partial^2 U}{\partial V\,\partial T}$$

Since U is a function of T and V, the second derivatives must be equal. However, $(\partial P/\partial T)_V$ is known from experiment to be nonzero in general; so (*3.4*) is not satisfied. Thus δQ_{rev} is not an exact differential, and Q_{rev} is not a state function. The sign δ is used with Q expressly to draw attention to this fact.

Example 3.5. Often an inexact differential expression can be made exact by dividing through the expression by some function of the independent variables. Such a function is called an *integrating denominator,* and it is known that for the case of two independent variables an integrating denominator always exists.

In Example 3.3 the expression $y\,dx - x\,dy$ was shown not to be exact. However, division by x^2 makes it exact:

$$dw = \frac{y}{x^2}dx - \frac{1}{x}dy$$

Application of (*3.4*) leads to $1/x^2 = 1/x^2$, and the criterion is satisfied. Moreover, it is readily confirmed that $w = -y/x$.

It was shown in Example 3.4 that the expression

$$\delta Q_{\text{rev}} = dU + P\,dV$$

is not exact. However, division by the absolute temperature T gives

$$dS = \frac{\delta Q_{\text{rev}}}{T} = \frac{dU}{T} + \frac{P}{T}dV \tag{3.7}$$

where use has been made of (*2.1*). Axiom 3 of Chapter 2 affirms that S is a property, and as such it should be functionally related to U and V, provided these variables characterize the system. Thus dS should be an exact differential and (*3.7*) should satisfy the exactness criterion (*3.4*).

Elimination of dU in (*3.7*) by means of (*3.6*) gives

$$\begin{aligned} dS &= \left[\frac{1}{T}\left(\frac{\partial U}{\partial V}\right)_T + \frac{P}{T}\right]dV + \frac{1}{T}\left(\frac{\partial U}{\partial T}\right)_V dT \\ &= M'\,dV + N'\,dT \end{aligned}$$

from which

$$\left(\frac{\partial M'}{\partial T}\right)_V = \frac{1}{T}\left(\frac{\partial^2 U}{\partial T\,\partial V}\right) - \frac{1}{T^2}\left(\frac{\partial U}{\partial V}\right)_T + \frac{1}{T}\left(\frac{\partial P}{\partial T}\right)_V - \frac{P}{T^2}$$

$$\left(\frac{\partial N'}{\partial V}\right)_T = \frac{1}{T}\left(\frac{\partial^2 U}{\partial V\,\partial T}\right)$$

If (*3.4*) is to be satisfied, then it is evident from these equations that

$$-\frac{1}{T^2}\left(\frac{\partial U}{\partial V}\right)_T + \frac{1}{T}\left(\frac{\partial P}{\partial T}\right)_V - \frac{P}{T^2} = 0$$

or

$$\left(\frac{\partial U}{\partial V}\right)_T = T\left(\frac{\partial P}{\partial T}\right)_V - P \tag{3.8}$$

Equation (*3.8*) is in fact a standard equation for single-phase PVT systems, and we have here derived it on the basis that the entropy S for such a system is a state function and that dS is an exact differential in the variables V and T. The validity of (*3.8*) is subject to experimental verification, and is known to be correct for such systems. (It is trivially true for an ideal gas.) This is simply another test of the validity of the fundamental axioms that form the basis for thermodynamics.

3.2 TRANSFORMATION RELATIONSHIPS FOR SYSTEMS OF TWO INDEPENDENT VARIABLES

This section is devoted to the development of some useful relationships among first partial derivatives for the thermodynamically important case of a system which can be fully specified by fixing two state variables. If these variables are designated y and z, then any other state function x is related to y and z by an equation having the functional form

$$f(x,y,z) = 0$$

Since any pair of the three variables may be selected as independent, this functional relationship may be expressed in three additional alternative forms

$$x = x(y,z) \qquad y = y(x,z) \qquad z = z(x,y)$$

Arbitrarily selecting the first two of these, we may write expressions for the total differentials of dx and dy:

$$dx = \left(\frac{\partial x}{\partial y}\right)_z dy + \left(\frac{\partial x}{\partial z}\right)_y dz \tag{3.9}$$

$$dy = \left(\frac{\partial y}{\partial x}\right)_z dx + \left(\frac{\partial y}{\partial z}\right)_x dz \tag{3.10}$$

Elimination of the differential dy between (*3.9*) and (*3.10*) gives

$$\left[\left(\frac{\partial x}{\partial y}\right)_z\left(\frac{\partial y}{\partial x}\right)_z - 1\right]dx + \left[\left(\frac{\partial x}{\partial y}\right)_z\left(\frac{\partial y}{\partial z}\right)_x + \left(\frac{\partial x}{\partial z}\right)_y\right]dz = 0 \tag{3.11}$$

Since x and z are independently variable, the coefficients of dx and dz must be identically zero if (*3.11*) is to be generally valid. Hence

$$\left(\frac{\partial x}{\partial y}\right)_z = \left(\frac{\partial y}{\partial x}\right)_z^{-1} \tag{3.12}$$

and

$$\left(\frac{\partial x}{\partial z}\right)_y = -\left(\frac{\partial x}{\partial y}\right)_z\left(\frac{\partial y}{\partial z}\right)_x$$

which in view of (*3.12*) can be written

$$\left(\frac{\partial x}{\partial y}\right)_z = -\left(\frac{\partial x}{\partial z}\right)_y\left(\frac{\partial z}{\partial y}\right)_x \tag{3.13}$$

If (*3.9*) is divided through by the differential of a fourth state variable, dw, then

$$\frac{dx}{dw} = \left(\frac{\partial x}{\partial y}\right)_z\frac{dy}{dw} + \left(\frac{\partial x}{\partial z}\right)_y\frac{dz}{dw}$$

Restriction of this equation to constant z reduces it to

$$\left(\frac{\partial x}{\partial w}\right)_z = \left(\frac{\partial x}{\partial y}\right)_z\left(\frac{\partial y}{\partial w}\right)_z$$

from which

$$\left(\frac{\partial x}{\partial y}\right)_z = \left(\frac{\partial x}{\partial w}\right)_z\left(\frac{\partial w}{\partial y}\right)_z \tag{3.14}$$

If x is now taken to be a function of y and w, then

$$dx = \left(\frac{\partial x}{\partial y}\right)_w dy + \left(\frac{\partial x}{\partial w}\right)_y dw$$

Division of this equation by dy with restriction to constant z gives

$$\left(\frac{\partial x}{\partial y}\right)_z = \left(\frac{\partial x}{\partial y}\right)_w + \left(\frac{\partial x}{\partial w}\right)_y\left(\frac{\partial w}{\partial y}\right)_z \tag{3.15}$$

The equations of this section, or variations on them, provide the basis for many of the transformations employed in the thermodynamics of systems described by two independent variables.

Example 3.6. If for a PVT system V is a function of T and P, then

$$dV = \left(\frac{\partial V}{\partial T}\right)_P dT + \left(\frac{\partial V}{\partial P}\right)_T dP \tag{3.16}$$

The partial differential coefficients in this equation are directly related to two properties commonly tabulated for pure substances:

(*a*) The *volume expansivity* β, where

$$\beta = \frac{1}{V}\left(\frac{\partial V}{\partial T}\right)_P \tag{3.17}$$

(*b*) The *isothermal compressibility* κ, where

$$\boxed{\kappa = -\frac{1}{V}\left(\frac{\partial V}{\partial P}\right)_T} \tag{3.18}$$

Substitution into (*3.16*) gives

$$\frac{dV}{V} = \beta\,dT - \kappa\,dP \tag{3.19}$$

Since dV is known to be an exact differential, (*3.4*) must be satisfied, and therefore

$$\left(\frac{\partial \beta}{\partial P}\right)_T = -\left(\frac{\partial \kappa}{\partial T}\right)_P \tag{3.20}$$

For the special case of an ideal gas, $PV = RT$, and by differentiation it is found that

$$\beta = \frac{1}{T} \quad \text{and} \quad \kappa = \frac{1}{P} \quad \text{(ideal gas)}$$

In this case (*3.19*) becomes

$$\frac{dV}{V} = \frac{dT}{T} - \frac{dP}{P} \quad \text{(ideal gas)} \tag{3.21}$$

If we let

$$\left(\frac{\partial x}{\partial y}\right)_z = \left(\frac{\partial P}{\partial T}\right)_V$$

then (*3.13*) becomes

$$\left(\frac{\partial P}{\partial T}\right)_V = -\left(\frac{\partial P}{\partial V}\right)_T\left(\frac{\partial V}{\partial T}\right)_P$$

or by (*3.12*)

$$\left(\frac{\partial P}{\partial T}\right)_V = -\frac{(\partial V/\partial T)_P}{(\partial V/\partial P)_T}$$

and in view of (*3.17*) and (*3.18*) this reduces to

$$\left(\frac{\partial P}{\partial T}\right)_V = \frac{\beta}{\kappa}$$

For an ideal gas this simplifies to $(\partial P/\partial T)_V = P/T$, which may be confirmed by direct differentiation of $PV = RT$.

Example 3.7. If we apply (*3.15*) to a PVT system and let

$$\left(\frac{\partial x}{\partial y}\right)_z = \left(\frac{\partial T}{\partial V}\right)_S$$

and if we identify w with U, then (*3.15*) becomes

$$\left(\frac{\partial T}{\partial V}\right)_S = \left(\frac{\partial T}{\partial V}\right)_U + \left(\frac{\partial T}{\partial U}\right)_V\left(\frac{\partial U}{\partial V}\right)_S$$

According to (*1.8*)

$$C_V = \left(\frac{\partial U}{\partial T}\right)_V$$

and therefore

$$\left(\frac{\partial T}{\partial U}\right)_V = \left(\frac{\partial U}{\partial T}\right)_V^{-1} = \frac{1}{C_V}$$

Furthermore, from (*2.6*) we have

$$\left(\frac{\partial U}{\partial V}\right)_S = -P$$

Therefore

$$\left(\frac{\partial T}{\partial V}\right)_S = \left(\frac{\partial T}{\partial V}\right)_U - \frac{P}{C_V} \quad \text{or} \quad \left(\frac{\partial T}{\partial V}\right)_U = \left(\frac{\partial T}{\partial V}\right)_S + \frac{P}{C_V}$$

This equation gives the derivative dT/dV for a process at constant internal energy. An example of such a process is the Joule expansion in which a gas initially confined to a portion of a rigid, insulated tank is allowed to expand to fill the entire tank. In this process Q and W are zero, and therefore $\Delta U = 0$. The quantity $(\partial T/\partial V)_S$ is the temperature-volume derivative for an *isentropic* or reversible, adiabatic process.

For an ideal gas U is a function of temperature only. Therefore if U is constant, so is T, and for an ideal gas $(\partial T/\partial V)_U = 0$. For this case

$$\left(\frac{\partial T}{\partial V}\right)_S = -\frac{P}{C_V}$$

which for a reversible, adiabatic process may be written

$$dT = -(P/C_V)\,dV$$

Division by T and substitution of R/V for P/T gives

$$\frac{dT}{T} = -\frac{R}{C_V}\frac{dV}{V}$$

This same equation was derived in Example 1.10 and again in Example 2.5 by entirely different methods, and leads to the relation

$$TV^{\gamma-1} = \text{constant}$$

for the reversible, adiabatic compression or expansion of an ideal gas with constant heat capacities.

3.3 LEGENDRE TRANSFORMATIONS

The fundamental property relation for a closed PVT system was developed in Chapter 2:

$$dU = T\,dS - P\,dV \tag{2.6}$$

This equation implies that U is always a function of the variables S and V in any closed system. However, the choice of S and V as variables is not always convenient. Other pairs of variables are often advantageously employed. It is therefore useful to define new functions whose total differentials are consistent with (*2.6*), but for which the natural variables are pairs other than S and V. Thus in Chapter 1 it was found convenient to define the enthalpy H by the equation

$$H = U + PV \tag{1.6}$$

Differentiation of (*1.6*) provides

$$dH = dU + P\,dV + V\,dP$$

which may be combined with (*2.6*) to give

$$dH = T\,dS + V\,dP$$

from which it is seen that S and P are the natural or special variables for the function H as it applies to a closed PVT system.

New thermodynamic functions cannot in general be defined by random combination of variables. There is, for example, the requirement of dimensional consistency. Fortunately, a standard mathematical method exists for the systematic definition of functions of the kind required: the *Legendre transformation*.

Recall the exact differential expression presented earlier in this chapter:

$$dY = C_1\,dX_1 + C_2\,dX_2 + \cdots + C_n\,dX_n \tag{3.2}$$

Legendre transformations define Y-related functions for which the sets of variables contain one or more of the C_i in place of the conjugate X_i. For a total differential expression exhibiting n variables there are $2^n - 1$ possible Legendre transformations, namely:

$$\begin{aligned}
\mathcal{T}_1 &= \mathcal{T}_1(C_1, X_2, X_3, \ldots, X_n) = Y - C_1X_1 \\
\mathcal{T}_2 &= \mathcal{T}_2(X_1, C_2, X_3, \ldots, X_n) = Y - C_2X_2 \\
&\vdots \\
\mathcal{T}_n &= \mathcal{T}_n(X_1, X_2, X_3, \ldots, C_n) = Y - C_nX_n \\
\mathcal{T}_{1,2} &= \mathcal{T}_{1,2}(C_1, C_2, X_3, \ldots, X_n) = Y - C_1X_1 - C_2X_2 \\
\mathcal{T}_{1,3} &= \mathcal{T}_{1,3}(C_1, X_2, C_3, \ldots, X_n) = Y - C_1X_1 - C_3X_3 \\
&\vdots \\
\mathcal{T}_{1,\ldots,n} &= \mathcal{T}_{1,\ldots,n}(C_1, C_2, C_3, \ldots, C_n) = Y - \Sigma\,(C_iX_i)
\end{aligned} \tag{3.22}$$

Each $\mathcal{T}$ in (*3.22*) represents a new *function*, and in each case the independent variables, shown in parentheses, are the *canonical*[1] variables for that function. Thus (*3.22*) provides a recipe for the definition of a set of new functions consistent with a particular exact differential expression, and it identifies the variables which are unique to each function. These have the following special property: When a transformation function $\mathcal{T}$ is known as a function of its n canonical variables, then the remaining n variables among those appearing in the original exact differential expression (the X_i and their conjugate C_i) can be recovered by differentiation of $\mathcal{T}$. This is not in general true for arbitrarily chosen sets of variables.

For example, if $Y = Y(X_1, X_2, X_3)$ then

$$dY = C_1\,dX_1 + C_2\,dX_2 + C_3\,dX_3 \tag{A}$$

From (*3.22*), $\mathcal{T}_1(C_1, X_2, X_3) = Y - C_1X_1$. To simplify notation, we will let this function be Z.

Thus

$$Z = Y - C_1X_1 \tag{B}$$

Differentiating $Z = Z(C_1, X_2, X_3)$ we have

$$dZ = \left(\frac{\partial Z}{\partial C_1}\right)_{X_2, X_3} dC_1 + \left(\frac{\partial Z}{\partial X_2}\right)_{C_1, X_3} dX_2 + \left(\frac{\partial Z}{\partial X_3}\right)_{C_1, X_2} dX_3 \tag{C}$$

The differential dZ can also be found from (B):

$$dZ = dY - C_1\,dX_1 - X_1\,dC_1$$

and substitution for dY by (A) gives:

$$dZ = -X_1\,dC_1 + C_2\,dX_2 + C_3\,dX_3 \tag{D}$$

Comparison of (C) and (D) shows that

$$X_1 = -\left(\frac{\partial Z}{\partial C_1}\right)_{X_2, X_3} \qquad C_2 = \left(\frac{\partial Z}{\partial X_2}\right)_{C_1, X_3} \qquad C_3 = \left(\frac{\partial Z}{\partial X_3}\right)_{C_1, X_2}$$

Thus from the original variables X_1, X_2, and X_3 and their conjugates C_1, C_2, and C_3 in the differential expression (A), we have Z as a function of C_1, X_2, and X_3, and the remaining variables X_1, C_2, and C_3, are given by derivatives of Z.

Example 3.8. The fundamental property relation (*2.6*) for a closed PVT system relates $U = U(V, S)$ to the two variables V and S:

$$dU = -P\,dV + T\,dS \tag{2.6}$$

In applying (*3.22*), we identify C_1 and C_2 with $-P$ and T, and X_1 and X_2 with V and S. There are $2^2 - 1 = 3$ possible U-related Legendre transforms, given by

[1]The adjective *canonical* means that the variables conform to a scheme that is both simple and clear.

$$\boxed{\begin{aligned} H &= H(P,S) &&= U + PV \\ A &= A(V,T) &&= U - TS \\ G &= G(P,T) &&= U + PV - TS \end{aligned}} \tag{3.23}$$

The first of these three functions is the enthalpy, already introduced. The functions A and G are the *Helmholtz function* and *Gibbs function*, respectively. (Both are sometimes called *free energies*, and in this case the modifiers *Helmholtz* and *Gibbs* are still appropriate to distinguish between them. A common European practice is to call A the *free energy* and G the *free enthalpy*.)

As a result of the definition of enthalpy, we earlier derived the equation

$$dH = T\,dS + V\,dP$$

Clearly, T and V must be identified with the partial differential coefficients of dS and dP. Thus

$$T = \left(\frac{\partial H}{\partial S}\right)_P \quad \text{and} \quad V = \left(\frac{\partial H}{\partial P}\right)_S$$

From these it is seen that knowledge of H as a function of S and P allows regeneration by differentiation of the two remaining properties that appear on the right side of the primitive equation (*2.6*).

Example 3.9. The properties T, S, P, and V assume the role of natural variables in the standard expositions of the classical thermodynamics of PVT systems. It is evident from Example 3.8 that this results from the particular form of the fundamental property relation (*2.6*), which is the basis for the definitions of H, A, and G. One could, however, proceed differently by rearranging the fundamental property relation as

$$dS = \frac{P}{T}dV + \frac{1}{T}dU$$

This equation was in fact used in Example 2.8. It expresses the dependent variable S as a function of V and U.

As in Example 3.8 there are but three possible Legendre transformations:

$$\begin{aligned} \Omega &= \Omega\left(\frac{P}{T}, U\right) &&= S - \frac{P}{T}V \\ \Psi &= \Psi\left(V, \frac{1}{T}\right) &&= S - \frac{1}{T}U \\ \Phi &= \Phi\left(\frac{P}{T}, \frac{1}{T}\right) &&= S - \frac{P}{T}V - \frac{1}{T}U \end{aligned} \tag{3.24}$$

The functions defined by (*3.24*) are called *Massieu functions*, and Φ is often called the *Planck function*. A complete network of thermodynamic equations can be developed from this formulation. However, the natural intensive variables (see Sec. 1.8) in this system are $1/T$ and P/T. These variables, as well as the Massieu functions, arise naturally in statistical mechanics and in irreversible thermodynamics, but they are less directly related to experience and are less useful in the description of real processes than are T and P, the intensive variables more commonly employed.

3.4 PRIMARY PROPERTY RELATIONSHIPS FOR PVT SYSTEMS OF VARIABLE COMPOSITION

As an important application of the principles of the preceding sections, we consider a homogeneous PVT system containing m chemical species present in mole numbers $n_1, n_2, \ldots, n_m$. The internal energy, entropy, and volume are extensive properties, so the total system properties are nU, nS, and nV, where U, S, and V are molar properties and n is the total number of moles of all chemical species. For the particular case of a reversible process in which all of the n_i are *constant*, we have

$$\delta Q_{\text{rev}} = T\,d(nS) \qquad \delta W_{\text{rev}} = P\,d(nV)$$

from which

$$d(nU) = T\,d(nS) - P\,d(nV) \tag{3.25}$$

Equation (*3.25*) is an exact differential expression and, according to (*3.3*),

$$\left[\frac{\partial(nU)}{\partial(nS)}\right]_{nV,\,n} = T \qquad (3.26)$$

$$\left[\frac{\partial(nU)}{\partial(nV)}\right]_{nS,\,n} = -P \qquad (3.27)$$

The additional subscript n indicates that *all* the n_i are held constant.

For the *general* case, however, nU must be considered a function of the n_i as well as of nS and nV. We can therefore write formally for the total differential of nU

$$d(nU) = \left[\frac{\partial(nU)}{\partial(nS)}\right]_{nV,\,n} d(nS) + \left[\frac{\partial(nU)}{\partial(nV)}\right]_{nS,\,n} d(nV) + \sum_{i=1}^{m} \left[\frac{\partial(nU)}{\partial n_i}\right]_{nS,\,nV,\,n_j} dn_i \qquad (3.28)$$

where the subscript n_j means that all mole numbers except n_i are held constant. Combining (*3.26*), (*3.27*), and (*3.28*), we obtain

$$\boxed{d(nU) = T\,d(nS) - P\,d(nV) + \Sigma\,\mu_i\,dn_i} \qquad (3.29)$$

where μ_i is the *chemical potential*, defined by

$$\mu_i = \left[\frac{\partial(nU)}{\partial n_i}\right]_{nS,\,nV,\,n_j}$$

and the plain summation symbol implies summation over all chemical species.

Equation (*3.29*) is the fundamental property relation for a *homogeneous PVT system of variable composition*, and is the basis for all derived property relations for such systems. The system may be open or closed, and composition changes may result either from chemical reaction or from the transport of matter or both. Equation (*3.25*) is a special case of (*3.29*), valid for systems of fixed mole numbers. In addition, (*3.25*) [or (*2.6*)] has separate validity for *all processes connecting equilibrium states in any closed PVT system*, whether homogeneous or heterogeneous and regardless of changes in the mole numbers on account of chemical reaction.

The equations of this section and those which follow could equally well be based on unit-mass (*specific*) properties rather than on molar properties. The symbol n would then represent mass rather than number of moles.

There are $m+2$ variables in (*3.29*), and therefore $2^{m+2}-1$ possible U-related Legendre transformations. However, only three of these find widespread use. They are the enthalpy, the Helmholtz function, and the Gibbs function, all of which were earlier defined in Example 3.8.

$$(nH) = (nU) + P(nV) \qquad (3.30)$$

$$(nA) = (nU) - T(nS) \qquad (3.31)$$

$$(nG) = (nU) + P(nV) - T(nS) \qquad (3.32)$$

By taking the total differentials of (*3.30*), (*3.31*), and (*3.32*) and employing (*3.29*) to eliminate $d(nU)$, one obtains differential expressions for $d(nH)$, $d(nA)$, and $d(nG)$ in forms which display their canonical variables:

$$d(nH) = T\,d(nS) + (nV)\,dP + \Sigma\,\mu_i\,dn_i \qquad (3.33)$$

$$d(nA) = -(nS)\,dT - P\,d(nV) + \Sigma\,\mu_i\,dn_i \qquad (3.34)$$

$$d(nG) = -(nS)\,dT + (nV)\,dP + \Sigma\,\mu_i\,dn_i \qquad (3.35)$$

A number of useful relationships follow from the fact that (3.29), (3.33), (3.34), and (3.35) are exact differential expressions. According to (3.3),

$$T = \left(\frac{\partial U}{\partial S}\right)_{V,n} = \left(\frac{\partial H}{\partial S}\right)_{P,n} \tag{3.36}$$

$$P = -\left(\frac{\partial U}{\partial V}\right)_{S,n} = -\left(\frac{\partial A}{\partial V}\right)_{T,n} \tag{3.37}$$

$$V = \left(\frac{\partial H}{\partial P}\right)_{S,n} = \left(\frac{\partial G}{\partial P}\right)_{T,n} \tag{3.38}$$

$$S = -\left(\frac{\partial A}{\partial T}\right)_{V,n} = -\left(\frac{\partial G}{\partial T}\right)_{P,n} \tag{3.39}$$

$$\mu_i = \left[\frac{\partial(nU)}{\partial n_i}\right]_{nS,nV,n_j} = \left[\frac{\partial(nH)}{\partial n_i}\right]_{nS,P,n_j} = \left[\frac{\partial(nA)}{\partial n_i}\right]_{nV,T,n_j} = \left[\frac{\partial(nG)}{\partial n_i}\right]_{T,P,n_j} \tag{3.40}$$

Equations (3.36) through (3.39) are written in terms of molar, rather than total, properties. This is permissible because all of the n_i and hence n are held constant in the definitions of the partial differential coefficients. The subscript n is retained to emphasize that (3.36) through (3.39) are valid for *constant-composition solutions.*

We may apply the exactness criterion (3.4) to (3.29), (3.33), (3.34), and (3.35), to obtain

$$\left(\frac{\partial T}{\partial V}\right)_{S,n} = -\left(\frac{\partial P}{\partial S}\right)_{V,n} \tag{3.41}$$

$$\left(\frac{\partial T}{\partial P}\right)_{S,n} = \left(\frac{\partial V}{\partial S}\right)_{P,n} \tag{3.42}$$

$$\left(\frac{\partial P}{\partial T}\right)_{V,n} = \left(\frac{\partial S}{\partial V}\right)_{T,n} \tag{3.43}$$

$$\left(\frac{\partial V}{\partial T}\right)_{P,n} = -\left(\frac{\partial S}{\partial P}\right)_{T,n} \tag{3.44}$$

$$\left(\frac{\partial \mu_i}{\partial T}\right)_{P,n} = -\left[\frac{\partial(nS)}{\partial n_i}\right]_{T,P,n_j} \tag{3.45}$$

$$\left(\frac{\partial \mu_i}{\partial P}\right)_{T,n} = \left[\frac{\partial(nV)}{\partial n_i}\right]_{T,P,n_j} \tag{3.46}$$

$$\left(\frac{\partial \mu_l}{\partial n_k}\right)_{T,P,n_j} = \left(\frac{\partial \mu_k}{\partial n_l}\right)_{T,P,n_j} \tag{3.47}$$

Equations (3.41) through (3.44) are called the *Maxwell equations,* and we have chosen to write them in terms of molar properties; as with (3.36) through (3.39), they apply to constant-composition solutions. Of the twelve possible equations of the form of (3.4) involving μ_i, we have only written the three given as (3.45), (3.46), and (3.47), which result from (3.35). It will be seen in Chapter 7 that they have special significance in the treatment of properties of solutions.

Example 3.10. One important use of the basic property relations given by (3.29), (3.33), (3.34), and (3.35) is in the development of expressions for a large number of partial derivatives. We illustrate the general technique with (3.29) applied to a closed system of fixed composition. For such a system n and all the n_i are constant, and (3.29) becomes

$$dU = T\,dS - P\,dV$$

We wish now to develop an expression for $(\partial U/\partial V)_T$. The procedure is to divide the equation through by dV and to restrict it to constant T. This gives

$$\left(\frac{\partial U}{\partial V}\right)_T = T\left(\frac{\partial S}{\partial V}\right)_T - P$$

The conversion to partial derivatives is based on the presumption that U and S are functions of T and V, and indeed this is so for single-phase PVT systems. By (*3.43*) we have

$$\left(\frac{\partial S}{\partial V}\right)_T = \left(\frac{\partial P}{\partial T}\right)_V$$

Substitution gives

$$\left(\frac{\partial U}{\partial V}\right)_T = T\left(\frac{\partial P}{\partial T}\right)_V - P$$

which is the same as (*3.8*), developed by a different and less direct method. It gives the required derivative in terms of P, V, and T, all measurable quantities.

As a further illustration, we determine an expression for $(\partial U/\partial T)_V$. Division of our initial equation by dT and restriction to constant V gives immediately

$$\left(\frac{\partial U}{\partial T}\right)_V = T\left(\frac{\partial S}{\partial T}\right)_V$$

3.5 HEAT CAPACITY RELATIONSHIPS FOR CONSTANT-COMPOSITION PVT SYSTEMS

The heat capacities C_V and C_P were defined in Chapter 1 by

$$C_V = \left(\frac{\partial U}{\partial T}\right)_V \tag{1.8}$$

$$C_P = \left(\frac{\partial H}{\partial T}\right)_P \tag{1.9}$$

For a constant-composition PVT system, (*3.29*) reduces to

$$dU = T\,dS - P\,dV \tag{2.6}$$

and (*3.33*) becomes

$$dH = T\,dS + V\,dP \tag{3.48}$$

From these, the derivatives appearing in (*1.8*) and (*1.9*) are directly obtained (see Example 3.10), and provide alternative expressions for the heat capacities:

$$C_V = T\left(\frac{\partial S}{\partial T}\right)_V \tag{3.49}$$

$$C_P = T\left(\frac{\partial S}{\partial T}\right)_P \tag{3.50}$$

Several equations will now be derived to demonstrate how heat-capacity data are used in conjunction with PVT data to evaluate changes in U, H, and S for changes in state of single-phase, constant-composition PVT systems. If U is considered a function of T and V, then

$$dU = \left(\frac{\partial U}{\partial T}\right)_V dT + \left(\frac{\partial U}{\partial V}\right)_T dV \tag{3.51}$$

According to (*3.8*)

$$\left(\frac{\partial U}{\partial V}\right)_T = T\left(\frac{\partial P}{\partial T}\right)_V - P \qquad (3.8)$$

Combination of (*1.8*), (*3.8*), and (*3.51*) gives

$$dU = C_V\,dT + \left[T\left(\frac{\partial P}{\partial T}\right)_V - P\right]dV \qquad (3.52)$$

An analogous equation can be derived for dH in terms of dT and dP:

$$dH = C_P\,dT + \left[V - T\left(\frac{\partial V}{\partial T}\right)_P\right]dP \qquad (3.53)$$

Two useful (and equivalent) expressions can be derived for the total differential of the entropy, depending on whether T and V or T and P are selected as the variables. In the former case

$$dS = \left(\frac{\partial S}{\partial T}\right)_V dT + \left(\frac{\partial S}{\partial V}\right)_T dV$$

and substitution for the partial derivatives by (*3.49*) and (*3.43*) gives

$$\boxed{dS = \frac{C_V}{T}\,dT + \left(\frac{\partial P}{\partial T}\right)_V dV} \qquad (3.54)$$

With T and P as variables

$$dS = \left(\frac{\partial S}{\partial T}\right)_P dT + \left(\frac{\partial S}{\partial P}\right)_T dP$$

and substitution by (*3.50*) and (*3.44*) yields

$$\boxed{dS = \frac{C_P}{T}\,dT - \left(\frac{\partial V}{\partial T}\right)_P dP} \qquad (3.55)$$

For a given change in state (*3.54*) and (*3.55*) must give the same value for dS, and they may therefore be equated. The resulting expression, upon rearrangement, becomes

$$(C_P - C_V)\,dT = T\left(\frac{\partial V}{\partial T}\right)_P dP + T\left(\frac{\partial P}{\partial T}\right)_V dV$$

Division by dT and restriction to either constant pressure or constant volume provides an equation for the *difference* in heat capacities

$$\boxed{C_P - C_V = T\left(\frac{\partial P}{\partial T}\right)_V \left(\frac{\partial V}{\partial T}\right)_P} \qquad (3.56)$$

An expression for the *ratio* of heat capacities follows directly from division of (*3.50*) by (*3.49*):

$$\frac{C_P}{C_V} = \left(\frac{\partial S}{\partial T}\right)_P \left(\frac{\partial T}{\partial S}\right)_V$$

Application of (*3.13*) separately to each of the partial derivatives gives

$$\frac{C_P}{C_V} = \left[-\left(\frac{\partial S}{\partial P}\right)_T \left(\frac{\partial P}{\partial T}\right)_S\right]\left[-\left(\frac{\partial T}{\partial V}\right)_S \left(\frac{\partial V}{\partial S}\right)_T\right] = \left[\left(\frac{\partial V}{\partial S}\right)_T \left(\frac{\partial S}{\partial P}\right)_T\right]\left[\left(\frac{\partial P}{\partial T}\right)_S \left(\frac{\partial T}{\partial V}\right)_S\right]$$

or as a result of (*3.14*)

$$\boxed{\frac{C_P}{C_V} = \left(\frac{\partial V}{\partial P}\right)_T \left(\frac{\partial P}{\partial V}\right)_S} \qquad (3.57)$$

Equations (*3.56*) and (*3.57*) provide alternative ways of relating C_P to C_V. The partial derivatives in both equations are obtained from PVT data, with the exception of $(\partial P/\partial V)_S$, which is related to the velocity of sound (see Problem 3.14).

Application of the exactness criterion (*3.4*) to the exact differential expressions (*3.54*) and (*3.55*) provides derivatives of the heat capacities that depend on PVT data only:

$$\left[\frac{\partial(C_V/T)}{\partial V}\right]_T = \left[\frac{\partial(\partial P/\partial T)_V}{\partial T}\right]_V \quad \text{and} \quad \left[\frac{\partial(C_P/T)}{\partial P}\right]_T = -\left[\frac{\partial(\partial V/\partial T)_P}{\partial T}\right]_P$$

These reduce to

$$\left(\frac{\partial C_V}{\partial V}\right)_T = T\left(\frac{\partial^2 P}{\partial T^2}\right)_V \qquad (3.58)$$

$$\left(\frac{\partial C_P}{\partial P}\right)_T = -T\left(\frac{\partial^2 V}{\partial T^2}\right)_P \qquad (3.59)$$

Example 3.11. The usefulness of the equations of this section depends on knowledge of a PVT relation for the particular substance to which the equations are applied. Such a relation may be provided by a specific equation of state or by tables of numerical data. For gases the simplest equation of state is the ideal-gas law, and we use it here to illustrate application of the equations just derived.

By differentiation of the ideal-gas equation, $PV = RT$, we find

$$\left(\frac{\partial P}{\partial T}\right)_V = \frac{R}{V} = \frac{P}{T}; \qquad \left(\frac{\partial^2 P}{\partial T^2}\right)_V = 0$$

$$\left(\frac{\partial V}{\partial T}\right)_P = \frac{R}{P} = \frac{V}{T}; \qquad \left(\frac{\partial^2 V}{\partial T^2}\right)_P = 0$$

$$\left(\frac{\partial V}{\partial P}\right)_T = \frac{-RT}{P^2} = \frac{-V}{P}$$

Substitution for these partial derivatives in (*3.52*) through (*3.59*) yields the following results valid for ideal gases:

$$dU = C_V\,dT \qquad (3.52\ ideal)\ \text{and}\ (1.13)$$

$$dH = C_P\,dT \qquad (3.53\ ideal)\ \text{and}\ (1.15)$$

$$dS = C_V\frac{dT}{T} + R\frac{dV}{V} \qquad (3.54\ ideal)\ \text{and}\ (2.7)$$

$$dS = C_P\frac{dT}{T} - R\frac{dP}{P} \qquad (3.55\ ideal)\ \text{and}\ (2.9)$$

$$C_P - C_V = R \qquad (3.56\ ideal)\ \text{and}\ (1.14)$$

$$\gamma = \frac{C_P}{C_V} = \frac{-V}{P}\left(\frac{\partial P}{\partial V}\right)_S \qquad (3.57\ ideal)$$

$$(\partial C_V/\partial V)_T = 0 \qquad (3.58\ ideal)$$

$$(\partial C_P/\partial P)_T = 0 \qquad (3.59\ ideal)$$

The first five of these equations have all been presented earlier as indicated by the equation numbers. Equation (*3.57 ideal*) also reduces to a familiar equation. For an isentropic or reversible, adiabatic process it may be written $\gamma\, dV/V = -dP/P$, which, for constant γ, integrates to $PV^\gamma = k$, a relation derived in Example 2.5 for the reversible, adiabatic expansion of an ideal gas with constant heat capacities. Equations (*3.58 ideal*) and (*3.59 ideal*) show again that the heat capacities of an ideal gas, like the internal energy and enthalpy, are functions of temperature only. When temperature is constant, so is the heat capacity.

Example 3.12. Just as the ideal-gas law is an idealization of gas-phase behavior, so the *incompressible liquid* is an idealization of liquid-phase behavior. Since the volume of liquid is usually quite insensitive to changes in temperature and pressure, liquids are sometimes assumed to be incompressible, and both $(\partial V/\partial T)_P$ and $(\partial V/\partial P)_T$ are taken to be zero. In this event both the volume expansivity β and the isothermal compressibility κ are zero (see Example 3.6). What this implies is that V is a *constant*, unrelated to T and P, and that there is no equation of state connecting P, V, and T.

Since V is a constant, dV is always zero, and as a result (*3.52*) reduces to $dU = C_V\,dT$. Thus the internal energy of an incompressible liquid is a function of temperature only. However, the enthalpy is a function of T and P, because (*3.53*) reduces to $dH = C_P\,dT + V\,dP$.

The entropy becomes a function of temperature only, because (*3.54*) reduces to

$$dS = C_V \frac{dT}{T}$$

since $dV = 0$, and (*3.55*) reduces to

$$dS = C_P \frac{dT}{T}$$

since $(\partial V/\partial T)_P = 0$. From these two equations it is seen that for an incompressible liquid

$$\gamma = \frac{C_P}{C_V} = 1$$

This same conclusion is reached from (*3.56*) and (*3.57*).

3.6 ATTAINMENT OF EQUILIBRIUM IN CLOSED, HETEROGENEOUS SYSTEMS

The preceding sections have dealt with relationships among thermodynamic properties in homogeneous systems presumed to be in equilibrium. We consider here the approach to equilibrium in heterogeneous systems which are not initially at equilibrium with respect to the distribution of the various chemical species among coexisting but separate phases. The simplest such system is one that is closed and in which the temperature and pressure are uniform (but not necessarily constant). The system is imagined to contain an arbitrary number of phases, the composition of each phase being uniform but not necessarily the same as the composition of any other phase. We further imagine that the system exchanges heat reversibly with its surroundings and that volume changes of the system occur in such a way that work exchange with the surroundings is also reversible. The process considered results from a change in the system *from* a nonequilibrium state *toward* an equilibrium state with respect to the distribution of species among the phases.

A total property of the system *as a whole* will be denoted by use of superscript t (for "total"). Thus S^t, U^t, and V^t are the entropy, internal energy, and volume, respectively, of the entire system.

For the reversible exchange of heat δQ, the entropy change of the surroundings is

$$dS_{\text{surr}} = \frac{-\delta Q}{T} \qquad (3.60)$$

where the minus sign arises because δQ is taken with reference to the system whereas dS_{surr} refers to the surroundings. The temperature T is that of the system and the surroundings, since the heat transfer is assumed reversible.

The second law applied to the process requires

$$dS^t + dS_{\text{surr}} \geqq 0$$

or by (*3.60*)

$$dS^t \geqq \frac{\delta Q}{T} \qquad (3.61)$$

The inequality sign signifies an irreversible process, here the transfer of mass between phases, since all heat and work effects have been taken to be reversible. Once the system reaches *phase equilibrium,* the equality of (*3.61*) applies.

The first law for a closed system is given by (*1.5*), $dU^t = \delta Q - \delta W$; and since the work of volume change is reversible, $\delta W = P\,dV^t$, where P is the pressure of the system. Therefore

$$dU^t = \delta Q - P\,dV^t \tag{3.62}$$

Combination of (*3.61*) and (*3.62*) gives

$$\boxed{dU^t - T\,dS^t + P\,dV^t \leqq 0} \tag{3.63}$$

This relation involves properties only and must be satisfied for changes of state *in any closed PVT system whatever,* without restriction to the conditions of reversibility assumed for its derivation. The equality is satisfied for *any* process that leads from one state of internal equilibrium to another or to variations around an equilibrium state, and in this case it is identical with (*2.6*). The inequality must be satisfied by *any* process that starts from an initial state of uniform T and P but otherwise not a state of internal equilibrium, and it dictates the direction of change that leads toward an equilibrium state. We are considering here nonequilibrium states with respect to the distribution of chemical species among the phases, but the treatment is in no way different for nonequilibrium states with respect to chemical reaction among the species or to a combination of the two.

Special forms of the general relation.

Equation (*3.63*) is of such generality that its implications are difficult to visualize. It is readily reduced to simpler forms that apply to processes restricted in various ways. Thus for processes restricted to occur at constant entropy and volume, dS^t and dV^t are zero, and (*3.63*) becomes

$$(dU^t)_{S^t, V^t} \leqq 0 \tag{3.64}$$

Similarly, for processes restricted to constant U^t and V^t

$$(dS^t)_{U^t, V^t} \geqq 0 \tag{3.65}$$

An isolated system, as discussed in Chapter 2, is a system constrained to constant internal energy and volume, and for such a system it follows immediately from the second law that (*3.65*) is valid.

Additional equations representing special cases of (*3.63*) follow from the definitions of enthalpy, the Helmholtz function, and the Gibbs function. Thus

$$H^t = U^t + PV^t \qquad \text{and} \qquad dH^t = dU^t + P\,dV^t + V^t\,dP$$

Combination with (*3.63*) gives

$$dH^t - V^t\,dP - T\,dS^t \leqq 0$$

from which

$$(dH^t)_{P, S^t} \leqq 0 \tag{3.66}$$

In an analogous fashion it is found that

$$(dA^t)_{T, V^t} \leqq 0 \tag{3.67}$$

and

$$\boxed{(dG^t)_{T, P} \leqq 0} \tag{3.68}$$

As an example of the meaning of these equations consider (*3.68*). The inequality requires that all irreversible processes occurring at constant T and P must proceed in such a direction as to cause the total Gibbs function of the system to decrease. Since the Gibbs function decreases in all changes toward equilibrium in a system at constant T and P, then the equilibrium state for a given T and P must be that state for which the Gibbs function has its minimum value with respect to all possible variations at the given T and P. At the equilibrium state the equality of (*3.68*) holds, and this means that differential variations may occur in the system at constant T and P without producing a change in G. Thus a criterion of equilibrium is given by the equation

$$(dG^t)_{T,P} = 0 \tag{3.69}$$

The same reasoning applies to (*3.64*) through (*3.67*); however (*3.69*) is the preferred relation on which to base equilibrium calculations, because it is more convenient to treat T and P as constant than it is the other pairs of state properties.

Consideration of (*3.68*) and (*3.69*) shows that there are two equivalent methods which may be used to identify equilibrium states in closed systems for a given T and P. First, one can develop an expression for $(dG^t)_{T,P}$ in terms of the composition variables of the system and set it equal to zero as indicated by (*3.69*). Alternatively, one can develop an expression for G^t as a function of the composition variables and then find the set of values for the compositions which minimizes G^t at a given T and P in accordance with (*3.68*). The former method is usually employed for phase equilibrium calculations, and is developed below. The latter method is used for chemical reaction equilibrium calculations, and is described in Chapter 7.

The application of (*3.69*) to phase equilibrium requires an expression for $(dG^t)_{T,P}$ that incorporates the compositions or mole numbers of the chemical species in the individual phases. We first note that the total Gibbs function G^t is the sum of the Gibbs functions of the phases present in the system:

$$G^t = \sum_{p=1}^{\pi} n^p G^p \tag{3.70}$$

where G^p is the molar Gibbs function of the pth phase and n^p is the total number of moles in the pth phase, and the summation is over all of the π phases. Differentiation of (*3.70*) provides

$$dG^t = \sum_{p=1}^{\pi} d(n^p G^p) \tag{3.71}$$

Equation (*3.35*) applied to phase p at constant T and P becomes

$$d(n^p G^p)_{T,P} = \sum_{i=1}^{m} (\mu_i^p \, dn_i^p) \tag{3.72}$$

where the summation is over all of the m species. Combination of (*3.71*) and (*3.72*) now gives

$$(dG^t)_{T,P} = \sum_{p=1}^{\pi} \sum_{i=1}^{m} (\mu_i^p \, dn_i^p) \tag{3.73}$$

In view of (*3.68*) we have the important result

$$\boxed{\sum_{p=1}^{\pi} \sum_{i=1}^{m} (\mu_i^p \, dn_i^p) \leqq 0} \tag{3.74}$$

The equality in (*3.74*) corresponds to (*3.69*), and therefore represents a criterion of equilibrium. It is the usefulness of this criterion in the solution of equilibrium problems that is the major justification for the introduction of the chemical potential μ_i as a thermodynamic property. The equality of (*3.74*) has many applications, but its use for the numerical calculation of equilibrium compositions must await the development of quantitative methods for the description of thermodynamic properties for solutions.

Example 3.13. Application of (*3.74*) to a system composed of two phases α and β and containing two chemical species or components 1 and 2 reduces it to

$$\mu_1^\alpha dn_1^\alpha + \mu_2^\alpha dn_2^\alpha + \mu_1^\beta dn_1^\beta + \mu_2^\beta dn_2^\beta \leqq 0$$

Since the system is closed and no chemical reaction occurs, the total number of moles of each chemical species must be constant:

$$dn_1^\beta = -dn_1^\alpha \qquad dn_2^\beta = -dn_2^\alpha$$

Combination of these three equations gives:

$$(\mu_1^\alpha - \mu_1^\beta)\,dn_1^\alpha + (\mu_2^\alpha - \mu_2^\beta)\,dn_2^\alpha \leqq 0 \tag{3.75}$$

Note that although the system as a whole is closed the individual phases are not. The differentials dn_1^α and dn_2^α arise because of the transfer of matter from one phase to the other.

Consider first the case where the inequality of (*3.75*) holds:

$$(\mu_1^\alpha - \mu_1^\beta)\,dn_1^\alpha + (\mu_2^\alpha - \mu_2^\beta)\,dn_2^\alpha < 0 \tag{3.75a}$$

The process represented by this equation is one during which the system changes from a nonequilibrium state toward the equilibrium state at constant T and P. The difference $(\mu_i^\alpha - \mu_i^\beta)$ and the differential dn_i^α can both be either positive or negative. However, (*3.75a*) will always be satisfied if

$$(\mu_i^\alpha - \mu_i^\beta)\,dn_i^\alpha < 0 \qquad (i = 1, 2)$$

and this will be true if and only if $(\mu_i^\alpha - \mu_i^\beta)$ and dn_i^α have opposite signs. It is clear upon a little thought that if the transfer of mass of species i is always in the direction of the smaller chemical potential then $(\mu_i^\alpha - \mu_i^\beta)$ and dn_i^α *will* have opposite signs, and (*3.75a*) *will always* be satisfied, as is required by the laws of thermodynamics. Thus a difference in chemical potential for a particular species represents a driving force for the transport of that species, just as temperature and pressure differences represent driving forces for heat and momentum transfer. It is a general principle that when driving forces become zero transport processes cease, and the condition characterized by the term *equilibrium* is established. In this case the equality of (*3.75*) holds, and

$$(\mu_1^\alpha - \mu_1^\beta)\,dn_1^\alpha + (\mu_2^\alpha - \mu_2^\beta)\,dn_2^\alpha = 0 \tag{3.75b}$$

This equation applies to differential variations around the equilibrium state, the particular type of equilibrium considered being phase equilibrium. In (*3.75b*) the differentials dn_1^α and dn_2^α are independent and arbitrary; thus the necessary and sufficient mathematical conditions for (*3.75b*) to be valid are:

$$\mu_1^\alpha = \mu_1^\beta \qquad \text{and} \qquad \mu_2^\alpha = \mu_2^\beta$$

Although this result has been developed for a two-phase, two-component system, it is readily generalized so as to apply to multiphase, multicomponent systems. For each additional component, (*3.75b*) would include an additional term on the left, and we would conclude immediately that

$$\mu_i^\alpha = \mu_i^\beta \qquad (i = 1, 2, \ldots, m)$$

For additional phases we would consider the equilibrium requirements for the possible pairs of phases, and would conclude that for multiple phases at the same T and P the equilibrium condition can be satisfied only when the chemical potential of each species in the system is the same in all phases. Mathematically, this is expressed by

$$\boxed{\mu_i^\alpha = \mu_i^\beta = \cdots = \mu_i^\pi} \qquad (i = 1, 2, \ldots, m) \tag{3.76}$$

Equation (*3.76*) is the practical or working equation that forms the basis for phase equilibrium calculations for PVT systems of uniform temperature and pressure.

Example 3.14. In addition to its quantitative role in the solution of phase equilibrium problems, (*3.76*) provides the basis for what is known as the *phase rule*, which allows one to calculate the number of independent variables that may be arbitrarily fixed in order to establish the *intensive* state of a PVT system. Such a state is established when its temperature, pressure, and the compositions of all phases are fixed. However, for equilibrium states these variables are not all independent, and fixing a limited number of them automatically establishes the others. This number of independent variables is called the *number of degrees of freedom* of the system, and is given by the phase rule. It is this number of variables that may be arbitrarily specified and which *must* be specified in order to fix the intensive state of a PVT system at equilibrium. This number is just the difference between the number of independent intensive variables associated with the system and the number of independent equations which may be written connecting these variables.

If m represents the number of chemical species in the system, then there are $(m-1)$ independent mole fractions for each phase. (The requirement that the sum of the mole fractions must equal unity makes one mole fraction dependent.) For π phases there is a total of $(m-1)(\pi)$ composition variables. In addition, the temperature and pressure, taken to be uniform throughout the system, are phase-rule variables, and this means that the total number of independent variables is $2+(m-1)(\pi)$. The masses of the phases are not phase-rule variables, because they have nothing to do with the intensive state of the system.

Equation (*3.76*) shows that one may write $(\pi-1)$ independent phase-equilibrium equations for each species and a total of $(\pi-1)(m)$ such equations for a nonreacting system. Since the μ_i's are functions of temperature, pressure, and the phase compositions, these equations represent relations among the phase-rule variables. Subtraction of the number of independent equations from the number of independent variables gives the number of degrees of freedom F as

$$F = 2 + (m-1)(\pi) - (\pi-1)(m)$$

or

$$\boxed{F = 2 - \pi + m} \tag{3.77}$$

which is the phase rule for a nonreacting PVT system.

Solved Problems

EXACT DIFFERENTIALS AND STATE FUNCTIONS (Sec. 3.1)

3.1. The fundamental property relationship for an electrochemical cell is

$$dU^t = T\,dS^t - P\,dV^t + \mathcal{E}\,dq$$

where U^t, S^t, and V^t are total properties of the cell, $\mathcal{E}$ is the cell emf (reversible cell voltage), and q is the charge of the cell. Write the consequences of (*3.3*) and of the exactness criterion (*3.4*).

By inspection we obtain from (*3.3*)

$$T = \left(\frac{\partial U^t}{\partial S^t}\right)_{V^t,q} \qquad P = -\left(\frac{\partial U^t}{\partial V^t}\right)_{S^t,q} \qquad \mathcal{E} = \left(\frac{\partial U^t}{\partial q}\right)_{S^t,V^t}$$

and from (*3.4*)

$$\left(\frac{\partial T}{\partial V^t}\right)_{S^t,q} = -\left(\frac{\partial P}{\partial S^t}\right)_{V^t,q} \qquad \left(\frac{\partial T}{\partial q}\right)_{S^t,V^t} = \left(\frac{\partial \mathcal{E}}{\partial S^t}\right)_{V^t,q} \qquad \left(\frac{\partial P}{\partial q}\right)_{S^t,V^t} = -\left(\frac{\partial \mathcal{E}}{\partial V^t}\right)_{S^t,q}$$

3.2. Find an equation which must be satisfied by any integrating denominator $D(x,y)$ for the inexact differential expression $\delta z = M(x,y)\,dx + N(x,y)\,dy$.

By definition the expression

$$\frac{\delta z}{D} = \frac{M}{D}dx + \frac{N}{D}dy$$

is exact. Equation (3.4) then requires that

$$\left(\frac{\partial MD^{-1}}{\partial y}\right)_x = \left(\frac{\partial ND^{-1}}{\partial x}\right)_y$$

or

$$\frac{1}{D}\left(\frac{\partial M}{\partial y}\right)_x - \frac{M}{D^2}\left(\frac{\partial D}{\partial y}\right)_x = \frac{1}{D}\left(\frac{\partial N}{\partial x}\right)_y - \frac{N}{D^2}\left(\frac{\partial D}{\partial x}\right)_y$$

Thus D must satisfy the partial differential equation

$$N\left(\frac{\partial D}{\partial x}\right)_y - M\left(\frac{\partial D}{\partial y}\right)_x = D\left[\left(\frac{\partial N}{\partial x}\right)_y - \left(\frac{\partial M}{\partial y}\right)_x\right]$$

3.3. By employing the definition of H, we can write (3.5) as $\delta Q_{rev} = dH - V\,dP$. Like (3.5), this expression is inexact. Considering H a function of T and P, find an expression for $(\partial H/\partial P)_T$ resulting from the knowledge that T is an integrating denominator for δQ_{rev}.

Division of the above expression by T gives

$$dS = \frac{\delta Q_{rev}}{T} = \frac{dH}{T} - \frac{V}{T}dP$$

But H is a function of T and P, so that

$$dH = \left(\frac{\partial H}{\partial P}\right)_T dP + \left(\frac{\partial H}{\partial T}\right)_P dT$$

from which

$$dS = \left[\frac{1}{T}\left(\frac{\partial H}{\partial P}\right)_T - \frac{V}{T}\right]dP + \frac{1}{T}\left(\frac{\partial H}{\partial T}\right)_P dT$$

$$= M\,dP + N\,dT$$

Then $\left(\frac{\partial M}{\partial T}\right)_P = \frac{1}{T}\frac{\partial^2 H}{\partial T\,\partial P} - \frac{1}{T^2}\left(\frac{\partial H}{\partial P}\right)_T - \frac{1}{T}\left(\frac{\partial V}{\partial T}\right)_P + \frac{V}{T^2}$ and $\left(\frac{\partial N}{\partial P}\right)_T = \frac{1}{T}\frac{\partial^2 H}{\partial P\,\partial T}$

Equation (3.4) requires that $\left(\frac{\partial M}{\partial T}\right)_P = \left(\frac{\partial N}{\partial P}\right)_T$, from which

$$-\frac{1}{T^2}\left(\frac{\partial H}{\partial P}\right)_T - \frac{1}{T}\left(\frac{\partial V}{\partial T}\right)_P + \frac{V}{T^2} = 0 \quad \text{or} \quad \left(\frac{\partial H}{\partial P}\right)_T = -T\left(\frac{\partial V}{\partial T}\right)_P + V$$

This equation can be obtained more directly from (3.48) by the method of Example 3.10.

TRANSFORMATION RELATIONSHIPS FOR SYSTEMS OF TWO INDEPENDENT VARIABLES (Sec. 3.2)

3.4. The differential coefficent $(\partial T/\partial P)_H$ is called the *Joule-Thomson coefficient,* and is of importance in refrigeration engineering. Show that it can be calculated from PVT and heat capacity data by

$$\left(\frac{\partial T}{\partial P}\right)_H = -\frac{1}{C_P}\left[V - T\left(\frac{\partial V}{\partial T}\right)_P\right]$$

Consider H a function of T and P and apply (3.13) to get

$$\left(\frac{\partial T}{\partial P}\right)_H = -\left(\frac{\partial T}{\partial H}\right)_P\left(\frac{\partial H}{\partial P}\right)_T$$

But from (*1.9*) and from Problem 3.3, respectively,

$$\left(\frac{\partial H}{\partial T}\right)_P = C_P \qquad \text{and} \qquad \left(\frac{\partial H}{\partial P}\right)_T = V - T\left(\frac{\partial V}{\partial T}\right)_P$$

Combining these three equations, we obtain the desired result.

3.5. The thermodynamic state of an axially stressed bar of constant volume can be described by the three coordinates T, σ, and ϵ, where σ is the stress and ϵ the natural strain (see Example 1.3). Any two of these coordinates may be taken as independent. The *linear expansivity* α and *Young's modulus* E of the material are defined as

$$\alpha = \left(\frac{\partial \epsilon}{\partial T}\right)_\sigma \qquad E = \left(\frac{\partial \sigma}{\partial \epsilon}\right)_T$$

Find a relationship between α and E.

Apply (*3.13*), letting

$$\left(\frac{\partial x}{\partial y}\right)_z = \left(\frac{\partial \sigma}{\partial T}\right)_\epsilon$$

Then

$$\left(\frac{\partial \sigma}{\partial T}\right)_\epsilon = -\left(\frac{\partial \sigma}{\partial \epsilon}\right)_T\left(\frac{\partial \epsilon}{\partial T}\right)_\sigma = -E\alpha$$

3.6. *Charles' law* states that, for a gas at low pressures, the volume of the gas is directly proportional to the temperature at constant pressure. *Boyle's law* asserts that, for a gas at low pressures, the pressure of the gas is inversely proportional to the volume at constant temperature. Derive the ideal-gas law from these two observations.

According to Charles' law, $V = C_1 T$ at constant P, where C_1 is a constant. Therefore

$$\left(\frac{\partial V}{\partial T}\right)_P = C_1 = \frac{V}{T}$$

According to Boyle's law, $P = C_2/V$ at constant T, where C_2 is a constant. Therefore

$$\left(\frac{\partial V}{\partial P}\right)_T = -\frac{C_2}{P^2} = -\frac{V}{P}$$

For a constant-composition PVT system, (*3.16*) applies:

$$dV = \left(\frac{\partial V}{\partial T}\right)_P dT + \left(\frac{\partial V}{\partial P}\right)_T dP$$

Inserting the expressions for $(\partial V/\partial T)_P$ and $(\partial V/\partial P)_T$ derived above, we obtain a differential equation for V:

$$\frac{dV}{V} = \frac{dT}{T} - \frac{dP}{P}$$

This is identical to (*3.21*), and gives on integration

$$V = \frac{RT}{P}$$

where R is the universal gas constant.

LEGENDRE TRANSFORMATIONS (Sec. 3.3)

3.7. Write the U^t-related Legendre transformations for an electrochemical cell.

The fundamental property relationship was given in Problem 3.1:

$$dU^t = T\,dS^t - P\,dV^t + \mathcal{E}\,dq$$

According to *(3.22)*, there are $2^3 - 1 = 7$ possible U^t-related Legendre transformations:

$$\mathcal{T}_1(T, V^t, q) = U^t - TS^t \equiv A^t$$

$$\mathcal{T}_2(S^t, P, q) = U^t + PV^t \equiv H^t$$

$$\mathcal{T}_3(S^t, V^t, \mathcal{E}) = U^t - \mathcal{E}q$$

$$\mathcal{T}_{1,2}(T, P, q) = U^t - TS^t + PV^t \equiv G^t$$

$$\mathcal{T}_{1,3}(T, V^t, \mathcal{E}) = U^t - TS^t - \mathcal{E}q$$

$$\mathcal{T}_{2,3}(S^t, P, \mathcal{E}) = U^t + PV^t - \mathcal{E}q$$

$$\mathcal{T}_{1,2,3}(T, P, \mathcal{E}) = U^t - TS^t + PV^t - \mathcal{E}q$$

The canonical variables for each transformation are displayed in parentheses. Because of their formal identity with the functions used in the analysis of PVT systems, the three transformations designated H^t, A^t, and G^t are those most commonly employed in the thermodynamic analysis of electrochemical cells.

3.8. It is sometimes convenient in the thermodynamic analysis of certain types of systems to normalize extensive properties with respect to the volume, rather than the mass, of the system. Thus one defines an *energy density* $\tilde{U}$ and an *entropy density* $\tilde{S}$ by

$$\tilde{U} = U^t/V^t \qquad \tilde{S} = S^t/V^t$$

where U^t, S^t, and V^t are total system properties. Rewrite the fundamental property relationship for a constant-composition PVT system, *(2.6)*, in terms of these functions and write the $\tilde{U}$-related Legendre transformations.

Equation *(2.6)* is $dU^t = T\,dS^t - P\,dV^t$. Since $U^t = \tilde{U}V^t$ and $S^t = \tilde{S}V^t$, then

$$dU^t = \tilde{U}\,dV^t + V^t\,d\tilde{U} \qquad dS^t = \tilde{S}\,dV^t + V^t\,d\tilde{S}$$

so that *(2.6)* becomes

$$\tilde{U}\,dV^t + V^t\,d\tilde{U} = T\tilde{S}\,dV^t + TV^t\,d\tilde{S} - P\,dV^t$$

Rearranging, we obtain

$$d\tilde{U} = T\,d\tilde{S} - \Gamma\,d\tau$$

where $\Gamma = \tilde{U} - T\tilde{S} + P$ and $d\tau = dV^t/V^t$. The function τ is called the *volume strain* and the differential $d\tau$ the *relative dilatation.* There are three Legendre transformations given by

$$\tilde{H}(\tilde{S}, \Gamma) = \tilde{U} + \Gamma\tau$$

$$\tilde{A}(T, \tau) = \tilde{U} - T\tilde{S}$$

$$\tilde{G}(T, \Gamma) = \tilde{U} - T\tilde{S} + \Gamma\tau$$

where the canonical variables are displayed in parentheses. A complete network of thermodynamic equations can be developed from this formulation.

PROPERTY RELATIONSHIPS FOR PVT SYSTEMS (Secs. 3.4 and 3.5)

3.9. The enthalpy can be related to the Gibbs function and its temperature derivatives through the *Gibbs-Helmholtz equation.* Prove that

$$H = G - T\left(\frac{\partial G}{\partial T}\right)_P = -T^2\left[\frac{\partial(G/T)}{\partial T}\right]_P$$

From *(3.30)* and *(3.32)*, $H = G + TS$. But from *(3.39)*

$$S = -\left(\frac{\partial G}{\partial T}\right)_P$$

so that
$$H = G - T\left(\frac{\partial G}{\partial T}\right)_P \qquad (1)$$

Now
$$\left[\frac{\partial(G/T)}{\partial T}\right]_P = \frac{1}{T}\left(\frac{\partial G}{\partial T}\right)_P - \frac{G}{T^2} = -\frac{1}{T^2}\left[G - T\left(\frac{\partial G}{\partial T}\right)_P\right] \qquad (2)$$

Combination of (1) and (2) gives
$$H = -T^2\left[\frac{\partial(G/T)}{\partial T}\right]_P$$

3.10. Statistical mechanics provides a link between the thermodynamic (macroscopic) and quantum mechanical (microscopic) descriptions of a system via a *partition function* $\mathcal{Z}$. One statistical mechanical formulation for a closed PVT system gives the following expression for the Helmholtz function:

$$A = -RT \ln \mathcal{Z}$$

where R is the universal gas constant and $\mathcal{Z}$ is a function of T and V only. Find expressions for P, S, U, H, and G in terms of T, V, and $\mathcal{Z}$.

From (3.37),
$$P = -\left(\frac{\partial A}{\partial V}\right)_T = RT\left(\frac{\partial \ln \mathcal{Z}}{\partial V}\right)_T$$

From (3.39),
$$S = -\left(\frac{\partial A}{\partial T}\right)_V = R \ln \mathcal{Z} + RT\left(\frac{\partial \ln \mathcal{Z}}{\partial T}\right)_V$$

From the definitions (3.30), (3.31), and (3.32),
$$U = A + TS = RT^2\left(\frac{\partial \ln \mathcal{Z}}{\partial T}\right)_V$$

$$H = U + PV = RT^2\left(\frac{\partial \ln \mathcal{Z}}{\partial T}\right)_V + RTV\left(\frac{\partial \ln \mathcal{Z}}{\partial V}\right)_T$$

$$G = A + PV = -RT \ln \mathcal{Z} + RTV\left(\frac{\partial \ln \mathcal{Z}}{\partial V}\right)_T$$

3.11. Show that in a PVT system $\mu_i = G_i$ for a pure material.

From (3.40)
$$\mu_i = \left[\frac{\partial(nG)}{\partial n_i}\right]_{T,P,n_j} = n\left(\frac{\partial G}{\partial n_i}\right)_{T,P,n_j} + G\left(\frac{\partial n}{\partial n_i}\right)_{T,P,n_j}$$

But for a pure material $n = n_i$ and $G = G_i$, and G_i is independent of n_i. Thus
$$\left(\frac{\partial G}{\partial n_i}\right)_{T,P,n_j} = \left(\frac{\partial G_i}{\partial n_i}\right)_{T,P,n_j} = 0$$

$$\left(\frac{\partial n}{\partial n_i}\right)_{T,P,n_j} = \left(\frac{\partial n_i}{\partial n_i}\right)_{T,P,n_j} = 1$$

and therefore $\mu_i = G_i$.

We will see in the next chapter that this identity is useful in studying the phase equilibrium of single-component systems; for such systems this result when combined with (3.76) requires that the molar Gibbs functions be the same for coexisting phases at equilibrium. Thus, $G_i^\alpha = G_i^\beta$ for equilibrium between two phases α and β of pure i.

3.12. Show that, for a constant-composition PVT system,

$$d\mu_i = -\bar{S}_i\,dT + \bar{V}_i\,dP$$

where $$\bar{S}_i = \left[\frac{\partial(nS)}{\partial n_i}\right]_{T,P,n_j} \qquad \bar{V}_i = \left[\frac{\partial(nV)}{\partial n_i}\right]_{T,P,n_j}$$

For a constant-composition PVT system, μ_i can be considered a function of T and P only. The chemical potential is an intensive property, so the restrictions of constant composition and constant mole numbers are equivalent for evaluating changes in μ_i. Thus for a change of state

$$d\mu_i = \left(\frac{\partial\mu_i}{\partial T}\right)_{P,n} dT + \left(\frac{\partial\mu_i}{\partial P}\right)_{T,n} dP$$

where the subscript n indicates that all mole numbers are held constant. But, by *(3.45)* and *(3.46)*,

$$\left(\frac{\partial\mu_i}{\partial T}\right)_{P,n} = -\left[\frac{\partial(nS)}{\partial n_i}\right]_{T,P,n_j} = -\bar{S}_i$$

$$\left(\frac{\partial\mu_i}{\partial P}\right)_{T,n} = \left[\frac{\partial(nV)}{\partial n_i}\right]_{T,P,n_j} = \bar{V}_i$$

and therefore

$$d\mu_i = -\bar{S}_i\,dT + \bar{V}_i\,dP$$

The quantities $\bar{S}_i$ and $\bar{V}_i$ are called the *partial molar entropy* and the *partial molar volume* of component i. Partial molar properties will be treated in more detail in Section 7.1.

3.13. In Example 3.12 some properties of U for an incompressible liquid were derived by considering U a function of T and V. Show that the same results obtain if instead U is considered a function of T and P.

Considering U a function of T and P, we write

$$dU = \left(\frac{\partial U}{\partial T}\right)_P dT + \left(\frac{\partial U}{\partial P}\right)_T dP$$

Expressions must now be found for $(\partial U/\partial T)_P$ and $(\partial U/\partial P)_T$. Division of *(2.6)* by dT and restriction to constant P gives

$$\left(\frac{\partial U}{\partial T}\right)_P = T\left(\frac{\partial S}{\partial T}\right)_P - P\left(\frac{\partial V}{\partial T}\right)_P$$

from which we obtain, employing *(3.50)*, *(3.56)*, and *(3.17)*,

$$\left(\frac{\partial U}{\partial T}\right)_P = C_P - PV\beta = C_V + \beta TV\left(\frac{\partial P}{\partial T}\right)_V - PV\beta$$

or

$$\left(\frac{\partial U}{\partial T}\right)_P = C_V + \beta V\left[T\left(\frac{\partial P}{\partial T}\right)_V - P\right] \tag{1}$$

An equation for $(\partial U/\partial P)_T$ follows from *(3.14)* and *(3.8)*:

$$\left(\frac{\partial U}{\partial P}\right)_T = \left(\frac{\partial U}{\partial V}\right)_T\left(\frac{\partial V}{\partial P}\right)_T = \left[T\left(\frac{\partial P}{\partial T}\right)_V - P\right]\left(\frac{\partial V}{\partial P}\right)_T$$

or

$$\left(\frac{\partial U}{\partial P}\right)_T = \kappa V\left[P - T\left(\frac{\partial P}{\partial T}\right)_V\right] \tag{2}$$

But $\beta = \kappa = 0$ for an incompressible liquid, so *(1)* and *(2)* become

$$\left(\frac{\partial U}{\partial T}\right)_P = C_V \qquad \text{and} \qquad \left(\frac{\partial U}{\partial P}\right)_T = 0$$

Thus for an incompressible liquid U is a function of T only and changes in U are calculated from $dU = C_V\,dT$, in agreement with Example 3.12.

3.14. At low frequencies the velocity of sound c in a fluid is related to the adiabatic compressibility κ_S by

$$c = \sqrt{\frac{g_c V}{M\kappa_S}}$$

where $\kappa_S = -\frac{1}{V}\left(\frac{\partial V}{\partial P}\right)_S$ and M is the molecular weight of the fluid. Show that:

(a) $c = \sqrt{\frac{\gamma g_c RT}{M}}$ in an ideal gas.

(b) $c = \infty$ in an incompressible liquid.

(a) From (*3.57 ideal*),

$$\kappa_S = \frac{1}{P}\frac{C_V}{C_P} = \frac{1}{P\gamma} \qquad \text{and thus} \qquad c = \sqrt{\frac{\gamma g_c PV}{M}}$$

But for an ideal gas $PV = RT$ and therefore

$$c = \sqrt{\frac{\gamma g_c RT}{M}}$$

(b) From (*3.57*), $\kappa_S = \kappa/\gamma$ and therefore

$$c = \sqrt{\frac{\gamma g_c V}{M\kappa}}$$

for *any* fluid. But for an incompressible fluid $\kappa = 0$ and therefore $c = \infty$.

These two results are idealizations, but they are in accord with the general observations that (a) the velocity of sound increases with the temperature of the medium, and (b) the sonic velocity is substantially greater in liquids than in gases.

ATTAINMENT OF EQUILIBRIUM IN CLOSED, HETEROGENEOUS SYSTEMS (Sec. 3.6)

3.15. How many degrees of freedom are there in a PVT system consisting of a vapor phase in equilibrium with a liquid phase and containing (a) a single component? (b) two nonreacting chemical species?

For both cases, there are two equilibrium phases and therefore $\pi = 2$. The phase rule (*3.77*) then gives $F = m$.

(a) For a single component, $m = 1$ and therefore $F = 1$. Thus, specification of T automatically fixes P (or vice versa), and also determines the other intensive properties of both phases.

(b) Here, $m = 2$ and therefore $F = 2$. Specification of any *two* independent intensive variables, such as T and P, determines *all* intensive properties of both phases.

3.16. Show that the maximum possible number of coexisting phases at equilibrium is three for a single-component PVT system.

There is one component and therefore $m = 1$. By the phase rule (*3.77*)

$$\pi = 2 + m - F = 3 - F$$

The minimum possible number degrees of freedom is zero, because it is generally impossible to solve an algebraic system for which the number of equations exceeds the number of variables. Thus the maximum possible number of phases is

$$\pi = 3 - 0 = 3$$

This is the condition that exists at the triple point of water, at which water vapor, liquid water, and "ice I" coexist in equilibrium at 0.01(°C) and 0.006113(bar).

3.17. Solve Problem 3.16 without directly appealing to the phase rule.

The condition of phase equilibrium for a π-phase single-component system follows from the equality of (*3.74*):

$$\sum_{p=1}^{\pi} \sum_{i=1}^{1} \mu_i^p \, dn_i^p = \sum_{p=1}^{\pi} \mu^p \, dn^p = 0 \tag{1}$$

where the subscript i has been dropped. The *total* number of moles n is fixed, and is equal to the summation of n^p over all phases:

$$\sum_{p=1}^{\pi} n^p = n \tag{2}$$

Thus, only $\pi - 1$ of the dn^p are independent and arbitrary, because differentiation of (*2*) gives

$$d\left(\sum_{p=1}^{\pi} n^p\right) = 0$$

or

$$\sum_{p=1}^{\pi} dn^p = 0 \tag{3}$$

We may choose the first $\pi - 1$ dn^p as independent. According to (*3*), they must satisfy the equation

$$-\sum_{p=1}^{\pi-1} dn^p = dn^\pi \tag{4}$$

Equation (*1*) may be written as

$$\sum_{p=1}^{\pi-1} \mu^p \, dn^p + \mu^\pi \, dn^\pi = 0$$

and combination of this expression with (*4*) gives

$$\sum_{p=1}^{\pi-1} (\mu^p - \mu^\pi) \, dn^p = 0 \tag{5}$$

Now, the dn^p are independent and arbitrary, so (*5*) can only hold in general if *each* of the terms $(\mu^p - \mu^\pi)$ is identically zero. Thus the single equation (*5*) is equivalent to the system of $\pi - 1$ equations

$$\begin{aligned} \mu^\alpha &= \mu^\pi \\ \mu^\beta &= \mu^\pi \\ &\vdots \\ \mu^{\pi-1} &= \mu^\pi \end{aligned} \tag{6}$$

For a single-component system, composition is not a variable and the μ^p are functions of T and P only. The maximum possible number of coexisting phases obtains when the number of equations of the form (*6*) just equals the number of possible variables, which is in this case two (T and P). Thus the maximum number of coexisting phases is *three*, and the equilibrium situation is defined by the two equations

$$\mu^\alpha(T,P) = \mu^\gamma(T,P)$$

$$\mu^\beta(T,P) = \mu^\gamma(T,P)$$

where the superscripts α, β, and γ identify the three phases in equilibrium.

3.18. **For a nonreacting PVT system at equilibrium, how many variables can be fixed arbitrarily in determining *completely* the state of the system if the total numbers of moles of each component are specified?**

The state of the system will be *completely* specified if we know, in addition to T, P, and the $m-1$ independent mole fractions in each phase, the total number of moles of each phase. The total number of variables is thus the sum of

	$(m-1)\pi$	independent mole fractions
plus	π	total mole numbers for the phases
plus	2	(T and P)
giving	$m\pi+2$	independent variables

Phase equilibrium is defined, as before, by the $m(\pi-1)$ equations among the μ_i^p which follow from (*3.76*). In addition to these equations there are m constraining equations of the form

$$\sum_{p=1}^{\pi} n_i^p \;=\; n_i$$

where $i = 1, 2, \ldots, m$, and the n_i are the total component mole numbers specified in the statement of the problem. The total number of available equations is thus the sum of

	$m(\pi-1)$	equations among the μ_i^p
plus	m	constraining equations on the n_i^p
giving	$m\pi$	independent equations

The number of variables which can be fixed arbitrarily is the number of independent variables less the number of independent equations, or

$$(m\pi+2) - (m\pi) \;=\; 2$$

Although this result was derived for phase equilibrium, it applies as well to systems which can undergo chemical reaction, and forms the basis for *Duhem's theorem*: For any closed PVT system formed from given initial amounts of prescribed chemical species, the equilibrium state is completely determined by any two properties of the whole system, provided that these properties are independently variable at equilibrium.

MISCELLANEOUS APPLICATIONS

3.19. **Set up the basic thermodynamic relationships for a constant-volume, axially stressed bar.**

Consider a process in which the bar is reversibly stressed and during which it reversibly exchanges heat with its surroundings. Then

$$\delta Q_{\text{rev}} \;=\; T\,dS \qquad\qquad \delta W_{\text{rev}} \;=\; -V\sigma\,d\epsilon$$

where S and V are the total entropy and volume of the bar. The equation for W_{rev} comes from Example 1.3. Applying the first law to this process, we get

$$dU \;=\; \delta Q_{\text{rev}} - \delta W_{\text{rev}} \;=\; T\,dS + V\sigma\,d\epsilon$$

But V is constant, so the above equation can be written

$$d\tilde{U} \;=\; T\,d\tilde{S} + \sigma\,d\epsilon \tag{1}$$

where $\tilde{U}$ and $\tilde{S}$ are the energy density and entropy density, respectively (see Problem 3.8). Equation (*1*) is the fundamental property relationship for our system and will form the basis for further derived relationships. There are three $\tilde{U}$-related Legendre transformations:

$$\tilde{H}(\tilde{S},\sigma) \;=\; \tilde{U} - \sigma\epsilon \tag{2}$$

$$\tilde{A}(T,\epsilon) \;=\; \tilde{U} - T\tilde{S} \tag{3}$$

$$\tilde{G}(T,\sigma) \;=\; \tilde{U} - T\tilde{S} - \sigma\epsilon \tag{4}$$

the total differentials of which are

$$d\tilde{H} = T\,d\tilde{S} - \epsilon\,d\sigma \tag{5}$$

$$d\tilde{A} = -\tilde{S}\,dT + \sigma\,d\epsilon \tag{6}$$

$$d\tilde{G} = -\tilde{S}\,dT - \epsilon\,d\sigma \tag{7}$$

A set of equations analogous to (*3.36*) through (*3.39*) follows from application of (*3.3*) to (*1*), (*5*), (*6*), and (*7*):

$$T = \left(\frac{\partial \tilde{U}}{\partial \tilde{S}}\right)_\epsilon = \left(\frac{\partial \tilde{H}}{\partial \tilde{S}}\right)_\sigma \tag{8}$$

$$\sigma = \left(\frac{\partial \tilde{U}}{\partial \epsilon}\right)_{\tilde{S}} = \left(\frac{\partial \tilde{A}}{\partial \epsilon}\right)_T \tag{9}$$

$$\epsilon = -\left(\frac{\partial \tilde{H}}{\partial \sigma}\right)_{\tilde{S}} = -\left(\frac{\partial \tilde{G}}{\partial \sigma}\right)_T \tag{10}$$

$$\tilde{S} = -\left(\frac{\partial \tilde{A}}{\partial T}\right)_\epsilon = -\left(\frac{\partial \tilde{G}}{\partial T}\right)_\sigma \tag{11}$$

Application of the exactness criterion (*3.4*) to (*1*), (*5*), (*6*), and (*7*) gives

$$\left(\frac{\partial T}{\partial \epsilon}\right)_{\tilde{S}} = \left(\frac{\partial \sigma}{\partial \tilde{S}}\right)_\epsilon \tag{12}$$

$$\left(\frac{\partial T}{\partial \sigma}\right)_{\tilde{S}} = -\left(\frac{\partial \epsilon}{\partial \tilde{S}}\right)_\sigma \tag{13}$$

$$\left(\frac{\partial \sigma}{\partial T}\right)_\epsilon = -\left(\frac{\partial \tilde{S}}{\partial \epsilon}\right)_T \tag{14}$$

$$\left(\frac{\partial \epsilon}{\partial T}\right)_\sigma = \left(\frac{\partial \tilde{S}}{\partial \sigma}\right)_T \tag{15}$$

which are analogues of the Maxwell equations (*3.41*) through (*3.44*).

The analogues of the volume expansivity and the isothermal compressibility were introduced in Problem 3.5. They are the linear expansivity α and Young's modulus E:

$$\alpha = \left(\frac{\partial \epsilon}{\partial T}\right)_\sigma \tag{16}$$

$$E = \left(\frac{\partial \sigma}{\partial \epsilon}\right)_T \tag{17}$$

Two other useful quantities are the constant-strain heat capacity and the constant-stress heat capacity:

$$C_\epsilon = \left(\frac{\partial \tilde{U}}{\partial T}\right)_\epsilon \tag{18}$$

$$C_\sigma = \left(\frac{\partial \tilde{H}}{\partial T}\right)_\sigma \tag{19}$$

Equations (*1*) through (*19*) form a basis for the thermodynamics of constant-volume stressed bars. Other formulations are possible (see Problem 3.39).

3.20. (*a*) For a reversible, isothermal process in a closed PVT system, show that $\delta W = -d(nA)$. (*b*) For a reversible, isothermal, constant-pressure process in a closed PVT system, show that $\delta W = -d(nG) + P\,d(nV)$.

For a closed system, not in motion, the first law requires that $\delta W = \delta Q - d(nU)$. For reversible processes, the second law requires $\delta Q = T\,d(nS)$. Thus

$$\delta W = T\,d(nS) - d(nU) \tag{1}$$

This basic equation applies to both (*a*) and (*b*).

(*a*) Differentiation of (*3.31*) gives

$$d(nU) = d(nA) + T\,d(nS) + (nS)\,dT$$

and substitution in (*1*) yields

$$\delta W = -d(nA) - (nS)\,dT$$

For an isothermal process this reduces to

$$\delta W = -d(nA) \tag{2}$$

(*b*) Again from (*3.31*) and (*3.32*) we see that

$$(nG) = (nA) + P(nV)$$

By differentiation

$$d(nA) = d(nG) - P\,d(nV) - (nV)\,dP$$

and substitution in (*2*) gives for an isothermal process

$$\delta W = -d(nG) + P\,d(nV) + (nV)\,dP$$

For a process that is isobaric as well, this reduces to

$$\delta W = -d(nG) + P\,d(nV) \tag{3}$$

It is seen from (*3*) that the work δW need not be merely $P\,d(nV)$, which is the work of expansion. The origin of additional work over and above expansion work is found by substituting for $d(nG)$ from (*3.35*), which for an isothermal, isobaric process becomes $d(nG) = \Sigma \mu_i\,dn_i$. Thus we have from (*3*)

$$\delta W = -\Sigma\,\mu_i\,dn_i + P\,d(nV) \tag{4}$$

From this we see that work other than expansion work can result from changes in the numbers of moles of the chemical species present. In a closed system such changes can only result from a chemical reaction. Work of this origin is most commonly manifested as electrical work, and the system in such a case is an electrochemical cell.

Equation (*2*) also reduces to (*4*) by substitution for $d(nA)$ from (*3.34*). Thus (*4*) gives the work for any reversible, isothermal process in a closed PVT system.

Supplementary Problems

EXACT DIFFERENTIALS AND STATE FUNCTIONS (Sec. 3.1)

3.21. Prove that the differential expression $M\,dx + N\,dy$ is exact if M is a function of x only and N is a function of y only.

3.22. Show that $D = Ax^{1-B}y^{1+B}$, where A and B are arbitrary constants, is an integrating denominator for $y\,dx - x\,dy$.

3.23. Prove that $D = AT + B$, where A and B are constants, is an integrating denominator for

$$\delta Q_{\text{rev}} = dU + P\,dV$$

if and only if B is identically zero. [*Hint*: Use the result of Problem 3.2.]

TRANSFORMATION RELATIONSHIPS FOR SYSTEMS OF TWO INDEPENDENT VARIABLES (Sec. 3.2)

3.24. Show that for a constant-volume stressed bar

$$\left(\frac{\partial \epsilon}{\partial T}\right)_{\tilde{S}} = \alpha + E^{-1}\left(\frac{\partial \sigma}{\partial T}\right)_{\tilde{S}}$$

where $\tilde{S} = S/V$. [*Hint*: See Problem 3.5.]

3.25. For a certain gas, β and κ are given by $\beta T = \kappa P = 1 - (b/V)$, where b is a constant. Find the equation of state of the gas if $PV/RT = 1$ at $P = 0$. *Ans.* $P(V-b) = RT$

3.26. Show that the Joule-Thomson coefficient defined in Problem 3.4 can be written in the alternate forms

$$\left(\frac{\partial T}{\partial P}\right)_H = T^2 \frac{\left[\dfrac{\partial(V/T)}{\partial T}\right]_P}{C_P}$$

$$= \frac{V(\beta T - 1)}{C_P}$$

Prove that $(\partial T/\partial P)_H = 0$ for an ideal gas.

LEGENDRE TRANSFORMATIONS (Sec. 3.3)

3.27. The fundamental property relation for a PVT system may be written

$$dV = \frac{T}{P}\,dS - \frac{1}{P}\,dU$$

Write the V-related Legendre transformations and show the canonical variables.

Ans. $\mathcal{T}_1\left(\frac{T}{P}, U\right) = V - \frac{T}{P}S$; $\mathcal{T}_2\left(S, \frac{1}{P}\right) = V + \frac{1}{P}U$; $\mathcal{T}_{1,2}\left(\frac{T}{P}, \frac{1}{P}\right) = V - \frac{T}{P}S + \frac{1}{P}U$

3.28. Show that the Massieu functions defined in Example 3.9 are related to the usual U-related transformations by

$$\Omega = \frac{U-G}{T} \qquad \psi = -\frac{A}{T} \qquad \Phi = -\frac{G}{T}$$

Prove that $\Omega =$ constant for an ideal gas undergoing an isentropic change of state.

PROPERTY RELATIONSHIPS FOR PVT SYSTEMS (Secs. 3.4 and 3.5)

3.29. Prove that for a PVT system

(*a*) $$U = -T^2\left[\frac{\partial(A/T)}{\partial T}\right]_V$$

(*b*) $$\left(\frac{\partial U}{\partial V}\right)_T = T^2\left[\frac{\partial(P/T)}{\partial T}\right]_V$$

(*c*) $$\left(\frac{\partial U}{\partial S}\right)_T = -P^2\left[\frac{\partial(T/P)}{\partial P}\right]_V$$

3.30. For a PVT system, show that if V is a function only of the ratio P/T, then U is a function only of T.

3.31. Prove that a substance for which $U = B + CV^{-R/\delta}e^{S/\delta}$ obeys the ideal gas law. B, C, and δ are constants.

3.32. In Problem 3.12 we defined the partial molar entropy $\overline{S}_i$ and the partial molar volume $\overline{V}_i$. Show that for a pure material $\overline{S}_i = S_i$ and $\overline{V}_i = V_i$.

3.33. Derive (*3.53*).

3.34. (*a*) Prove that $$dS = \frac{C_V}{T}\left(\frac{\partial T}{\partial P}\right)_V dP + \frac{C_P}{T}\left(\frac{\partial T}{\partial V}\right)_P dV.$$

(*b*) For an ideal gas with constant heat capacities, use the above equation to derive the PV relation for an isentropic process: $PV^\gamma = \text{constant}$.

(*c*) Show that the equation of part (*a*) may also be written

$$T\,dS = C_V\left(\frac{\partial T}{\partial P}\right)_V \left[dP - \gamma\left(\frac{\partial P}{\partial V}\right)_T dV\right]$$

(*d*) Derive (*3.57*) from the result of part (*c*).

3.35. Prove that

(*a*) $$\left(\frac{\partial U}{\partial T}\right)_S = C_V\left(\frac{\partial \ln T}{\partial \ln P}\right)_V$$

(*b*) $$\left(\frac{\partial H}{\partial T}\right)_S = C_P\left(\frac{\partial \ln T}{\partial \ln V}\right)_P$$

ATTAINMENT OF EQUILIBRIUM IN CLOSED, HETEROGENEOUS SYSTEMS (Sec. 3.6)

3.36. Derive (*3.67*) and (*3.68*).

3.37. Deduce (*3.74*) from (*3.63*) *without* specifically considering a process at constant T and P.

3.38. Which of the following phase equilibrium situations are compatible with the phase rule for non-reacting systems?

(*a*) Equilibrium among three allotropic forms of ice.

(*b*) Equilibrium among water vapor, liquid water, and two allotropic forms of ice.

(*c*) Equilibrium between two dense gas phases in a system with an arbitrary number of components.

(*d*) Equilibrium among an A-rich liquid phase, a B-rich liquid phase, and a vapor phase containing A and B in an AB binary system.

(*e*) Equilibrium among $m+3$ phases in a system containing m components.

Ans. (*a*), (*c*), (*d*)

MISCELLANEOUS APPLICATIONS

3.39. In Problem 3.19 we outlined the derivation of the primary thermodynamic property relationships for an axially stressed bar, and in that problem the special variables were T, $\tilde{S}$, σ, and ϵ. In that development, however, we could equally as well have employed the reversible work expression $\delta W_{\text{rev}} = -F\,dl$, rather than the equivalent expression $\delta W_{\text{rev}} = -\sigma V\,d\epsilon$ (see Example 1.3), in which case the fundamental property relationship would be

$$dU = T\,dS + F\,dl$$

For this formulation, the special variables are T, S, F, and l, and a network of equations analogous to those in Problem 3.19 can be written down by formally substituting S for $\tilde{S}$, F for σ, and l for ϵ

in equations (2) through (15). In this alternate system quantities analogous to α, E, C_ϵ, and C_σ are defined as follows:

$$\alpha' = \frac{1}{l}\left(\frac{\partial l}{\partial T}\right)_F \qquad E' = \frac{l^2}{V}\left(\frac{\partial F}{\partial l}\right)_T$$

$$C_l = \left(\frac{\partial U}{\partial T}\right)_l \qquad C_F = \left(\frac{\partial H}{\partial T}\right)_F$$

Recalling that $\sigma = Fl/V$ and that V is assumed constant, derive the following equations relating the expansivities, Young's moduli, and heat capacities for the two formulations:

$$\alpha' = \frac{\alpha E}{E - \sigma} \qquad E' = E - \sigma$$

$$C_l = VC_\epsilon \qquad C_F = V\left[C_\sigma + \frac{E\alpha\sigma}{E - \sigma}(\alpha T - \epsilon)\right]$$

Chapter 4

Properties of Pure Substances

It is evident from the preceding chapter that thermodynamics provides a multitude of equations interrelating the properties of substances. The properties themselves depend on the nature of the substance, and they differ from one substance to the next. Thermodynamics is in no sense a model or description of the behavior of matter; rather, it depends for its usefulness on the availability of experimental or theoretical values for a minimum number of properties. Given appropriate data, thermodynamics allows the development of a complete set of thermodynamic property values from which one can subsequently calculate the heat and work effects of various processes and determine equilibrium conditions in a variety of systems.

Thermodynamic properties such as internal energy, enthalpy, and entropy, are not directly measurable. Values must be calculated by the equations of thermodynamics from experimental values for measurable quantities, such as temperature, pressure, volume, and heat capacity. Although the behavior of substances is not inherently a part of thermodynamics, knowledge of the properties of substances is essential to any effective application of thermodynamics. This chapter therefore treats qualitatively the general behavior of substances in equilibrium states and quantitatively the methods used for correlation of experimental data and for calculation of property values.

4.1 PVT BEHAVIOR OF A PURE SUBSTANCE

The relationship of specific or molar volume to temperature and pressure for a pure substance in equilibrium states can be represented by a surface in three dimensions, as shown in Fig. 4-1. The surfaces marked S, L, and G represent respectively the solid, liquid, and gas regions of the diagram. The unshaded surfaces are the regions of coexistence of two phases in equilibrium, and there are three such regions: solid-gas (S-G), solid-liquid (S-L), and liquid-gas (L-G). Heavy lines separate the various regions and form boundaries of the surfaces representing the individual phases. The heavy line passing through points A and B marks the intersections of the two-phase regions, and is the three-phase line, along which solid, liquid, and gas phases exist in three-phase equilibrium. According to the phase rule, such systems have zero degrees of freedom; they exist for a given pure substance at but one temperature and one pressure. For this reason the projection of this line on the PT plane (shown to the left of the main diagram of Fig. 4-1) is a point, known as the *triple point*. The phase rule also requires that systems made up of two phases in equilibrium have just one degree of freedom, and therefore the two-phase regions must project as lines on the PT plane, forming a PT diagram of three lines, fusion (or melting), sublimation, and vaporization, which meet at the triple point. The PT projection of Fig. 4-1 provides no information about the volumes of the systems represented. This is, however, given explicitly by the other projection of Fig. 4-1, where all surfaces of the three-dimensional diagram show up as areas on the PV plane.

The PT projection of Fig. 4-1 is shown to a larger scale in Fig. 4-2, and the liquid and gas regions of the PV projection are shown in more detail by Fig. 4-3.

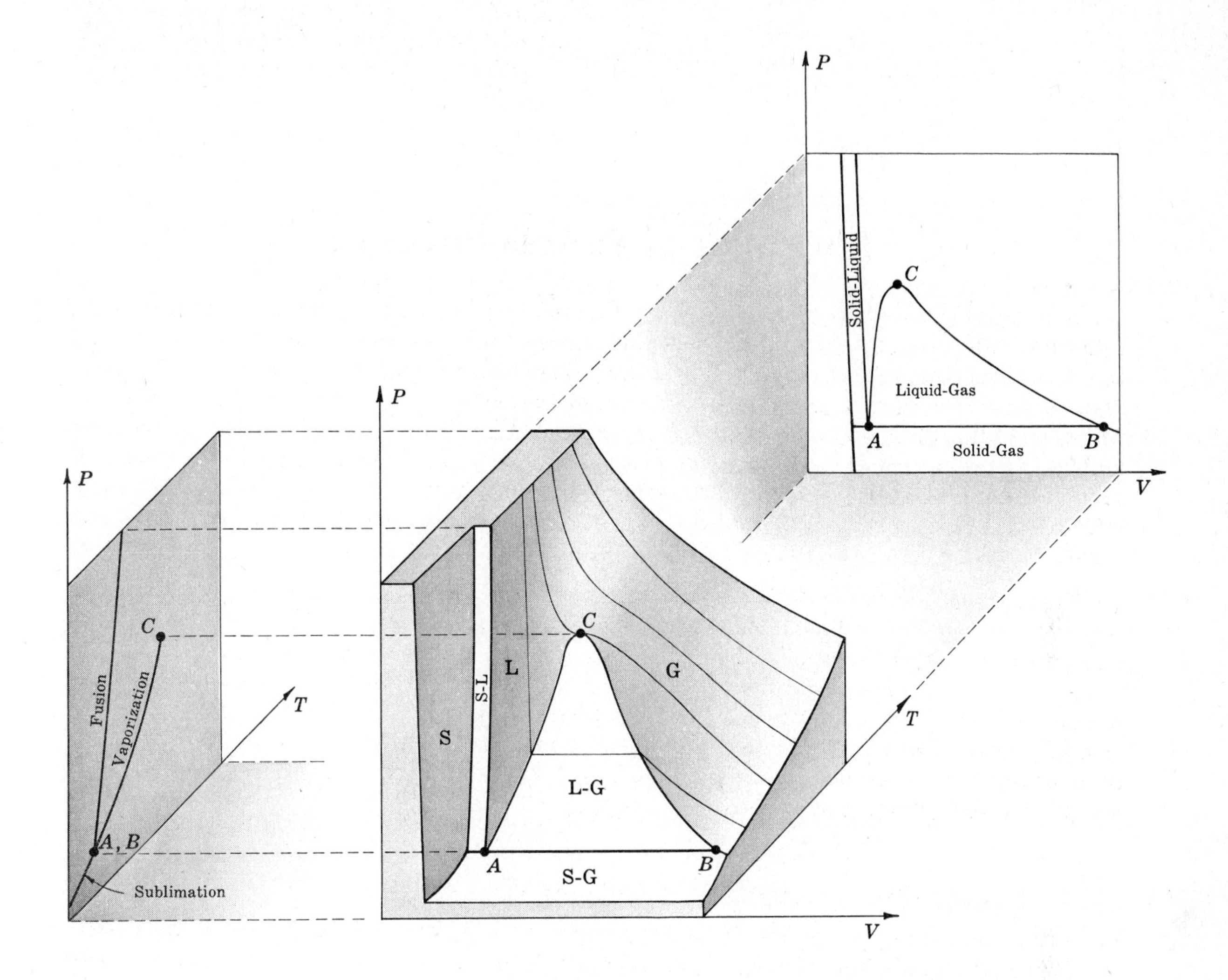
P
T
C
Fusion
Vaporization
A, B
Sublimation
P
C
S
S-L
L
G
L-G
A
B
S-G
V
T
P
Solid-Liquid
C
Liquid-Gas
A
Solid-Gas
B
V

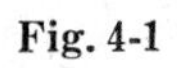

Fig. 4-1

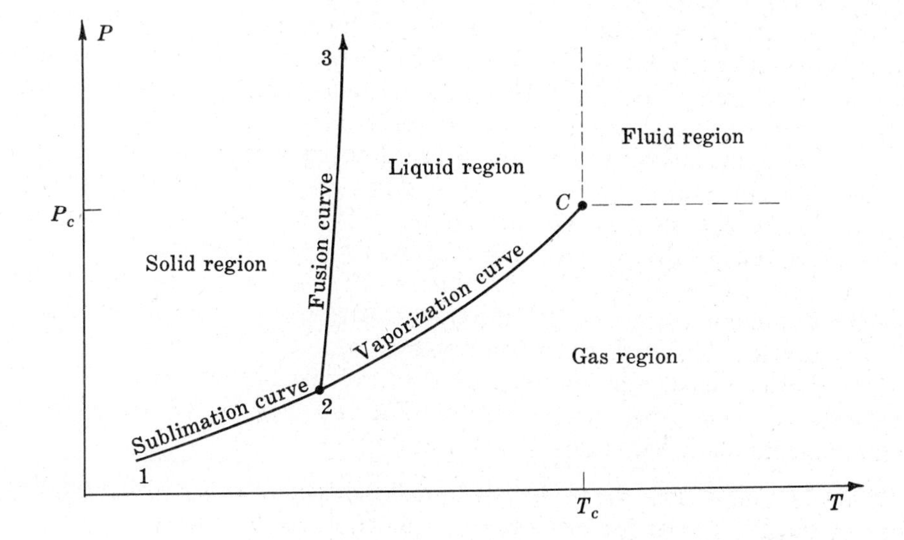
P
3
Fluid region
Liquid region
Fusion curve
P_c
C
Solid region
Vaporization curve
Gas region
Sublimation curve
2
1
T_c
T

Fig. 4-2

The solid lines of Fig. 4-2 clearly represent phase boundaries. The fusion curve (line 2-3) normally has a positive slope, but for a few substances (water is the best known) it has a negative slope. This line is believed to continue upward indefinitely. The two curves 1-2 and 2-C represent the *vapor pressures of* solid and liquid respectively. The terminal point C is the *critical point,* which represents the highest pressure and highest temperature at which liquid and gas can coexist in equilibrium. The termination of the vapor-pressure curve at point C means that at higher temperatures and pressures no clear distinction can be drawn between what is called liquid and what is called gas. Thus there is a region extending indefinitely upward from and indefinitely to the right of the critical point that is called simply the fluid region. It is bounded by dashed lines that do not represent phase transitions, but which conform to arbitrary definitions of what is considered a liquid and what is considered a gas. The region designated as liquid in Fig. 4-2 lies above the vaporization curve. Thus a liquid can always be vaporized by a sufficient reduction in pressure at constant temperature. The gas region of Fig. 4-2 lies to the right of the sublimation and vaporization curves. Thus a gas can always be condensed by a sufficient reduction in temperature at constant pressure. A substance at a temperature above its critical temperature T_c *and* at a pressure above its critical pressure P_c is said to be a *fluid* because it cannot be caused either to liquefy by temperature reduction at constant P or to vaporize by pressure reduction at constant T.

A *vapor* is a gas existing at temperatures below T_c, and it may therefore be condensed either by reduction of temperature at constant P or by increase of pressure at constant T.

Example 4.1. Show how it is possible to change a vapor into a liquid without condensation.

When a vapor condenses it undergoes an abrupt change of phase, characterized by the appearance of a meniscus. This can only occur when the vaporization curve of Fig. 4-2 is crossed. Figure 4-4 shows two points A and B on either side of the vaporization curve. Point A represents a vapor state and point B a liquid state. If the pressure on the vapor at A is raised at constant T, then the vertical path from A to B is followed and this path crosses the vaporization curve, at which point condensation occurs at a fixed pressure. However, it is also possible to follow a path from A to B that goes around the critical point C and which then does not cross the vaporization curve, as can be seen in Fig. 4-4. When this path is followed the transition from vapor to liquid is gradual, and at no point is an abrupt change in properties observed.

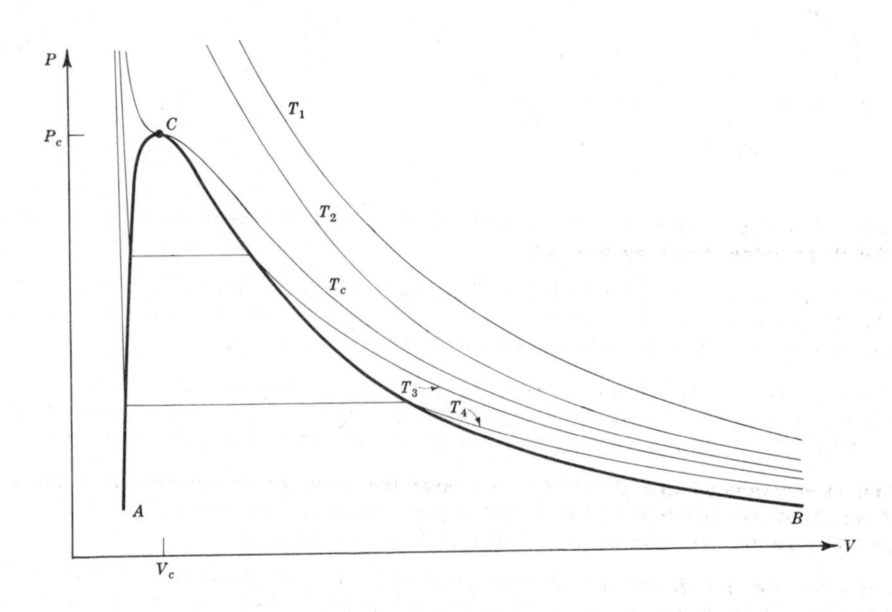

Fig. 4-3

The primary curves of Fig. 4-3 give the pressure-volume relations for *saturated liquid* (A to C) and for *saturated vapor* (C to B). The area lying below the curve ACB represents the two-phase region where saturated liquid and saturated vapor coexist at equilibrium. Point C is the critical point, for which the coordinates are P_c and V_c. A number of constant-temperature lines or *isotherms* are included on Fig. 4-3. The critical isotherm at temperature T_c passes through the critical point. Isotherms for higher temperatures T_1 and T_2 lie above the critical isotherm. Lower-temperature isotherms lie below the critical isotherm, and are made up of three sections. The middle section traverses the two-phase region, and is horizontal, because equilibrium mixtures of liquid and vapor at a fixed temperature have a fixed pressure, the vapor pressure, regardless of the proportions of liquid and vapor present. Points along this horizontal line represent different proportions of liquid and vapor, ranging from all liquid at the left end to all vapor at the right. These horizontal line segments become progressively shorter with increasing temperature, until at the critical point the isotherm exhibits a horizontal inflection. As this point is approached the liquid and vapor phases become more and more nearly alike, eventually becoming unable to sustain a meniscus between them, and ultimately becoming indistinguishable.

The sections of isotherms to the left of line AC traverse the liquid region, and are very steep, because liquids are not very *compressible*; that is, it takes large changes in pressure to effect a small volume change. A liquid which is not saturated (not at its boiling point, along line AC) is often called subcooled liquid or compressed liquid.

The sections of isotherms to the right of line CB lie in the vapor region. Vapor in this region is called superheated vapor in order to distinguish it from saturated vapor as represented by the line CB.

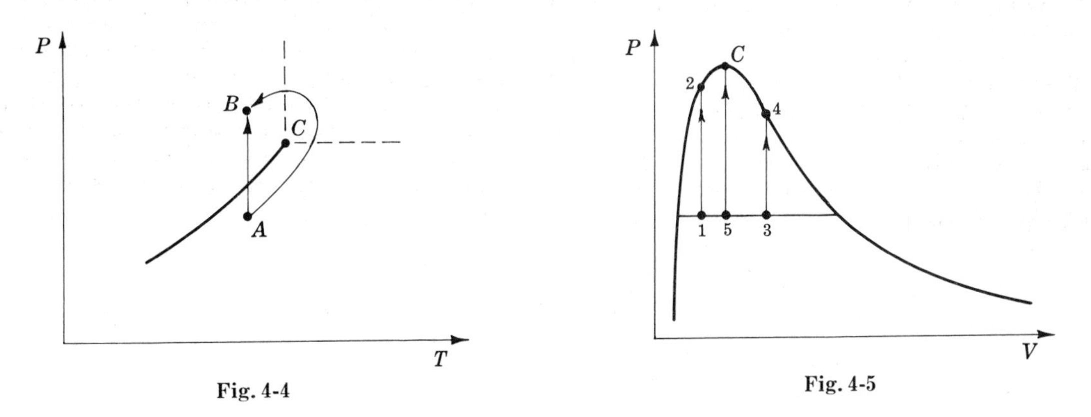

Fig. 4-4

Fig. 4-5

Example 4.2. Describe what would be observed if a pure substance were heated in a sealed tube as indicated by the three paths shown by Fig. 4-5.

Points 1, 3, and 5 represent three different fillings of the tube. Point 1 indicates the filling for which the most liquid is present and point 3, the least. The remainder of the tube contains vapor. (We are considering *pure* substances, and there must be no impurities such as air.)

Heating at constant volume along line 1-2 causes the meniscus to rise until at point 2 the meniscus reaches the top of the tube, and the tube is filled with liquid. Thus starting from point 1 we see that heating causes the liquid to expand sufficiently to fill the tube, causing the vapor to condense in the process.

Heating at constant volume along line 3-4 causes the meniscus to fall until at point 4 it reaches the bottom of the tube, and the tube is filled with vapor. Thus in this process the liquid evaporates more rapidly than it expands, and eventually the tube contains only vapor.

Heating from the appropriate intermediate point along line 5-C produces a different result. The liquid expands and evaporates at compensating rates, and the meniscus does not move much either way. As the critical point is approached the meniscus becomes indistinct, then hazy, and finally disappears. The tube is then filled with neither liquid nor vapor, but with fluid.

4.2 PHASE CHANGES OF PURE SUBSTANCES. CLAPEYRON'S EQUATION

A phase change or transition occurs whenever one of the curves of Fig. 4-2 is crossed, and it is evident from this diagram that phase changes for pure substances occur at constant temperature and pressure. However, the extensive thermodynamic functions change abruptly as a result of a phase transition. Thus the molar or specific volume of a saturated liquid is very different from that for the saturated vapor at the same T and P. Discrete changes of internal energy, enthalpy, and entropy also result from a phase transition. The exception is the Gibbs function, which does not change during melting, vaporization, sublimation, or an allotropic transformation that occurs at constant T and P.

Since a phase change represents a transition between two phases that coexist in equilibrium, (*3.69*) applies to the process of change:

$$(dG^t)_{T,P} = 0 \tag{3.69}$$

For a pure substance T and P are indeed constant, and according to (*3.69*) the total Gibbs function for the closed system in which the transition occurs must also be constant. This can be true only if the molar Gibbs function is the same for both phases. Thus for two coexisting phases α and β of a pure substance we must have $G^\alpha = G^\beta$. This result was obtained in Problem 3.11 by a different method. If T and P are now simultaneously changed by dT and dP in such a way as to maintain equilibrium between the two phases, then for this change

$$dG^\alpha = dG^\beta \tag{4.1}$$

Equation (*3.35*) applied to one mole of a homogeneous pure material reduces immediately to

$$dG = -S\,dT + V\,dP$$

Substitution on both sides of (*4.1*) gives

$$-S^\alpha\,dT + V^\alpha\,dP = -S^\beta\,dT + V^\beta\,dP \qquad \text{or} \qquad \frac{dP}{dT} = \frac{S^\beta - S^\alpha}{V^\beta - V^\alpha}$$

Since the pressure in this equation is always a saturation pressure on the appropriate phase boundary, we will write P^{sat} in place of P to make this explicit. Furthermore $S^\beta - S^\alpha$ and $V^\beta - V^\alpha$, representing property changes of phase transition, are most conveniently written $\Delta S^{\alpha\beta}$ and $\Delta V^{\alpha\beta}$, and our equation is expressed most simply as

$$\frac{dP^{\text{sat}}}{dT} = \frac{\Delta S^{\alpha\beta}}{\Delta V^{\alpha\beta}} \tag{4.2}$$

A phase transition occurring at constant T and P requires an exchange of heat between the substance taken as the system and its surroundings. When the transition is carried out reversibly this heat is known as the *latent heat* and is equal to the enthalpy change. That is, $Q = \Delta H^{\alpha\beta}$. This result for a constant-pressure process was found in Example 1.7. In addition, for a reversible process at constant temperature, $\Delta S^{\alpha\beta} = Q/T$. Since the phase transition is carried out at constant T and P, we may combine these two equations to get $\Delta S^{\alpha\beta} = \Delta H^{\alpha\beta}/T$. This equation involves properties only and therefore does not depend on the assumption of reversibility made in its derivation. Combination with (*4.2*) gives the Clapeyron equation:

$$\boxed{\frac{dP^{\text{sat}}}{dT} = \frac{\Delta H^{\alpha\beta}}{T\,\Delta V^{\alpha\beta}}} \tag{4.3}$$

This equation applies to any phase change of a pure substance and relates the slope of the appropriate PT phase-boundary curve at a given T and P to the enthalpy change (latent heat) and the volume change of phase transition at the same T and P.

Example 4.3. The notation commonly used by engineers is that of the *steam tables*, an extensive compilation of the thermodynamic properties of the substance water. This notation employs subscripts rather than superscripts to designate the saturated phases and double subscripts without the prefixed Δ to denote phase changes:

s = saturated solid

f = saturated liquid (from the German, *flüssig*)

g = saturated vapor or gas

sf = fusion

fg = vaporization

sg = sublimation

Let us write the Clapeyron equation with this notation for each of the three phase transitions.

Fusion
$$\frac{dP^{\text{sat}}}{dT} = \frac{H_{sf}}{TV_{sf}} = \frac{H_f - H_s}{T(V_f - V_s)} \tag{4.3a}$$

Vaporization
$$\frac{dP^{\text{sat}}}{dT} = \frac{H_{fg}}{TV_{fg}} = \frac{H_g - H_f}{T(V_g - V_f)} \tag{4.3b}$$

Sublimation
$$\frac{dP^{\text{sat}}}{dT} = \frac{H_{sg}}{TV_{sg}} = \frac{H_g - H_s}{T(V_g - V_s)} \tag{4.3c}$$

Example 4.4. For sublimation and vaporization processes at low pressure one may introduce reasonable approximations into the Clapeyron equation by assuming that the vapor phase is an ideal gas and that the molar volume of the condensed phase is negligible compared with the molar volume of the vapor phase. What form does the Clapeyron equation assume under these approximations?

Consider the case of vaporization. If $V_g \gg V_f$ then $V_g - V_f \cong V_g$, and by the ideal-gas law, $V_g = RT/P^{\text{sat}}$. Then (4.3b) becomes

$$\frac{dP^{\text{sat}}}{dT} = \frac{H_{fg}}{RT^2/P^{\text{sat}}}$$

or

$$\frac{dP^{\text{sat}}/P^{\text{sat}}}{dT/T^2} = \frac{H_{fg}}{R}$$

or

$$H_{fg} = -R\,\frac{d(\ln P^{\text{sat}})}{d(1/T)} \tag{4.4}$$

An identical expression is obtained for H_{sg}, the latent heat of sublimation. This equation is known as the Clausius-Clapeyron equation, and it allows an approximate value for the latent heat to be determined from data for the saturation pressure alone. No volumetric data are required.

More specifically, (4.4) indicates that the value of H_{fg} (or of H_{sg}) is given by the slope of a plot of ln P^{sat} versus $1/T$. It is a fact, however, that such plots made of experimental data produce lines that are nearly straight, which implies through (4.4) that the latent heat of vaporization H_{fg} is independent of temperature. This is far from true. In fact, H_{fg} must become zero at the critical point, where the phases become identical. The difficulty is that (4.4) is based on assumptions which have approximate validity only at low pressures, at pressures well below the critical pressure for any pure substance considered.

4.3 VAPOR PRESSURES AND LATENT HEATS

The Clapeyron equation provides a vital connection between the properties of different phases. It is usually applied to the calculation of latent heats of vaporization and sublimation from vapor-pressure and volumetric data:

$$\Delta H^{\alpha\beta} = T\,\Delta V^{\alpha\beta}\frac{dP^{\text{sat}}}{dT}$$

For this, we need accurate representations of P^{sat} as a function of T. As stated following Example 4.4, a plot of $\ln P^{\text{sat}}$ versus $1/T$ generally yields a line that is very nearly straight. This suggests a vapor-pressure equation of the form

$$\ln P^{\text{sat}} = A - \frac{B}{T} \tag{4.5}$$

where A and B are constants. This equation is useful for many purposes, but it does not represent data sufficiently well to provide accurate values of derivatives.

The Antoine equation is more satisfactory, and has found wide use:

$$\ln P^{\text{sat}} = A - \frac{B}{T+C} \tag{4.6}$$

where A, B, and C are constants.

When extensive vapor-pressure data of high accuracy are available, it is found difficult to represent them faithfully by any simple equation. An equation that is usually satisfactory has the form

$$\ln P^{\text{sat}} = A - \frac{B}{T+C} + DT + E\ln T \tag{4.7}$$

where A, B, C, D, and E are constants.

Equations for the representation of vapor-pressure data are empirical. Again, thermodynamics provides no model for the behavior of substances, either in general or individually.

4.4 PROPERTIES OF TWO-PHASE SYSTEMS

When a phase change of a pure substance is brought about at constant T and P, the molar properties of the individual phases do not change. We start with a given amount of a substance in phase α, for which the molar (or unit-mass) properties are V^α, U^α, H^α, S^α, etc. If we gradually carry out a change at constant T and P to give finally the same quantity of the substance in phase β, then it will have the molar (or unit-mass) properties V^β, U^β, H^β, S^β, etc. Intermediate states of the system will be made up of the two phases α and β in varying amounts, but each phase will always have the same set of molar (or unit-mass) properties. If we want an average molar (or unit-mass) property over the entire two-phase system, we need a weighted sum of the properties of the individual phases. If we let x be the fraction of the total number of moles (or of the total mass) of the system that exists in phase β, then the average properties of the two-phase system are given by

$$V = (1-x)V^\alpha + xV^\beta \qquad U = (1-x)U^\alpha + xU^\beta$$

$$H = (1-x)H^\alpha + xH^\beta \qquad S = (1-x)S^\alpha + xS^\beta$$

These equations can be represented in general by

$$\boxed{M = (1-x)M^\alpha + xM^\beta} \tag{4.8}$$

where M represents any molar (or unit-mass) thermodynamic property. When $x = 0$, $M = M^\alpha$, and when $x = 1$, $M = M^\beta$. In between these limits M is a linear function of x.

This is most clearly seen by rewriting (4.8) in the form

$$M = M^{\alpha} + x(M^{\beta} - M^{\alpha})$$

or

$$\boxed{M = M^{\alpha} + x\,\Delta M^{\alpha\beta}} \tag{4.9}$$

Example 4.5. Rewrite (4.8) and (4.9) in the subscript notation of the steam tables for the case of the volume of a two-phase system of liquid and vapor.

In this case M becomes V, and the superscripts α and β become subscripts f and g. Then (4.8) and (4.9) are written as

$$V = (1-x)V_f + xV_g \qquad \text{and} \qquad V = V_f + xV_{fg}$$

The last equation has a simple, physical interpretation: the mixture has at least the volume of saturated liquid but has in addition the volume gained by vaporization of the fraction of the system represented by x. For liquid-vapor systems the mass fraction x which is vapor is called the *quality*.

4.5 VOLUME EXPANSIVITY AND ISOTHERMAL COMPRESSIBILITY OF SOLIDS AND LIQUIDS

The volume expansivity β and isothermal compressibility κ were defined by (3.17) and (3.18):

$$\beta = \frac{1}{V}\left(\frac{\partial V}{\partial T}\right)_P \qquad \kappa = -\frac{1}{V}\left(\frac{\partial V}{\partial P}\right)_T$$

Experimental values for these quantities are tabulated in handbooks of data, and for solids and liquids such tabulations take the place of an equation of state when used in conjunction with

$$\frac{dV}{V} = \beta\,dT - \kappa\,dP \tag{3.19}$$

Both β and κ are generally positive numbers; however β can take on negative values in unusual cases, as it does for liquid water between 0 and 4(°C). Both β and κ are in general functions of temperature and pressure. Normally, both β and κ increase with increasing temperature, though there are exceptions. The effect of pressure is usually quite small, an increase in pressure invariably causing κ to decrease and usually, but not always, causing β also to decrease.

When both β and κ are weak functions of T and P or where the changes in T and P are relatively small, β and κ may be treated as constants, and (3.19) may be integrated to give:

$$\ln\frac{V_2}{V_1} = \beta(T_2 - T_1) - \kappa(P_2 - P_1) \tag{4.10}$$

This is a different order of approximation than the assumption of incompressibility, for which β and κ are taken to be zero, as discussed in Example 3.12.

When β and κ cannot be considered constant as a reasonable approximation, (3.19) must be integrated more generally:

$$\ln\frac{V_2}{V_1} = \int_{T_1}^{T_2}\beta\,dT - \int_{P_1}^{P_2}\kappa\,dP \tag{4.11}$$

Since V is a state function, the value of $\ln(V_2/V_1)$ must be independent of the path of integration for given states 1 and 2. This means that one may choose any path at all leading from state 1 to state 2, and the choice is made on the basis of convenience. Different paths

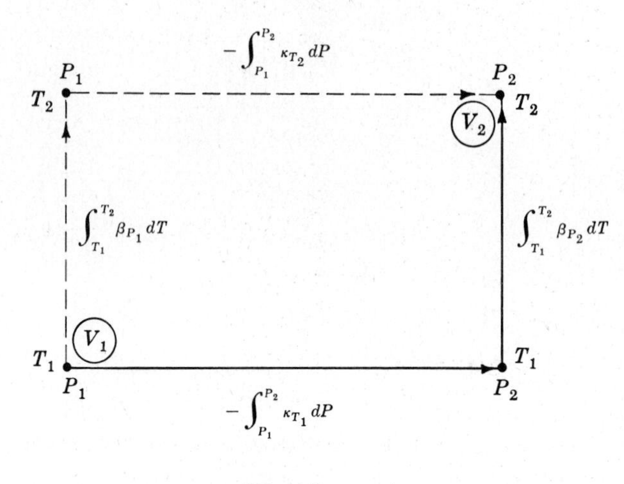

Fig. 4-6

will give different values for the individual integrals of (*4.11*), but the difference between the two integrals must always be the same.

The two most obvious (and most useful) paths are shown by Fig. 4-6, where each path consists of two steps, one at constant T and the other at constant P. Only one term of (*4.11*) applies to each step as shown. The path indicated by dashed lines consists of a first step at the constant pressure P_1 and a second step at the constant temperature T_2. The two integrals require data for β *as a function of* T *at pressure* P_1 and data for κ *as a function of* P *at temperature* T_2. The path described by the solid lines takes the steps in the opposite order, and requires different data for β and κ as indicated. All other paths would require data for both β and κ *as functions of both* T *and* P, and an exact T versus P relation would have to be mapped out. Fortunately, it is never necessary to complicate the problem in this way. The choice of which rectangular path to use depends entirely upon what data are available.

Example 4.6. The following values of β and κ for liquid water are available:

T(°C)	P(atm)	β(°C)$^{-1}$	κ(atm)$^{-1}$
20	1	208×10^{-6}	46.4×10^{-6}
20	1000	—	35.3×10^{-6}
30	1	304×10^{-6}	45.2×10^{-6}
30	1000	—	34.3×10^{-6}

Water undergoes a change of state from 20(°C) and 1(atm) to 30(°C) and 1000(atm). Determine the percentage change in the specific volume of water caused by this change of state.

Since data are available for β just at $P_1 = 1$(atm), the only reasonable path to adopt for this calculation is the one described by the dashed lines of Fig. 4-6. For this path (*4.11*) becomes

$$\ln \frac{V_2}{V_1} = \int_{T_1}^{T_2} \beta_{P_1}\, dT - \int_{P_1}^{P_2} \kappa_{T_2}\, dP$$

Since no data are given for β and κ at intermediate values of T and P, we can do no better than to take β and κ out from under the integral signs as arithmetic average values. Thus

$$\ln \frac{V_2}{V_1} = \beta_{P_1}^{\text{ave}}(T_2 - T_1) - \kappa_{T_2}^{\text{ave}}(P_2 - P_1)$$

where for $P_1 = 1(\text{atm})$

$$\beta_{P_1}^{\text{ave}} = \frac{208 + 304}{2} \times 10^{-6} = 256 \times 10^{-6}(°\text{C})^{-1}$$

and for $T_2 = 30(°\text{C})$

$$\kappa_{T_2}^{\text{ave}} = \frac{45.2 + 34.3}{2} \times 10^{-6} = 39.75 \times 10^{-6}(\text{atm})^{-1}$$

Substitution of numerical values gives

$$\ln \frac{V_2}{V_1} = [256 \times 10^{-6}(°\text{C})^{-1} \times 10(°\text{C})] - [39.75 \times 10^{-6}(\text{atm})^{-1} \times 999(\text{atm})] = -0.03715$$

and

$$\frac{V_2}{V_1} = 0.9635$$

By subtracting unity from both sides of this equation, we get $(V_2 - V_1)/V_1 = -0.0365$, which means that the change of state caused a 3.65% decrease in the specific volume of the water.

Example 4.7. Using the data given with Example 4.6, determine the average value of β for water at 1000(atm) between the temperatures of 20 and 30(°C).

Figure 4-6 shows two paths from point 1 to point 2. Using arithmetic averages for β and κ for the two paths, we get two equally valid reductions of (*4.11*):

$$\ln \frac{V_2}{V_1} = \beta_{P_1}^{\text{ave}}(T_2 - T_1) - \kappa_{T_2}^{\text{ave}}(P_2 - P_1)$$

$$\ln \frac{V_2}{V_1} = -\kappa_{T_1}^{\text{ave}}(P_2 - P_1) + \beta_{P_2}^{\text{ave}}(T_2 - T_1)$$

Since both expressions must give the same result, we have

$$(\beta_{P_1}^{\text{ave}} - \beta_{P_2}^{\text{ave}})(T_2 - T_1) = (\kappa_{T_2}^{\text{ave}} - \kappa_{T_1}^{\text{ave}})(P_2 - P_1)$$

where as before

$$\beta_{P_1}^{\text{ave}} = 256 \times 10^{-6}(°\text{C})^{-1} \quad \text{and} \quad \kappa_{T_2}^{\text{ave}} = 39.75 \times 10^{-6}(\text{atm})^{-1}$$

For $T_1 = 20(°\text{C})$

$$\kappa_{T_1}^{\text{ave}} = \frac{46.4 + 35.3}{2} \times 10^{-6} = 40.85 \times 10^{-6}(\text{atm})^{-1}$$

Thus

$$(256 \times 10^{-6} - \beta_{P_2}^{\text{ave}})(10) = (39.75 \times 10^{-6} - 40.85 \times 10^{-6})(999)$$

Solution for $\beta_{P_2}^{\text{ave}}$ gives

$$\beta_{P_2}^{\text{ave}} = 366 \times 10^{-6}(°\text{C})^{-1}$$

The procedure followed in this example illustrates the integral counterpart of the reciprocity relation for an exact differential, given for the present case by (*3.20*).

4.6 HEAT CAPACITIES OF SOLIDS AND LIQUIDS

In general, heat capacities must be determined by experiment for each substance of interest. Data for solids and liquids are usually taken at atmospheric pressure and are reported as functions of temperature by equations of the form

$$C_P = a + bT + cT^2$$

or

$$C_P = a + bT + cT^{-2}$$

where in either case the set of constants a, b, and c must be specific for the substance. Heat capacities usually increase with increasing temperature. The effect of pressure on the heat capacities of liquids and solids is normally very small, and unless the pressure is very high can be neglected.

The difference in heat capacities as given by (*3.56*) may be expressed in terms of the volume expansivity and the isothermal compressibility by substitution for the derivatives:

$$\left(\frac{\partial V}{\partial T}\right)_P = V\beta \quad \text{and} \quad \left(\frac{\partial P}{\partial T}\right)_V = \frac{\beta}{\kappa}$$

This gives

$$C_P - C_V = \frac{TV\beta^2}{\kappa} \tag{4.12}$$

This difference is usually significant, except at very low temperatures.

Example 4.8. For metallic copper at 300(K) the following values are known:

$$C_P = 24.5\text{(J)/(g mole)(°C)}$$
$$\beta = 50.4 \times 10^{-6}\text{(°C)}^{-1}$$
$$\kappa = 0.788 \times 10^{-6}\text{(atm)}^{-1}$$
$$V = 7.06\text{(cm)}^3\text{/(g mole)}$$

Determine C_V.

Direct substitution into (*4.12*) gives

$$C_P - C_V = \frac{300\text{(K)} \times 7.06\text{(cm)}^3\text{/(g mole)} \times [50.4 \times 10^{-6}\text{(°C)}^{-1}]^2}{0.788 \times 10^{-6}\text{(atm)}^{-1}}$$

$$= 6.828\text{(cm}^3\text{-atm)(K)(°C)}^{-2}\text{/(g mole)}$$

In this result we have units of (K) and (°C). Since (°C) here refers to a temperature *change*, from $(\partial V/\partial T)_P$, (K) may be substituted for (°C), because temperature *changes* are the same in kelvins as in degrees Celsius. This gives

$$C_P - C_V = 6.828 \frac{\text{(cm}^3\text{-atm)}}{\text{(g mole)(°C)}}$$

The composite unit (cm^3-atm) is a unit of energy, representing the product of pressure and volume. The necessary conversion factor is

$$9.869\text{(cm}^3\text{-atm)} = 1\text{(J)}$$

Thus

$$C_P - C_V = \frac{6.828\text{(cm}^3\text{-atm)/(g mole)(°C)}}{9.869\text{(cm}^3\text{-atm)/(J)}} = 0.69\text{(J)/(g mole)(°C)}$$

As a result

$$C_V = C_P - 0.69 = 24.5 - 0.69 \cong 23.8\text{(J)/(g mole)(°C)}$$

Example 4.9. The heat capacity of quartz (SiO_2) at atmospheric pressure between 298 and 848(K) is given by the equation

$$C_P = 11.22 + (8.20 \times 10^{-3})T - (2.70 \times 10^5)T^{-2}$$

where T is in kelvins and C_P is in (cal)/(g mole)(°C). If a metric ton, 10^3(kg), of quartz is heated from 300(K) to 700(K) at atmospheric pressure, how much heat is required?

For a constant-pressure process it was shown in Example 1.7 that $\Delta H = Q$. Furthermore, (*1.11*) gives $dH = C_P\, dT$ for a constant-pressure process. As a result, we have

$$Q = \int_{T_1}^{T_2} C_P\, dT$$

Substitution for C_P and integration provide the equation:

$$Q = 11.22(T_2 - T_1) + \frac{8.20 \times 10^{-3}}{2}(T_2^2 - T_1^2) - (2.70 \times 10^5)\left(\frac{1}{T_1} - \frac{1}{T_2}\right)$$

Substitution of numerical values for the temperatures gives the result:

$$Q = 5614\text{(cal)/(g mole)}$$

The molecular weight of SiO_2 is 52, and therefore

$$Q = \frac{5614}{52} = 107.96\text{(cal)/(g)}$$

This gives a total heat requirement for 10^6(g) of

$$Q = 107.96 \times 10^6(\text{cal}) \quad \text{or} \quad 451.7 \times 10^6(\text{J})$$

Example 4.10. Introducing the definition of β into (*3.53*), we get the general equation

$$dH = C_P\,dT + V(1-\beta T)\,dP$$

Determine the enthalpy change of mercury for a change of state from 1(atm) and 100(°C) to 1000(atm) and 0(°C).

The following data for mercury are available:

T(°C)	P(atm)	V(cm)^3/(g mole)	β(°C)$^{-1}$	C_P(cal)/(g mole)(°C)
0	1	14.72	181×10^{-6}	6.69
0	1000	14.67	174×10^{-6}	6.69
100	1	—	—	6.57

Enthalpy is a state function, and enthalpy changes are independent of path. Thus we may choose a convenient path for calculation just as was done for the volume in Example 4.6. Integration of the equation for dH gives

$$\Delta H = H_2 - H_1 = \int_{T_1}^{T_2} C_P\,dT + \int_{P_1}^{P_2} V(1-\beta T)\,dP$$

With the given data the most convenient path for integration is as in Fig. 4-7.

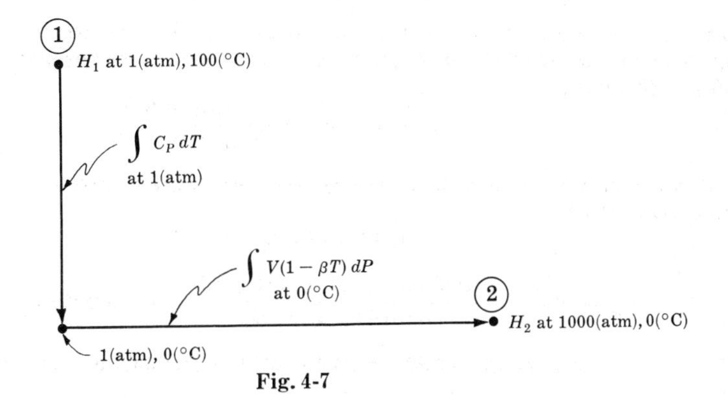

Fig. 4-7

The data show that C_P, V, and β are all very weak functions of T and P, and therefore the use of average values should give accurate results. The equation for ΔH then becomes

$$\Delta H = \underbrace{C_P^{\text{ave}}(T_2 - T_1)}_{\text{for } P = 1(\text{atm})} + \underbrace{V^{\text{ave}}(1-\beta^{\text{ave}}T)(P_2 - P_1)}_{\text{for } T = 0(°\text{C})}$$

where

$$C_P^{\text{ave}} = \frac{6.69 + 6.57}{2} = 6.63(\text{cal})/(\text{g mole})(°\text{C})$$

$$V^{\text{ave}} = \frac{14.72 + 14.67}{2} = 14.695(\text{cm})^3/(\text{g mole})$$

$$\beta^{\text{ave}} = \frac{181 + 174}{2} \times 10^{-6} = 177.5 \times 10^{-6}(°\text{C})^{-1}$$

Substitution of these values into the two terms of the equation for ΔH gives:

$$\Delta H = 6.63(\text{cal})/(\text{g mole})(°\text{C}) \times [0 - 100](°\text{C})$$

$$+ \frac{14.695(\text{cm})^3/(\text{g mole}) \times [1 - 177.5 \times 10^{-6}(°\text{C})^{-1} \times 273(\text{K})] \times 999(\text{atm})}{41.293(\text{cm}^3\text{-atm})/(\text{cal})}$$

$$= -663.0 + 338.3 = -324.7(\text{cal})/(\text{g mole}) \quad \text{or} \quad -1358.5(\text{J})/(\text{g mole})$$

4.7 HEAT CAPACITIES OF GASES

The heat capacities of gases are also represented by empirical functions of temperature, usually of the same form as the equations employed for solids and liquids. The heat capacities of gases are also strong functions of pressure. However, the effect of pressure on the thermodynamic properties of gases is determined in a way that does not require knowledge of heat capacities as a function of pressure. Instead, use is made of heat capacities of gases in the *ideal-gas state,* as was discussed in Chapter 1. These heat capacities, designated C'_P and C'_V, are independent of pressure, and are always related by the equation

$$C'_P - C'_V = R$$

The temperature dependence of C'_P is usually expressed by equations having one or the other of the forms:

$$C'_P = a + bT + cT^2 \qquad C'_P = a + bT + cT^{-2}$$

where the sets of constants must be determined specifically for each gas. Since gases at low pressures usually are approximately ideal, the ideal-gas heat capacities are suitable for almost all calculations for real gases at atmospheric pressure.

Example 4.11. One gram mole of nitrogen is contained in a constant-volume tank at low pressure. Assuming nitrogen to be an ideal gas, calculate the amount of heat required to raise its temperature from 300 to 1000(K). For nitrogen

$$C'_P = 6.529 + 1.250 \times 10^{-3}T$$

where T is in kelvins and C'_P is in (cal)/(g mole)(K).

For a constant-volume process there is no work and the first law becomes $\Delta U = Q$. Since we have taken nitrogen to be an ideal gas,

$$Q = \Delta U = \int_{T_1}^{T_2} C'_V\, dT$$

To get C'_V, we use the relation $C'_P - C'_V = R$, and our equation becomes

$$Q = \int_{T_1}^{T_2} (C'_P - R)\, dT$$

where R has the value 1.987(cal)/(g mole)(K). Substitution for C'_P and R gives

$$Q = \int_{300}^{1000} (6.529 - 1.987 + 0.00125\, T)\, dT$$

Integrating,

$$Q = 4.542(1000 - 300) + \frac{0.00125}{2}(1000^2 - 300^2) = 3748\text{(cal)} \quad \text{or} \quad 15{,}682\text{(J)}$$

4.8 EVALUATION OF THERMODYNAMIC PROPERTIES FOR PVT SYSTEMS

A general procedure for the calculation of the thermodynamic properties of gases or vapors makes use of what are called *residual functions.* These are defined by the generic equation

$$\Delta M' = M' - M \tag{4.13}$$

where the M's may represent any extensive thermodynamic property. The residual function $\Delta M'$ is the difference between M', which represents the property for the *ideal-gas state,* and M, which represents the actual property, both for the same temperature and pressure.

The ideal-gas state at temperature T and pressure P for a given gas is the state that would be reached if the gas at temperature T and at a pressure approaching zero, where $PV = RT$, were to remain an ideal gas when compressed isothermally to pressure P. This state is, of course, imaginary, except at pressures approaching zero; nevertheless it is of considerable practical use. Calculations of property changes may be made along any arbitrary path, and hypothetical states are as acceptable as real states. Property changes for ideal gases are given by very simple equations, and the use of residual functions allows conversion back and forth between real-gas and ideal-gas properties.

Residual enthalpy, entropy, and volume.

We now develop equations for the residual enthalpy $\Delta H'$ and the residual entropy $\Delta S'$. By definition (*4.13*), $\Delta H' = H' - H$. Differentiation with respect to P at constant T gives

$$\left(\frac{\partial\,\Delta H'}{\partial P}\right)_T = \left(\frac{\partial H'}{\partial P}\right)_T - \left(\frac{\partial H}{\partial P}\right)_T$$

For an ideal gas the enthalpy is a function of temperature only, independent of P. Thus $(\partial H'/\partial P)_T$ is zero. An expression for $(\partial H/\partial P)_T$ was developed in Problem 3.3:

$$\left(\frac{\partial H}{\partial P}\right)_T = V - T\left(\frac{\partial V}{\partial T}\right)_P$$

Thus

$$\left(\frac{\partial\,\Delta H'}{\partial P}\right)_T = T\left(\frac{\partial V}{\partial T}\right)_P - V$$

For a constant-temperature change, this equation may be written

$$d(\Delta H') = \left[T\left(\frac{\partial V}{\partial T}\right)_P - V\right]dP \qquad \text{(const. } T\text{)}$$

Integration from P^* to P gives

$$\Delta H' - (\Delta H')^* = \int_{P^*}^{P}\left[T\left(\frac{\partial V}{\partial T}\right)_P - V\right]dP \qquad \text{(const. } T\text{)}$$

When $P^* \to 0$, $(\Delta H')^*$ becomes the residual enthalpy at zero pressure. Since this is an ideal-gas state, $(\Delta H')^*$ is assumed to be zero, and we get finally

$$\boxed{\Delta H' = \int_0^P\left[T\left(\frac{\partial V}{\partial T}\right)_P - V\right]dP} \qquad \text{(const. } T\text{)} \qquad (4.14)$$

For the residual entropy $\Delta S' = S' - S$ and differentiation gives

$$\left(\frac{\partial\,\Delta S'}{\partial P}\right)_T = \left(\frac{\partial S'}{\partial P}\right)_T - \left(\frac{\partial S}{\partial P}\right)_T$$

Proceeding as before, we have from (*2.9*) and (*3.44*) that

$$\left(\frac{\partial S'}{\partial P}\right)_T = -\frac{R}{P} \qquad \text{and} \qquad \left(\frac{\partial S}{\partial P}\right)_T = -\left(\frac{\partial V}{\partial T}\right)_P$$

Therefore

$$\left(\frac{\partial\,\Delta S'}{\partial P}\right)_T = -\frac{R}{P} + \left(\frac{\partial V}{\partial T}\right)_P$$

and

$$d(\Delta S') = \left[\left(\frac{\partial V}{\partial T}\right)_P - \frac{R}{P}\right]dP \qquad \text{(const. } T\text{)}$$

Integration from a pressure approaching zero to P gives

$$\boxed{\Delta S' = \int_0^P\left[\left(\frac{\partial V}{\partial T}\right)_P - \frac{R}{P}\right]dP} \qquad \text{(const. } T\text{)} \qquad (4.15)$$

where the residual entropy at $P = 0$ has been assumed zero.

It is also possible to define a residual volume $\Delta V'$. According to (*4.13*)

$$\Delta V' = V' - V = (RT/P) - V \qquad (4.16)$$

Thus $\Delta V'$ is experimentally accessible through PVT measurements. Moreover, it is convenient to deal with $\Delta V'$ rather than with V itself, because it has a much more limited range of values. Whereas V becomes infinite as $P \to 0$, $\Delta V'$ remains finite. Thus it is useful to recast (*4.14*) and (*4.15*) so as to replace V by $\Delta V'$. Since $V = V' - \Delta V'$,

$$\left(\frac{\partial V}{\partial T}\right)_P = \left(\frac{\partial V'}{\partial T}\right)_P - \left(\frac{\partial \Delta V'}{\partial T}\right)_P$$

and from $V' = RT/P$ we have

$$\left(\frac{\partial V'}{\partial T}\right)_P = \frac{R}{P} = \frac{V'}{T}$$

Therefore
$$\left(\frac{\partial V}{\partial T}\right)_P = \frac{R}{P} - \left(\frac{\partial \Delta V'}{\partial T}\right)_P = \frac{V'}{T} - \left(\frac{\partial \Delta V'}{\partial T}\right)_P$$

Appropriate substitution in (*4.14*) and (*4.15*) for $(\partial V/\partial T)_P$ and for V leads directly to

$$\Delta H' = \int_0^P \left[\Delta V' - T\left(\frac{\partial \Delta V'}{\partial T}\right)_P\right] dP \qquad \text{(const. } T\text{)} \tag{4.17}$$

and
$$\Delta S' = -\int_0^P \left(\frac{\partial \Delta V'}{\partial T}\right)_P dP \qquad \text{(const. } T\text{)} \tag{4.18}$$

For an ideal gas $\Delta V'$ is always zero, and in this case (*4.17*) and (*4.18*) immediately reduce to $\Delta H' = \Delta S' = 0$. We shall illustrate the use of (*4.17*) and (*4.18*) in Example 4.12.

Property calculations from residual functions.

The manner in which these residual functions are incorporated into equations for calculation of enthalpy and entropy is as follows. For the ideal-gas state (*1.15*) and (*2.9*) may be integrated to give

$$H' - H_0' = \int_{T_0}^T C_P' \, dT \qquad \text{and} \qquad S' - S_0' = \int_{T_0}^T C_P' \frac{dT}{T} - R \ln \frac{P}{P_0}$$

These integrations are from an initial ideal-gas *reference state* designated by the subscript (0) to the ideal-gas state at the T and P of interest. Combination of these equations with the defining equations $H = H' - \Delta H'$ and $S = S' - \Delta S'$ gives

$$\boxed{H = H_0' + \int_{T_0}^T C_P' \, dT - \Delta H'} \tag{4.19}$$

$$\boxed{S = S_0' + \int_{T_0}^T C_P' \frac{dT}{T} - R \ln \frac{P}{P_0} - \Delta S'} \tag{4.20}$$

The reference state at T_0 and P_0 may be selected arbitrarily and values may be assigned to H_0' and S_0' arbitrarily. This is another manifestation of the characteristic of exact differentials mentioned in Chapter 3, namely, that the function (here H or S) can be defined only to within an additive constant. For the determination of numerical values of H and S the data needed are ideal-gas heat capacities as a function of T and PVT data to allow evaluation of the residual functions. Once H and S are known (along with the PVT data), then the other thermodynamic functions are calculated by the defining equations (*3.23*): $U = H - PV$, $A = U - TS$, $G = H - TS$.

The extension of property calculations from the vapor phase to the liquid phase is made by use of the Clapeyron equation (*4.3b*):

$$H_g - H_f = H_{fg} = TV_{fg}\frac{dP^{sat}}{dT} \qquad (4.21)$$

In addition

$$S_g - S_f = S_{fg} = \frac{H_{fg}}{T} \qquad (4.22)$$

The additional data needed for these calculations are vapor pressures as a function of T and volumetric data for the saturated liquid and saturated vapor.

Once property values are known for the saturated liquid, the calculations may be extended into the liquid phase by application of (3.53) and (3.55), integrated at constant T from the saturation pressure to any higher pressure:

$$H - H_f = \int_{P^{sat}}^{P}\left[V - T\left(\frac{\partial V}{\partial T}\right)_P\right]dP \qquad \text{(const. } T\text{)} \qquad (4.23)$$

$$S - S_f = -\int_{P^{sat}}^{P}\left(\frac{\partial V}{\partial T}\right)_P dP \qquad \text{(const. } T\text{)} \qquad (4.24)$$

PVT data for the liquid phase are required for these calculations.

Example 4.12. The data needed for the calculations outlined in this section consist first of numerical values determined from a large number of experiments. These must be smoothed and correlated to produce tables of consistent data, or graphs, or equations that faithfully represent the data. Calculations are possible by numerical methods, graphical methods, or analytical methods, and the procedure is exacting. For purposes of illustration we will use analytical methods with equations that represent data for an imaginary substance. This allows demonstration of the procedures without the distraction of complex computational methods.

Let us suppose that our substance has ideal-gas heat capacities as given by the equation

$$C'_P = 100 + 0.01\,T$$

where T is in kelvins and C'_P is in (J)/(g mole)(°C). Further, the PVT data for the vapor phase are given by the equation of state

$$V = \frac{RT}{P} + b - \frac{a}{RT}$$

where $a = 20 \times 10^6(\text{cm})^6(\text{atm})/(\text{g mole})^2$ and $b = 140(\text{cm})^3/(\text{g mole})$. With these constants, V is in $(\text{cm})^3/(\text{g mole})$ and P is in atmospheres.

The vapor pressure is given by

$$\ln P^{sat} = 10.0 - \frac{3550}{T}$$

where T is in kelvins and P^{sat} is in atmospheres.

For the liquid phase the molar volumes of saturated liquid are given by

$$V_f = 60 + 0.10\,T$$

where T is in kelvins and V_f is in $(\text{cm})^3/(\text{g mole})$. The volume expansivity of the liquid is constant at

$$\beta = 0.0012\,(°\text{C})^{-1}$$

and the liquid volume is independent of pressure.

We wish now to determine the numerical relationship of H and S to the pressure for our substance at a temperature of 400(K) at pressures from 1 to 10(atm).

For the application of (4.19) and (4.20) we need to evaluate $\Delta H'$ and $\Delta S'$ for the gas phase by means of (4.17) and (4.18). Since we are given that

$$V = \frac{RT}{P} + b - \frac{a}{RT}$$

and since $V' = RT/P$ it follows that

$$\Delta V' = V' - V = -b + \frac{a}{RT} \qquad \text{and} \qquad \left(\frac{\partial\, \Delta V'}{\partial T}\right)_P = \frac{-a}{RT^2}$$

Both $\Delta V'$ and $(\partial\, \Delta V'/\partial T)_P$ are seen to be independent of P, and substitution into (4.17) and (4.18) gives after integration

$$\Delta H' = \left(\frac{2a}{RT} - b\right) P \qquad \text{and} \qquad \Delta S' = \frac{aP}{RT^2}$$

Putting in numerical values for a, b, T, and R, we get

$$\Delta H' = 1080\, P(\text{cm}^3\text{-atm})/(\text{g mole}) \qquad \text{or} \qquad 109.4\, P(\text{J})/(\text{g mole})$$

Similarly,

$$\Delta S' = 0.1544\, P(\text{J})/(\text{g mole})(\text{K})$$

With respect to (4.19) and (4.20) we must now assign values to H_0' and S_0' for the ideal-gas state at some selected T_0 and P_0. We do this arbitrarily, and set

$$H_0' = 1000(\text{J})/(\text{g mole}) \qquad S_0' = 10(\text{J})/(\text{g mole})(\text{K})$$

at $T_0 = 300(K)$ and $P_0 = 1(\text{atm})$. We can now substitute for C_P', H_0', S_0', $\Delta H'$, $\Delta S'$, P_0, T_0, and T in (4.19) and (4.20) and carry out the indicated integrations:

$$H = 1000 + \int_{300}^{400} (100 + 0.01\, T)\, dT - 109.4\, P$$

$$= 11{,}350 - 109.4\, P(\text{J})/(\text{g mole})$$

$$S = 10 + \int_{300}^{400} \left(\frac{100 + 0.01\, T}{T}\right) dT - 8.314 \ln P - 0.1544\, P$$

$$= 39.77 - 8.314 \ln P - 0.1544\, P(\text{J})/(\text{g mole})(\text{K})$$

Substitution of numerical values for P provides values of H and S. The initial value of P is 1(atm), and the final value is P^{sat} at 400(K), which is given by

$$\ln P^{\text{sat}} = 10.0 - \frac{3550}{400} = 1.125 \qquad \text{or} \qquad P^{\text{sat}} = 3.081(\text{atm})$$

Calculations for several pressures at 400(K) give the following results.

P(atm)	H(J)/(g mole)	S(J)/(g mole)(K)
1	11,241	39.62
2	11,131	33.70
3	11,022	30.17
$P^{\text{sat}} = 3.081$	$H_g = 11{,}013$	$S_g = 29.94$

These property values are plotted in Figs. 4-8 and 4-9.

Application of the Clapeyron equation requires the derivative dP^{sat}/dT, evalued at 400(K). Differentiation of the vapor-pressure equation gives

$$\frac{dP^{\text{sat}}}{dT} = \frac{3550 P^{\text{sat}}}{T^2} = \frac{(3550)(3.081)}{(400)^2} = 0.06836(\text{atm})/(\text{K})$$

The molar volume of saturated liquid at 400(K) is

$$V_f = 60 + 0.10\, T = 60 + (0.10)(400) = 100(\text{cm})^3/(\text{g mole})$$

The molar volume of saturated vapor is computed from the vapor-phase equation of state at $T = 400(\text{K})$ and $P^{\text{sat}} = 3.081(\text{atm})$:

$$V_g = \frac{(82.05)(400)}{3.081} + 140 - \frac{20 \times 10^6}{(82.05)(400)} = 10{,}183(\text{cm})^3/(\text{g mole})$$

This gives $V_{fg} = V_g - V_f = 10{,}083(\text{cm})^3/(\text{g mole})$. Substitution of numerical values into (*4.21*) gives

$$H_{fg} = TV_{fg}\frac{dP^{\text{sat}}}{dT} = \frac{400 \times 10{,}083 \times 0.06836(\text{cm}^3\text{-atm})/(\text{g mole})}{9.869(\text{cm}^3\text{-atm})/(\text{J})}$$

$$= 27{,}937(\text{J})/(\text{g mole})$$

Also, by (*4.22*),

$$S_{fg} = \frac{H_{fg}}{T} = \frac{27{,}937}{400} = 69.84(\text{J})/(\text{g mole})(\text{K})$$

The enthalpy and entropy of saturated liquid at 400(K) and $P^{\text{sat}} = 3.081(\text{atm})$ are therefore

$$H_f = H_g - H_{fg} = 11{,}013 - 27{,}937 = -16{,}924(\text{J})/(\text{g mole})$$

$$S_f = S_g - S_{fg} = 29.94 - 69.84 = -39.90(\text{J})/(\text{g mole})(\text{K})$$

For the liquid phase (*4.23*) and (*4.24*) can be combined with the definition of the volume expansivity β to give

$$H = H_f + \int_{P^{\text{sat}}}^{P} V[1 - T\beta]\,dP \qquad S = S_f - \int_{P^{\text{sat}}}^{P} V\beta\,dP$$

Since V, T, and β are all constant, integration yields:

$$H = H_f + V_f(1 - T\beta)(P - P^{\text{sat}}) \qquad S = S_f - V_f\beta(P - P^{\text{sat}})$$

Substitution of known numerical values gives

$$H = -16{,}924 + \frac{100[1 - 400 \times 0.0012][P - 3.081](\text{cm}^3\text{-atm})/(\text{g mole})}{9.869(\text{cm}^3\text{-atm})/(\text{J})}$$

$$= -16{,}924 + 5.269[P - 3.081](\text{J})/(\text{g mole})$$

$$S = -39.90 - \frac{100 \times 0.0012[P - 3.081]}{9.869}$$

$$= -39.90 - 0.01216[P - 3.081](\text{J})/(\text{g mole})(\text{K})$$

Values of H and S for compressed liquid are now readily calculated at various pressures, and are tabulated as follows:

P(atm)	H(J)/(g mole)	S(J)/(g mole)(K)
$P^{\text{sat}} =$ 3.081	$H_f = -16{,}924$	$S_f = -39.90$
4	−16,919	−39.91
6	−16,909	−39.94
8	−16,898	−39.96
10	−16,888	−39.98

All results are shown in Figs. 4-8 and 4-9 on the following page.

The magnitudes of the values for H and S depend upon the choice made at the beginning for the values of H_0' and S_0'. Were different choices made, all values of H and of S would be changed by constant amounts. This would in no way change *differences* in values of H or S, and in application it is always differences that are needed. Therefore, once calculations are made on some particular basis, there is no need to abide by the initial arbitrary choices of constants. One can adjust the scale of values by adding a constant to all values of H or S, and in this way the zero point of the scale can be placed wherever one wishes. This amounts to no more than a shifting of scales on the graphs of Figs. 4-8 and 4-9, and an example of such a scale shift is shown at the bottom of each graph. The enthalpy scale has been shifted by the addition of 17.0×10^3 to each value and the entropy scale by the addition of 40.0.

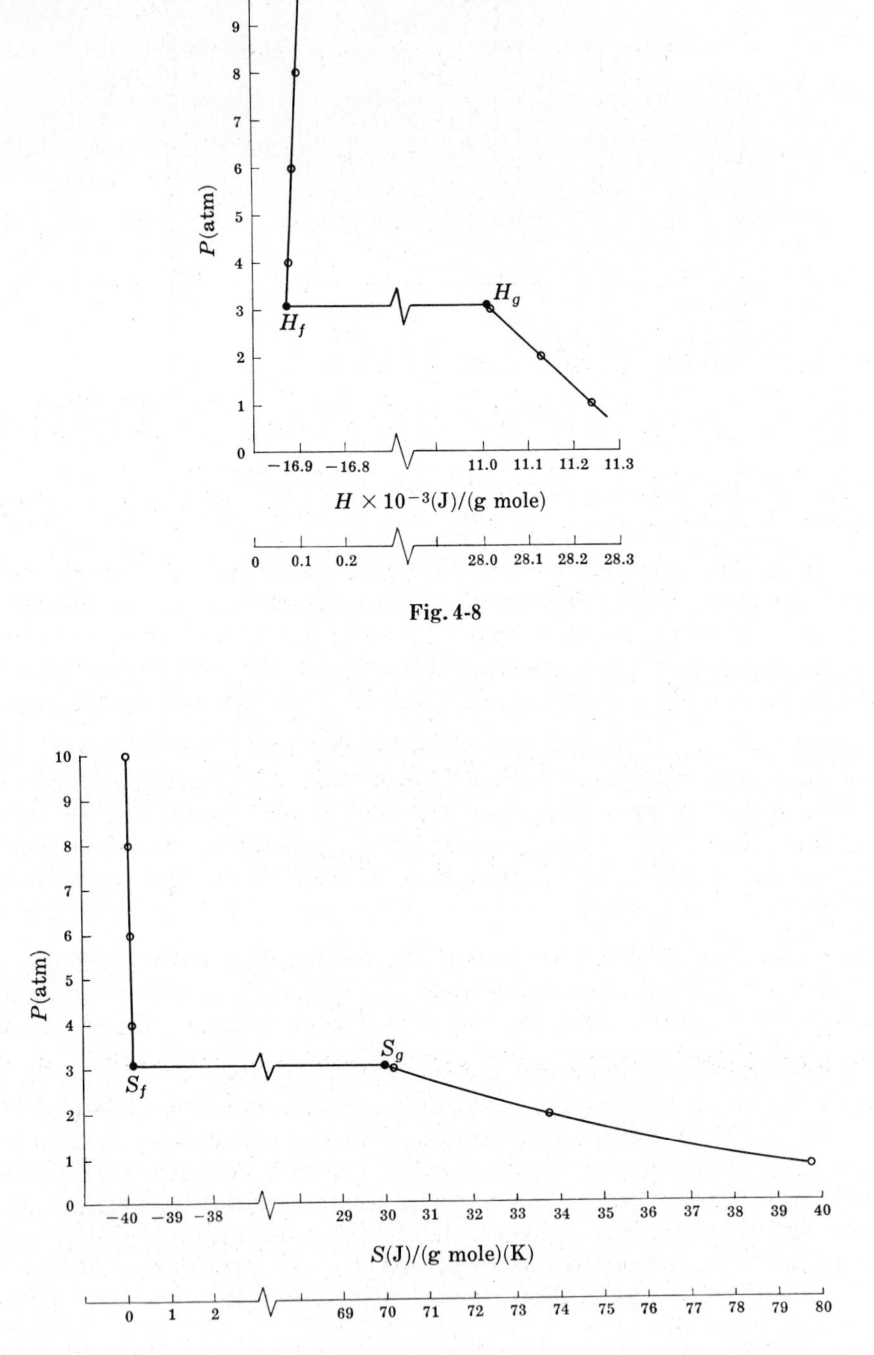

Fig. 4-8

Fig. 4-9

4.9 THERMODYNAMIC DIAGRAMS AND TABLES FOR PVT SYSTEMS

Calculations such as those illustrated in Example 4.12 may be carried out for a pure substance for a series of different temperatures. If the results are plotted as in Example 4.12, the graphs will display a set of isotherms rather than just one. The PH and PS sketches in Fig. 4-10 indicate the nature of the graphs obtained. Also shown is the corresponding PV diagram, which represents the data upon which the calculations are based.

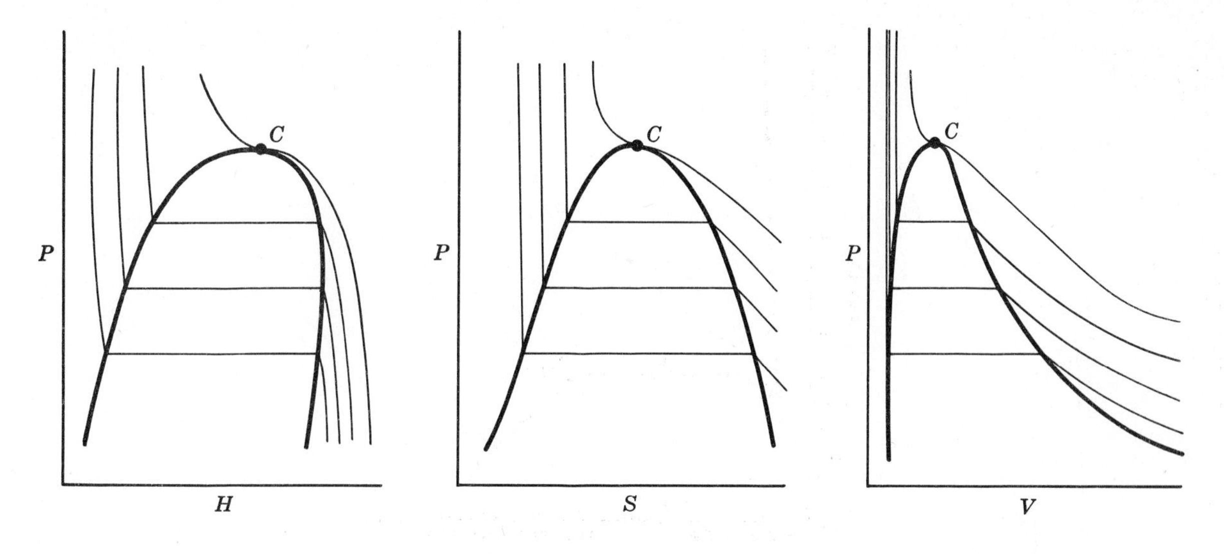

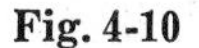
Fig. 4-10

The lighter lines represent isotherms at various temperatures, whereas the heavy curves drawn through the points where the isotherms abruptly change slope represent saturated liquid (to the left of C) and saturated vapor (to the right of C). Point C is in each case the critical point, and the isotherm passing through C is for temperature T_c. Isotherms for temperatures greater than T_c are smooth and lie above the critical isotherm.

Once diagrams such as these (or equivalent tabular data) are available, information may be taken from one graph and included on another. For example, vertical lines on the PS and PV diagrams represent lines of constant S and of constant V, and values of P and T along these lines may be used to locate lines of constant S and of constant V on the PH diagram. In this way the PH diagram may be made to include the relations among all the properties P, V, T, H, and S.

One can also use other pairs of coordinates. For example, one can take data from the PS diagram and plot a TS diagram showing lines of constant P (isobars). In addition, data from the PV and PH diagrams can be included as lines of constant V and of constant H.

The PH diagram and the TS diagram are in common use along with a third diagram, employed almost exclusively for steam and known as the *Mollier diagram.* It displays H and S as coordinates and shows lines of constant pressure. Data for volume are not usually included, but in the two-phase (liquid-vapor) region lines of constant moisture (percentage of liquid by weight) appear. In the vapor or gas region lines of constant temperature and of constant *superheat* are shown. Superheat is a term used to designate the difference between the temperature and the saturation temperature at the same pressure. In fact, the entire vapor region exclusive of the saturation curve is often called the superheat region.

The three diagrams just discussed are outlined in Figs. 4-11 through 4-13 on page 118. These figures are based on data for water, and will vary in detail from substance to substance. In all cases the heavy lines represent phase boundaries, and the lighter lines are examples of curves which show the relation between coordinates for a constant value of a particular property.

The information presented by means of graphs can, of course, also be given in tables. The most widely used such tables are the *steam tables,* and they represent extensive tabulations of data for solid, liquid, and vapor H_2O. Similar, but usually less detailed, tables can be found in data compilations and in the literature for other common substances.

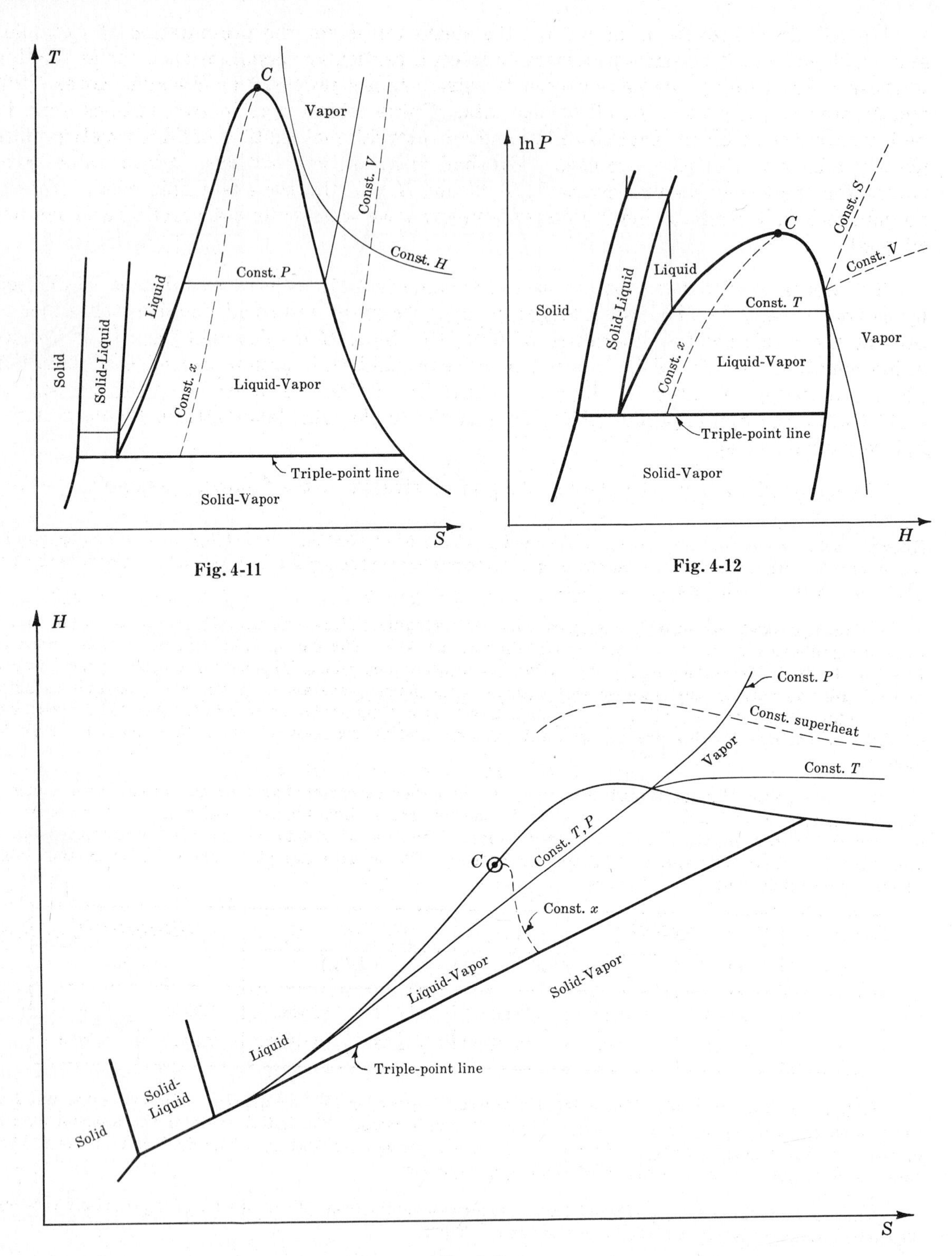

Fig. 4-11

Fig. 4-12

Fig. 4-13

In the steam tables data for the saturated phases in equilibrium are given separately from data for the single-phase regions. Thus there is a table for saturated liquid and saturated vapor, a table for the vapor region, a table for the liquid region, and a table for saturated solid and saturated vapor.

We will draw heavily on data from the steam tables for the presentation of examples and problems, and it becomes necessary to select a particular version of such tables for this purpose. Accordingly, we have chosen the steam tables prepared by Keenan, Keyes, Hill, and Moore* as the source for all of our data. These tables come in two editions – one in which data are given in English units, and an International Edition in which metric units (closely related to SI units) are used. We shall draw on both editions. Steam tables give values for the extensive properties V, U, H, and S, and the basis is a unit mass, either a pound mass or a gram. Their notation employs lower-case letters, a practice we have not adopted.

By international agreement the basis for numerical values in steam tables is established by setting U_f and S_f, the internal energy and entropy of saturated liquid water, equal to zero at the triple-point temperature of 0.01(°C). Since H is calculated from U by the defining equation $H = U + PV$, it is not possible in addition to assign an arbitrary value for H_f, the enthalpy of saturated liquid, at 0.01(°C). Rather, H_f must be calculated, and at 0.01(°C) is given by the product PV. Thus at the triple-point temperature, H_f has a finite, but very small, value.

The use of data from the steam tables is illustrated in the following example.

Example 4.13. Consider two tanks, each having a volume of $1000(\text{cm})^3$, one filled with saturated liquid water and the other filled with saturated vapor (steam), both at a pressure of 10(bar). If both tanks explode, which will do the greater damage?

We assume that the tanks undergo sudden, catastrophic failure and that the contents of the tanks expand rapidly to a pressure of 1(bar) in an adiabatic process. The damage that can result is assumed to be proportional to the work done by the system. The work depends on the path followed by the system, and in order to specify the path we will assume the expansion process to be reversible as well as adiabatic. This assumption is justified on the basis that only a comparative answer is sought. The reversible, adiabatic expansion is also one of constant entropy, and it produces the maximum possible work or damage.

It is seen from Fig. 4-11 that the expansion of either saturated liquid or saturated vapor along a line of constant S will lead to a final state in the two-phase (liquid-vapor) region, a state made up of a mixture of saturated liquid and saturated vapor. Thus the only data needed are for saturated liquid and saturated vapor at pressures of 10(bar) and 1(bar). These data are taken from the saturation table of the steam tables and are as follows:

P	T	V_f	V_g	U_f	U_g	S_f	S_g
(bar)	(°C)	$(\text{cm})^3/(\text{g})$		(J)/(g)		(J)/(g)(K)	
1	99.63	1.0432	1694.0	417.36	2506.1	1.3026	7.3594
10	179.91	1.1273	194.44	761.68	2583.6	2.1387	6.5865

By the first law the work of an adiabatic process is given by $W = -m(\Delta U)$, where m is the mass of the system and ΔU is the internal energy change for unit mass. The initial value of the internal energy is given in either case directly in the above table for a pressure of 10(bar). The final value of U in both cases is determined by the condition of isentropic expansion.

Consider *Case A*, the expansion of 1(g) of saturated liquid from 10(bar) to 1(bar) at constant entropy. The initial entropy is given directly in the table. Thus

$$S_1 = 2.1387(\text{J})/(\text{g})(\text{K})$$

Now $S_2 = S_1$ and by *(4.8)*

$$S_2 = (1-x)S_{f_2} + xS_{g_2}$$

---

*Joseph H. Keenan, Frederick G. Keyes, Philip G. Hill, and Joan G. Moore, "Steam Tables", John Wiley and Sons, Inc., New York, 1969.

where x is the quality, or fraction which is vapor. Combination of these last two equations gives

$$S_1 = (1-x)S_{f_2} + xS_{g_2}$$

and substituting numerical values, we get

$$2.1387 = (1-x)(1.3026) + x(7.3594) \quad \text{or} \quad x = 0.1380$$

We now find the final internal energy by again applying (*4.8*):

$$U_2 = (1-x)U_{f_2} + xU_{g_2} = (1-0.1380)(417.36) + (0.1380)(2506.1) = 705.61$$

Thus for *Case A*,

$$\Delta U = U_2 - U_1 = 705.61 - 761.68 = -56.07(\text{J})/(\text{g})$$

For *Case B*, the expansion of 1(g) of saturated vapor from 10(bar) to 1(bar) at constant entropy, the above calculation is repeated:

$$S_2 = S_1 = 6.5865(\text{J})/(\text{g})(\text{K})$$

Also

$$S_2 = (1-x)S_{f_2} + xS_{g_2}$$

Thus

$$6.5865 = (1-x)(1.3026) + x(7.3594) \quad \text{or} \quad x = 0.8724$$

For the internal energy

$$U_2 = (1-0.8724)(417.36) + (0.8724)(2506.1) = 2239.7(\text{J})/(\text{g})$$

Thus for *Case B*,

$$\Delta U = U_2 - U_1 = 2239.7 - 2583.6 = -343.9(\text{J})/(\text{g})$$

The expansion of a gram of saturated vapor clearly results in a much greater decrease of internal energy than the expansion of a gram of saturated liquid. However, the mass contained in the tank is very different in the two cases. This is determined as follows:

For *Case A*, the specific volume of saturated liquid at 10(bar) is 1.1273(cm)^3/(g). Thus a 1000(cm)^3 tank contains a mass of

$$m = \frac{1000(\text{cm})^3}{1.1273(\text{cm})^3/(\text{g})} = 887.1(\text{g})$$

The total work of expansion is therefore

$$W = -m(\Delta U) = -(887.1)(-56.07) = 49{,}740(\text{J})$$

For *Case B*, the specific volume of saturated vapor at 10(bar) is 194.44(cm)^3/(g), and the 1000(cm)^3 tank holds

$$m = \frac{1000}{194.44} = 5.143(\text{g})$$

In this case the total work is

$$W = -(5.143)(-343.9) = 1770(\text{J})$$

These results may be summarized as follows:

	m(g)	ΔU(J)/(g)	W(J)
Case A, expansion of saturated liquid	887.1	−56.07	49,740
Case B, expansion of saturated vapor	5.143	−343.9	1,770

The surprising conclusion is that for a tank of a given volume the destructive potential upon explosion is nearly 30 times greater when the tank is filled with saturated liquid water than when it is filled with saturated steam at the same pressure. The same considerations apply when one is concerned with the storage of energy in a given volume. Saturated liquid at elevated pressure is a much more effective medium than saturated vapor at the same pressure. The reason is that a far greater mass of liquid than of vapor can be put in a given volume.

Solved Problems

PVT BEHAVIOR OF A PURE SUBSTANCE (Sec. 4.1)

4.1. On a PV diagram show the location of (*a*) the fluid region, (*b*) the vapor region.

The fluid region and the vapor region are both indicated in Fig. 4-14. The fluid region includes all states for which the temperature *and* the pressure are above their critical values. The vapor region includes those states from which condensation can be caused *both* by compression at constant T *and* by cooling at constant P.

4.2. The PT diagram of Fig. 4-2 gives no information on volumes. For the liquid and gas regions, what would be the general appearance of constant-volume lines (isometrics) on this diagram?

An examination of Fig. 4-3 shows that vertical lines on a PV diagram, i.e. lines of constant volume, intersect the saturation line ACB throughout the full range of saturation pressures for liquid and vapor. For small volumes these lines intersect the saturated liquid curve AC, and for larger volumes, the saturated vapor curve CB. Thus there are two constant-volume lines that intersect curve ACB at a given pressure, one for liquid and one for vapor. Of course there is an exception for the line at the critical volume which passes through point C at the maximum of ACB. On a PT diagram the two branches of the saturation curve AC and BC are identical, and the coordinates (P, T) of the intersections of constant-volume lines for both liquid and vapor lie along the vaporization curve 2-C of Fig. 4-2. Moreover, they represent terminal points for these curves on the PT diagram. From these terminal points the constant-volume lines (isometrics) extend upward into the liquid region for the set of liquid isometrics, and to the right into the gas region for the set of gas isometrics.

An idea of the general shape of the isometrics can be obtained by consideration of two idealizations: the ideal gas and the liquid for which β and κ are constant. For the ideal gas, $PV = RT$, and

$$\left(\frac{\partial P}{\partial T}\right)_V = \frac{R}{V}$$

Thus at constant volume the slope of an isometric for an ideal gas is constant and positive, and the slope decreases with increasing V.

For the liquid phase we have from *(3.19)* $dV/V = \beta\,dT - \kappa\,dP$, which at constant V becomes

$$\left(\frac{\partial P}{\partial T}\right)_V = \frac{\beta}{\kappa}$$

For constant and positive values of β and κ, the slope of isometrics is constant and positive, independent of V.

These results for idealized gas and liquid phases hold qualitatively for real materials. Isometrics do in fact show curvature, but it is small, and PT diagrams look essentially like Fig. 4-15.

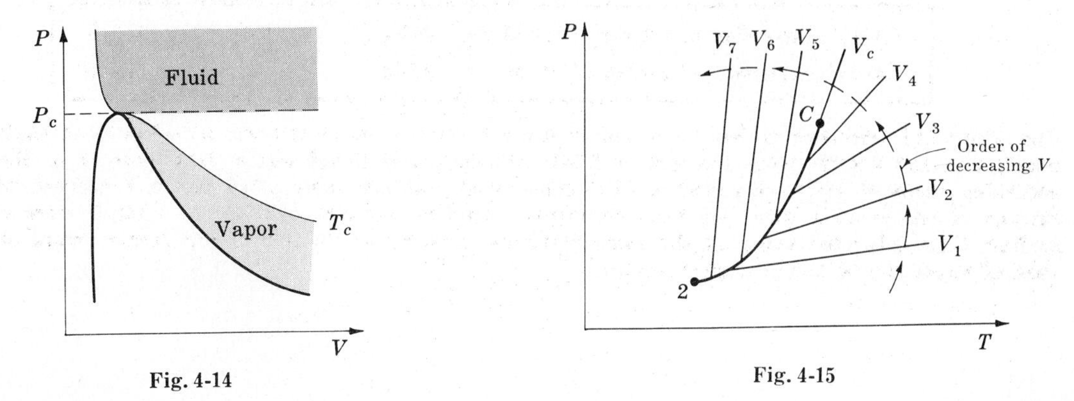

Fig. 4-14

Fig. 4-15

VAPOR PRESSURES AND LATENT HEATS (Sec. 4.3)

4.3. The vapor pressure P^{sat} of a given liquid is represented by an equation of the form

$$\ln P^{sat} = A - \frac{B}{T}$$

where A and B are constants, and T is absolute temperature. Show that for this material

$$S_{fg} = V_{fg}\frac{BP^{sat}}{T^2}$$

Equation (*4.2*) is directly applicable, once it is rewritten in the subscript notation of the steam tables:

$$\frac{dP^{sat}}{dT} = \frac{S_{fg}}{V_{fg}}$$

This may also be written

$$\frac{d\ln P^{sat}}{dT} = \frac{S_{fg}}{P^{sat}V_{fg}}$$

and differentiating the given expression for $\ln P^{sat}$, we find upon substitution:

$$\frac{B}{T^2} = \frac{S_{fg}}{P^{sat}V_{fg}} \qquad \text{or} \qquad S_{fg} = V_{fg}\frac{BP^{sat}}{T^2}$$

4.4. From the data given below determine the latent heat of vaporization H_{fg} of Freon-11 (a refrigerant) at 91(°C).

T(°F)	P^{sat}(psia)	V_g (ft)³/(lb$_m$)	V_f (ft)³/(lb$_m$)
90	19.69	2.09	0.01
91	20.06	2.06	0.01
92	20.43	2.02	0.01

The problem calls for direct application of the Clapeyron equation (*4.3*), which in subscript notation becomes

$$H_{fg} = TV_{fg}\frac{dP^{sat}}{dT}$$

The temperature and the volume change of vaporization are

$$T = 91 + 460 = 551(\text{R}) \qquad V_{fg} = 2.06 - 0.01 = 2.05(\text{ft})^3/(\text{lb}_m)$$

The derivative dP^{sat}/dT is obtained by noting from the data given that P^{sat} is linear in T. Thus

$$\frac{dP^{sat}}{dT} = \frac{\Delta P^{sat}}{\Delta T} = \frac{[20.43 - 19.69](\text{psia})}{2(\text{R})} = 0.37(\text{psia})/(\text{R})$$

Note that $\Delta T = 2(°\text{F}) = 2(\text{R})$. Substitution into the Clapeyron equation gives

$$H_{fg} = 551(\text{R}) \times 2.05(\text{ft})^3/(\text{lb}_m) \times 0.37(\text{psia})/(\text{R}) = 417.9(\text{ft})^3(\text{psia})/(\text{lb}_m)$$

The product of (ft)³(psia) or (ft)³(lb$_f$/in²) is an energy unit, as the product of pressure and volume always is. The usual procedure with these units is to convert them to (ft-lb$_f$) or to (Btu). Thus

$$H_{fg} = 417.9(\text{ft})^3(\text{lb}_f/\text{in}^2)/(\text{lb}_m) \times 144(\text{in})^2/(\text{ft})^2 = 60{,}180(\text{ft-lb}_f)/(\text{lb}_m)$$

and

$$H_{fg} = \frac{60{,}180(\text{ft-lb}_f)/(\text{lb}_m)}{778(\text{ft-lb}_f)/(\text{Btu})} = 77.35(\text{Btu})/(\text{lb}_m)$$

TWO-PHASE SYSTEMS (Sec. 4.4)

4.5. Steam at a pressure of 35(bar) is known to have a specific volume of 50(cm)³/(g). What is its enthalpy?

Examination of the steam tables for values of V at a pressure of 35(bar) shows the given value of V lies between $V_f = 1.2347$ and $V_g = 57.07(\text{cm})^3/(\text{g})$. The steam is therefore "wet," consisting of a mixture of saturated vapor and saturated liquid. In order to find the enthalpy we must determine the quality of the steam, and this is done through use of the known volume data. From Example 4.5 we have

$$V = V_f + xV_{fg} \qquad \text{where } V_{fg} = V_g - V_f$$

Solving for the quality (mass fraction of vapor) x,

$$x = \frac{V - V_f}{V_{fg}} = \frac{50.000 - 1.2347}{57.07 - 1.2347} = 0.8733$$

The enthalpy is given by

$$H = H_f + xH_{fg}$$

From the saturation steam tables at 35(bar)

$$H_f = 1049.7 \quad \text{and} \quad H_{fg} = 1753.7(\text{J})/(\text{g})$$

Thus

$$H = 1049.7 + (0.8733)(1753.7) = 2581.2(\text{J})/(\text{g})$$

4.6. A tank contains exactly one pound mass of H_2O consisting of liquid and vapor in equilibrium at 100(psia). If the liquid and the vapor each occupy one-half the volume of the tank, what is the enthalpy of the contents of the tank?

For saturated liquid and vapor at 100(psia) the steam tables provide the values

$$V_f = 0.017736 \qquad V_g = 4.434(\text{ft})^3/(\text{lb}_\text{m})$$

$$H_f = 298.61 \qquad H_g = 1187.8(\text{Btu})/(\text{lb}_\text{m})$$

If x is the mass of vapor, then the volume of vapor is xV_g and similarly the volume of liquid is $(1-x)V_f$. However, these two volumes are equal, and therefore

$$xV_g = (1-x)V_f$$

or

$$\frac{x}{1-x} = \frac{V_f}{V_g} = \frac{0.017736}{4.434}$$

Solution for x gives

$$x = 0.003984(\text{lb}_\text{m}) \quad \text{and} \quad 1 - x = 0.996016(\text{lb}_\text{m})$$

Then by *(4.8)*

$$H = (1-x)H_f + xH_g = (0.996016)(298.61) + (0.003984)(1187.8) = 301.70(\text{Btu})$$

4.7. A rigid tank having a total volume of 1(m)³ contains 0.05(m)³ of saturated liquid water and 0.95(m)³ of saturated vapor water at 1(bar) pressure. How much heat must be added to the water so that the liquid is just vaporized?

The system is closed, and the first-law energy equation is simply $Q = \Delta U^t = m\,\Delta U$, because there is no work. In its initial state the system contains saturated liquid and saturated vapor in equilibrium at 1(bar). Property values from the steam tables are:

$$V_{f_1} = 1.0432 \qquad V_{g_1} = 1694.0(\text{cm})^3/(\text{g})$$

$$U_{f_1} = 417.36 \qquad U_{g_1} = 2506.1 \qquad U_{fg_1} = 2088.7(\text{J})/(\text{g})$$

The final state is saturated vapor at a pressure yet to be determined. This is done through use of the given volumes. The initial mass of liquid is

$$m_{f_1} = \frac{V_{\text{liquid}}}{V_{f_1}} = \frac{0.05(\text{m})^3 \times (100)^3(\text{cm})^3/(\text{m})^3}{1.0432(\text{cm})^3/(\text{g})} = 47{,}929(\text{g})$$

Similarly

$$m_{g_1} = \frac{V_{\text{vapor}}}{V_{g_1}} = \frac{0.95 \times (100)^3}{1694.0} = 560.8(\text{g})$$

The total mass of the system is

$$m = m_{f_1} + m_{g_1} = 47{,}929 + 561 = 48{,}490(\text{g})$$

The final state of the system is saturated vapor occupying the full $1(\text{m})^3$ of tank volume and consisting of the total mass of the system. Thus

$$V_{g_2} = \frac{V_{\text{tank}}}{m} = \frac{1(\text{m})^3 \times (100)^3(\text{cm})^3/(\text{m})^3}{48{,}490(\text{g})} = 20.622(\text{cm})^3/(\text{g})$$

Saturated vapor with this specific volume occurs at just one pressure, which may be found by interpolation in the steam tables for saturated vapor. This pressure is $P_2 = 89.5(\text{bar})$ and at this pressure the internal energy of saturated vapor is

$$U_2 = U_{g_2} = 2558.4(\text{J})/(\text{g})$$

The initial internal energy is

$$U_1 = U_{f_1} + x_1 U_{fg_1}$$

where x_1 is the initial quality, given by

$$x_1 = \frac{m_{g_1}}{m} = \frac{560.8}{48{,}490} = 0.011565$$

(Note that although the system is 95% vapor by volume initially, it is only a little more than 1% vapor on a mass basis.) Substitution of values in the expression for U_1 gives

$$U_1 = 417.36 + (0.011565)(2088.7) = 441.52(\text{J})/(\text{g})$$

We may now solve for Q by our original equation:

$$Q = m\,\Delta U = 48{,}490(\text{g}) \times [2558.4 - 441.5](\text{J})/(\text{g}) = 1.0265 \times 10^8(\text{J})$$

4.8. A rigid tank of $10(\text{ft})^3$ capacity contains saturated liquid and vapor in equilibrium at 200(psia). If 99% of the mass is liquid, how much heat must be added before the tank becomes just full of liquid?

The reason the tank fills with liquid in this case and with vapor in the preceding problem is explained in Example 4.2. The energy equation is $Q = \Delta U^t = m\,\Delta U$ and initial property values from the steam tables are:

$$V_{f_1} = 0.018387 \qquad V_{g_1} = 2.289(\text{ft})^3/(\text{lb}_\text{m})$$

$$U_{f_1} = 354.9 \qquad U_{g_1} = 1114.6(\text{Btu})/(\text{lb}_\text{m})$$

The initial specific volume of the contents of the tank is

$$V_1 = (1 - x_1)V_{f_1} + x_1 V_{g_1}$$

where x_1 is the quality or mass fraction vapor, which by the problem statement is 0.01. Thus

$$V_1 = (0.99)(0.018387) + (0.01)(2.289) = 0.041093(\text{ft})^3/(\text{lb}_\text{m})$$

(Note that the fraction *by volume* of the tank contents that is vapor is $0.02289/0.041093 = 0.557$. Thus it is 55.7% vapor by volume although only 1% vapor on a mass basis.) The mass of the system is

$$m = \frac{V_{\text{tank}}}{V_1} = \frac{10(\text{ft})^3}{0.041093(\text{ft})^3/(\text{lb}_\text{m})} = 243.35(\text{lb}_\text{m})$$

The initial internal energy is

$$U_1 = (1 - x_1)U_{f_1} + x_1 U_{g_1}$$

$$= (0.99)(354.9) + (0.01)(1114.6) = 362.5(\text{Btu})/(\text{lb}_\text{m})$$

The final state just as the last vapor condenses is one of saturated liquid, containing the entire mass of the system and occupying the entire tank volume. Neither the tank volume nor the total mass of the system has changed. Therefore the final specific volume of the contents of the tank is unchanged, and $V_{f_2} = 0.041093(\text{ft})^3/(\text{lb}_m)$. The pressure at which saturated liquid has this specific volume is found by interpolation in the table for saturated liquid to be $P_2 = 3159.6(\text{psia})$ and at this pressure the final internal energy is

$$U_2 = U_{f_2} = 827.8(\text{Btu})/(\text{lb}_m)$$

We may now solve for Q:

$$Q = m\,\Delta U = 243.35(\text{lb}_m) \times [827.8 - 362.5](\text{Btu})/(\text{lb}_m) = 113{,}230(\text{Btu})$$

HEAT CAPACITIES (Secs. 4.6 and 4.7)

4.9. The following data are available for liquid water at 25(°C) and atmospheric pressure: $\beta = 256 \times 10^{-6}(°\text{C})^{-1}$, $(\partial\beta/\partial T)_P = 9.6 \times 10^{-6}(°\text{C})^{-2}$, $V = 1.003(\text{cm})^3/(\text{g})$. Determine the effect of pressure on C_P, i.e. calculate the value of $(\partial C_P/\partial P)_T$, for water at these conditions.

By (*3.59*) we have

$$\left(\frac{\partial C_P}{\partial P}\right)_T = -T\left(\frac{\partial^2 V}{\partial T^2}\right)_P$$

which in view of (*3.17*) may also be written

$$\left(\frac{\partial C_P}{\partial P}\right)_T = -T\left[\frac{\partial(\beta V)}{\partial T}\right]_P = -T\left[\beta\left(\frac{\partial V}{\partial T}\right)_P + V\left(\frac{\partial \beta}{\partial T}\right)_P\right]$$

Again using (*3.17*) we obtain

$$\left(\frac{\partial C_P}{\partial P}\right)_T = -TV\left[\beta^2 + \left(\frac{\partial \beta}{\partial T}\right)_P\right]$$

Substitution of numerical values gives

$$\left(\frac{\partial C_P}{\partial P}\right)_T = -298(\text{K}) \times 1.003(\text{cm})^3/(\text{g})[(256 \times 10^{-6})^2(\text{K})^{-2} + 9.6 \times 10^{-6}(\text{K})^{-2}]$$

$$= -0.00288(\text{cm}^3\text{-atm})/(\text{g})(\text{K})(\text{atm}) \quad \text{or} \quad -2.92 \times 10^{-4}(\text{J})/(\text{g})(\text{K})(\text{atm})$$

Thus a pressure increase of 1(atm) at the stated conditions causes a very small decrease in C_P. Since the heat capacity of water at 25(°C) is about 4.18(J)/(g)(K), the percentage change for a pressure increase of 1(atm) is

$$\frac{-2.92 \times 10^{-4}}{4.18}(100) = -0.007\%$$

4.10. The general defining equation for a heat capacity is given in Chapter 1 as

$$C_X = \left(\frac{\delta Q}{dT}\right)_X$$

where X indicates a reversible and fully specified path. We have so far considered but two heat capacities, C_V and C_P, but many others could be defined. Two that are sometimes found useful are the heat capacities of saturated liquid and of saturated vapor. These represent $\delta Q/dT$ for reversible changes along the appropriate saturation curves, and we will designate them C_{sat} in general. Since the pressure, temperature, and volume all vary along a saturation curve, C_{sat} is different from C_V and C_P. Nor is it given by $(dU/dT)_{\text{sat}}$ or by $(dH/dT)_{\text{sat}}$, derivatives taken along a saturation curve. These quantities, however, are all related to one another. Derive equations that show these relationships.

Since the defining equation for C_X restricts consideration to reversible processes, we may always make the substitution $\delta Q = T\,dS$. Thus for the case considered here

$$C_{sat} = T\left(\frac{dS}{dT}\right)_{sat}$$

where $(dS/dT)_{sat}$ is the change of S with T along a saturation curve, i.e. the reciprocal of the slope of the saturated liquid or saturated vapor curve on the TS diagram of Fig. 4-11. In this form the equation is written in terms of properties only, and no restriction to reversible processes is now necessary.

If we consider S to be a function of T and V, then

$$dS = \left(\frac{\partial S}{\partial T}\right)_V dT + \left(\frac{\partial S}{\partial V}\right)_T dV$$

Division by dT and restriction to changes along a saturation curve give

$$\left(\frac{dS}{dT}\right)_{sat} = \left(\frac{\partial S}{\partial T}\right)_V + \left(\frac{\partial S}{\partial V}\right)_T \left(\frac{dV}{dT}\right)_{sat} \qquad (1)$$

This equation is very similar to (*3.15*) except that restriction is not to a constant value of a property but to the set of states prescribed by a saturation curve. By (*3.49*) and (*3.43*):

$$\left(\frac{\partial S}{\partial T}\right)_V = \frac{C_V}{T} \quad \text{and} \quad \left(\frac{\partial S}{\partial V}\right)_T = \left(\frac{\partial P}{\partial T}\right)_V$$

Hence

$$C_{sat} = T\left(\frac{dS}{dT}\right)_{sat} = T\left[\frac{C_V}{T} + \left(\frac{\partial P}{\partial T}\right)_V \left(\frac{dV}{dT}\right)_{sat}\right]$$

But from Example 3.6, $(\partial P/\partial T)_V = \beta/\kappa$, and by substitution we get a final expression relating C_{sat} and C_V:

$$C_{sat} = C_V + \frac{T\beta}{\kappa}\left(\frac{dV}{dT}\right)_{sat} \qquad (2)$$

An equation entirely analogous to (*1*) is

$$\left(\frac{dV}{dT}\right)_{sat} = \left(\frac{\partial V}{\partial T}\right)_P + \left(\frac{\partial V}{\partial P}\right)_T \left(\frac{dP}{dT}\right)_{sat}$$

which by (*3.17*) and (*3.18*) becomes

$$\left(\frac{dV}{dT}\right)_{sat} = \beta V - \kappa V\left(\frac{dP}{dT}\right)_{sat} \qquad (3)$$

Combination of (*2*) and (*3*) gives another expression relating C_{sat} and C_V:

$$C_{sat} = C_V + \frac{T\beta^2 V}{\kappa} - T\beta V\left(\frac{dP}{dT}\right)_{sat} \qquad (4)$$

However, by (*4.12*), $T\beta^2 V/\kappa = C_P - C_V$. Therefore

$$C_{sat} = C_P - T\beta V\left(\frac{dP}{dT}\right)_{sat} \qquad (5)$$

and we have a relation between C_{sat} and C_P. Note that $(dP/dT)_{sat}$ is just the slope of the vaporization curve of Fig. 4-2.

Another equation analogous to (*1*) is

$$\left(\frac{dH}{dT}\right)_{sat} = \left(\frac{\partial H}{\partial T}\right)_P + \left(\frac{\partial H}{\partial P}\right)_T \left(\frac{dP}{dT}\right)_{sat}$$

However, $(\partial H/\partial T)_P = C_P$ and by substitution of the definition of β from (*3.17*) into the result of Problem 3.3, we have $(\partial H/\partial P)_T = V(1 - \beta T)$. Therefore

$$\left(\frac{dH}{dT}\right)_{sat} = C_P - T\beta V\left(\frac{dP}{dT}\right)_{sat} + V\left(\frac{dP}{dT}\right)_{sat}$$

According to (5) the first two terms on the right equal C_{sat}. Thus

$$C_{sat} = \left(\frac{dH}{dT}\right)_{sat} - V\left(\frac{dP}{dT}\right)_{sat} \tag{6}$$

which shows the relation between C_{sat} and $(dH/dT)_{sat}$.

Since by definition of enthalpy $U = H - PV$, we have

$$\left(\frac{dU}{dT}\right)_{sat} = \left(\frac{dH}{dT}\right)_{sat} - V\left(\frac{dP}{dT}\right)_{sat} - P\left(\frac{dV}{dT}\right)_{sat}$$

However (6) shows the first two terms on the right to be equal to C_{sat}. Therefore

$$C_{sat} = \left(\frac{dU}{dT}\right)_{sat} + P\left(\frac{dV}{dT}\right)_{sat} \tag{7}$$

an equation which relates C_{sat} to $(dU/dT)_{sat}$.

In summary, we have from (2) and (7)

$$C_{sat} = C_V + \frac{T\beta}{\kappa}\left(\frac{dV}{dT}\right)_{sat} = \left(\frac{dU}{dT}\right)_{sat} + P\left(\frac{dV}{dT}\right)_{sat}$$

and from (5) and (6)

$$C_{sat} = C_P - T\beta V\left(\frac{dP}{dT}\right)_{sat} = \left(\frac{dH}{dT}\right)_{sat} - V\left(\frac{dP}{dT}\right)_{sat}$$

It is clear from these equations that C_{sat}, C_V, C_P, $(dU/dT)_{sat}$, and $(dH/dT)_{sat}$ are all different, although related, quantities.

An idea of the magnitudes of these quantities can be obtained by use of data from the steam tables. Consider the following data for saturated liquid and saturated vapor:

T	P	V_f	V_g	H_f	H_g	S_f	S_g
(°C)	(bar)	(cm)³/(g)		(J)/(g)		(J)/(g)(K)	
99	0.9778	1.0427	1729.9	414.83	2674.5	1.2956	7.3669
100	1.0135	1.0435	1672.9	419.04	2676.1	1.3069	7.3549
101	1.0502	1.0443	1618.2	423.26	2677.6	1.3181	7.3429

The most direct calculation that can be made for C_{sat} makes use of the entropy data in accord with the original equation

$$C_{sat} = T\left(\frac{dS}{dT}\right)_{sat}$$

Applied to saturated liquid we may write this

$$C_f = T\frac{dS_f}{dT}$$

and for saturated vapor

$$C_g = T\frac{dS_g}{dT}$$

We assume as a reasonable approximation that $(dS/dT)_{sat}$ at 100(°C) is equal to $(\Delta S/\Delta T)_{sat}$ applied to the interval between 99 and 101(°C). Thus for 100(°C),

$$\frac{dS_f}{dT} \cong \frac{\Delta S_f}{\Delta T} = \frac{1.3181 - 1.2956}{2} = 0.01125\text{(J)/(g)(K)}^2$$

and

$$C_f = 373.15\text{(K)} \times 0.01125\text{(J)/(g)(K)}^2 = 4.198\text{(J)/(g)(K)}$$

Similarly

$$\frac{dS_g}{dT} \cong \frac{\Delta S_g}{\Delta T} = \frac{7.3429 - 7.3669}{2} = -0.0120\text{(J)/(g)(K)}^2$$

and $$C_g = (373.15)(-0.0120) = -4.478(\text{J})/(\text{g})(\text{K})$$

We have here the surprising result that C_g, the heat capacity of saturated vapor, is *negative*; that is, as the temperature rises, heat must be *withdrawn*. The explanation is that along the saturation curve both T and P change and both affect the entropy. An increase in temperature causes the entropy to increase, whereas the simultaneous pressure increase causes the entropy to decrease. When the latter effect overbalances the former, then the entropy decreases and the heat capacity is negative.

By comparison, C_P for vapor at 100(°C) and 1.0135(bar) may be computed from data in the steam tables. This requires values of the enthalpy for superheated vapor as indicated by the equation

$$C_P = \left(\frac{\partial H}{\partial T}\right)_P \cong \left(\frac{\Delta H}{\Delta T}\right)_P$$

The enthalpy change between the temperatures of 100 and 110(°C) at $P = 1$(bar) and at $P = 1.1$(bar) is +20.3(J)/(g) in each case, and this is therefore presumed to be the enthalpy change at $P = 1.0135$(bar) as well. Thus at 100(°C)

$$C_P = \left(\frac{\Delta H}{\Delta T}\right)_{P=1.0135(\text{bar})} = \frac{20.3}{10} = 2.03(\text{J})/(\text{g})(\text{K})$$

The enthalpy change along the saturation curve is quite different from this and is given by

$$\frac{dH_g}{dT} \cong \frac{\Delta H_g}{\Delta T} = \frac{2677.6 - 2674.5}{2} = 1.55(\text{J})/(\text{g})(\text{K})$$

PROPERTY VALUES FOR PVT SYSTEMS (Sec. 4.8)

4.11. Property values for subcooled or compressed liquid are calculated from property values for saturated liquid at the same temperature by integration of the equations

$$dH = \left[V - T\left(\frac{\partial V}{\partial T}\right)_P\right] dP \quad \text{and} \quad dS = -\left(\frac{\partial V}{\partial T}\right)_P dP$$

Often data are not available for V and $(\partial V/\partial T)_P$ as functions of P in the liquid region, and the assumption is made that these values are constant at the saturation values. Integration then gives

$$H - H_f = \left(V_f - T\frac{dV_f}{dT}\right)(P - P^{\text{sat}}) \qquad (\text{const. } T)$$

$$S - S_f = -\frac{dV_f}{dT}(P - P^{\text{sat}}) \qquad (\text{const. } T)$$

Since volumes are assumed independent of P, we have switched to the total derivative dV_f/dT. Apply these equations to liquid water, using values of the saturation properties from the steam tables, and calculate H and S for water at (*a*) $T = 100$(°C) and $P = 25$(bar), (*b*) $T = 100$(°C) and $P = 200$(bar). Compare results with values taken from the steam tables.

At 100(°C) or 373(K), we have the following values from the steam tables:

$$P^{\text{sat}} = 1.0135(\text{bar}) \qquad H_f = 419.04(\text{J})/(\text{g})$$

$$S_f = 1.3069(\text{J})/(\text{g})(\text{K}) \qquad V_f = 1.0435(\text{cm})^3/(\text{g})$$

$$\frac{dV_f}{dT} \cong \frac{V_{f_{101}} - V_{f_{99}}}{101 - 99} = \frac{1.0443 - 1.0427}{2} = 0.0008(\text{cm})^3/(\text{g})(\text{K})$$

(*a*) Direct substitution of values in the enthalpy formula gives

$$H - H_f = [1.0435 - (373)(0.0008)](25 - 1.0135) = 17.87(\text{cm}^3\text{-bar})/(\text{g})$$

Since $10(\text{cm}^3\text{-bar}) = 1(\text{J})$, we have $H - H_f = 1.79(\text{J})/(\text{g})$ and

$$H = H_f + 1.79 = 419.04 + 1.79 = 420.83(\text{J})/(\text{g})$$

The value given in the steam tables is 420.85(J)/(g).

Similarly, by the entropy formula,

$$S - S_f = -0.01919(\text{cm}^3\text{-bar})/(\text{g})(\text{K}) \quad \text{or} \quad -0.0019(\text{J})/(\text{g})(\text{K})$$

whence

$$S = 1.3050$$

which is exactly the steam-table value.

(*b*) The same procedure for a pressure of 200(bar) yields the following results:

$$H - H_f = 14.83(\text{J})/(\text{g})$$

$$H = 419.04 + 14.83 = 433.87(\text{J})/(\text{g})$$

[from the steam tables $H = 434.06(\text{J})/(\text{g})$]

$$S - S_f = -0.0159(\text{J})/(\text{g})(\text{K})$$

$$S = 1.3069 - 0.0159 = 1.2910(\text{J})/(\text{g})(\text{K})$$

[from the steam tables $S = 1.2917(\text{J})/(\text{g})(\text{K})$]

These results show that the approximations employed lead to rather small errors at the conditions considered. However, the method would be entirely inappropriate in the critical region where liquid properties change rapidly with pressure.

4.12. Very pure liquid water can be subcooled at atmospheric pressure to temperatures well below 0(°C). Assume that a mass of water has been cooled as a liquid to −5(°C). A small crystal of ice (whose mass is negligible) is added to "seed" the subcooled liquid. If the subsequent change of state occurs adiabatically and at constant (atmospheric) pressure, what fraction of the system solidifies? What is the entropy change of the system?

The following data are known:

The latent heat of fusion of water at 0(°C) = 333.4(J)/(g)

The heat capacity of water between 0 and −5(°C) = 4.22(J)/(g)(°C)

The final state of the system is presumed to be the equilibrium state of a mixture of ice and water at atmospheric pressure. This state exists at 0(°C) or 273.15(K). The change of state occurs in a closed system at constant pressure, and for such a process (see Example 1.7) $Q = \Delta H^t = m\,\Delta H$. However, this process is adiabatic, and therefore $\Delta H = 0$.

Since ΔH is a property change, its value is independent of path, and therefore for purposes of calculation we may consider an arbitrary path made up of the two following steps. (1) The subcooled liquid is warmed as a liquid from −5(°C) to 0(°C). (2) The heat added during step (1) is now withdrawn at 0(°C) causing part of the liquid to freeze. The amounts of heat for the two steps just compensate, so that overall $Q = \Delta H = 0$ for each gram of water.

For step (1)

$$\Delta H_1 = C_P\,\Delta T = 4.22(\text{J})/(\text{g})(°\text{C}) \times 5(°\text{C}) = 21.1(\text{J})/(\text{g})$$

For step (2)

$$\Delta H_2 = -z\,\Delta H_{\text{fusion}} = -333.4\,z(\text{J})/(\text{g})$$

where z is the fraction of the system that freezes. We can now write

$$-333.4\,z + 21.1 = 0 \quad \text{or} \quad z = 0.0633$$

Thus 6.33% of the system solidifies.

The process as it occurs is obviously irreversible. This can be confirmed by showing that the entropy change of the system is positive. (Since the process is adiabatic, no entropy change occurs in the surroundings.) The entropy is a property, and its change may be calculated along the same path as the enthalpy. For step (1)

$$\Delta S_1 = C_P \ln \frac{T_2}{T_1} = 4.22 \ln \frac{273.15}{268.15} = 0.07796(\text{J})/(\text{g})(\text{K})$$

For step (2) the temperature is constant, and

$$\Delta S_2 = \frac{Q_2}{T_2} = \frac{-21.1}{273.15} = -0.07725(\text{J})/(\text{g})(\text{K})$$

Hence $$\Delta S_{\text{total}} = +0.00071(\text{J})/(\text{g})(\text{K})$$

4.13. The heat capacity at constant pressure of a certain gas in the ideal-gas state is given as a function of temperature by the equation

$$C_P' = 12.0 + 3.0 \times 10^{-3}T - \frac{2.1 \times 10^5}{T^2}$$

where T is in kelvins and C_P' is in (cal)/(g mole)(K). Determine $\Delta H_{12}'$ and $\Delta S_{12}'$ for a temperature change at constant P from $T_1 = 300$ to $T_2 = 700$(K).

In the ideal-gas state enthalpy and entropy changes at constant pressure are given by

$$\Delta H_{12}' = \int_{T_1}^{T_2} C_P'\,dT \qquad \text{and} \qquad \Delta S_{12}' = \int_{T_1}^{T_2} \frac{C_P'\,dT}{T}$$

Substituting for C_P' and integrating, we get

$$\Delta H_{12}' = 12.0(T_2 - T_1) + 1.5 \times 10^{-3}(T_2^2 - T_1^2) + 2.1 \times 10^5\left(\frac{1}{T_2} - \frac{1}{T_1}\right)$$

$$= \left[12.0 + 1.5 \times 10^{-3}(T_2 + T_1) - \frac{2.1 \times 10^5}{T_1 T_2}\right](T_2 - T_1)$$

and $$\Delta S_{12}' = 12.0 \ln \frac{T_2}{T_1} + 3.0 \times 10^{-3}(T_2 - T_1) - 1.05 \times 10^5 \frac{T_2^2 - T_1^2}{(T_1 T_2)^2}$$

With temperature in kelvins, the units of $\Delta H_{12}'$ will be (cal)/(g mole) and those of $\Delta S_{12}'$ will be (cal)/(g mole)(K). Substitution of numerical values for T_1 and T_2 gives

$$\Delta H_{12}' = 5000(\text{cal})/(\text{g mole}) \qquad \Delta S_{12}' = 10.415(\text{cal})/(\text{g mole})(\text{K})$$

4.14. Equation (*4.16*) gives the definition of residual volume as

$$\Delta V' = \frac{RT}{P} - V$$

and it is stated following this equation that experimental data show that as $P \to 0$, $\Delta V'$ remains finite. Although $\Delta V'$ becomes zero at a unique temperature for any particular substance, in general it approaches a nonzero limiting value. How is this experimental evidence reconciled with the well-known fact that the ideal-gas law becomes valid as $P \to 0$?

The question is, how can

$$\Delta V' = \frac{RT}{P} - V$$

be anything but zero if we have the simultaneous requirement that

$$V = \frac{RT}{P}$$

In general, of course, these two equations together require $\Delta V'$ to be zero. However, when $P \to 0$, both V and RT/P aproach infinity. Thus in this limit

$$\Delta V' = \frac{RT}{P} - V = \infty - \infty$$

which is an indeterminate form. The indeterminacy can be resolved only by experiment, and this shows $\Delta V'$ to be generally nonzero but finite in the limit as $P \to 0$. Thus, strictly speaking, we cannot write

$$V = \frac{RT}{P}$$

for a real gas at zero pressure, but must write instead

$$V + \Delta V' = \frac{RT}{P}$$

However, as $P \to 0$, $V \to \infty$, and $\Delta V'$ becomes entirely negligible in comparison with V; so for all practical purposes the ideal-gas law is valid.

Consider a numerical example. At 283(K) the limiting value of $\Delta V'$ for methane is 50(cm)^3/(g mole). At this temperature and $P = 0.0001$(atm) we calculate

$$\frac{RT}{P} = \frac{82.05(\text{cm}^3\text{-atm})/(\text{g mole})(\text{K}) \times 283(\text{K})}{0.0001(\text{atm})} = 232{,}201{,}500(\text{cm})^3/(\text{g mole})$$

The volume is given by

$$V = \frac{RT}{P} - \Delta V' = 232{,}201{,}500 - 50 = 232{,}201{,}450(\text{cm})^3/(\text{g mole})$$

Had we set $V = RT/P$ we would have been in error by less than 1 part in 4 million, and as P gets smaller and smaller the fractional error becomes less and less, approaching zero in the limit. Nevertheless, the *difference* between RT/P and V remains essentially constant at 50(cm)^3/(g mole). It does not become zero.

4.15. Sketch graphs of V, S, G, and C_P versus T for a pure substance along a constant-pressure path that includes a phase change from liquid to vapor.

The volume of a liquid normally increases slowly with temperature. During vaporization at constant T and P the volume increases enormously (except in the critical region). The vapor then expands fairly rapidly with further temperature increase. Thus a plot of V versus T usually has the general characteristics shown in Fig. 4-16.

The change of entropy with temperature for a homogeneous system is given by (*3.50*) as

$$\left(\frac{\partial S}{\partial T}\right)_P = \frac{C_P}{T}$$

The heat capacity C_P is always positive, both for liquids and vapors. Therefore a plot of S versus T will have a positive slope in both regions. During vaporization the entropy increases by an amount equal to the latent heat divided by the absolute temperature. Thus an S versus T plot looks generally like Fig. 4-17. The slope of the vapor line is indicated to be less than that of the liquid line because the heat capacity of a vapor is generally less than that of the liquid. Since C_P normally increases as the temperature rises, C_P/T may either increase or decrease, and therefore no curvature is shown for these lines.

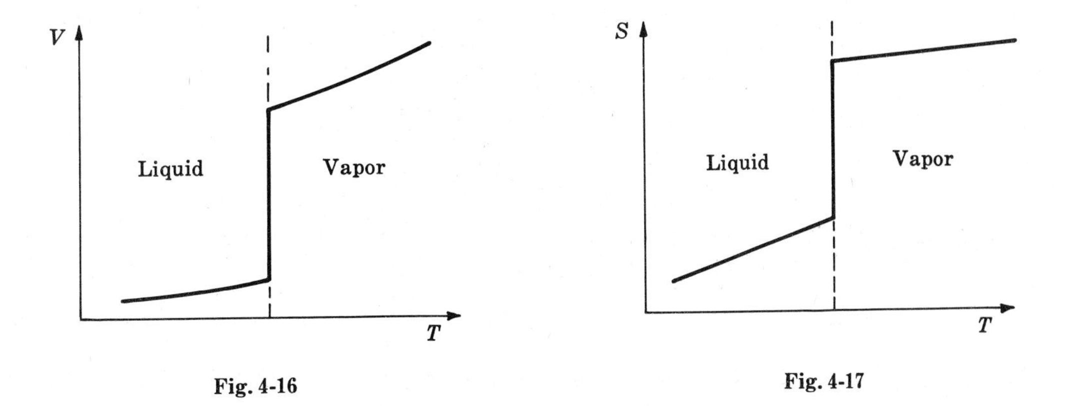

Fig. 4-16

Fig. 4-17

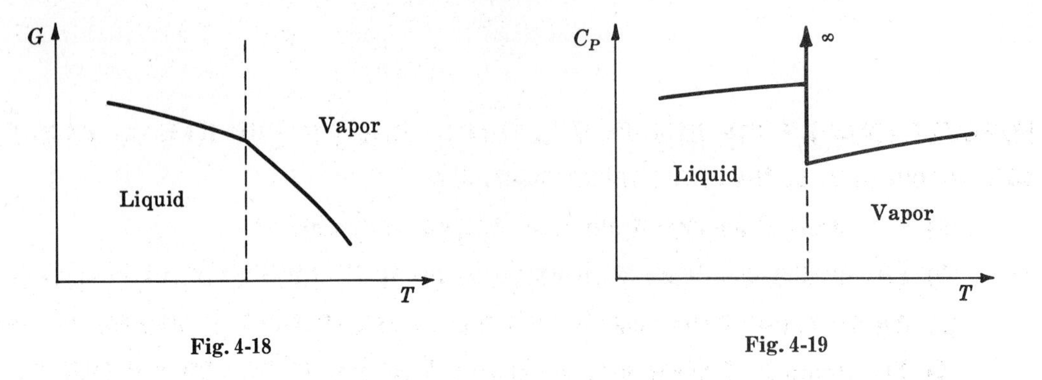

Fig. 4-18

Fig. 4-19

The Gibbs function is related to T and P by (*3.35*), which for 1 mole of a pure material becomes $dG = -S\,dT + V\,dP$. At constant T and P, G must also be constant; thus we arrive again at the conclusion reached in Problem 3.11 that no change in the Gibbs function occurs during phase change at constant T and P. For the single phases we have from the above equation

$$\left(\frac{\partial G}{\partial T}\right)_P = -S$$

From this and Fig. 4-17 we construct Fig. 4-18 for G versus T.

As already mentioned C_P is positive and generally increases with T. Moreover the value for saturated vapor is generally less than for saturated liquid, as shown in Fig. 4-19, which also indicates that C_P becomes infinite during the vaporization process. This follows from the defining equation

$$\left(\frac{\delta Q}{dT}\right)_P = C_P$$

and the fact that the temperature change is zero during the vaporization process.

4.16. Develop the three general expressions which give the average molar properties of a three-phase mixture of a pure material at conditions of its triple point in terms of the molar property of one phase and the property changes of fusion and vaporization.

Let x = fraction of system that is vapor, y = fraction that is liquid, z = fraction that is solid; where $x + y + z = 1$. Then an average molar property of the system, M, is given by

$$M = zM_s + yM_f + xM_g \tag{1}$$

Substitution of $z = 1 - x - y$ into (*1*) gives

$$M = M_s + y(M_f - M_s) + x(M_g - M_s) = M_s + yM_{sf} + xM_{sg}$$

However, $M_{sg} = M_{sf} + M_{fg}$. This relation is general, and its validity for $M = V, S,$ and H can be seen by examination of Figs. 4-1, 4-11, and 4-12. Making the substitution, we have

$$M = M_s + (x + y)M_{sf} + xM_{fg} \tag{2}$$

In words this equation says that any three-phase, pure-component system has the property of the solid phase increased by the property change of fusion for the total fraction of the system that is not solid plus the property change of vaporization for that fraction of the system that is vapor.

A second equation is obtained by substitution of $y = 1 - x - z$ into (*1*):

$$M = z(M_s - M_f) + M_f + x(M_g - M_f)$$

But $$M_s - M_f = -M_{sf} \quad \text{and} \quad M_g - M_f = M_{fg}$$

Therefore $$M = M_f + xM_{fg} - zM_{sf} \tag{3}$$

The final equation results in similar fashion by substitution of $x = 1 - z - y$ into (*1*):

$$M = M_g - zM_{sg} - yM_{fg}$$

Again we make use of the equation $M_{sg} = M_{sf} + M_{fg}$ and obtain

$$M = M_g - (z+y)M_{fg} - zM_{sf} \qquad (4)$$

MISCELLANEOUS PROBLEMS REQUIRING USE OF THE STEAM TABLES

4.17. From data in the steam tables determine:

(a) The internal energy of steam at 50(psia) and 400(°F).

(b) The quality of steam with an enthalpy of 1074.0(Btu)/(lb$_m$) at 50(psia).

(c) An estimate of the specific heat capacity C_P of steam at 30(psia) and 300(°F).

(d) The number of degrees of superheat for steam at 50(psia) and 400(°F).

(e) Which one of Maxwell's equations, (3.41)–(3.44), would be most directly and easily applied for testing the consistency of the data in the superheat tables? Illustrate the use of the equation selected.

(a) Steam at 50(psia) and 400(°F) is superheated, and its internal energy is read directly from the tables as $U = 1141.9$(Btu)/(lb$_m$).

(b) At 50(psia): $H_f = 250.24$, $H_g = 1174.4$, and $H_{fg} = 924.2$(Btu)/(lb$_m$). Since $H = H_f + xH_{fg}$,

$$x = \frac{H - H_f}{H_{fg}} = \frac{1074.0 - 250.2}{924.2} = 0.891$$

(c) By definition, $C_P = (\partial H/\partial T)_P$. For a small temperature interval we can write $C_P \cong (\Delta H/\Delta T)_P$. We can apply this equation over the temperature interval from 290 to 310(°F). This gives

$$C_P \cong \frac{1194.1 - 1184.2}{20} = 0.495(\text{Btu})/(\text{lb}_\text{m})(°\text{F})$$

(d) The saturation temperature at a pressure of 50(psia) is 281.03(°F). Thus

$$\text{Degrees of superheat} = 400 - 281.03 = 118.97(°\text{F})$$

(e) The equation most directly and easily used is (3.44), because it requires only T and P to be held constant, and these are the primary variables of the steam tables. For small intervals of T and P (3.44) may be expressed as

$$\left(\frac{\Delta V}{\Delta T}\right)_P \cong -\left(\frac{\Delta S}{\Delta P}\right)_T$$

As an example of the application of this equation, consider the following data taken from the superheat tables. The data refer to 1(lb$_m$) of steam.

T(°F)	V(ft)^3 $P = 28$(psia)	V(ft)^3 $P = 30$(psia)	S(Btu)/(R) $P = 28$(psia)	S(Btu)/(R) $P = 30$(psia)
300	15.890	14.812	1.7415	1.7334
310	16.115	15.023	1.7479	1.7399
	at 28(psia) $\Delta V = 0.225$	at 30(psia) $\Delta V = 0.211$	at 300(°F) $\Delta S = -0.0081$; at 310(°F) $\Delta S = -0.0080$	
	$\Delta V^{\text{ave}} = 0.218$		$\Delta S^{\text{ave}} = -0.00805$	
	$\left(\frac{\Delta V}{\Delta T}\right)_P = \frac{0.218(\text{ft})^3}{10(\text{R})} = 0.0218(\text{ft})^3/(\text{R})$		$\left(\frac{\Delta S}{\Delta P}\right)_T = \frac{-0.00805(\text{Btu})/(\text{R})}{2(\text{psia})} = -0.004025(\text{Btu})/(\text{R})(\text{psia})$	

Thus we have average values of $(\Delta V/\Delta T)_P$ and $(\Delta S/\Delta P)_T$ centered around 305(°F) and 29(psia). The remaining problem is to reconcile units. We can write

$$\left(\frac{\Delta S}{\Delta P}\right)_T = \frac{-0.004025(\text{Btu})/(\text{R})(\text{lb}_f/\text{in}^2) \times 778.16(\text{ft-lb}_f)/(\text{Btu})}{144(\text{in})^2/(\text{ft})^2}$$

$$= -0.02175(\text{ft})^3/(\text{R})$$

The requirement of consistency as expressed by the relation

$$\left(\frac{\Delta V}{\Delta T}\right)_P \cong -\left(\frac{\Delta S}{\Delta P}\right)_T$$

is indeed obeyed, since $0.0218 \cong -(-0.02175)$.

4.18. Superheated steam at 500(psia) and 650(°F) expands to 100(psia). Determine the final state of the steam if the expansion is (*a*) at constant enthalpy, (*b*) at constant entropy.

The following initial property values are given in the steam tables:

$$H_1 = 1328.0(\text{Btu})/(\text{lb}_m) \qquad S_1 = 1.5860(\text{Btu})/(\text{lb}_m)(\text{R})$$

(*a*) We find from the steam tables that superheated vapor at 100(psia) has an enthalpy equal to H_1 at a temperature just below 600(°F). Linear interpolation in the tables gives the value $T_2 = 597.4(°\text{F})$, at which

$$H_2 = H_1 = 1328.0(\text{Btu})/(\text{lb}_m) \quad \text{and} \quad S_2 = 1.7570(\text{Btu})/(\text{lb}_m)(\text{R})$$

(*b*) We find from the steam tables that no stable vapor state at 100(psia) has an entropy as low as S_1. We find further that the value of

$$S_2 = S_1 = 1.5860(\text{Btu})/(\text{lb}_m)(\text{R})$$

lies between the entropy of saturated liquid and saturated vapor at 100(psia). The final state must therefore be a mixture of these, or "wet" steam, for which we can determine the quality by

$$S_2 = S_{f_2} + x_2 S_{fg_2} \quad \text{or} \quad x_2 = \frac{S_2 - S_{f_2}}{S_{fg_2}}$$

At 100(psia): $S_{f_2} = 0.4744$ and $S_{fg_2} = 1.1290(\text{Btu})/(\text{lb}_m)(\text{R})$. Thus

$$x_2 = \frac{1.5860 - 0.4744}{1.1290} = 0.9846$$

and

$$H_2 = H_{f_2} + x_2 H_{fg_2}$$

$$= 298.6 + (0.9846)(889.2) = 1174.1(\text{Btu})/(\text{lb}_m)$$

Note that in (*a*) the entropy increased at constant enthalpy and in (*b*) the enthalpy decreased at constant entropy. Both of these processes are easily followed on a Mollier (H versus S) diagram.

4.19. Ten pounds mass of steam initially at a pressure P_1 of 20(psia) and a temperature T_1 of 300(°F) is compressed isothermally and reversibly in a piston-cylinder device to a final pressure such that the steam is just saturated (i.e. saturated vapor). Calculate Q and W for the process.

In its initial state the steam is a superheated vapor, whereas it is a saturated vapor at the end of the process. From the appropriate steam tables we have the following values:

$$\begin{aligned} T_1 &= 300 & T_2 &= 300(°\text{F}) \\ P_1 &= 20 & P_2 &= 66.98(\text{psia}) \\ V_1 &= 22.36 & V_2 &= 6.472(\text{ft})^3/(\text{lb}_m) \\ U_1 &= 1108.7 & U_2 &= 1100.0(\text{Btu})/(\text{lb}_m) \\ S_1 &= 1.7805 & S_2 &= 1.6356(\text{Btu})/(\text{lb}_m)(\text{R}) \end{aligned}$$

The first-law energy equation for a closed system is $\Delta U^t = Q - W$ or $m\,\Delta U = Q - W$. The value of ΔU is readily calculated:

$$\Delta U = U_2 - U_1 = 1100.0 - 1108.7 = -8.7(\text{Btu})/(\text{lb}_\text{m})$$

Note that this quantity is not zero as it would be for an ideal gas; it is, however, small.

It is possible to determine W by the integration

$$W = \int_{V_1}^{V_2} P\,dV$$

but this would require preparation of a table or graph of values of P and the corresponding values of V selected from the steam tables at the constant $T = 300(°\text{F})$, and subsequent numerical or graphical integration. Since ΔU is known, a much simpler procedure is to determine Q first and then to evaluate W by the first law: $W = Q - m\,\Delta U$.

The value of Q is obtained from the entropy change of the system. Since the process is reversible and T is constant, we have $\Delta S^t = Q/T$ or $Q = Tm\,\Delta S$. Since $T = 300 + 460 = 760(\text{R})$,

$$Q = 760(\text{R}) \times 10(\text{lb}_\text{m}) \times [1.6356 - 1.7805](\text{Btu})/(\text{lb}_\text{m})(\text{R}) = -1101.2(\text{Btu})$$

Substitution of numbers into the equation for W gives

$$W = -1101.2 - 10(-8.7) = -1014.2(\text{Btu})$$

Thus Q and W are not equal as they would be for an ideal gas.

4.20. Select data from the steam tables to show that for a phase transition from vapor to liquid they conform to the requirements of *(3.68)*, $(dG^t)_{T,P} \leqq 0$.

For a finite change of state at constant T and P *(3.68)* may be integrated to give $(\Delta G^t)_{T,P} \leqq 0$. The inequality applies to an irreversible process, whereas the equality applies to a reversible process, i.e. one where equilibrium conditions are always satisfied. Consider now 1(g) of water undergoing a phase change from saturated vapor to saturated liquid. At saturation the liquid and vapor phases are always at equilibrium and our equation becomes $\Delta G_{fg} = 0$.

Let us test the following data from the steam tables at 25(bar) and $T = 223.99(°\text{C})$ or 497.14(K):

$$H_f = 962.1 \qquad H_g = 2803.1(\text{J})/(\text{g})$$

$$S_f = 2.5547 \qquad S_g = 6.2575(\text{J})/(\text{g})(\text{K})$$

Since $G = H - TS$, we find

$$G_f = 962.1 - (497.14)(2.5547) = -307.9(\text{J})/(\text{g})$$

$$G_g = 2803.1 - (497.14)(6.2575) = -307.8(\text{J})/(\text{g})$$

Thus for all practical purposes ΔG_{fg} is zero. That it is not exactly zero results from the fact that steam tables and other tables of data are quite different from tables of logarithms and other mathematical functions, which can be made as accurate as one chooses. Tables of data depend on experimental measurements and on the correlation of results by empirical equations. Every effort is made to force conformity of the calculated results with the requirements of exact thermodynamic equations, but this is possible only within the limits of an inherent uncertainty that comes from the treatment of experimental measurements.

Now let us try another set of data, for which G_f and G_g should not be the same. Data are given in the steam tables for subcooled vapor, i.e. vapor cooled to an unstable state at a temperature below its saturation temperature. For example, data are given for subcooled vapor at 25(bar) and 220(°C). Since the saturation temperature is 223.99(°C), this vapor has been cooled almost 4(°C) below the temperature at which it would normally condense. The stable state at these conditions is liquid water, data for which are also available. Thus we have at 25(bar) and $T = 220(°\text{C})$ or 493.15(K):

$$H^{\text{vap}} = 2790.2(\text{J})/(\text{g}) \qquad S^{\text{vap}} = 6.2315(\text{J})/(\text{g})(\text{K})$$

$$H^{\text{liq}} = 943.7(\text{J})/(\text{g}) \qquad S^{\text{liq}} = 2.5174(\text{J})/(\text{g})(\text{K})$$

From these we calculate $G = H - TS$:

$$G^{vap} = 2790.2 - (493.15)(6.2315) = -282.9(\text{J})/(\text{g})$$

$$G^{liq} = 943.7 - (493.15)(2.5174) = -297.8(\text{J})/(\text{g})$$

These two states are at the same T and P, and the change that occurs from one to the other at this T and P is from the unstable state to the stable state, i.e. from vapor to liquid. Thus for this case

$$(\Delta G^t)_{T,P} = G^{liq} - G^{vap} = -14.9(\text{J})/(\text{g})$$

which conforms with the requirement that $(\Delta G^t)_{T,P} < 0$ for irreversible processes. Any change from an unstable state to a stable one is of course inherently irreversible.

4.21. Show how the saturation or vapor pressure of a spherical droplet of water at 25(°C) depends on the radius of the droplet. At 25(°C) the surface tension σ of water is $\sigma = 69.4(\text{dyne})/(\text{cm})$.

A free-body diagram (Fig. 4-20) showing the forces on a droplet hemisphere of radius r provides the basis for a force balance.

Force acting to the left as a result of the internal pressure P_i: $P_i(\pi r^2)$

Force acting to the right as a result of the external pressure P_e: $P_e(\pi r^2)$

Force acting to the right as a result of the surface tension at the surface of the droplet: $\sigma(2\pi r)$

Thus
$$P_i(\pi r^2) = P_e(\pi r^2) + \sigma(2\pi r)$$

or
$$P_i - P_e = 2\sigma/r$$

Using the conversion $1(\text{dyne})/(\text{cm})^2 = 10^{-6}(\text{bar})$, we calculate:

r(cm)	$P_i - P_e$(bar)
1	0.00014
0.01	0.0139
10^{-4}(1 micron)	1.388
0	∞

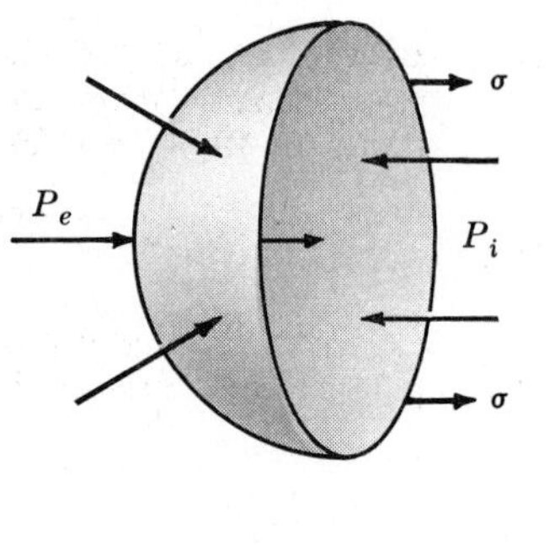

Fig. 4-20

The properties of the liquid in a droplet depend, of course, on the *interior* pressure, which is always higher than the exterior pressure. We have already developed equations which allow us to determine the influence of this pressure increase on the thermodynamic properties of the droplet. In particular, we have from (*3.35*) at constant T that

$$dG = V\,dP \qquad (\text{constant } T) \tag{1}$$

Consider a pool of liquid water with a flat surface at 25(°C) in equilibrium with water vapor at its saturation pressure P_0^{sat} of 0.0317(bar) as given by the steam tables. If a small amount of liquid is taken from this pool and formed into a droplet of 1-micron radius at 25(°C), there would be a pressure increase $P_i - P_e = P_i - P_0^{sat}$ of 1.388(bar) as determined above. This causes a change in the Gibbs function which we determine by integration of (*1*), with V taken to be constant for liquid water at $1.003(\text{cm})^3/(\text{g})$ for $T = 25(°\text{C})$. Thus

$$\Delta G = V(P_i - P_0^{sat}) = \frac{1.003(\text{cm})^3/(\text{g}) \times 1.388(\text{bar})}{10(\text{cm}^3\text{-bar})/(\text{J})} = 0.139(\text{J})/(\text{g})$$

If the droplet is to remain in equilibrium with water vapor, then the Gibbs function for the water vapor must also increase by exactly this amount, in accord with (*4.1*). Integration of (*1*) from P_0^{sat} to the new vapor pressure P^{sat}, with V for the vapor given by the ideal-gas law, gives

$$\Delta G = \int_{P_0^{sat}}^{P^{sat}} \frac{RT}{P}\,dP = RT \ln \frac{P^{sat}}{P_0^{sat}}$$

Noting that $\Delta P^{sat} = P^{sat} - P_0^{sat}$ is very small, we have

$$\Delta G = RT \ln \frac{P_0^{sat} + \Delta P^{sat}}{P_0^{sat}} = RT \ln\left(1 + \frac{\Delta P^{sat}}{P_0^{sat}}\right) \cong RT \frac{\Delta P^{sat}}{P_0^{sat}}$$

Substitution of numerical values gives

$$0.139(\text{J})/(\text{g}) \times 18(\text{g})/(\text{g mole}) = 8.314(\text{J})/(\text{g mole})(\text{K}) \times 298(\text{K}) \times \frac{\Delta P^{sat}}{0.0317}$$

from which

$$\Delta P^{sat} = 0.000032(\text{bar})$$

This increase of vapor pressure with total pressure is known as the Poynting effect.

Thus the saturation pressure of water vapor in equilibrium with liquid water droplets of 1-micron radius is higher by 0.000032(bar) or 0.024(mm Hg) than the value recorded in the steam tables. Moreover, the smaller the droplet, the greater this difference will be, approaching infinity as $r \to 0$. This raises the question of how a droplet such as a fog droplet or a rain drop could ever begin to form from water vapor. The answer is that they form only on dust particles, which provide a finite radius from which the droplet can grow. Another question is how a fog can be stable with respect to ground water, since the fog exerts a vapor pressure higher than does the ground water, and therefore the fog droplets should evaporate. The answer is that vapor pressure is a strong function of temperature, and a slight reduction of temperature will reduce the vapor pressure of a fog to that of the ground water. We can calculate the required temperature decrease from (*4.4*), the Clausius-Clapeyron equation, written

$$\frac{d \ln P^{sat}}{dT} = \frac{H_{fg}}{RT^2}$$

or approximately

$$\Delta T = \frac{RT^2 \, \Delta P^{sat}}{H_{fg} P_0^{sat}}$$

Here ΔP^{sat} is -0.000032(bar), the negative of the increase calculated above, because we seek the temperature change that *compensates* the Poynting effect. Taking the value of H_{fg} from the steam tables at 25(°C), we have

$$\Delta T = \frac{8.314(\text{J})/(\text{g mole})(\text{K}) \times 298^2(\text{K})^2 \times [-0.000032(\text{bar})]}{2442.3(\text{J})/(\text{g}) \times 18(\text{g})/(\text{g mole}) \times 0.0317(\text{bar})} = -0.016(\text{K})$$

or a temperature decrease of 0.016(°C). We note that the vapor pressure is also lowered by impurities dissolved in the liquid droplets, and fogs stabilized in this way become smogs.

Supplementary Problems

VAPOR PRESSURES AND LATENT HEATS (Sec. 4.3)

4.22. The vapor pressure of a certain pure liquid is given by the equation

$$\ln P^{sat} = 9.3781 - \frac{3456.80}{T - 78.67}$$

where P^{sat} is in atmospheres and T is in kelvins. Making the assumptions of Example 4.4, calculate a value for H_{fg} at 25(°C). Are these assumptions justified in this case?

Ans. $H_{fg} = 12{,}675$(cal)/(g mole) or 53,033(J)/(g mole). At 25(°C) P^{sat} is 0.001709(atm), a pressure so low that the assumptions made should introduce negligible error.

TWO-PHASE SYSTEMS (Sec. 4.4)

4.23. A rigid tank having a total volume of 100(ft)^3 contains 1(ft)^3 of saturated liquid water and 99(ft)^3 of saturated vapor water at 1(atm). How much heat must be added to the system so that the liquid is just vaporized?

Data: Initial properties are

$$P_1 = 1(\text{atm}) \quad \text{or} \quad 14.696(\text{psia}),$$

$$V_{f_1} = 0.016715, \quad V_{g_1} = 26.80(\text{ft})^3/(\text{lb}_m),$$

$$U_{f_1} = 180.1, \quad U_{g_1} = 1077.6(\text{Btu})/(\text{lb}_m)$$

Final properties are

$$P_2 = 294.15(\text{psia}) \quad (\text{calculated})$$

$$V_{g_2} = 1.5743(\text{ft})^3/(\text{lb}_\text{m}), \quad U_{g_2} = 1118.0(\text{Btu})/(\text{lb}_\text{m})$$

Ans. $Q = 56{,}260(\text{Btu})$

4.24. One kilogram of water, 20% vapor by weight, is contained in a rigid vessel at a pressure of 2(bar). It is heated until the liquid is just vaporized. Determine the final temperature and pressure, the property changes ΔU^t, ΔH^t, and ΔS^t for the kilogram of water, and the heat added to the system.

Data: Initial properties are

$$P_1 = 2(\text{bar}), \quad T_1 = 120.23(°\text{C})$$

$$V_{f_1} = 1.0605 \qquad V_{g_1} = 885.7(\text{cm})^3/(\text{g})$$

$$U_{f_1} = 504.5 \qquad U_{g_1} = 2529.5(\text{J})/(\text{g})$$

$$H_{f_1} = 504.7 \qquad H_{g_1} = 2706.7(\text{J})/(\text{g})$$

$$S_{f_1} = 1.5301 \qquad S_{g_1} = 7.1271(\text{J})/(\text{g})(\text{K})$$

Final properties are

$$P_2 = 10.97(\text{bar}) \quad (\text{calculated}) \qquad T_2 = 183.97(°\text{C})$$

$$U_{g_2} = 2586.3(\text{J})/(\text{g}), \quad H_{g_2} = 2781.6(\text{J})/(\text{g}), \quad S_{g_2} = 6.5545(\text{J})/(\text{g})(\text{K})$$

Ans. $\Delta U^t = 1{,}676{,}800(\text{J})$, $\Delta H^t = 1{,}836{,}500(\text{J})$, $\Delta S^t = 3905(\text{J})/(\text{K})$, $Q = 1{,}676{,}800(\text{J})$

HEAT CAPACITIES (Secs. 4.6 and 4.7)

4.25. (*a*) In the equation

$$\int_{T_1}^{T_2} C_P\,dT = C_{P_m}(T_2 - T_1)$$

where C_{P_m} is the appropriate mean heat capacity, show that when $C_P = \alpha + \beta T$, C_{P_m} is the heat capacity evaluated at the arithmetic average of T_1 and T_2.

(*b*) In the equation

$$\int_{T_1}^{T_2} C_P \frac{dT}{T} = C_{P_m} \ln \frac{T_2}{T_1}$$

show that when $C_P = \alpha + \beta T$, C_{P_m} is the heat capacity evaluated at the logarithmic mean of T_1 and T_2.

Note that α and β are constants and that the logarithmic mean of T_1 and T_2 is

$$\frac{T_2 - T_1}{\ln (T_2/T_1)}$$

4.26. Making use of the Clapeyron equation and of (*6*) of Problem 4.10 as applied to both saturated liquid and to saturated vapor, show that

$$C_g - C_f = \frac{dH_{fg}}{dT} - \frac{H_{fg}}{T}$$

PROPERTY VALUES FOR PVT SYSTEMS (Sec. 4.8)

4.27. A mass of saturated liquid water at a pressure of 1(bar) fills a container. The saturation temperature is 99.63(°C). Heat is added to the water until its temperature reaches 120(°C). If the volume of the container does not change, what is the final pressure?

Data: The average value of β between 100 and 120(°C) is $80.8 \times 10^{-5}(°\text{C})^{-1}$. The value of κ at 1(bar) and 120(°C) is $4.93 \times 10^{-5}\ (\text{bar})^{-1}$, and may be assumed independent of P. The volume of saturated liquid water at 1(bar) is $1.0432(\text{cm})^3/(\text{g})$.

Ans. 335(bar) [Note: It can be found from the steam tables for liquid that the pressure at 120(°C), where water has a specific volume of 1.0432, is 326(bar).]

4.28. The steam tables for superheated steam at 400(°F) and 200(psia) give:

$$V = 2.361(\text{ft})^3/(\text{lb}_m) \qquad H = 1210.8(\text{Btu})/(\text{lb}_m) \qquad S = 1.5600(\text{Btu})/(\text{lb}_m)(\text{R})$$

The enthalpy of water vapor at a pressure approaching zero and at 400(°F) is given as 1241.9(Btu)/(lb_m), and the entropy of water vapor as an ideal gas at 1(psia) and 400(°F) is given as 2.1721(Btu)/(lb_m)(R). Determine $\Delta V'$, $\Delta H'$, and $\Delta S'$ for steam at 400(°F) and 200(psia).

Ans. $\Delta V' = 0.199(\text{ft})^3/(\text{lb}_m)$, $\Delta H' = 31.1(\text{Btu})/(\text{lb}_m)$, $\Delta S' = 0.0281(\text{Btu})/(\text{lb}_m)(\text{R})$

4.29. Data for liquid mercury at 0(°C) and 1(atm) are:

$$V = 14.72(\text{cm})^3/(\text{g mole}) \qquad \beta = 181 \times 10^{-6}(\text{K})^{-1} \qquad \kappa = 3.94 \times 10^{-6}(\text{atm})^{-1}$$

Taking these values to be essentially independent of P, calculate:

(*a*) The pressure increase above 1(atm) at 0(°C) required to cause a 0.1% decrease in V. (For a 0.1% change, V may be considered "essentially" constant.)

(*b*) ΔU, ΔH, and ΔS for 1(g mole) of mercury for the change described in (*a*).

(*c*) Q and W if the change is carried out reversibly.

Ans. (*a*) 254(atm); (*b*) $\Delta U = -18.53(\text{J})$, $\Delta H = 360.1(\text{J})$, $\Delta S = -0.06857(\text{J})/(\text{K})$;
(*c*) $Q = -18.73(\text{J})$, $W = -0.20(\text{J})$

4.30. For a single-phase PVT system consisting of one mole (or a unit mass) of a pure material, (*3.33*) becomes $dH = T\,dS + V\,dP$. Starting with this equation, show that

$$(a)\quad (\partial H/\partial P)_S = V \qquad (b)\quad (\partial H/\partial T)_S = V(\partial P/\partial T)_S = C_P/\beta T$$

$$(c)\quad (\partial H/\partial V)_S = V(\partial P/\partial V)_S = -\gamma/\kappa \qquad (d)\quad \left(\frac{\partial H}{\partial S}\right)_P = T$$

$$(e)\quad \left(\frac{\partial H}{\partial S}\right)_T = T - V\left(\frac{\partial T}{\partial V}\right)_P = T - \frac{1}{\beta} \qquad (f)\quad \left(\frac{\partial H}{\partial S}\right)_V = T - V\left(\frac{\partial T}{\partial V}\right)_S = T + \frac{\gamma - 1}{\beta}$$

(*Hint*: See Example 3.10 and Problem 4.32.)

4.31. For a PVT system consisting of one mole (or a unit mass) of a pure material, show that

$$(a)\quad \left(\frac{\partial^2 G}{\partial T^2}\right)_P = \frac{-C_P}{T} \qquad (b)\quad \left(\frac{\partial^2 G}{\partial P^2}\right)_T = -\kappa V \qquad (c)\quad \frac{\partial^2 G}{\partial P\,\partial T} = \beta V$$

Show also that all three quantities become infinite during a phase transition.

4.32. For a single-phase PVT system consisting of one mole (or a unit mass) of a pure material, show that

$$(a)\quad \left(\frac{\partial V}{\partial T}\right)_S = -\frac{C_V\kappa}{\beta T} \qquad (b)\quad \left(\frac{\partial P}{\partial T}\right)_S = \frac{C_P}{V\beta T} \qquad (c)\quad \left(\frac{\partial V}{\partial P}\right)_S = -\frac{\kappa V}{\gamma}$$

Show that for an ideal gas these equations lead to the results of Example 2.5.

4.33. For a single-phase PVT system consisting of one mole (or a unit mass) of a pure material, the following table indicates various derivatives *at constant T*, $(\partial y/\partial x)_T$. Those given elsewhere are indicated by an equation or problem number. For the others the proper expression is listed, and is to be verified. In addition, show that all derivatives reduce to their proper values for an ideal gas.

$\partial y \downarrow / \partial x \rightarrow$	∂P	∂V
∂P	—	(3.18)
∂V	(3.18)	—
∂U	$-V(T\beta - P\kappa)$	$\frac{T\beta}{\kappa} - P$
∂H	$V(1-\beta T)$	$\frac{1}{\kappa}(T\beta - 1)$
∂S	$-\beta V$	$\frac{\beta}{\kappa}$
∂C_V	—	$\frac{T}{\kappa}\left[\left(\frac{\partial \beta}{\partial T}\right)_V - \frac{\beta}{\kappa}\left(\frac{\partial \kappa}{\partial T}\right)_V\right]$
∂C_P	Problem 4.9	—

MISCELLANEOUS PROBLEMS REQUIRING USE OF THE STEAM TABLES

4.34. A closed rigid vessel having a volume of $20(\text{ft})^3$ is filled with steam at 100(psia) and 600(°F). Heat is transferred from the steam until its temperature reaches 350(°F). Determine Q.

Data: Initial condition is superheated steam for which

$$V_1 = 6.216(\text{ft})^3/(\text{lb}_\text{m}), \quad U_1 = 1214.2(\text{Btu})/(\text{lb}_\text{m})$$

Final condition is found to be superheated steam at 74.88(psia) and 350(°F), with

$$V_2 = 6.216(\text{ft})^3/(\text{lb}_\text{m}), \quad U_2 = 1119.3(\text{Btu})/(\text{lb}_\text{m})$$

Ans. $Q = -305.3(\text{Btu})$

4.35. Steam at 300(°F) and 1(atm) is compressed isothermally in a reversible process until it reaches a final state of saturated liquid. Determine Q and W for $1(\text{lb}_\text{m})$ of steam.

Data: Initial properties are

$$U_1 = 1109.6(\text{Btu})/(\text{lb}_\text{m}), \quad S_1 = 1.8157(\text{Btu})/(\text{lb}_\text{m})(\text{R})$$

Final properties are

$$P_2 = 66.98(\text{psia}), \quad U_2 = 269.5(\text{Btu})/(\text{lb}_\text{m}), \quad S_2 = 0.4372(\text{Btu})/(\text{lb}_\text{m})(\text{R})$$

Ans. $Q = -1047.7(\text{Btu})/(\text{lb}_\text{m}), \quad W = -207.6(\text{Btu})/(\text{lb}_\text{m})$

Review Questions for Chapters 1 through 4

For each of the following statements indicate whether it is true or false.

______ 1. For a closed system the value of $\int P\,dV$ for the change of a gas from one given state to another is a constant regardless of the path so long as all processes are reversible.

______ 2. All ideal gases have the same molar heat capacity at constant pressure (C_P).

______ 3. The molar heat capacity at constant volume (C_V) of an ideal gas is independent of temperature.

______ 4. The molar heat capacity at constant pressure (C_P) of an ideal gas is independent of pressure.

______ 5. The enthalpy of an ideal gas is a function of temperature only.

______ 6. The entropy of an ideal gas is a function of temperature only.

______ 7. Work is *always* given by the integral $\int P\,dV$.

______ 8. The first law of thermodynamics requires that the total energy of any system be conserved within the system.

______ 9. For any gas at constant temperature as the pressure approaches zero, the product PV approaches zero.

______ 10. The energy of an isolated system must be constant.

______ 11. The entropy of an isolated system must be constant.

______ 12. The equation $PV^{\gamma} = \text{const.}$ is valid for any adiabatic process involving ideal gases.

______ 13. If a system undergoes a reversible adiabatic change of state, it is correct to say that the entropy of the system does not change.

______ 14. There is but a single degree of freedom for a three-phase PVT system at equilibrium made up of three nonreacting chemical species.

______ 15. For wet steam it is true in general that $V = V_g - x'V_{fg}$, where V is the specific volume of the mixture and x' is the mass fraction of liquid.

______ 16. If a given amount of an ideal gas undergoes a process during which $PV^2 = k$, where k is a constant, then it is also true that $T/P^{1/2} = k'$, where k' is another constant.

______ 17. If a system undergoes an *irreversible* change from an initial equilibrium state i to a final equilibrium state f, the entropy change of the surroundings must be less (algebraically) than it would be if the system changed from i to f reversibly.

______ 18. The heat capacity at constant volume of a single-component system consisting of liquid and vapor in equilibrium is infinite.

_____ 19. The heat capacity at constant pressure of a single-component system consisting of liquid and vapor in equilibrium is infinite.

_____ 20. At the critical point the internal energy of saturated liquid is equal to the internal energy of saturated vapor.

_____ 21. If a system undergoes a process during which its entropy does not change, it is necessarily true that the process is reversible and adiabatic.

_____ 22. Heat is *always* given by the integral $\int T\,dS$.

_____ 23. The equation $dH = T\,dS + V\,dP$ can be applied *only* to reversible processes.

_____ 24. It is true in general that $H_{fg} = TS_{fg}$, where T is the absolute temperature at which H_{fg} and S_{fg} are evaluated.

_____ 25. For any process the second law of thermodynamics requires that the entropy change of the system be zero or positive.

_____ 26. Cyclic processes get work from heat in defiance of the qualitative expression of the second law.

_____ 27. If stirring work is done on an ideal gas in a closed system during a constant-volume process, then $\delta Q \neq C_V\,dT$.

_____ 28. In view of the equation for a bar stressed at constant volume:

$$d\tilde{U} = T\,d\tilde{S} + \sigma\,d\epsilon$$

we may write

$$\left(\frac{\partial \sigma}{\partial \tilde{S}}\right)_\epsilon = \left(\frac{\partial T}{\partial \epsilon}\right)_{\tilde{S}}$$

_____ 29. The change in the Gibbs function for vaporization at constant T and P, i.e. G_{fg}, is always positive.

_____ 30. When a molten salt crystallizes, the atoms arrange themselves in a highly ordered lattice structure; since increasing order is associated with decreasing entropy, we must conclude that the entropy of the universe decreases as a result of this process.

_____ 31. If a saturated liquid undergoes a reversible adiabatic expansion to a lower pressure, some of the liquid will vaporize.

_____ 32. For any single-phase PVT system at constant pressure, it is always true that $\Delta H = \int_{T_1}^{T_2} C_P\,dT$.

_____ 33. The Hilsch vortex tube holds special interest for scientists and engineers because it operates in violation of the second law of thermodynamics.

_____ 34. Nuclear fission power plants present a special problem to society because they cause thermal pollution, whereas conventional steam power plants do not.

Ans.

1	2	3	4	5	6	7	8	9	10	11	12	13	14	15
F	F	F	T	T	F	F	F	F	T	F	F	T	F	T

16	17	18	19	20	21	22	23	24	25	26	27	28	29	30
T	F	F	T	T	F	F	F	T	F	F	T	T	F	F

31	32	33	34
T	T	F	F

For the following multiple-choice questions indicate your answers by the appropriate numbers: 1, 2, 3, or 4.

(a) An inventor claims to have devised an engine that produces 1200(Btu) of work while receiving 1000(Btu) of heat from a single heat reservoir during a complete cycle of the engine. Such an engine would violate:

(1) The first law

(2) The second law

(3) Both the first and second laws

(4) Neither the first nor the second law

(b) The inventor also claims to have constructed a device that rejects 100(Btu) of heat to a single heat reservoir while absorbing 100(Btu) of work during a single cycle of the device. This device violates:

(1), (2), (3), (4) as in part (a).

(c) A system is changed from a single initial equilibrium state to the same final equilibrium state by two different processes, one reversible, and one irreversible. Which of the following is true, where ΔS refers to the system?

(1) $\Delta S_{\text{irr}} = \Delta S_{\text{rev}}$

(2) $\Delta S_{\text{irr}} > \Delta S_{\text{rev}}$

(3) $\Delta S_{\text{irr}} < \Delta S_{\text{rev}}$

(4) No decision is possible with respect to (1), (2), and (3).

(d) For any process at all, the second law requires that the entropy change of the *system* be:

(1) Positive or zero

(2) Zero

(3) Negative or zero

(4) Positive or zero or negative, but does not say which.

(e) A hypothetical substance has the following volume expansivity and isothermal compressibility: $\beta = a/V$ and $\kappa = b/V$, where a and b are constants. The equation of state of such a substance would be:

(1) $V = aT + bP + \text{const.}$

(2) $V = aT - bP + \text{const.}$

(3) $V = bT + aP + \text{const.}$

(4) $V = bT - aP + \text{const.}$

(f) An exact differential expression relating thermodynamic variables is given by

$$dB = C\,dE - F\,dG + H\,dJ$$

Which of the following would *not* be a new thermodynamic function consistent with the above expression?

(1) $B - FG - CE$

(2) $B - CE$

(3) $B - HJ$

(4) $B - HJ + FG - CE$

(g) Given the same exact differential expression as in (f) we conclude that:

(1) $(\partial C/\partial G)_E = (\partial F/\partial E)_G$

(2) $(\partial C/\partial J)_{E,G} = (\partial H/\partial E)_{J,G}$

(3) $(\partial F/\partial G)_{E,J} = -(\partial E/\partial C)_{G,J}$

(4) None of the above

(h) For a PVT system, the expression $T\left(\frac{\partial S}{\partial T}\right)_P - T\left(\frac{\partial S}{\partial T}\right)_V$ is *always* equal to

(1) Zero

(2) $\gamma = C_P/C_V$

(3) R

(4) $T\left(\frac{\partial P}{\partial T}\right)_V\left(\frac{\partial V}{\partial T}\right)_P$

(i) The expression $\left(\frac{\partial P}{\partial V}\right)_T\left(\frac{\partial T}{\partial P}\right)_S\left(\frac{\partial S}{\partial T}\right)_P$ is equivalent to

(1) $(\partial S/\partial V)_T$

(2) $(\partial P/\partial T)_V$

(3) $(\partial V/\partial T)_S$

(4) $-(\partial P/\partial T)_V$

(j) A system consisting of a liquid phase and a vapor phase in equilibrium contains three chemical species: water, ethanol, and methanol. The number of degrees of freedom for the system is:

(1) Zero

(2) One

(3) Two

(4) Three

Ans.

a	b	c	d	e	f	g	h	i	j
(3)	(4)	(1)	(4)	(2)	(1)	(2)	(4)	(4)	(4)

Chapter 5

Equations of State and Corresponding-States Correlations for PVT Systems

It was noted in Chapter 4 that the usefulness of thermodynamic property relations depends on the availability of a certain amount of data. In fact, data for a minimum number of properties must be provided as input to the network of thermodynamic equations if they are to yield any quantitative output. Such data may be stored and displayed as tables or graphs, but the most concise and generally useful kind of representation is through mathematical *equations of state.*

For a thermodynamic system (not necessarily a PVT system) characterized by *n independent* variables, we define an equation of state to be an algebraic expression connecting $n+1$ state variables. Such an equation describes the inherent material behavior of a system or substance, and is an example of what in continuum mechanics is called a constitutive equation.

As a practical matter, the material behavior of real systems is often too complicated to be faithfully described by equations of convenient simplicity, except over very limited ranges of the state variables. As a result, the relatively simple expressions which usually serve as equations of state describe models of material behavior which approximate real behavior more or less well depending on the material and on the conditions. Thus the equation of state represented by the ideal-gas law $PV = RT$ correctly describes real-gas behavior only in the limit as $P \to 0$, although it is an exact expression for a model gas made up of molecules which have no volume and which exert no forces on one another.

Strictly, any $n+1$ state variables may appear in an equation of state. However, we will restrict ourselves in this chapter to equations which apply to PVT systems and which relate the measurable variables P, V, T, and composition. Such equations are sometimes called *thermal equations of state.*

5.1 THE COMPRESSIBILITY FACTOR

It was stated in Section 4.8 that the residual volume $\Delta V'$ is a convenient function with which to describe the volumetric behavior of PVT systems, because although V can take on an enormous range of values, $\Delta V'$ is observed to vary over a much more limited range. The use of $\Delta V'$ still presents a minor disadvantage, however, in that as shown by Problem 4.14 it remains finite for real gases in the limit as $P \to 0$, even though for an ideal gas $\Delta V'$ is identically zero. Moreover, the value of $\Delta V'$ at zero pressure is different for different gases, and is in addition a function of temperature.

This difficulty can be avoided by employing a different volumetric auxiliary function, the *compressibility factor Z.* Whereas $\Delta V'$ was defined as the difference between V' and V at the same T and P, Z is defined as the ratio of V to V' at the same T and P: $Z = V/V'$. But $V' = RT/P$, so that

$$\boxed{Z = \frac{PV}{RT}} \tag{5.1}$$

According to (5.1), $Z = 1$ identically for an ideal gas. From the definition (4.16) of $\Delta V'$, Z can also be written

$$Z = 1 - \frac{P\,\Delta V'}{RT}$$

or

$$Z = 1 - \frac{\Delta V'}{V + \Delta V'}$$

It follows from the above two equations that, if $\Delta V'$ remains finite as $P \to 0$ (or $V \to \infty$), then Z also approaches unity for a real gas in the limit as $P \to 0$ (or $V \to \infty$). The existence of a universal limiting value of Z for all real gases at zero pressure is a major motivation for its use. As with $\Delta V'$, Z also exhibits a much more limited range of values than does V, and thus is a convenient function for representation of the volumetric properties of real fluids. Since Z is a state function and is defined in terms of T, P, and V, equations for Z as functions of P, T, and composition (or V, T, and composition, or P, V, and composition) constitute thermal equations of state.

Example 5.1. By using volumetric data from the steam tables, calculate and plot Z for water as a function of P for several isotherms. Show the vapor-liquid saturation region, the low-pressure gas region, the critical isotherm, and extend a few isotherms to 1000(bar), the maximum steam-table pressure for the vapor phase.

Values of Z are calculated from (5.1), and the results are shown on Figs. 5-1 and 5-2. Figure 5-1 covers the pressure range 0 to 1000(bar) and shows the saturation region and selected isotherms from 300(°C) to 1300(°C), the upper temperature limit of the tables. Figure 5-2 shows vapor-phase Z's from 0 to 4(bar), and at temperatures from 80(°C) to 1300(°C).

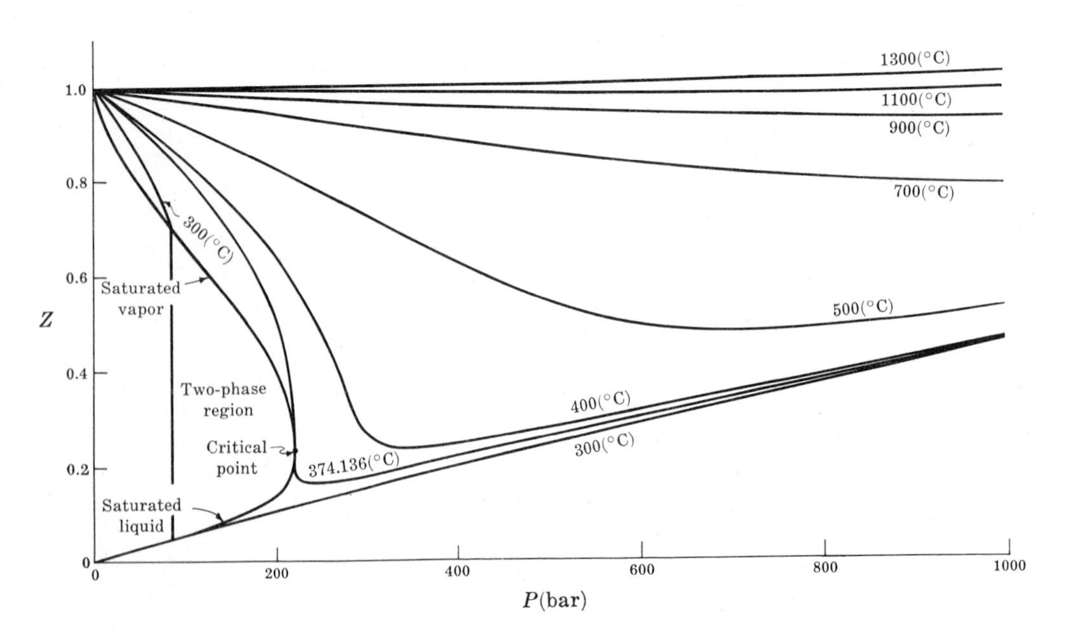

Fig. 5-1

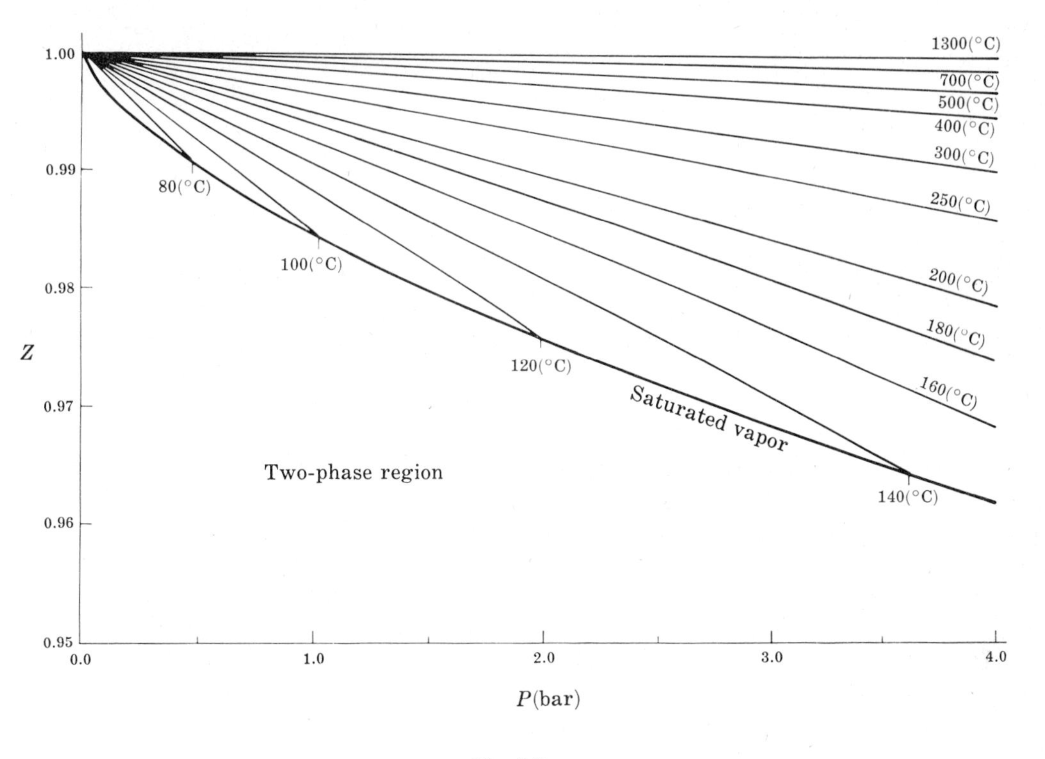

Fig. 5-2

Although the behavior of water differs in detail from that of other substances, the gross characteristics of Figs. 5-1 and 5-2 are representative of many materials. Typical features of most ZP diagrams include (*a*) the relatively small area occupied by the two-phase region, (*b*) the pronounced hook in the critical isotherm [374.136(°C) on Fig. 5-1], (*c*) the gradual flattening with increasing temperature of the supercritical isotherms at moderate pressures, and (*d*) the increase of slope (in the algebraic sense) with increasing temperature of the isotherms at zero pressure. A feature of *all* ZP diagrams is the trend toward linearity of the isotherms at low pressures, shown more clearly in Fig. 5-2. This behavior suggests that at low pressures simple polynomial expressions for Z in terms of P, containing only a few terms, should suffice as empirical equations of state.

Example 5.2. Find expressions for $(\partial V/\partial T)_P$, $(\partial V/\partial P)_T$, and $(\partial P/\partial T)_V$ in terms of T, P, and Z and its derivatives.

We first find the total differential dZ by logarithmic differentiation of the definition (*5.1*):

$$dZ = \frac{Z}{P}dP + \frac{Z}{V}dV - \frac{Z}{T}dT \tag{5.2}$$

Division of (*5.2*) by dT and restriction to constant P gives

$$\left(\frac{\partial Z}{\partial T}\right)_P = \frac{Z}{V}\left(\frac{\partial V}{\partial T}\right)_P - \frac{Z}{T} = \frac{P}{RT}\left(\frac{\partial V}{\partial T}\right)_P - \frac{Z}{T}$$

from which

$$\left(\frac{\partial V}{\partial T}\right)_P = \frac{RZ}{P} + \frac{RT}{P}\left(\frac{\partial Z}{\partial T}\right)_P \tag{5.3}$$

Similarly,

$$\left(\frac{\partial Z}{\partial P}\right)_T = \frac{Z}{V}\left(\frac{\partial V}{\partial P}\right)_T + \frac{Z}{P} = \frac{P}{RT}\left(\frac{\partial V}{\partial P}\right)_T + \frac{Z}{P}$$

from which

$$\left(\frac{\partial V}{\partial P}\right)_T = -\frac{ZRT}{P^2} + \frac{RT}{P}\left(\frac{\partial Z}{\partial P}\right)_T \tag{5.4}$$

The remaining derivative follows from (*5.3*), (*5.4*), and the relationship

$$\left(\frac{\partial P}{\partial T}\right)_V = -\left(\frac{\partial P}{\partial V}\right)_T\left(\frac{\partial V}{\partial T}\right)_P = -\left(\frac{\partial V}{\partial T}\right)_P \Big/ \left(\frac{\partial V}{\partial P}\right)_T$$

Thus

$$\left(\frac{\partial P}{\partial T}\right)_V = \frac{\dfrac{Z}{T} + \left(\dfrac{\partial Z}{\partial T}\right)_P}{\dfrac{Z}{P} - \left(\dfrac{\partial Z}{\partial P}\right)_T} \qquad (5.5)$$

Example 5.3. Considering Z a function of T and P, derive equations giving the residual enthalpy $\Delta H'$ and the residual entropy $\Delta S'$ in terms of T, P, and Z and its derivatives.

We use (*4.14*) for $\Delta H'$:

$$\Delta H' = \int_0^P \left[T\left(\frac{\partial V}{\partial T}\right)_P - V\right] dP \qquad \text{(const. } T) \qquad (4.14)$$

Combination of (*5.1*), (*5.3*), and (*4.14*) gives

$$\Delta H' = \int_0^P \left\{T\left[\frac{RZ}{P} + \frac{RT}{P}\left(\frac{\partial Z}{\partial T}\right)_P\right] - \frac{ZRT}{P}\right\} dP \qquad \text{(const. } T)$$

or

$$\frac{\Delta H'}{RT} = T\int_0^P \left(\frac{\partial Z}{\partial T}\right)_P \frac{dP}{P} \qquad \text{(const. } T) \qquad (5.6)$$

The desired expression for $\Delta S'$ follows similarly from combination of (*5.3*) and (*4.15*):

$$\frac{\Delta S'}{R} = T\int_0^P \left(\frac{\partial Z}{\partial T}\right)_P \frac{dP}{P} + \int_0^P (Z-1)\frac{dP}{P} \qquad \text{(const. } T) \qquad (5.7)$$

Note that we have expressed (*5.6*) and (*5.7*) in dimensionless form.

5.2 VIRIAL EQUATIONS OF STATE

Since $Z = 1$ for any gas at $P = 0$ (or $1/V = 0$), empirical description of the vapor state is conveniently accomplished by considering Z a function of T and P (or T and $1/V$) and expanding Z in a power series about $P = 0$ (or $1/V = 0$):

$$Z = 1 + B'P + C'P^2 + D'P^3 + \cdots \qquad (5.8)$$

$$Z = 1 + \frac{B}{V} + \frac{C}{V^2} + \frac{D}{V^3} + \cdots \qquad (5.9)$$

Equations (*5.8*) and (*5.9*) are called *virial equations of state*, or virial equations, and the coefficients B, B', C, C', etc., are called *virial coefficients*. Thus B and B' are second virial coefficients, C and C' are third virial coefficients, etc. The virial coefficients of pure gases are functions of temperature only; for mixtures they are functions both of temperature and composition. Comparison of the two infinite series (*5.8*) and (*5.9*) shows that the coefficients are simply related (see Problem 5.6):

$$B' = \frac{B}{RT}$$

$$C' = \frac{C - B^2}{(RT)^2} \qquad (5.10)$$

$$D' = \frac{D + 2B^3 - 3BC}{(RT)^3}$$

etc.

Example 5.4. We wish to show how virial coefficients can in practice be obtained from PVT data. Consider the volumetric virial equation (*5.9*). Rearrangement of this equation gives

$$(Z-1)V = B + \frac{C}{V} + \frac{D}{V^2} + \cdots \tag{5.11}$$

Equation (5.11) asserts that an isothermal plot of $(Z-1)V$ versus $1/V$ should yield B for that temperature as the intercept at $1/V = 0$:

$$B = \lim_{1/V \to 0} [(Z-1)V]_T \tag{5.12}$$

In addition, C is obtained from this plot as the limiting slope at $1/V = 0$:

$$C = \lim_{1/V \to 0} \left[\frac{\partial(Z-1)V}{\partial(1/V)}\right]_T$$

C can be found by an alternative method. Rearrangement of (5.11) gives

$$[(Z-1)V - B]V = C + \frac{D}{V} + \cdots \tag{5.13}$$

If B is known, say from (5.12), then (5.13) shows that C is equal to the intercept at $1/V = 0$ on a $[(Z-1)V - B]V$ versus $1/V$ plot:

$$C = \lim_{1/V \to 0} \{[(Z-1)V - B]V\}_T$$

and D is given by the limiting slope at $1/V = 0$:

$$D = \lim_{1/V \to 0} \left\{\frac{\partial[(Z-1)V - B]V}{\partial(1/V)}\right\}_T$$

We illustrate the use of the above equations by applying them to the volumetric data for water vapor at 260(°C) given in the steam tables.

Figure 5-3 is the $(Z-1)V$ versus $1/V$ plot. The points shown represent pressures ranging from 0.25(bar) to the saturation pressure of 46.88(bar). On the scale of this graph, there is a substantial amount of scatter up to a density of about 20×10^{-5}(g mole)/(cm)^3, but a much smoother trend at higher densities. The curve drawn through the points yields an intercept of -142.2(cm)^3/(g mole), and by (5.12) this is the second virial coefficient B. Depending on how the limiting tangent to the curve at $1/V = 0$ is drawn, values for C between about -6500 and -7500(cm)^6/(g mole)^2 are obtained.

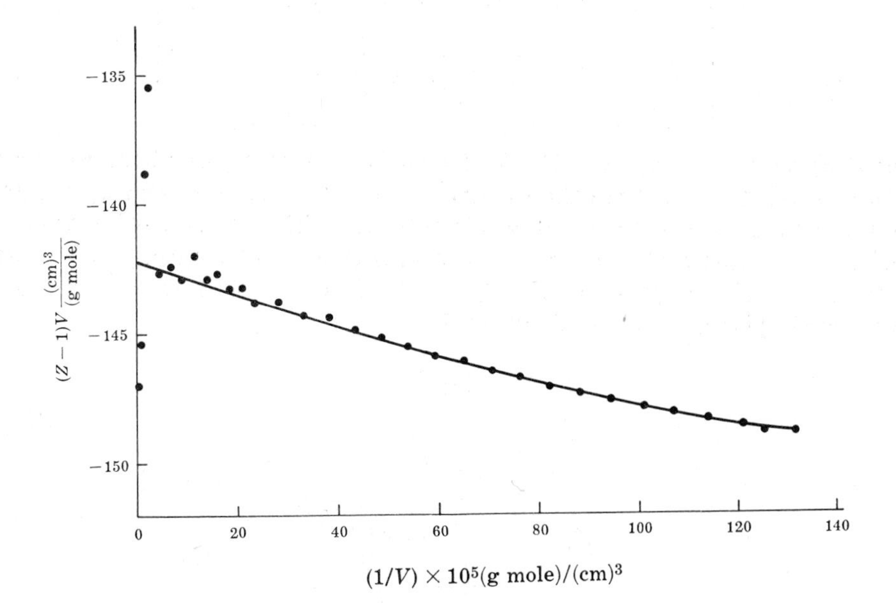

Fig. 5-3

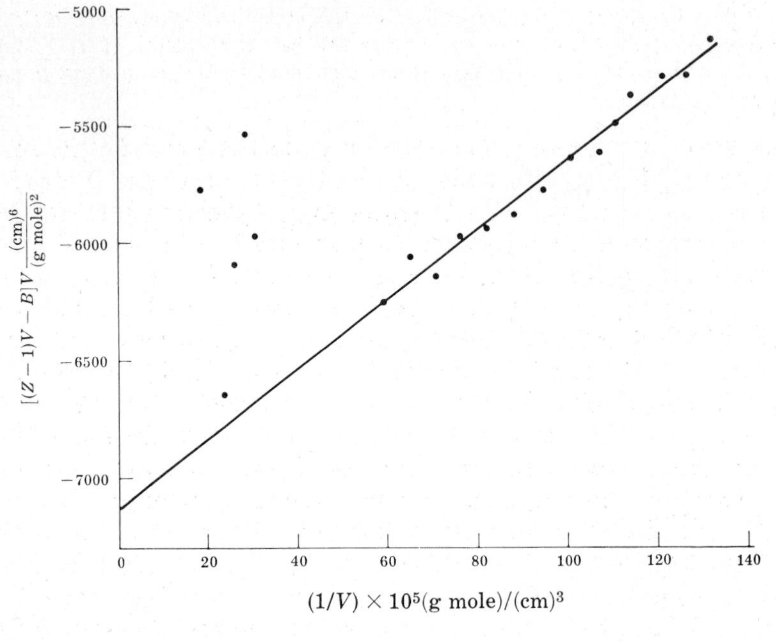

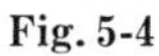
Fig. 5-4

Another estimate for C results from the $[(Z-1)V-B]V$ versus $1/V$ plot, shown in Fig. 5-4. The B of -142.2(cm)^3/(g mole) from Fig. 5-3 was used to calculate these points. Very significant scatter is obtained up to a density of about 50×10^{-5}(g mole)/(cm)^3, with many of the points off scale on Fig. 5-4. Above this density, a linear trend is observed, with less scatter. At 260(°C) and below 50×10^{-5}(g mole)/(cm)^3, Z for water vapor is insensitive to C within the range of values covered by the scatter, so for purposes of determining the intercept an extrapolation (as shown) from higher densities is entirely satisfactory. This extrapolation gives a C of -7140(cm)^6/(g mole)^2, and the slope of the line yields $D = 1.51 \times 10^6$(cm)^9/(g mole)^3. To summarize:

$$B = -142.2(\text{cm})^3/(\text{g mole}) \qquad C = -7140(\text{cm})^6/(\text{g mole})^2 \qquad D = 1.51 \times 10^6(\text{cm})^9/(\text{g mole})^3$$

for water vapor at 260(°C). No indication has been given of the uncertainties in these coefficients, but we note that the determination of the latter two, particularly D, is subject to rather large errors. Estimates of the higher virial coefficients cannot be made from the steam table data, but it is apparent from Figs. 5-3 and 5-4 that coefficients beyond D are unnecessary at this temperature.

It is instructive to compare the relative contributions to Z of the terms containing B, C, and D, and we do this in the table below. Also shown are volumes calculated with the derived virial coefficients and the corresponding volumes from the steam tables.

P(bar)	B/V	C/V^2	D/V^3	$V_{\text{calculated}}$ (cm)^3/(g)	$V_{\text{steam tables}}$ (cm)^3/(g)
0.01	−0.00003	-4×10^{-10}	2×10^{-14}	246,046	246,046
0.1	−0.00032	-4×10^{-8}	2×10^{-11}	24,598	24,598
1	−0.00322	-4×10^{-6}	2×10^{-8}	2452.6	2453
10	−0.03318	−0.00039	0.00002	237.80	237.8
25	−0.08816	−0.00274	0.00036	89.511	89.51
46.88	−0.18695	−0.01234	0.00343	42.206	42.21

Up to a pressure of about 5(bar), the virial equation truncated to two terms gives essentially a perfect fit to the 260(°C) steam-table data, and even at 25(bar) the contribution of the C and D terms to Z is only about 0.4%. This is why the extrapolation to $1/V = 0$ from higher densities in Fig. 5-4 was sufficient for determination of C. The excellent agreement between calculated and steam-table volumes merely illustrates the appropriateness of a four-term polynomial in $1/V$ as a curve-fitting device for this isotherm.

Although we have justified (*5.8*) and (*5.9*) on empirical grounds, the real utility of the virial equations derives from the fact that the coefficients also have theoretical significance. If we picture a gas as consisting of a large number of discrete particles (molecules), then we find that the macroscopic PVT behavior is determined on the microscopic scale by interactions among molecular force fields. At least two factors determine the effects of these molecular interactions: (1) the nature of the molecules themselves, which determines the types of forces in action, and (2) the intermolecular separations. At very large intermolecular separations (low densities), the molecules exert negligible forces on each other, and the gas approaches ideal behavior ($Z = 1$). As the density of the system is increased, the average distance between molecules decreases, and intermolecular interactions become increasingly important. If we view the overall behavior of the collection of molecules as a combination of effects due to interactions among various average *numbers* of molecules, we conclude that at very low densities only "one-body" interactions are significant (i.e., the molecules act independently of each other), that at slightly higher densities "two-body" interactions must also be considered (i.e., interactions involving molecules taken two at a time are important), that at still higher densities "three-body" interactions are also important, etc. Statistical mechanics provides the interpretation of the terms of the virial equation B/V, C/V^2, etc., as representing the successive contributions of N-body interactions to the deviation of a real gas from ideal-gas behavior, and in fact provides recipes for the calculation of B, C, etc., from mechanical models of the intermolecular force fields. The most important result for our purposes is that statistical mechanics also yields *exact* expressions for the composition dependence of the virial coefficients for mixtures. In fact the virial equation is the only equation of state for which rigorous "mixing rules" are available. The mixing rules for an m-component mixture are

$$B = \sum_{i=1}^{m} \sum_{j=1}^{m} y_i y_j B_{ij} \tag{5.14}$$

$$C = \sum_{i=1}^{m} \sum_{j=1}^{m} \sum_{k=1}^{m} y_i y_j y_k C_{ijk} \tag{5.15}$$

etc.

where the y's are mole fractions. The coefficients B_{ij}, C_{ijk}, etc., are functions of temperature only, and have the properties

$$\begin{gathered} B_{ij} = B_{ji} \\ C_{ijk} = C_{ikj} = C_{jik} = C_{jki} = C_{kij} = C_{kji} \end{gathered} \tag{5.16}$$

etc.

When the series formulae (*5.14*) and (*5.15*) are expanded in full, two types of coefficients appear in the result: those in which the successive subscripts are identical, and those for which at least one of the subscripts differs from the others. The first kind of coefficient refers to a pure component; the second type, called a *cross-coefficient*, is a mixture property.

Example 5.5. Write (*5.14*) and (*5.15*) in expanded form for a binary mixture.

Equation (*5.14*) becomes

$$B = \sum_{i=1}^{2} \sum_{j=1}^{2} y_i y_j B_{ij}$$

$$= y_1^2 B_{11} + y_1 y_2 B_{12} + y_2 y_1 B_{21} + y_2^2 B_{22}$$

But by (*5.16*), $B_{12} = B_{21}$; therefore

$$B = y_1^2 B_{11} + 2y_1 y_2 B_{12} + y_2^2 B_{22} \tag{5.17}$$

Similarly,

$$C = \sum_{i=1}^{2} \sum_{j=1}^{2} \sum_{k=1}^{2} y_i y_j y_k C_{ijk}$$

$$= y_1^3 C_{111} + y_2 y_1^2 C_{211} + y_1^2 y_2 C_{121} + y_1 y_2^2 C_{221} + y_1^2 y_2 C_{112} + y_2^2 y_1 C_{212} + y_1 y_2^2 C_{122} + y_2^3 C_{222}$$

But by (*5.16*)

$$C_{211} = C_{121} = C_{112}$$

$$C_{221} = C_{212} = C_{122}$$

Therefore

$$C = y_1^3 C_{111} + 3y_1^2 y_2 C_{112} + 3y_1 y_2^2 C_{221} + y_2^3 C_{222}$$

If we repeat Example 5.5 for a ternary vapor mixture, we find that the expanded equation for B includes the additional coefficients B_{33}, B_{13}, and B_{23}. However, these coefficients can all be determined from data on pure components and binary mixtures, and *in general* only pure-component and binary data are required to calculate a mixture B. To generalize, the Nth virial coefficient for a mixture containing any number of components is a function of coefficients calculable from measurements on systems containing at most N components. This useful property derives from the statistical mechanical significance of the Nth virial coefficient as deriving from N-body molecular interactions. Thus B_{11} and B_{22} take into account the averaged effects of two-body interactions between identical molecules, whereas B_{12} similarly accounts for interactions between pairs of unlike molecules.

5.3 TRUNCATED FORMS OF THE VIRIAL EQUATIONS

It is evident from Fig. 5-2 and from the numerical comparisons presented in Example 5.4 that for low pressures either of the virial equations truncated after two terms

$$Z = 1 + B'P \tag{5.18}$$

or

$$Z = 1 + \frac{B}{V} \tag{5.19}$$

should in general provide a good representation of isothermal volumetric data. These equations do in fact produce comparable results for gases at low pressures.

At higher pressures one must employ the three-term virial equations:

$$Z = 1 + B'P + C'P^2 \tag{5.20}$$

$$\boxed{Z = 1 + \frac{B}{V} + \frac{C}{V^2}} \tag{5.21}$$

Comparison of the curve-fitting capabilities of (*5.20*) and (*5.21*) shows that (*5.21*) generally provides an adequate representation of the data to considerably higher pressures than does (*5.20*). Experimental volumetric data are therefore usually fitted to (*5.21*), producing values for the coefficients B and C, rather than values of B' and C'.

In order to combine the computational convenience of (*5.18*) for low pressures with the availability of values for B rather than B', substitution is made in (*5.18*) for B' as given by

(5.10), i.e. $B' = B/RT$. Thus (5.18) becomes

$$\boxed{Z = 1 + \frac{BP}{RT}} \tag{5.22}$$

and this is the preferred form of the virial equation for the low-pressure region.

An approximation has been made in passing from (5.18) to (5.22), because the relations of (5.10) are strictly valid only when they relate to the two *infinite* series given by (5.8) and (5.9). However, this approximation introduces negligible error for most practical purposes, and (5.22) finds very extensive use.

Example 5.6. Compare the expressions for molar volume obtained by solving (5.19) and (5.22) for V.

Since $Z = PV/RT$, (5.19) is quadratic in volume. Solution for V gives

$$V = \frac{RT}{2P}\left(1 + \sqrt{1 + \frac{4BP}{RT}}\right) \tag{5.23}$$

The plus sign before the radical is chosen because for $B = 0$, we must obtain the ideal-gas volume $V = RT/P$. Solving (5.22) for V, we obtain

$$V = \frac{RT}{P} + B \tag{5.24}$$

We note that for sufficiently small values of the argument BP/RT, (5.23) and (5.24) become nearly identical, because

$$\sqrt{1 + \frac{4BP}{RT}} \cong 1 + \frac{1}{2}\left(\frac{4BP}{RT}\right)$$

for small BP/RT. A numerical example of the practical equivalence of (5.23) and (5.24) at low pressures can be given for water vapor at 260(°C) with the second virial coefficient derived in Example 5.4. The results are given in the table below, and show that even at a pressure of 10(bar) a difference of only 0.1% results in the calculated volumes. The pressure level at which significant differences in the calculated volumes appear will of course depend in general on the magnitudes of B and T.

	V(cm)^3/(g mole)	
P(bar)	Equation (5.23)	Equation (5.24)
0.01	4,433,752	4,433,752
0.1	443,247	443,247
1.0	44,196.3	44,196.7
10.0	4,286.8	4,291.7

Example 5.7. Derive expressions for the residual enthalpy and the residual entropy of a gas described by the truncated virial equation of state (5.22).

We use the results of Example 5.3:

$$\frac{\Delta H'}{RT} = T\int_0^P \left(\frac{\partial Z}{\partial T}\right)_P \frac{dP}{P} \qquad \text{(const. } T) \tag{5.6}$$

$$\frac{\Delta S'}{R} = T\int_0^P \left(\frac{\partial Z}{\partial T}\right)_P \frac{dP}{P} + \int_0^P (Z-1)\frac{dP}{P} \qquad \text{(const. } T) \tag{5.7}$$

Substituting

$$Z = 1 + \frac{BP}{RT} \qquad \left(\frac{\partial Z}{\partial T}\right)_P = \frac{P}{R}\left[\frac{1}{T}\frac{dB}{dT} - \frac{B}{T^2}\right]$$

and carrying out the integrations, we obtain

$$\boxed{\frac{\Delta H'}{RT} = \frac{P}{R}\left(\frac{dB}{dT} - \frac{B}{T}\right)} \tag{5.25}$$

$$\boxed{\frac{\Delta S'}{R} = \frac{P}{R}\frac{dB}{dT}} \tag{5.26}$$

Note again the dimensionless forms in which (*5.25*) and (*5.26*) are expressed.

5.4 EMPIRICAL EQUATIONS OF STATE

Although either of the virial equations (*5.8*) or (*5.9*) can in principle be used to fit gas-phase isotherms to any required accuracy, experimentally determined virial coefficients beyond the third are scarce because very extensive and precise PVT data are required for their evaluation. As a result (*5.21*), which is generally valid only up to densities of the order of the critical density, is for all practical purposes the best one can do with the virial equations. Unfortunately, simultaneous description of both the liquid and vapor portions of subcritical isotherms cannot be accomplished with a truncated virial equation and a single set of virial coefficients. Therefore, prediction and correlation of the volumetric properties of fluids at high densities and in the liquid region is usually done with empirical equations of state.

Literally scores of such equations have been proposed. They range from relatively simple expressions containing a few arbitrary constants to complex equations suitable only for computerized calculations and involving as many as twenty or more constants. We shall not consider these complex equations, but will restrict ourselves in this outline to the simplest realistic class of empirical equations of state: those containing only two constants. The prototype for all equations of this kind is the *van der Waals equation*:

$$Z = \frac{V}{V-b} - \frac{a}{RTV} \tag{5.27}$$

A modern modification of the van der Waals equation is the *Redlich-Kwong equation of state*:

$$Z = \frac{V}{V-b} - \frac{a}{RT^{3/2}(V+b)} \tag{5.28}$$

In both (*5.27*) and (*5.28*), a and b are taken as positive constants (different for the two equations) which vary from substance to substance. For mixtures, the following empirical mixing rules are employed:

$$a = \sum_{i=1}^{m}\sum_{j=1}^{m} y_i y_j a_{ij} \tag{5.29}$$

$$b = \sum_{i=1}^{m} y_i b_i \tag{5.30}$$

The cross-coefficients in (*5.29*) are usually assumed to follow the combination rule

$$a_{ij} = \sqrt{a_i a_j} \tag{5.31}$$

Equations (*5.29*), (*5.30*), and (*5.31*) are taken to hold for any mixture containing any number of components. Unlike (*5.14*) and (*5.15*), however, they have little theoretical basis, and occasionally lead to mixture property estimates in very poor agreement with experiment. They are most realistic for calculation of properties of mixtures containing components of similar chemical nature.

No equation of state is known that with fixed values for its constants can describe the PVT behavior of a substance throughout the complete range of temperatures and pressures of practical interest. Thus if a and b in (*5.27*) or (*5.28*) are regarded as curve-fitting coefficients, then their values clearly depend on the set of data used to derive them. It is possible to arrive at optimal values of a and b for a particular range of T and P through numerical regression techniques which provide the best fit of a set of experimental data points. Alternatively, one can determine values of a and b, as shown in Example 5.8, by requiring the equation of state to satisfy two arbitrarily chosen conditions.

Example 5.8. We wish to find the values of a and b for the van der Waals equation which result when the following two conditions which apply at the critical point (see Section 4.1) are satisfied:

$$\left(\frac{\partial P}{\partial V}\right)_{T,\,cr} = 0 \tag{5.32}$$

$$\left(\frac{\partial^2 P}{\partial V^2}\right)_{T,\,cr} = 0 \tag{5.33}$$

First we solve (*5.27*) for P:

$$P = \frac{RT}{V-b} - \frac{a}{V^2} \tag{5.34}$$

from which, according to (*5.32*) and (*5.33*),

$$\left(\frac{\partial P}{\partial V}\right)_{T,\,cr} = \frac{-RT_c}{(V_c-b)^2} + \frac{2a}{V_c^3} = 0 \qquad \left(\frac{\partial^2 P}{\partial V^2}\right)_{T,\,cr} = \frac{2RT_c}{(V_c-b)^3} - \frac{6a}{V_c^4} = 0$$

where T_c and V_c are the critical temperature and critical volume. Solution of these last equations for a and b gives

$$a = \frac{9}{8}RT_cV_c \qquad b = \frac{V_c}{3} \tag{5.35}$$

Equations (*5.35*) show that values for a and b can be determined for any real fluid from experimentally determined values for T_c and V_c. We note however that combination of (*5.35*) with (*5.34*) as written to apply at the critical point will generally yield an incorrect value for P_c (or Z_c):

$$P_c = \frac{RT_c}{V_c-b} - \frac{a}{V_c^2} = \frac{RT_c}{V_c - V_c/3} - \frac{9/8RT_cV_c}{V_c^2} = \frac{3}{8}\frac{RT_c}{V_c}$$

or

$$Z_c = 3/8 \tag{5.36}$$

Thus the van der Waals equation gives $Z_c = 0.375$ for *all* fluids if a and b are determined from (*5.32*) and (*5.33*). But Z_c is known to vary from substance to substance, and in fact is usually between about 0.23 and 0.29, or about 30% lower than the van der Waals Z_c. Similarly, *all* two-constant equations of state will give a single value of Z_c if the constants are determined from (*5.32*) and (*5.33*); the value of Z_c so determined will, however, be different for different equations of state.

Example 5.9. Calculate and plot on pressure-volume coordinates several van der Waals isotherms. It is convenient to use *reduced* (or *generalized*) coordinates. These are designated T_r, P_r, and V_r, and are dimensionless quantities normalized with respect to T_c, P_c, and V_c:

$$T_r = \frac{T}{T_c} \qquad P_r = \frac{P}{P_c} \qquad V_r = \frac{V}{V_c}$$

When written in terms of these variables, empirical equations of state usually assume particularly simple forms. Thus, when (*5.34*), (*5.35*), and (*5.36*) are combined, we obtain

$$P = \frac{RT}{V - \dfrac{V_c}{3}} - \frac{9/8RT_cV_c}{V^2}$$

from which

$$\frac{P}{P_c} = \frac{3}{8}\frac{RT_c}{P_cV_c}\left[\frac{8(T/T_c)}{3(V/V_c)-1} - \frac{3}{(V/V_c)^2}\right]$$

or
$$P_r = \frac{8T_r}{3V_r - 1} - \frac{3}{V_r^2} \tag{5.37}$$

The van der Waals equation written in the form (*5.37*) is thus applicable to *all* fluids, because both arbitrary constants have been absorbed in the dimensionless variables. Figure 5-5 is a P_rV_r diagram of (*5.37*) evaluated at $T_r = 0.9$, 1.0, 1.1, and 1.2.

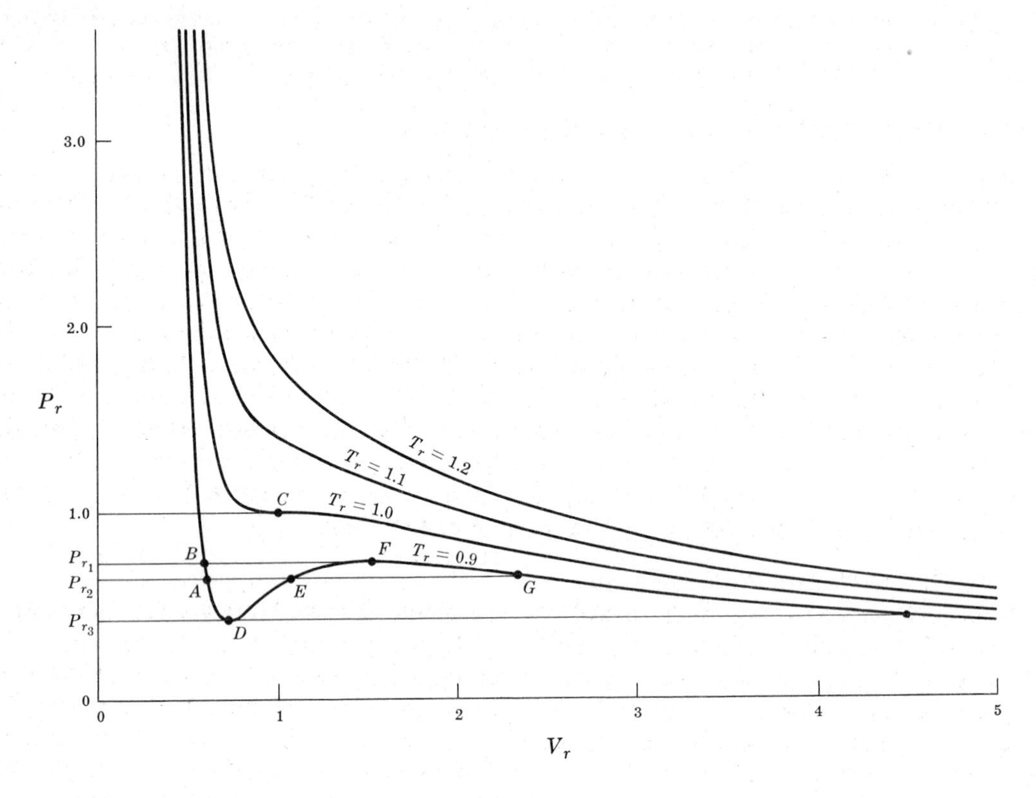

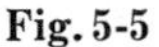
Fig. 5-5

It is useful to relate the algebraic characteristics of (*5.37*) as shown on Fig. 5-5 to observed PVT behavior. Since (*5.37*) is cubic in V_r, there are either one or three real roots which satisfy the equation for every P_r at fixed T_r. If only one real root exists it will correspond either to a gas or fluid state (e.g. to every point on the $T_r = 1.1$ and 1.2 isotherms) or to a liquid state (e.g. to every point above B on the $T_r = 0.9$ isotherm). When three real roots exist, there are three possible situations.

If all three roots are different, as on $T_r = 0.9$ at P_{r_2}, the smallest (at point A) is a liquid volume and the largest (at point G) is a vapor volume. The middle root for V_r (at point E) has no physical significance, as discussed later. Thus the liquid state at A and the vapor state at G exist at the same T_r (and T) and the same P_r (and P), and are therefore presumably the states of liquid and vapor which exist in vapor-liquid equilibrium at $T_r = 0.9$. However, it is seen from Fig. 5-5 that there is a range of values for P_r (from P_{r_3} to P_{r_1}) for which both liquid-like and vapor-like states are shown for the $T_r = 0.9$ isotherm. On the other hand, the phase rule applied to a system consisting of vapor and liquid in equilibrium requires that for a given T_r there be but one P_r, namely P_r^{sat}. This apparent contradiction is resolved by *Maxwell's equal-area rule* (Problem 5.16), which states that P_r^{sat} is that pressure for which the horizontal line AEG cuts off equal areas $ADEA$ and $EFGE$ below and above the isobar. The line meeting this condition on Fig. 5-5 is the one at P_{r_2}. Thus points A and G correspond to saturated liquid and vapor respectively and exist at $P_r^{\text{sat}} = P_{r_2} \cong 0.65$. The portions of the $T_r = 0.9$ isotherm above A and to the right of G represent subcooled liquid and superheated vapor respectively.

If two of the three roots of (*5.37*) are equal, as at P_{r_1} and P_{r_3}, then a minimum or maximum occurs in the isotherm at the double root. Between these two extremes, i.e. between points D and F, $(\partial P_r/\partial V_r)_{T_r}$

is positive. However, it is known that this quantity must be negative for all real materials, and we must conclude that the portion of the isotherm between D and F has no physical significance. This is the reason we earlier disregarded the middle root at E. Between A and D and between F and G, $(\partial P_r/\partial V_r)_{T_r}$ is negative as required, although the values of P_r do not equal the equilibrium pressure P_r^{sat}. These portions of the isotherm represent *metastable* liquid and vapor states, respectively, and correspond to superheated liquid and subcooled vapor, nonequilibrium physical states that can be produced and maintained under carefully controlled conditions in the laboratory.

For three equal real roots, (*5.37*) predicts a horizontal inflection point. This is an exceptional point and occurs only on the critical isotherm $T_r = 1$ at the critical pressure where $P_r = 1$. This is the gas-liquid critical point and is designated as point C on Fig. 5-5.

5.5 CORRESPONDING-STATES CORRELATIONS

Although the PVT behavior of one fluid is qualitatively similar to that of another, gross differences may be observed when comparisons are made of the same thermodynamic property of two different fluids at, say, the same T and P. For convenience in correlation and prediction of thermodynamic properties it is desirable to reduce available data for different substances to some common basis. The classical approach to this goal is motivated by Example 5.9, where it was shown that the van der Waals equation could be put into a form containing only pure numbers and the dimensionless variables P_r, V_r, and T_r. The resulting equation (*5.37*) asserts that all substances, when compared at the same T_r and P_r, should have the same V_r. This is the *theorem of corresponding states.* It is found that only a few substances actually conform to this principle, but simple extensions of the corresponding-states concept have led to the development of correlations which unify to a reasonable degree the volumetric behavior of a large number of fluids.

The acentric factor.

These modern correlations are based on the premise that all fluids having the same value of some dimensionless parameter characteristic of the fluid should have the same V_r when compared at the same T_r and P_r; they are sometimes called *three-parameter* corresponding-states correlations. The most obvious choice for a dimensionless material parameter is Z_c (see Problem 5.18) and correlations employing Z_c have appeared in the literature. However, determination of Z_c requires a value for V_c, a difficult quantity to measure experimentally. A material parameter subject to more precise determination is the *acentric factor* ω.

In Sec. 5.2 we noted that interactions among molecular force fields are manifested as PVT behavior on the macroscopic level. The simplest type of molecular force field is spherical, and we expect that the behavior of fluids consisting of molecules having fields of this type should conform to some simple scheme against which we could compare the behavior of more complex substances. Indeed it is found that these *simple fluids*, exemplified by the heavier rare gases, argon, krypton, and xenon, conform very closely to the theorem of corresponding states. The basis for the definition of the acentric factor is the observation that, for Ar, Kr, and Xe,

$$P_r^{\text{sat}} \cong 0.1 \quad \text{at} \quad T_r = 0.7$$

It is also observed that $P_r^{\text{sat}} < 0.1$ at $T_r = 0.7$ for most other fluids, and that P_r^{sat} at $T_r = 0.7$ generally decreases with increasing asymmetry of the molecular force field. Accordingly, ω is defined as

$$\omega = \log_{10}(P_r^{\text{sat}})_{\substack{\text{simple fluid}\\ \text{at } T_r = 0.7}} - \log_{10}(P_r^{\text{sat}})_{T_r = 0.7}$$

Since P_r^{sat} for the simple fluids averages out to be 0.1 at $T_r = 0.7$, this definition reduces to

$$\omega = -1 - \log_{10}(P_r^{\text{sat}})_{T_r = 0.7} \tag{5.38}$$

Thus ω can be determined from T_c, P_c, and a single vapor-pressure measurement made at $T_r = 0.7$. A list of values for the acentric factor for some common fluids is given in Appendix 3, along with data for T_c, P_c, V_c, and Z_c.

The Pitzer correlation.

An extensive corresponding-states correlation for gases and liquids employing ω, T_r, and P_r has been developed by K. S. Pitzer and coworkers.* They designate substances which conform to their correlation as *normal fluids,* and criteria are presented for determining whether a fluid is normal. It is found that highly polar materials, such as ammonia, water, and alcohols, cannot be considered normal; other investigators have proposed the use of a second dimensionless material parameter to characterize such materials and to allow their inclusion within the framework of a *four-parameter* corresponding-states theory.

We shall consider in this outline only that portion of Pitzer's correlation which is suitable for low-pressure gas- or vapor-phase property calculations. It was shown in Example 5.7 that, in regions where (*5.22*) is valid, $\Delta H'$ and $\Delta S'$ can be calculated if B is known as a function of temperature. The Pitzer correlation for B is of the form

$$\frac{BP_c}{RT_c} = B^0 + \omega B^1 \tag{5.39}$$

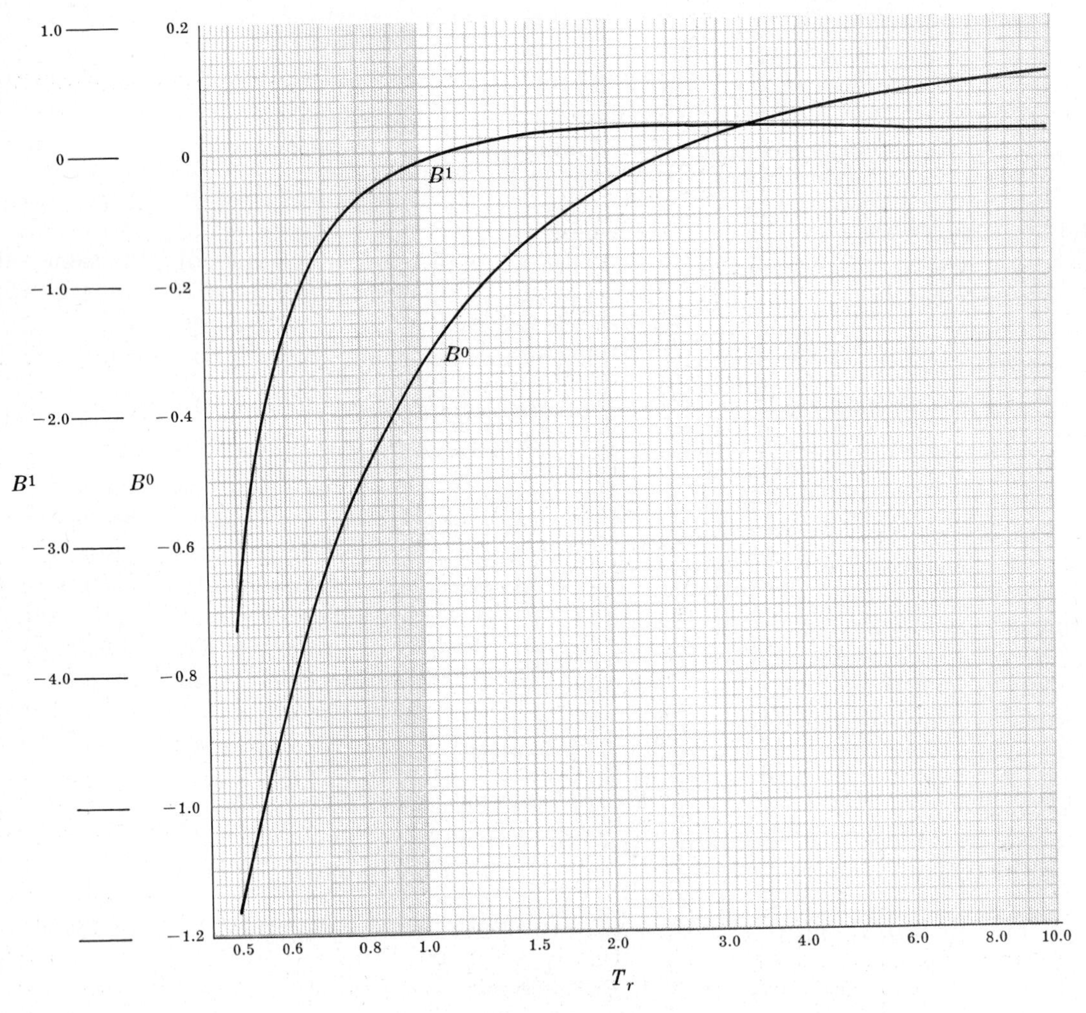

Fig. 5-6

*G. N. Lewis, M. Randall, K. S. Pitzer, and L. Brewer, "Thermodynamics," Appendix 1, McGraw-Hill, New York, 1961.

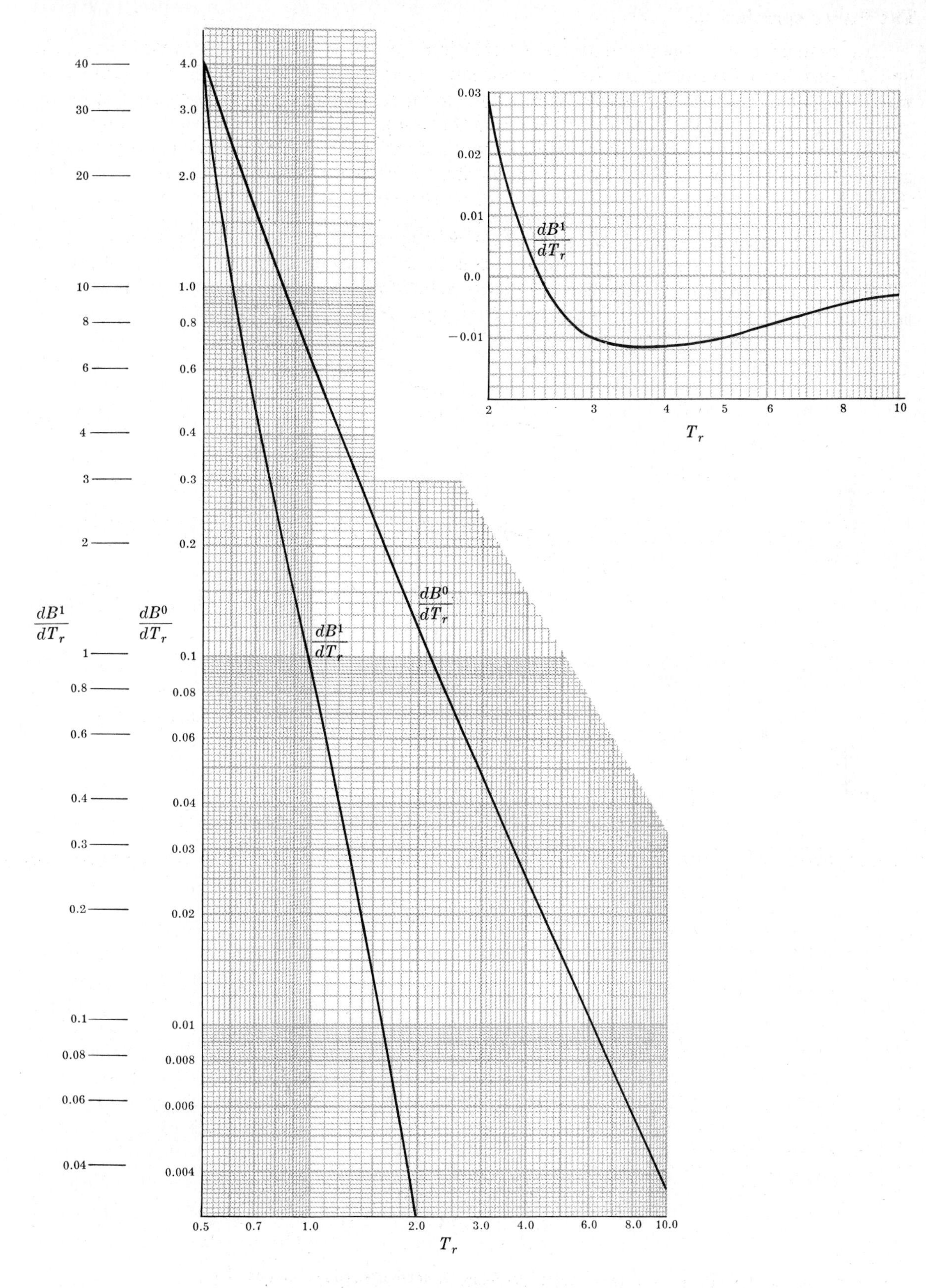

40
30
20
10
8
6
4
3
2
1
0.8
0.6
0.4
0.3
0.2
0.1
0.08
0.06
0.04
$\frac{dB^1}{dT_r}$
4.0
3.0
2.0
1.0
0.8
0.6
0.4
0.3
0.2
0.1
0.08
0.06
0.04
0.03
0.02
0.01
0.008
0.006
0.004
$\frac{dB^0}{dT_r}$
$\frac{dB^1}{dT_r}$
$\frac{dB^0}{dT_r}$
0.5
0.7
1.0
2.0
3.0
4.0
6.0
8.0
10.0
T_r
0.03
0.02
0.01
0.0
−0.01
$\frac{dB^1}{dT_r}$
2
3
4
5
6
8
10
T_r

Fig. 5-7

where B^0 and B^1 are functions of T_r only. Differentiation of (5.39) with respect to T gives

$$\frac{P_c}{RT_c}\frac{dB}{dT} = \frac{dB^0}{dT} + \omega\frac{dB^1}{dT}$$

Since $T = T_c T_r$ then $dT = T_c dT_r$, and substitution for dT on the right-hand side of the above equation reduces it to

$$\frac{P_c}{R}\frac{dB}{dT} = \frac{dB^0}{dT_r} + \omega\frac{dB^1}{dT_r} \tag{5.40}$$

The four dimensionless functions B^0, B^1, dB^0/dT_r, and dB^1/dT_r are plotted as functions of T_r in Figs. 5-6 and 5-7.

The correlation just described was developed for pure gases. It has been extended to apply to gas mixtures by Prausnitz* through the following considerations. Equation (5.14) gives the exact composition dependence of the second virial coefficient for a mixture:

$$B = \sum_i \sum_j y_i y_j B_{ij} \tag{5.14}$$

Differentiation with respect to T gives

$$\frac{dB}{dT} = \sum_i \sum_j y_i y_j \frac{dB_{ij}}{dT} \tag{5.41}$$

We clearly need a correlation for B_{ij}, in particular for the case where $i \neq j$. Prausnitz proposed that Pitzer's correlation be used, and rewrote (5.39) as

$$\frac{P_{c_{ij}} B_{ij}}{RT_{c_{ij}}} = B^0 + \omega_{ij} B^1 \tag{5.42}$$

An analogous equation can be written to replace (5.40). All that is required in addition is a set of rules for determination of the "pseudo-parameters" $T_{c_{ij}}$, $P_{c_{ij}}$, and ω_{ij}. These are given as follows. For $T_{c_{ij}}$:

$$T_{c_{ij}} = (T_{c_i} T_{c_j})^{1/2} \tag{5.43}$$

For $P_{c_{ij}}$ we write

$$P_{c_{ij}} = \frac{Z_{c_{ij}} R T_{c_{ij}}}{V_{c_{ij}}} \tag{5.44}$$

where

$$Z_{c_{ij}} = \frac{Z_{c_i} + Z_{c_j}}{2} \tag{5.45}$$

and

$$V_{c_{ij}} = \left(\frac{V_{c_i}^{1/3} + V_{c_j}^{1/3}}{2}\right)^3 \tag{5.46}$$

Finally,

$$\omega_{ij} = \frac{\omega_i + \omega_j}{2} \tag{5.47}$$

When $i = j$ all of these equations reduce to the appropriate values for a pure material. Furthermore, (5.42) becomes identical to (5.39) for $i = j$. This extension of Pitzer's correlation to mixtures is essentially empirical and can be expected in general to give only estimates of second virial cross-coefficients. It is most reliable for mixtures of molecules of similar size and chemical type.

In general the correlation described is useful only in the range of T_r and P_r where the two-term virial equation is appropriate. This limits applicability to the region where $V_r \geqq 2$.

---

*J. M. Prausnitz, "Molecular Thermodynamics of Fluid-Phase Equilibria," Prentice-Hall, Englewood Cliffs, 1969.

Example 5.10. Find expressions for Z, the dimensionless residual enthalpy $\Delta H'/RT$, and the dimensionless residual entropy $\Delta S'/R$ in terms of T_r, P_r, and the Pitzer correlation for B.

Combination of (*5.22*) and (*5.39*) gives:

$$Z = 1 + \frac{BP}{RT} = 1 + \frac{P}{RT}\frac{RT_c}{P_c}(B^0 + \omega B^1)$$

or

$$Z = 1 + \frac{P_r}{T_r}(B^0 + \omega B^1) \tag{5.48}$$

From (*5.25*), (*5.39*), and (*5.40*),

$$\frac{\Delta H'}{RT} = \frac{P}{R}\left(\frac{dB}{dT} - \frac{B}{T}\right) = \frac{P}{R}\left(\frac{R}{P_c}\frac{dB^0}{dT_r} + \omega\frac{R}{P_c}\frac{dB^1}{dT_r} - \frac{1}{T}\frac{RT_c}{P_c}B^0 - \frac{\omega}{T}\frac{RT_c}{P_c}B^1\right)$$

or

$$\frac{\Delta H'}{RT} = P_r\left[\left(\frac{dB^0}{dT_r} - \frac{B^0}{T_r}\right) + \omega\left(\frac{dB^1}{dT_r} - \frac{B^1}{T_r}\right)\right] \tag{5.49}$$

From (*5.26*) and (*5.40*),

$$\frac{\Delta S'}{R} = \frac{P}{R}\frac{dB}{dT} = \frac{P}{R}\frac{R}{P_c}\left(\frac{dB^0}{dT_r} + \omega\frac{dB^1}{dT_r}\right)$$

or

$$\frac{\Delta S'}{R} = P_r\left(\frac{dB^0}{dT_r} + \omega\frac{dB^1}{dT_r}\right) \tag{5.50}$$

Solved Problems

COMPRESSIBILITY FACTOR (Sec. 5.1)

5.1. Show that the heat of vaporization H_{fg} of a pure fluid is related to P^{sat} by

$$H_{fg} = -R(Z_g^{sat} - Z_f^{sat})\frac{d\ln P^{sat}}{d(1/T)}$$

where Z_g^{sat} and Z_f^{sat} are the compressibility factors of the equilibrium vapor and liquid phases, respectively.

We start with the Clapeyron equation (*4.3b*):

$$\frac{dP^{sat}}{dT} = \frac{H_{fg}}{TV_{fg}} \tag{4.3b}$$

But

$$V_{fg} = V_g - V_f = \frac{Z_g^{sat}RT}{P^{sat}} - \frac{Z_f^{sat}RT}{P^{sat}}$$

or

$$V_{fg} = \frac{RT}{P^{sat}}(Z_g^{sat} - Z_f^{sat}) \tag{1}$$

Also, we have the mathematical identity:

$$\frac{d\ln P^{sat}}{d(1/T)} = -\frac{T^2}{P^{sat}}\frac{dP^{sat}}{dT} \tag{2}$$

Combination of (*4.3b*), (*1*), and (*2*) and rearrangement gives the desired expression. Note the similarity between this exact result and the approximate Clausius-Clapeyron equation (*4.4*).

5.2. Find expressions for $(\partial P/\partial V)_T$, $(\partial P/\partial T)_V$, and $(\partial V/\partial T)_P$ in terms of T, V, and Z and its derivatives.

Division of (*5.2*) by dV and restriction to constant T gives

$$\left(\frac{\partial Z}{\partial V}\right)_T = \frac{Z}{P}\left(\frac{\partial P}{\partial V}\right)_T + \frac{Z}{V} = \frac{V}{RT}\left(\frac{\partial P}{\partial V}\right)_T + \frac{Z}{V}$$

or
$$\left(\frac{\partial P}{\partial V}\right)_T = \frac{RT}{V}\left(\frac{\partial Z}{\partial V}\right)_T - \frac{ZRT}{V^2} \tag{1}$$

Similarly,
$$\left(\frac{\partial Z}{\partial T}\right)_V = \frac{Z}{P}\left(\frac{\partial P}{\partial T}\right)_V - \frac{Z}{T} = \frac{V}{RT}\left(\frac{\partial P}{\partial T}\right)_V - \frac{Z}{T}$$

or
$$\left(\frac{\partial P}{\partial T}\right)_V = \frac{RT}{V}\left(\frac{\partial Z}{\partial T}\right)_V + \frac{RZ}{V} \tag{2}$$

The remaining derivative follows from (1), (2), and the relationship [see (3.13)]

$$\left(\frac{\partial V}{\partial T}\right)_P = -\left(\frac{\partial V}{\partial P}\right)_T\left(\frac{\partial P}{\partial T}\right)_V = -\left(\frac{\partial P}{\partial T}\right)_V \Big/ \left(\frac{\partial P}{\partial V}\right)_T$$

Thus
$$\left(\frac{\partial V}{\partial T}\right)_P = \frac{\dfrac{Z}{T} + \left(\dfrac{\partial Z}{\partial T}\right)_V}{\dfrac{Z}{V} - \left(\dfrac{\partial Z}{\partial V}\right)_T} \tag{3}$$

5.3. Considering Z a function of T and V, derive equations giving $\Delta H'/RT$ and $\Delta S'/R$ in terms of T, V, and Z and its derivatives.

At constant T, changes in $\Delta H'$ can be evaluated from changes in V only. Thus, integration from the ideal-gas state at zero pressure (where $\Delta H' = 0$ and $V = \infty$) to any real state at the same T gives

$$\Delta H' = \int_\infty^V \left(\frac{\partial\,\Delta H'}{\partial V}\right)_T dV \qquad \text{(const. } T) \tag{1}$$

But
$$\left(\frac{\partial\,\Delta H'}{\partial V}\right)_T = \left(\frac{\partial\,\Delta H'}{\partial P}\right)_T\left(\frac{\partial P}{\partial V}\right)_T \tag{2}$$

and, from Sec. 4.8,
$$\left(\frac{\partial\,\Delta H'}{\partial P}\right)_T = T\left(\frac{\partial V}{\partial T}\right)_P - V \tag{3}$$

Combination of (2) and (3) and application of (3.13) gives

$$\left(\frac{\partial\,\Delta H'}{\partial V}\right)_T = -T\left(\frac{\partial P}{\partial T}\right)_V - V\left(\frac{\partial P}{\partial V}\right)_T \tag{4}$$

The required expressions for $(\partial P/\partial V)_T$ and $(\partial P/\partial T)_V$ are given by equations (1) and (2) of Problem 5.2. Substitution of these expressions in (4) gives

$$\left(\frac{\partial\,\Delta H'}{\partial V}\right)_T = -\frac{RT^2}{V}\left(\frac{\partial Z}{\partial T}\right)_V - RT\left(\frac{\partial Z}{\partial V}\right)_T \tag{5}$$

and the desired expression for $\Delta H'/RT$ follows from combination of (1) and (5):

$$\frac{\Delta H'}{RT} = -T\int_\infty^V \left(\frac{\partial Z}{\partial T}\right)_V \frac{dV}{V} - \int_\infty^V \left(\frac{\partial Z}{\partial V}\right)_T dV \qquad \text{(const. } T)$$

or
$$\frac{\Delta H'}{RT} = -T\int_\infty^V \left(\frac{\partial Z}{\partial T}\right)_V \frac{dV}{V} - (Z-1) \qquad \text{(const. } T) \tag{6}$$

The equation for $\Delta S'/R$ is found in an entirely analogous manner:

$$\frac{\Delta S'}{R} = -T\int_\infty^V \left(\frac{\partial Z}{\partial T}\right)_V \frac{dV}{V} - \int_\infty^V \frac{(Z-1)}{V}\,dV - \ln Z \qquad \text{(const. } T) \tag{7}$$

Equations (*6*) and (*7*) are merely alternatives to (*5.6*) and (*5.7*), but are more convenient for use with equations of state which cannot be solved easily for Z in terms of P and T.

5.4. Derive expressions for the Joule-Thomson coefficient (see Problem 3.4) in terms of (*a*) T, P, C_P, and Z and its derivatives, and (*b*) T, V, C_P, and Z and its derivatives.

From Problem 3.4,

$$\left(\frac{\partial T}{\partial P}\right)_H = -\frac{1}{C_P}\left[V - T\left(\frac{\partial V}{\partial T}\right)_P\right] \qquad (1)$$

(*a*) From Example 5.2,

$$\left(\frac{\partial V}{\partial T}\right)_P = \frac{RZ}{P} + \frac{RT}{P}\left(\frac{\partial Z}{\partial T}\right)_P \qquad (5.3)$$

Combination of (*1*) and (*5.3*) gives

$$\left(\frac{\partial T}{\partial P}\right)_H = -\frac{1}{C_P}\left[V - \frac{ZRT}{P} - \frac{RT^2}{P}\left(\frac{\partial Z}{\partial T}\right)_P\right]$$

or

$$\left(\frac{\partial T}{\partial P}\right)_H = \frac{RT^2}{C_P P}\left(\frac{\partial Z}{\partial T}\right)_P \qquad (2)$$

(*b*) From Problem 5.2,

$$\left(\frac{\partial V}{\partial T}\right)_P = \frac{\dfrac{Z}{T} + \left(\dfrac{\partial Z}{\partial T}\right)_V}{\dfrac{Z}{V} - \left(\dfrac{\partial Z}{\partial V}\right)_T} \qquad (3)$$

Combining (*1*) and (*3*), we obtain

$$\left(\frac{\partial T}{\partial P}\right)_H = \frac{V}{C_P}\left[\frac{V\left(\dfrac{\partial Z}{\partial V}\right)_T + T\left(\dfrac{\partial Z}{\partial T}\right)_V}{Z - V\left(\dfrac{\partial Z}{\partial V}\right)_T}\right] \qquad (4)$$

The *inversion curve* of a fluid separates the region for which $(\partial T/\partial P)_H$ is positive from the region where it is negative. The equation defining this curve, $(\partial T/\partial P)_H = 0$, can be put in a particularly simple form by use of (*2*). Thus,

$$\left(\frac{\partial Z}{\partial T}\right)_P = 0 \qquad (5)$$

is a practical definition which permits direct determination of the inversion curve from charts or tables of compressibility-factor data. Most *equations of state*, on the other hand, are of the form $Z = Z(V, T)$, and (*5*) is inconvenient to use. In such cases, (*4*) gives an alternative useful definition of the inversion curve as the locus of points for which

$$V\left(\frac{\partial Z}{\partial V}\right)_T + T\left(\frac{\partial Z}{\partial T}\right)_V = 0 \qquad (6)$$

VIRIAL EQUATIONS (Secs. 5.2 and 5.3)

5.5. Interpret the virial coefficients of a constant-composition mixture as derivatives of the compressibility factor Z. Show that, for such a mixture, they are functions of T only.

As defined in the text, the virial equations are series expansions of Z about $P = 0$ or $1/V = 0$. In these equations, Z is a function of T as well as of P or V. Taylor's formula for the infinite series expansion of $f(x, y)$ in powers of x about $x = x_0$ is

$$f(x, y) = f(x_0, y) + \frac{1}{1!}\left(\frac{\partial f}{\partial x}\right)_{y,\, x=x_0}(x - x_0) + \frac{1}{2!}\left(\frac{\partial^2 f}{\partial x^2}\right)_{y,\, x=x_0}(x - x_0)^2 + \cdots$$

Letting $f \equiv Z$, $y \equiv T$, $x \equiv P$ or $1/V$, $x_0 = 0$, and noting that $Z(0, T) = 1$, we obtain on comparing the above equation with (*5.8*) and (*5.9*)

$$B' = \left(\frac{\partial Z}{\partial P}\right)_{T,\, P=0} \qquad B = \left[\frac{\partial Z}{\partial(1/V)}\right]_{T,\, 1/V=0}$$

$$C' = \frac{1}{2}\left(\frac{\partial^2 Z}{\partial P^2}\right)_{T,\, P=0} \qquad C = \frac{1}{2}\left[\frac{\partial^2 Z}{\partial(1/V)^2}\right]_{T,\, 1/V=0} \tag{1}$$

etc. etc.

Thus, B and B' are the limiting slopes of isotherms on Z versus $1/V$ and Z versus P diagrams, C and C' are proportional to the limiting second derivatives on the same diagrams, etc.

That the virial coefficients are functions of T only follows from equations (*1*), because they can be interpreted as derivatives of Z at a *specified* pressure or density. Thus they are independent of the actual *system* pressure or density.

5.6. Derive equations (*5.10*).

Solve (*5.9*) for P:

$$P = RT\left(\frac{1}{V} + \frac{B}{V^2} + \frac{C}{V^3} + \frac{D}{V^4} + \cdots\right) \tag{1}$$

Substitute (*1*) in (*5.8*):

$$Z = 1 + B'RT\left(\frac{1}{V} + \frac{B}{V^2} + \frac{C}{V^3} + \frac{D}{V^4} + \cdots\right) + C'(RT)^2\left(\frac{1}{V} + \frac{B}{V^2} + \frac{C}{V^3} + \frac{D}{V^4} + \cdots\right)^2$$

$$+ D'(RT)^3\left(\frac{1}{V} + \frac{B}{V^2} + \frac{C}{V^3} + \frac{D}{V^4} + \cdots\right)^3 + \cdots$$

Thus
$$Z = 1 + \frac{B'RT}{V} + \frac{BB'RT + C'(RT)^2}{V^2} + \frac{CB'RT + 2BC'(RT)^2 + D'(RT)^3}{V^3} + \cdots \tag{2}$$

Term-by-term comparison of (*2*) with (*5.9*) gives the desired results. Note that (*5.10*) is strictly valid only when the *infinite* series are compared.

5.7. Derive an expression for the residual volume $\Delta V'$ as a function of T, P, and the virial coefficients in the volumetric virial equation (*5.9*).

Combine (*4.16*), (*5.8*), and (*5.10*):

$$\Delta V' = V' - V = -\frac{RT}{P}(Z - 1)$$

$$= -RT\left[\frac{B}{RT} + \frac{C - B^2}{(RT)^2}P + \cdots\right]$$

or
$$\Delta V' = -\left(B + \frac{C - B^2}{RT}P + \cdots\right) \tag{1}$$

In agreement with the discussion of Problem 4.14, (*1*) shows that, in the limit as $P \to 0$, $\Delta V'$ remains finite and generally nonzero, i.e. $\lim_{P \to 0} \Delta V' = -B$. This limiting $\Delta V'$ is also clearly a function of T.

5.8. Derive an expression for the Joule-Thomson coefficient from the virial equation (*5.9*).

From Problem 5.4, we have

$$\left(\frac{\partial T}{\partial P}\right)_H = \frac{V}{C_P}\left[\frac{V\left(\frac{\partial Z}{\partial V}\right)_T + T\left(\frac{\partial Z}{\partial T}\right)_V}{Z - V\left(\frac{\partial Z}{\partial V}\right)_T}\right] \tag{1}$$

From (5.9),

$$\left(\frac{\partial Z}{\partial V}\right)_T = -\frac{B}{V^2} - \frac{2C}{V^3} - \frac{3D}{V^4} - \cdots \tag{2}$$

$$\left(\frac{\partial Z}{\partial T}\right)_V = \frac{1}{V}\frac{dB}{dT} + \frac{1}{V^2}\frac{dC}{dT} + \frac{1}{V^3}\frac{dD}{dT} + \cdots \tag{3}$$

Combination of (1), (2), and (3) and rearrangement gives the desired result:

$$\left(\frac{\partial T}{\partial P}\right)_H = \frac{1}{C_P}\left[\frac{\left(T\frac{dB}{dT} - B\right) + \left(T\frac{dC}{dT} - 2C\right)\frac{1}{V} + \cdots}{1 + \frac{2B}{V} + \frac{3C}{V^2} + \cdots}\right] \tag{4}$$

The Joule-Thomson coefficient of an ideal gas is identically zero (Problem 3.26). However, we see from (4) that, for a real gas in the limit as $P \to 0$ $(1/V \to 0)$,

$$\left(\frac{\partial T}{\partial P}\right)_H = \frac{T}{C_P'}\left(\frac{dB}{dT} - \frac{B}{T}\right) \tag{5}$$

where C_P' is the constant-pressure ideal-gas heat capacity. Thus the Joule-Thomson coefficient of a real gas at zero pressure is in general nonzero and is also a function of temperature.

5.9. The second virial coefficient of an equimolar binary vapor mixture of methane and n-hexane is $-517(\text{cm})^3/(\text{g mole})$ at 50(°C). What is B at the same temperature for a mixture containing 25 mole % methane and 75 mole % n-hexane? At 50(°C), $B = -33(\text{cm})^3/(\text{g mole})$ for methane and $-1512(\text{cm})^3/(\text{g mole})$ for n-hexane.

Let methane $\equiv 1$ and n-hexane $\equiv 2$. B_{11} and B_{22} are given, and B_{12} must be determined from the given mixture B. For a binary mixture

$$B = y_1^2B_{11} + 2y_1y_2B_{12} + y_2^2B_{22} \tag{5.17}$$

from which

$$B_{12} = \frac{B - y_1^2B_{11} - y_2^2B_{22}}{2y_1y_2} = \frac{-517 - (0.5)^2(-33) - (0.5)^2(-1512)}{(2)(0.5)(0.5)}$$

$$= -262(\text{cm})^3/(\text{g mole})$$

For the 25%/75% mixture B can now be calculated from (5.17):

$$B = (0.25)^2(-33) + (2)(0.25)(0.75)(-262) + (0.75)^2(-1512) = -951(\text{cm})^3/(\text{g mole})$$

5.10. The mixing rule (5.14) for B is quadratic in mole fraction. Rewrite this expression as the sum of two terms, one linear in mole fraction and the other quadratic, such that the linear part reduces to B_{ii} at $y_i = 1$ for all i.

The only linear expression satisfying the stated conditions is $\sum_{i=1}^{m} y_iB_{ii}$. We therefore write (5.14) as

$$B = \sum_i y_iB_{ii} + \sum_i\sum_j y_iy_jB_{ij} - \sum_i y_iB_{ii} \tag{1}$$

where we have abbreviated the summation notation. We seek to express the last term of (1) in a quadratic form. First note that

$$\sum_i y_iB_{ii} \equiv \frac{1}{2}\sum_i y_iB_{ii} + \frac{1}{2}\sum_j y_jB_{jj} \tag{2}$$

The two sums on the right-hand side of (2) remain unaltered if we multiply them by $\sum_j y_j$ and $\sum_i y_i$, respectively, because the mole fractions sum to unity. Thus

$$\frac{1}{2}\sum_i y_i B_{ii} = \frac{1}{2}\left(\sum_j y_j\right)\sum_i y_i B_{ii} = \frac{1}{2}\sum_i\sum_j y_i y_j B_{ii} \tag{3}$$

$$\frac{1}{2}\sum_j y_j B_{jj} = \frac{1}{2}\left(\sum_i y_i\right)\sum_j y_j B_{jj} = \frac{1}{2}\sum_i\sum_j y_i y_j B_{jj} \tag{4}$$

Combination of (*1*), (*2*), (*3*), and (*4*) gives the desired result:

$$B = \sum_i y_i B_{ii} + \frac{1}{2}\sum_i\sum_j y_i y_j \delta_{ij} \tag{5}$$

where

$$\delta_{ij} = 2B_{ij} - B_{ii} - B_{jj} \tag{6}$$

Note that when $i \equiv j$, (*6*) reduces to $\delta_{ij} = 0$. Also, $\delta_{ji} = \delta_{ij}$ because $B_{ji} = B_{ij}$. Thus for a binary system,

$$B = y_1 B_{11} + y_2 B_{22} + y_1 y_2 \delta_{12} \qquad \text{where} \qquad \delta_{12} = 2B_{12} - B_{11} - B_{22}$$

If $\delta_{ij} = 0$ for all i and j, then from (*6*),

$$B_{ij} = \tfrac{1}{2}(B_{ii} + B_{jj}) \tag{7}$$

and (*5*) assumes the simple linear form $B = \sum_i y_i B_{ii}$. We show in Problem 7.3 that (*7*) is one of the properties of vapor mixtures that conform to *ideal-solution* behavior. In general, however, $\delta_{ij} \neq 0$.

5.11. Ten grams of steam at 260(°C) expands reversibly and isothermally in a piston-and-cylinder apparatus from an initial volume of 2(liter) to a final volume of 20(liter). How much work is done by the steam?

We use (*5.21*) and the B and C for steam at 260(°C) determined in Example 5.4. The reversible work is calculated from

$$W_{\text{rev}} = \int_{V_1^t}^{V_2^t} P\,dV^t$$

or

$$W_{\text{rev}} = n\int_{V_1}^{V_2} P\,dV \tag{1}$$

where V^t is the total system volume, V is the molar volume of steam, and n is the total number of moles of steam. From (*5.21*),

$$P = RT\left(\frac{1}{V} + \frac{B}{V^2} + \frac{C}{V^3}\right) \tag{2}$$

Combination of (*1*) and (*2*) and integration from V_1 to V_2 give

$$W_{\text{rev}} = nRT\left[\ln\frac{V_2}{V_1} + \frac{V_2 - V_1}{V_1 V_2}\left(B + \frac{C}{2}\frac{V_1 + V_2}{V_1 V_2}\right)\right] \tag{3}$$

From the statement of the problem,

$$n = 10(\text{g}) \times \frac{1}{18.02}\frac{(\text{g mole})}{(\text{g})} = 0.5549(\text{g mole})$$

$$V_1 = \frac{2(\text{l}) \times 1000(\text{cm})^3/(\text{l})}{0.5549(\text{g mole})} = 3604\,\frac{(\text{cm})^3}{(\text{g mole})}$$

$$V_2 = \frac{20 \times 1000}{0.5549} = 36{,}040\,\frac{(\text{cm})^3}{(\text{g mole})}$$

and from Example 5.4

$$B = -142.2(\text{cm})^3/(\text{g mole}) \qquad C = -7140(\text{cm})^6/(\text{g mole})^2$$

Substitution of these quantities into (*3*), with $R = 8.314(\text{J})/(\text{g mole})(\text{K})$ and $T = 533.15(\text{K})$, yields $W_{\text{rev}} = 5575(\text{J})$.

We may check the validity of (*2*) for this application by calculating the initial and final pressures. Thus, (2) gives

$$P_1 = 11.65(\text{atm}) = 11.80(\text{bar}) \qquad P_2 = 1.209(\text{atm}) = 1.225(\text{bar})$$

Comparison of these pressures with those in the table of Example 5.4 shows that the three-term virial equation is entirely adequate in this range.

5.12. Derive expressions for $\Delta H'/RT$ and $\Delta S'/R$ for a gas described by the truncated virial equation (5.21).

We use the results of Problem 5.3. From (5.21),

$$\left(\frac{\partial Z}{\partial T}\right)_V = \frac{1}{V}\left(\frac{dB}{dT}\right) + \frac{1}{V^2}\left(\frac{dC}{dT}\right)$$

$$\left(\frac{\partial Z}{\partial V}\right)_T = -\frac{B}{V^2} - \frac{2C}{V^3}$$

$$\frac{Z-1}{V} = \frac{B}{V^2} + \frac{C}{V^3}$$

By (6) of Problem 5.3,

$$\frac{\Delta H'}{RT} = -T\int_\infty^V \left(\frac{\partial Z}{\partial T}\right)_V \frac{dV}{V} - (Z-1) \qquad (\text{const. } T)$$

$$= -T\int_\infty^V \left(\frac{1}{V}\frac{dB}{dT} + \frac{1}{V^2}\frac{dC}{dT}\right)\frac{dV}{V} - \left(\frac{B}{V} + \frac{C}{V^2}\right) \qquad (\text{const. } T)$$

from which

$$\frac{\Delta H'}{RT} = \frac{T}{V}\left(\frac{dB}{dT} - \frac{B}{T}\right) + \frac{T}{V^2}\left(\frac{1}{2}\frac{dC}{dT} - \frac{C}{T}\right) \qquad (1)$$

Similarly, by (7) of Problem 5.3,

$$\frac{\Delta S'}{R} = -T\int_\infty^V \left(\frac{\partial Z}{\partial T}\right)_V \frac{dV}{V} - \int_\infty^V \frac{Z-1}{V}\,dV - \ln Z \qquad (\text{const. } T)$$

$$= -T\int_\infty^V \left(\frac{1}{V}\frac{dB}{dT} + \frac{1}{V^2}\frac{dC}{dT}\right)\frac{dV}{V} - \int_\infty^V \left(\frac{B}{V^2} + \frac{C}{V^3}\right)dV - \ln Z \qquad (\text{const. } T)$$

or

$$\frac{\Delta S'}{R} = \frac{T}{V}\left(\frac{dB}{dT} + \frac{B}{T}\right) + \frac{T}{2V^2}\left(\frac{dC}{dT} + \frac{C}{T}\right) - \ln Z \qquad (2)$$

Equations (1) and (2) are the counterparts of (5.25) and (5.26), respectively.

5.13. One gram mole of methane gas is mixed with one gram mole of n-hexane gas under conditions of constant total volume and constant temperature. If the initial temperature and pressure of both pure gases is 75(°C) and 1(atm), what is the pressure change on mixing? For methane(1) and n-hexane(2) at 75(°C), $B_{11} = -26(\text{cm})^3/(\text{g mole})$, $B_{22} = -1239(\text{cm})^3/(\text{g mole})$, and $B_{12} = -180(\text{cm})^3/(\text{g mole})$.

For the two pure gases we may write

$$PV_1^t = n_1Z_1RT \qquad (1)$$

$$PV_2^t = n_2Z_2RT \qquad (2)$$

Similarly, for the gas mixture

$$P_fV_f^t = nZ_fRT \qquad (3)$$

where the subscript f denotes the final, mixed condition and $n = n_1 + n_2$. But by the statement of the problem

$$V_f^t = V_1^t + V_2^t$$

and (3) becomes
$$P_f(V_1^t + V_2^t) = nZ_fRT$$

Substitution of V_1^t and V_2^t from (1) and (2) into this last equation yields on rearrangement

$$P_f = \frac{nZ_fP}{n_1Z_1 + n_2Z_2} \tag{4}$$

The pressure change on mixing is $\Delta P = P_f - P$ and the mole fractions of 1 and 2 in the mixture are $y_1 = n_1/n$ and $y_2 = n_2/n$. Thus we obtain from (4) an exact expression for the pressure change on mixing at constant temperature and constant total volume:

$$\Delta P = \left[\frac{Z_f - (y_1Z_1 + y_2Z_2)}{y_1Z_1 + y_2Z_2}\right]P \tag{5}$$

For this problem, the gases are adequately described by the truncated virial equation (5.22). Thus

$$Z_f = 1 + \frac{BP_f}{RT} = 1 + \frac{B(P + \Delta P)}{RT} \qquad Z_1 = 1 + \frac{B_{11}P}{RT} \qquad Z_2 = 1 + \frac{B_{22}P}{RT}$$

where B is the second virial coefficient of the mixture. Substitution of these expressions into (5) and solution for ΔP yields

$$\Delta P = \frac{(B - y_1B_{11} - y_2B_{22})P^2}{RT - (B - y_1B_{11} - y_2B_{22})P}$$

This equation can be abbreviated through the use of δ-notation (see Problem 5.10). Thus for a binary mixture

$$B - y_1B_{11} - y_2B_{22} = y_1y_2\delta_{12} \qquad \text{where} \qquad \delta_{12} = 2B_{12} - B_{11} - B_{22}$$

and we obtain finally

$$\Delta P = \frac{y_1y_2\delta_{12}P^2}{RT - y_1y_2\delta_{12}P} \tag{6}$$

From the statement of the problem

$$\delta_{12} = 2(-180) - (-26) - (-1239) = 905(\text{cm})^3/(\text{g mole})$$

and $y_1 = y_2 = 0.5$, $P = 1(\text{atm})$, $T = 75(°\text{C}) = 348.2(\text{K})$. Therefore we have from (6)

$$\Delta P = \frac{(0.5)(0.5)(905)(1)^2}{(82.05)(348.2) - (0.5)(0.5)(905)(1)} = 0.008(\text{atm})$$

The small but significant pressure increase resulting from the mixing of the gases is about 1% of the initial pressure. Since all the B's and hence δ_{12} are zero for ideal gases, no pressure change would occur if the gases were ideal.

EMPIRICAL EQUATIONS OF STATE (Sec. 5.4)

5.14. We noted in Sec. 5.4 the necessity of using empirical equations of state to describe the volumetric properties of liquids, and we showed in Example 5.9 that even the relatively simple van der Waals equation is capable of providing a qualitative picture of liquid-phase PVT behavior. *Quantitative* correlation and prediction of liquid volumes requires special methods, however, and for precise calculations one must generally use equations of state developed specifically for liquids. One such equation of state is the following version of the *Tait equation*:

$$V = V_0 - D\ln\left(\frac{P + E}{P_0 + E}\right) \tag{1}$$

where the parameters D and E are functions of temperature only, and V_0 and P_0 are the liquid volume and pressure in some reference state. For convenience, V_0 and P_0 are often taken to be V^{sat} and P^{sat}, the saturation values at the temperature of interest. For this choice of reference state, a complete description of the PVT behavior of a liquid follows if D, E, V^{sat}, and P^{sat} are known as functions of T. Published values of these parameters for liquid water at 60(°C) are:

$$P^{sat} = 0.1994(\text{bar}) \qquad V^{sat} = 1.0172(\text{cm})^3/(\text{g})$$

$$D = 0.1255(\text{cm})^3/(\text{g}) \qquad E = 2712(\text{bar})$$

(a) Show that for pressures not too far removed from P_0 the Tait equation predicts a linear dependence of V on P. Using the *exact* form of the Tait equation (*1*) and the values for P^{sat}, V^{sat}, D, and E given above, calculate V for liquid water at 60(°C) and at pressures between 25 and 1400(bar). Compare these calculated values with steam-table data.

(b) Find an expression for the isothermal compressibility κ, and calculate κ at 60(°C) for liquid water at saturation and at pressures of 250(bar) and 500(bar).

(a) The Tait equation can be written as

$$V = V_0 - D\ln\left(1 + \frac{P - P_0}{P_0 + E}\right) \tag{2}$$

and a series expansion for $\ln(1+x)$, valid for $-1 < x < 1$, is

$$\ln(1+x) = x - \tfrac{1}{2}x^2 + \tfrac{1}{3}x^3 - \cdots$$

Thus for sufficiently small x, $\ln(1+x) \cong x$, and for $P - P_0 \ll P_0 + E$, (*2*) becomes

$$V \cong V_0 - \frac{D(P-P_0)}{P_0 + E} \tag{3}$$

which is the required linear approximation.

If the reference state is chosen as the saturated liquid, then the range of validity of (*3*) is relatively small: for liquid water at 60(°C) a linear fit of volumetric data near saturation yields extrapolated values of $V - V^{sat}$ which are in error by 15% at 100(bar) and 25% at 500(bar). However, as shown in the table below, the *exact* form of the Tait equation provides an excellent representation of the steam-table data for liquid water at 60(°C), giving an essentially perfect fit up to about 150(bar) and an error of only 0.1% at 1400(bar), which is the upper pressure limit of the steam tables.

P(bar)	V(cm)^3/(g)	
	Equation (*1*)	Steam Tables
25	1.0161	1.0160
50	1.0149	1.0149
100	1.0127	1.0127
150	1.0105	1.0105
200	1.0083	1.0084
500	0.9960	0.9962
1000	0.9778	0.9777
1200	0.9712	0.9708
1400	0.9650	0.9640

(b) By definition,

$$\kappa = -\frac{1}{V}\left(\frac{\partial V}{\partial P}\right)_T$$

All the parameters in (*1*) are functions of T only. Thus we obtain by differentiation of (*1*) at constant T:

$$\left(\frac{\partial V}{\partial P}\right)_T = -\frac{D}{P+E}$$

and the required expression for κ is

$$\kappa = \frac{D}{V(P+E)} \qquad (4)$$

with V given by (1).

For the calculation of κ for water, we choose the saturated liquid as the reference state. Then (4) becomes

$$\kappa = \frac{D}{(P+E)\left[V^{\text{sat}} - D\ln\left(\frac{P+E}{P^{\text{sat}}+E}\right)\right]} \qquad (5)$$

The required values of κ at 60(°C) calculated from the above equation at $P = P^{\text{sat}} =$ 0.1994(bar) and at 250(bar) and 500(bar) are shown in the following table, together with values derived from the steam-table data. The agreement is seen to be good.

P(bar)	$\kappa \times 10^5$(bar)$^{-1}$	
	Equation (5)	Steam Tables
0.1994	4.55	4.55
250	4.21	4.15
500	3.92	3.93

5.15. Write the van der Waals and Redlich-Kwong equations of state in virial form.

Write the van der Waals equation (5.27) as

$$Z = \frac{1}{1-(b/V)} - \frac{a}{RTV} \qquad (1)$$

The binomial expansion for $[1-(b/V)]^{-1}$ is

$$\frac{1}{1-(b/V)} = 1 + \frac{b}{V} + \frac{b^2}{V^2} + \frac{b^3}{V^3} + \cdots \qquad (2)$$

This series converges for $b < V$, which is always the case. Combination of (1) and (2) gives the desired result:

$$Z = 1 + \frac{b-(a/RT)}{V} + \frac{b^2}{V^2} + \frac{b^3}{V^3} + \cdots \qquad (3)$$

Similarly, write the Redlich-Kwong equation (5.28) as

$$Z = \frac{1}{1-(b/V)} - \frac{a/VRT^{3/2}}{1+(b/V)} \qquad (4)$$

The binomial expansion for $[1+(b/V)]^{-1}$ is

$$\frac{1}{1+(b/V)} = 1 - \frac{b}{V} + \frac{b^2}{V^2} - \frac{b^3}{V^3} + \cdots \qquad (5)$$

Again convergence is obtained for $b < V$, a condition that always holds.

Combination of (2), (4), and (5) gives

$$Z = 1 + \frac{b-(a/RT^{3/2})}{V} + \frac{b^2+(ab/RT^{3/2})}{V^2} + \frac{b^3-(ab^2/RT^{3/2})}{V^3} + \cdots \qquad (6)$$

Term-by-term comparison of (3) and (6) with (5.9) shows that the virial coefficients are

$$B_{vdW} = b - (a/RT) \qquad B_{RK} = b - (a/RT^{3/2})$$

$$C_{vdW} = b^2 \qquad C_{RK} = b^2 + (ab/RT^{3/2})$$

$$D_{vdW} = b^3 \qquad\qquad D_{RK} = b^3 - (ab^2/RT^{3/2})$$

etc. etc.

Neither of these predictions of the functional forms of the virial coefficients is in general correct. However, the Redlich-Kwong coefficients are more realistic than those of the van der Waals equation.

5.16. Derive Maxwell's *equal-area rule* for subcritical isotherms having the general shape of van der Waals isotherms (see Fig. 5-5 and Example 5.9).

A typical subcritical isotherm is shown in Fig. 5-8. By assumption, P^{sat} is the vapor pressure to be found from the equation of state. We construct a horizontal line intersecting the isotherm at A, C, and E. The states A and E, corresponding to saturated liquid and vapor volumes V^l and V^g, are then equilibrium states, for which (see Problem 3.11) $G^l = G^g$. But, from (*3.23*), $G = A + PV$. Thus

$$A^l + P^lV^l = A^g + P^gV^g$$

But $P^l = P^g = P^{sat}$, and we obtain on rearrangement

$$A^l - A^g = P^{sat}(V^g - V^l) \qquad (1)$$

An alternative equation can be found for $A^l - A^g$ by evaluating the integral $\int dA$ along the path $EDCBA$:

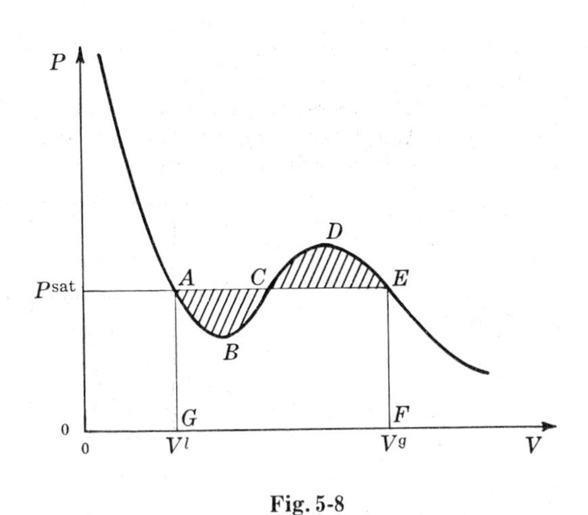

Fig. 5-8

$$A^l - A^g = \int_{EDCBA} dA$$

But this path is an isotherm for which, by (*3.34*), $dA = -P\,dV$. Thus

$$A^l - A^g = -\int_{EDCBA} P\,dV$$

or

$$A^l - A^g = \int_{ABCDE} P\,dV \qquad (2)$$

where P is the pressure calculated from the equation of state. Equations (*1*) and (*2*) apply to the same change of state, and they may be combined to give

$$P^{sat}(V^g - V^l) = \int_{ABCDE} P\,dV \qquad (3)$$

But

$$P^{sat}(V^g - V^l) = \text{area } AEFGA \qquad (4)$$

and

$$\int_{ABCDE} P\,dV = \text{area } ABCDEFGA \qquad (5)$$

Also, by inspection of the figure,

$$\text{area } ABCDEFGA = \text{area } AEFGA - \text{area } ABCA + \text{area } CDEC \qquad (6)$$

Combination of (*3*), (*4*), (*5*), and (*6*) gives the desired result:

$$\text{area } ABCA = \text{area } CDEC$$

Thus, if P^{sat} is the vapor pressure consistent with the equation of state, then the shaded areas are equal. The converse is also true.

5.17. (*a*) Find expressions for $\Delta H'/RT$ and $\Delta S'/R$ for a gas described by the van der Waals equation. (*b*) Using the results of (*a*) calculate $\Delta H'$ and $\Delta S'$ for carbon dioxide gas at 100(°C) and 40(atm).

(*a*) Since van der Waals' equation

$$Z = \frac{V}{V-b} - \frac{a}{RTV} \tag{5.27}$$

gives Z as a function of T and V, the results of Problem 5.3 are most conveniently used to find the required expressions. From (*6*) of that problem

$$\frac{\Delta H'}{RT} = -T\int_\infty^V \left(\frac{\partial Z}{\partial T}\right)_V \frac{dV}{V} + (1-Z)$$

By differentiation of van der Waals' equation

$$\left(\frac{\partial Z}{\partial T}\right)_V = \frac{a}{RT^2V}$$

Integration then provides

$$\frac{\Delta H'}{RT} = \frac{a}{RTV} + 1 - Z$$

From (*7*) of Problem 5.3

$$\frac{\Delta S'}{R} = -T\int_\infty^V \left(\frac{\partial Z}{\partial T}\right)_V \frac{dV}{V} - \int_\infty^V \frac{(Z-1)}{V}\,dV - \ln Z$$

The first integral on the right has just been evaluated as a/RTV. The second integral is treated as follows: Subtraction of unity from both sides of van der Waals' equation gives

$$Z - 1 = \frac{V}{V-b} - 1 - \frac{a}{RTV} = \frac{b}{V-b} - \frac{a}{RTV}$$

Therefore

$$\int_\infty^V \left(\frac{Z-1}{V}\right) dV = b\int_\infty^V \frac{dV}{V(V-b)} - \frac{a}{RT}\int_\infty^V \frac{dV}{V^2} = \ln\frac{V-b}{V} + \frac{a}{RTV}$$

Thus

$$\frac{\Delta S'}{R} = \frac{a}{RTV} - \ln\frac{V-b}{V} - \frac{a}{RTV} - \ln Z$$

or finally

$$\frac{\Delta S'}{R} = \ln\frac{V}{Z(V-b)} = \ln\frac{RT}{P(V-b)}$$

(*b*) In order to make use of these equations for the calculation of numerical results, we must have values for the van der Waals constants a and b. These may be obtained from critical constants through (*5.35*) of Example 5.8. Thus

$$a = \frac{9}{8}RT_cV_c \quad \text{or} \quad \frac{a}{RT} = \frac{9}{8}\frac{T_cV_c}{T}$$

$$b = \frac{V_c}{3}$$

Numerical values for T_c and V_c from Appendix 3 give

$$\frac{a}{RT} = \frac{9}{8} \times \frac{304.2}{373.15} \times 94.0(\text{cm})^3/(\text{g mole}) = 86.21(\text{cm})^3/(\text{g mole})$$

$$b = \frac{94.0}{3} = 31.3(\text{cm})^3/(\text{g mole})$$

Our expressions for $\Delta H'$ and $\Delta S'$ include V and Z, and values for these quantities must first be determined from the van der Waals equation itself. Since this equation is cubic in volume, solution for V is best done by trial. With the above constants and for a pressure of 40(atm) the result is

$$V = 707.7(\text{cm})^3/(\text{g mole}) \qquad \text{from which} \qquad Z = 0.9245$$

Evaluation of $\Delta H'$ and $\Delta S'$ is now a matter of simple substitution:

$$\frac{\Delta H'}{RT} = \frac{a}{RTV} + (1-Z) = \frac{86.21}{707.7} + 1 - 0.9245 = 0.1973$$

Hence $$\Delta H' = 0.1973\,RT = 0.1973 \times 8.314(\text{J})/(\text{g mole})(\text{K}) \times 373.15(\text{K})$$

$$= 612(\text{J})/(\text{g mole})$$

$$\frac{\Delta S'}{R} = \ln\frac{RT}{P(V-b)} = \ln\frac{(82.05)(373.15)}{(40)(707.7-31.3)} = \ln 1.132 = 0.124$$

and $$\Delta S' = 0.124\,R = (0.124)(8.314) = 1.03(\text{J})/(\text{g mole})(\text{K})$$

The above calculations were based on values of the van der Waals constants determined by use of (*5.35*). There is, however, an alternative and more widely used pair of equations, which are developed by elimination of V_c from (*5.35*) by means of (*5.36*):

$$Z_c = \frac{P_cV_c}{RT_c} = \frac{3}{8}$$

Thus $V_c = \frac{3}{8}\frac{RT_c}{P_c}$ and combination with (*5.35*) yields

$$a = \frac{27}{64}\frac{R^2T_c^2}{P_c} \quad \text{or} \quad \frac{a}{RT} = \frac{27}{64}\frac{RT_c^2}{P_cT}$$

$$b = \frac{1}{8}\frac{RT_c}{P_c}$$

These expressions for the van der Waals constants are based on values for T_c and P_c, and do not require knowledge of V_c, which is much more difficult to determine experimentally. Substitution of numerical values gives

$$\frac{a}{RT} = \frac{27 \times 82.05(\text{cm}^3\text{-atm})/(\text{g mole})(\text{K}) \times 304.2^2(\text{K})^2}{64 \times 72.9(\text{atm}) \times 373.15(\text{K})} = 117.8(\text{cm})^3/(\text{g mole})$$

$$b = \frac{82.05 \times 304.2}{8 \times 72.9} = 42.8(\text{cm})^3/(\text{g mole})$$

These values of van der Waals constants are different from those determined earlier, because the true critical constants T_c, P_c, and V_c used do not satisfy the artificial requirement of (*5.36*) that $Z_c = 3/8$. Use of this second pair of constants gives a second set of results as shown in the second row of the following table.

	Z	V (cm)^3/(g mole)	$\Delta H'$ (J)/(g mole)	$\Delta S'$ (J)/(g mole)(K)
a and b from T_c and V_c	0.9245	707.7	612	1.03
a and b from T_c and P_c	0.8947	684.9	860	1.46
Experimental	0.9012	689.9	1150	2.25

The calculated results of this problem can be compared with the experimental values listed in the last row of the table. The calculated volumes are not too far in error. However, the derived quantities $\Delta H'$ and $\Delta S'$ are seen to be in poor agreement with experiment. It is generally true that the prediction of satisfactory values for derived quantities requires an equation of state that provides a very accurate representation of volumetric behavior.

5.18. One form of a three-constant equation of state is

$$P = \frac{RT}{V-b} - \frac{a}{T(V+c)^2} \qquad (1)$$

If one requires this equation to satisfy (*5.32*) and (*5.33*) and in addition to produce the correct value of Z_c for any given fluid, show that a particular form of three-parameter corresponding-states correlation results.

Differentiation of (*1*) and imposition of (*5.32*) and (*5.33*) yields:

$$\left(\frac{\partial P}{\partial V}\right)_{T,cr} = -\frac{RT_c}{(V_c-b)^2} + \frac{2a}{T_c(V_c+c)^3} = 0 \tag{2}$$

$$\left(\frac{\partial^2 P}{\partial V^2}\right)_{T,cr} = \frac{2RT_c}{(V_c-b)^3} - \frac{6a}{T_c(V_c+c)^4} = 0 \tag{3}$$

Equation (*1*) may be put in the form

$$Z = \frac{V}{V-b} - \frac{aV}{RT^2(V+c)^2}$$

Therefore

$$Z_c = \frac{V_c}{V_c-b} - \frac{aV_c}{RT_c^2(V_c+c)^2} \tag{4}$$

Combination of (*2*) and (*3*) and solution for b gives

$$b = \frac{V_c-2c}{3} \tag{5}$$

and combination of (*2*) and (*5*) gives

$$a = \frac{9}{8}RT_c^2(V_c+c) \tag{6}$$

Substituting (*5*) and (*6*) in (*4*) and solving for c, we obtain

$$c = V_c\left(\frac{3-8Z_c}{8Z_c}\right) \tag{7}$$

from which

$$a = \frac{27}{64}\frac{RT_c^2V_c}{Z_c} \tag{8}$$

and

$$b = V_c\left(\frac{4Z_c-1}{4Z_c}\right) \tag{9}$$

The original equation of state can now be written in dimensionless form. Noting that

$$P_r = \frac{P}{P_c} = \frac{PV_c}{Z_cRT_c} \tag{10}$$

we obtain, after combination of (*1*), (*7*), (*8*), (*9*), and (*10*),

$$P_r = \frac{4T_r}{1+4Z_c(V_r-1)} - \frac{3}{T_r[1+\frac{8}{3}Z_c(V_r-1)]^2} \tag{11}$$

According to (*11*), all fluids having the same Z_c will have the same P_r when compared at the same V_r and T_r. All three-constant equations of state in which the constants are determined as above will yield the same conclusion, although the final dimensionless equations corresponding to (*11*) will in general be different.

CORRESPONDING-STATES CORRELATIONS (Sec. 5.5)

5.19. Derive an expression for ω for a fluid obeying the vapor-pressure equation (*4.5*). For the same fluid, show how ω is related to the slope of the reduced-vapor-pressure curve plotted with $\log_{10} P_r$ and $1/T_r$ as coordinates.

Equation (*4.5*) is

$$\ln P^{sat} = A - \frac{B}{T} \tag{4.5}$$

where A and B are constants (B is not to be confused with the second virial coefficient). Since the gas-liquid critical point is the terminus of the vapor-pressure curve,

$$\ln P_c = A - \frac{B}{T_c} \qquad (1)$$

and subtraction of (1) from (4.5) gives

$$\ln P_r^{\text{sat}} = \frac{B}{T_c}\left(1 - \frac{1}{T_r}\right) \qquad (2)$$

Letting $T_r = 0.7$ and changing to common logarithms,

$$\log_{10} P_r^{\text{sat}} = -\frac{3 \log_{10} e}{7} \frac{B}{T_c} \quad \text{at } T_r = 0.7$$

from which we obtain, according to the definition of ω, (5.38),

$$\omega = \frac{3 \log_{10} e}{7} \frac{B}{T_c} - 1 \qquad (3)$$

Equation (2) is linear in $1/T_r$, so the slope $\mathcal{S}$ of a $\log_{10} P_r^{\text{sat}}$ versus $1/T_r$ plot is

$$\mathcal{S} = -(\log_{10} e)\frac{B}{T_c} \qquad (4)$$

Combination of (3) and (4) yields

$$\mathcal{S} = -\frac{7}{3}(\omega + 1)$$

For a simple fluid ($\omega = 0$) obeying (4.5), $\mathcal{S} = -7/3$, and for non-simple fluids $\mathcal{S}$ is increasingly negative with increasing ω. Equation (4.5) is only a first approximation to the behavior of real simple fluids, and the first of these conclusions is only in qualitative agreement with experiment; the second, however, is generally valid.

5.20. A 10(ft)³ tank to contain propane has a bursting pressure of 400(psia). Safety considerations dictate that the tank be charged with no more propane than would exert half the bursting pressure at a temperature of 260(°F). How many pounds mass of propane may be charged to the tank? The molecular weight of propane is 44.1.

At 260(°F) = 399.8(K) propane is (see Appendix 3) above its critical temperature, so the system is certain to contain only gas at this temperature. The allowable mass m of propane is found from the equation

$$PV^t = (m/M)ZRT$$

or

$$m = MPV^t/ZRT \qquad (1)$$

where V^t is the system volume and M is the molecular weight of propane. We now estimate Z from the Pitzer correlation and (5.48). From Appendix 3, $T_c = 369.9(\text{K})$, $P_c = 42.0(\text{atm})$, and $\omega = 0.152$. At the stated conditions, then,

$$T_r = \frac{399.8(\text{K})}{369.9(\text{K})} = 1.08$$

$$P_r = \frac{200(\text{psia})}{42.0(\text{atm}) \times 14.7\dfrac{(\text{psia})}{(\text{atm})}} = 0.324$$

From Fig. 5-6, at $T_r = 1.08$

$$B^0 = -0.285 \qquad B^1 = -0.03$$

so, by (5.48),

$$Z = 1 + \frac{P_r}{T_r}(B^0 + \omega B^1) = 1 + \frac{0.324}{1.08}[(-0.285) + (0.152)(-0.03)] = 0.913$$

The mass m can now be calculated from (1):

$$m = \frac{44.1 \frac{(\text{lb}_\text{m})}{(\text{lb mole})} \times 200(\text{psia}) \times 10(\text{ft})^3}{0.913 \times 10.73 \frac{(\text{psia})(\text{ft})^3}{(\text{lb mole})(\text{R})} \times 719.7(\text{R})} = 12.51(\text{lb}_\text{m})$$

5.21. Estimate Z, $\Delta H'$, and $\Delta S'$ for hydrogen sulfide gas at 10(atm) and 200(°C).

We use the Pitzer correlation and the results of Example 5.10. From Appendix 3, $T_c = 373.6(\text{K})$, $P_c = 88.9(\text{atm})$, and $\omega = 0.100$. Then

$$T_r = \frac{200 + 273.2}{373.6} = 1.267 \qquad P_r = \frac{10}{88.9} = 0.112$$

and from Figs. 5-6 and 5-7

$$B^0 = -0.21 \qquad B^1 = 0.06$$

$$\frac{dB^0}{dT_r} = 0.36 \qquad \frac{dB^1}{dT_r} = 0.31$$

By (*5.48*) and (*5.49*)

$$Z = 1 + \frac{P_r}{T_r}(B^0 + \omega B^1) = 1 + \frac{0.112}{1.267}[-0.21 + (0.100)(0.06)] = 0.982$$

$$\frac{\Delta H'}{RT} = P_r\left[\left(\frac{dB^0}{dT_r} - \frac{B^0}{T_r}\right) + \omega\left(\frac{dB^1}{dT_r} - \frac{B^1}{T_r}\right)\right]$$

$$= 0.112\left[\left(0.36 + \frac{0.21}{1.267}\right) + 0.100\left(0.31 - \frac{0.06}{1.267}\right)\right] = 0.06183$$

Thus $$\Delta H' = 0.06183 \times 8.314 \frac{(\text{J})}{(\text{g mole})(\text{K})} \times 473.2(\text{K}) = 243.3(\text{J})/(\text{g mole})$$

and, by (*5.50*),

$$\frac{\Delta S'}{R} = P_r\left(\frac{dB^0}{dT_r} + \omega\frac{dB^1}{dT_r}\right) = 0.112[0.36 + (0.100)(0.31)] = 0.0438$$

from which $$\Delta S' = 0.0438 \times 8.314 = 0.364(\text{J})/(\text{g mole})(\text{K})$$

5.22. Estimate the second virial coefficient of air at 300(K).

We use the Pitzer correlation and the combination rules (*5.43*) through (*5.47*). Take air to contain 79 mole % nitrogen and 21 mole % oxygen. Identify nitrogen as component 1 and oxygen as component 2. From Appendix 3:

$$T_{c_1} = 126.2(\text{K}) \qquad T_{c_2} = 154.8(\text{K})$$

$$P_{c_1} = 33.5(\text{atm}) \qquad P_{c_2} = 50.1(\text{atm})$$

$$Z_{c_1} = 0.291 \qquad Z_{c_2} = 0.293$$

$$V_{c_1} = 90.1(\text{cm})^3/(\text{g mole}) \qquad V_{c_2} = 74.4(\text{cm})^3/(\text{g mole})$$

$$\omega_1 = 0.040 \qquad \omega_2 = 0.021$$

By (*5.43*): $$T_{c_{12}} = \sqrt{(126.2)(154.8)} = 139.8(\text{K})$$

By (*5.45*): $$Z_{c_{12}} = \frac{0.291 + 0.293}{2} = 0.292$$

By (*5.46*): $$V_{c_{12}} = \left(\frac{90.1^{1/3} + 74.4^{1/3}}{2}\right)^3 = 82.0(\text{cm})^3/(\text{g mole})$$

By (*5.47*): $$\omega_{12} = \frac{0.040 + 0.021}{2} = 0.0305$$

Finally by (*5.44*)

$$P_{c_{12}} = \frac{(0.292)(82.05)(139.8)}{82.0} = 40.9(\text{atm})$$

At 300(K), then,

$$T_{r_1} = \frac{300}{126.2} = 2.38 \qquad T_{r_2} = \frac{300}{154.8} = 1.94$$

$$T_{r_{12}} = \frac{T}{T_{c_{12}}} = \frac{300}{139.8} = 2.15$$

From Fig. 5-6,

$$B_{11}^0 = -0.02 \qquad B_{11}^1 = 0.18$$

$$B_{22}^0 = -0.065 \qquad B_{22}^1 = 0.17$$

$$B_{12}^0 = -0.04 \qquad B_{12}^1 = 0.175$$

and, from (*5.39*) and (*5.42*),

$$B_{11} = \frac{RT_{c_1}}{P_{c_1}}[B_{11}^0 + \omega_1 B_{11}^1] = \frac{(82.05)(126.2)}{(33.5)}[-0.02 + (0.040)(0.18)]$$

$$= -4.0(\text{cm})^3/(\text{g mole})$$

$$B_{22} = \frac{RT_{c_2}}{P_{c_2}}[B_{22}^0 + \omega_2 B_{22}^1] = \frac{(82.05)(154.8)}{(50.1)}[-0.065 + (0.021)(0.17)]$$

$$= -15.6(\text{cm})^3/(\text{g mole})$$

$$B_{12} = \frac{RT_{c_{12}}}{P_{c_{12}}}[B_{12}^0 + \omega_{12} B_{12}^1] = \frac{(82.05)(139.8)}{(40.9)}[-0.04 + (0.0305)(0.175)]$$

$$= -9.7(\text{cm})^3/(\text{g mole})$$

For a binary mixture

$$B = y_1^2 B_{11} + 2y_1 y_2 B_{12} + y_2^2 B_{22} \tag{5.17}$$

so that

$$B_{\text{air}} = (0.79)^2(-4.0) + (2)(0.79)(0.21)(-9.7) + (0.21)^2(-15.6) = -6.4(\text{cm})^3/(\text{g mole})$$

at 300(K). The experimental value is $-7.5(\text{cm})^3/(\text{g mole})$.

5.23. Propane gas undergoes a change of state from an initial condition of 5(atm) and 105(°C) to 25(atm) and 190(°C). Using the virial equation of state (*5.22*) and the generalized correlation of Sec. 5.5, determine values of ΔH and ΔS for this change of state. The molar heat capacity of propane in the ideal-gas state is given by $C_P' = 5.49 + 0.0424\,T$, where T is in (K) and C_P' is in (cal)/(g mole)(K).

Where tabular data for derived thermodynamic functions are not available, one must devise a *calculational path* connecting the states of interest which allows use of what information is available. The property changes ΔH and ΔS are independent of path, and therefore the choice of path is made solely on the basis of convenience. The residual functions $\Delta H'$ and $\Delta S'$ connect the real state of a gas with the ideal-gas state at the same T and P. For the ideal-gas state, property changes are readily calculated by the equations for ideal gases. Thus the most convenient calculational path consists of three steps:

(1) We imagine the real gas in its initial state to be transformed into an ideal gas at the same conditions of T_1 and P_1. Associated with this step are the property changes $\Delta H_1'$ and $\Delta S_1'$.

(2) The *ideal gas* is now changed from a state at T_1 and P_1 to a state at T_2 and P_2, the final conditions of temperature and pressure. Associated with this step are enthalpy and entropy changes resulting from the change from T_1, P_1 to T_2, P_2.

(3) We now imagine the ideal gas to be transformed back into a real gas at T_2 and P_2.

Associated with this step are the property changes $-\Delta H_2'$ and $-\Delta S_2'$. The minus signs are required because the change of state is from ideal to real, whereas the quantities $\Delta H'$ and $\Delta S'$ represent the difference (ideal) − (real).

The required values of $\Delta H'$ and $\Delta S'$ can be calculated by *(5.49)* and *(5.50)* of Example 5.10:

$$\frac{\Delta H'}{RT} = P_r\left[\left(\frac{dB^0}{dT_r} - \frac{B^0}{T_r}\right) + \omega\left(\frac{dB^1}{dT_r} - \frac{B^1}{T_r}\right)\right] \tag{5.49}$$

$$\frac{\Delta S'}{R} = P_r\left(\frac{dB^0}{dT_r} + \omega\frac{dB^1}{dT_r}\right) \tag{5.50}$$

The quantities B^0, B^1, dB^0/dT_r, and dB^1/dT_r are found from Figs. 5-6 and 5-7.

For propane we have from Appendix 3:

$$T_c = 369.9(\text{K}) \qquad P_c = 42.0(\text{atm}) \qquad \omega = 0.152$$

Thus at the initial state

$$T_{r_1} = \frac{105 + 273}{369.9} = 1.022 \qquad P_{r_1} = \frac{5}{42.0} = 0.119$$

From Figs. 5-6 and 5-7 we find

$$B^0 = -0.32 \qquad B^1 = -0.07$$

$$dB^0/dT_r = 0.62 \qquad dB^1/dT_r = 0.84$$

Thus by *(5.49)*

$$\frac{\Delta H_1'}{RT_1} = 0.119\left[\left(0.62 - \frac{-0.32}{1.022}\right) + 0.152\left(0.84 - \frac{-0.07}{1.022}\right)\right]$$

$$= 0.119[0.93 + 0.152(0.91)] = 0.127$$

from which

$$\Delta H_1' = 0.127\,RT_1 = 0.127 \times 8.314(\text{J})/(\text{g mole})(\text{K}) \times 378(\text{K}) = 400(\text{J})/(\text{g mole})$$

From *(5.50)*

$$\frac{\Delta S_1'}{R} = 0.119(0.62 + 0.152 \times 0.84) = 0.089$$

and

$$\Delta S_1' = 0.089 \times 8.314(\text{J})/(\text{g mole})(\text{K}) = 0.74(\text{J})/(\text{g mole})(\text{K})$$

At the final state

$$T_{r_2} = \frac{190 + 273}{369.9} = 1.25 \qquad P_{r_2} = \frac{25}{42.0} = 0.595$$

From Figs. 5-6 and 5-7 we get

$$B^0 = -0.21 \qquad B^1 = 0.06$$

$$dB^0/dT_r = 0.37 \qquad dB^1/dT_r = 0.33$$

These values allow us to compute

$$\frac{\Delta H_2'}{RT_2} = 0.345 \quad \text{or} \quad \Delta H_2' = 0.345 \times 8.314 \times 463 = 1330(\text{J})/(\text{g mole})$$

and

$$\frac{\Delta S_2'}{R} = 0.25 \quad \text{or} \quad \Delta S_2' = 0.25 \times 8.314 = 2.08(\text{J})/(\text{g mole})(\text{K})$$

The calculational path to be used is represented in Fig. 5-9. Since we have already evaluated $\Delta H_1'$, $\Delta S_1'$, $\Delta H_2'$, and $\Delta S_2'$, we have only to determine $\Delta H_{12}'$ and $\Delta S_{12}'$:

$$\Delta H_{12}' = \int_{T_1}^{T_2} C_P'\,dT = \int_{378}^{463} (5.49 + 0.0424\,T)\,dT$$

$$= 5.49(463 - 378) + \frac{0.0424}{2}(463^2 - 378^2)$$

$$= 1980(\text{cal})/(\text{g mole}) \quad \text{or} \quad 8290(\text{J})/(\text{g mole})$$

$$\Delta S'_{12} = \int_{T_1}^{T_2} C'_P \frac{dT}{T} - R \ln \frac{P_2}{P_1} = \int_{378}^{463} \left(\frac{5.49}{T} + 0.0424\right) dT - R \ln \frac{25}{5}$$

$$= 5.49 \ln \frac{463}{378} + 0.0424(463 - 378) - 1.987 \ln 5$$

$$= 1.52(\text{cal})/(\text{g mole})(\text{K}) \quad \text{or} \quad 6.36(\text{J})/(\text{g mole})(\text{K})$$

For the entire process

$$\Delta H = H_2 - H_1 = \Delta H'_1 + \Delta H'_{12} - \Delta H'_2$$

$$= 400 + 8290 - 1330 = 7360(\text{J})/(\text{g mole})$$

$$\Delta S = S_2 - S_1 = \Delta S'_1 + \Delta S'_{12} - \Delta S'_2$$

$$= 0.74 + 6.36 - 2.08 = 5.02(\text{J})/(\text{g mole})(\text{K})$$

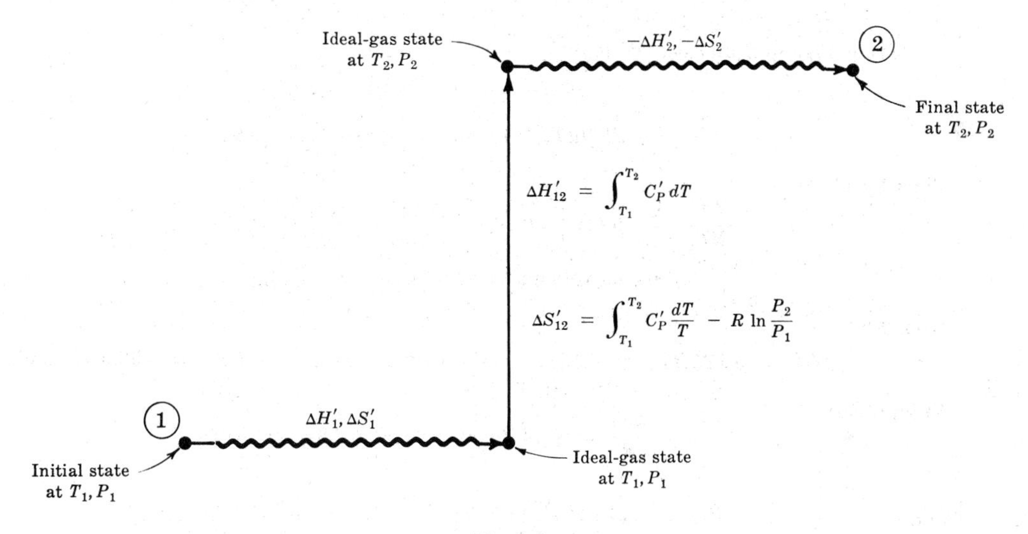

Fig. 5-9

Supplementary Problems

THE COMPRESSIBILITY FACTOR (Sec. 5.1)

5.24. Derive expressions for the volume expansivity β and the isothermal compressibility κ in terms of T, P, and Z and its derivatives.

Ans. $\beta = \dfrac{1}{T} + \dfrac{1}{Z}\left(\dfrac{\partial Z}{\partial T}\right)_P$, $\quad \kappa = \dfrac{1}{P} - \dfrac{1}{Z}\left(\dfrac{\partial Z}{\partial P}\right)_T$

5.25. (*a*) Show that in the two-phase vapor-liquid region $Z = Z_f + xZ_{fg}$, where the notation is that of Sec. 4.4. (*b*) Show that at the gas-liquid critical point

$$\left(\frac{\partial Z}{\partial V}\right)_{T,cr} = \frac{Z_c}{V_c} \qquad \left(\frac{\partial^2 Z}{\partial V^2}\right)_{T,cr} = 0$$

5.26. Prove:

(a) $$C_P - C_V = \frac{\dfrac{RT^2}{Z}\left(\dfrac{\partial Z}{\partial T}\right)_V^2 + 2RT\left(\dfrac{\partial Z}{\partial T}\right)_V + RZ}{1 - \dfrac{V}{Z}\left(\dfrac{\partial Z}{\partial V}\right)_T}$$

(b) $$\left(\frac{\partial C_V}{\partial V}\right)_T = \frac{RT^2}{V}\left(\frac{\partial^2 Z}{\partial T^2}\right)_V + \frac{2RT}{V}\left(\frac{\partial Z}{\partial T}\right)_V$$

(c) $$\left(\frac{\partial C_P}{\partial P}\right)_T = -\frac{RT^2}{P}\left(\frac{\partial^2 Z}{\partial T^2}\right)_P - \frac{2RT}{P}\left(\frac{\partial Z}{\partial T}\right)_P$$

5.27. Analogous to the quantities $\Delta V'$, $\Delta H'$, and $\Delta S'$, we may define a residual internal energy $\Delta U'$, a residual Helmholtz function $\Delta A'$, and a residual Gibbs function $\Delta G'$. Show that the dimensionless forms of these functions are calculable from T, P, and Z and its derivatives as follows:

$$\frac{\Delta U'}{RT} = T\int_0^P \left(\frac{\partial Z}{\partial T}\right)_P \frac{dP}{P} + (Z-1) \qquad \text{(const. } T\text{)}$$

$$\frac{\Delta A'}{RT} = (Z-1) - \int_0^P (Z-1)\frac{dP}{P} \qquad \text{(const. } T\text{)}$$

$$\frac{\Delta G'}{RT} = -\int_0^P (Z-1)\frac{dP}{P} \qquad \text{(const. } T\text{)}$$

VIRIAL EQUATIONS (Secs. 5.2 and 5.3)

5.28. Prove: $$B' = \lim_{P\to 0}\left[\frac{Z-1}{P}\right]_T \quad \text{and} \quad C' = \lim_{P\to 0}\left[\frac{\partial\left(\dfrac{Z-1}{P}\right)}{\partial P}\right]_T = \lim_{P\to 0}\left[\frac{\dfrac{Z-1}{P} - B'}{P}\right]_T$$

5.29. Calculate B for an equimolar ternary mixture of methane(1), propane(2), and n-pentane(3) at 100(°C). At 100(°C), in (cm)³/(g mole) units,

$$B_{11} = -20 \qquad B_{22} = -241 \qquad B_{33} = -621$$
$$B_{12} = -75 \qquad B_{13} = -122 \qquad B_{23} = -399$$

Ans. $B = -230$(cm)³/(g mole)

5.30. Estimate Z, $\Delta H'$, and $\Delta S'$ for steam at 300(°C) and 5(atm) from the following values for the second virial coefficient of water vapor:

T(°C)	B(cm)³/(g mole)
290	−125
300	−119
310	−113

Ans. $Z = 0.987$, $\Delta H' = 235$(J)/(g mole), $\Delta S' = 0.304$(J)/(g mole)(K)

EMPIRICAL EQUATIONS OF STATE (Sec. 5.4)

5.31. Find $\Delta H'/RT$ and $\Delta S'/R$ for the Redlich-Kwong equation.

Ans. $$\Delta H'/RT = \frac{3a}{2bRT^{3/2}}\ln\left(\frac{V+b}{V}\right) + (1-Z)$$

$$\Delta S'/R = \ln\left(\frac{V}{V-b}\right) + \frac{a}{2bRT^{3/2}}\ln\left(\frac{V+b}{V}\right) - \ln Z$$

5.32. (*a*) Proceeding as in Example 5.8, find a and b in terms of T_c and V_c for the Redlich-Kwong equation. What is the Redlich-Kwong Z_c? (*b*) Write the Redlich-Kwong equation in reduced form from the results of part (*a*). (*c*) Derive an expression for the reduced Redlich-Kwong second virial coefficient $B_{RK}P_c/RT_c$.

Ans. (*a*) $$a = \frac{RT_c^{3/2}V_c}{3A}, \quad b = AV_c, \; Z_c = \tfrac{1}{3}$$

where $A = 2^{1/3} - 1$

(*b*) $$P_r = \frac{3T_r}{V_r - A} - \frac{1}{AT_r^{1/2}(V_r + A)V_r}$$

(*c*) $$\frac{B_{RK}P_c}{RT_c} = \frac{A}{3} - \frac{1}{9AT_r^{3/2}}$$

5.33. (*a*) Show that a point on the inversion curve for a van der Waals fluid must satisfy the equation

$$(2a - bRT)V^2 - 4abV + 2ab^2 = 0$$

(*b*) The *Boyle curve* of a fluid is the locus of points for which $(\partial Z/\partial P)_T = 0$. Show that a point on the Boyle curve for a van der Waals fluid must satisfy the equation

$$(a - bRT)V^2 - 2abV + ab^2 = 0$$

CORRESPONDING-STATES CORRELATIONS (Sec. 5.5)

5.34. What is the acentric factor of a fluid which obeys the reduced van der Waals equation (*5.37*)? [*Hint:* Use the definition (*5.38*) and the result of Problem 5.16.]

Ans. $\omega = -0.303$

5.35. The *Boyle temperature* of a real gas is the temperature for which $(\partial Z/\partial P)_T = 0$ at $P = 0$, i.e. the temperature at which the Boyle curve intersects the Z axis on a ZP diagram [Problem 5.33, part (*b*)]. Show that $B = 0$ at the Boyle temperature, and estimate the Boyle temperature for oxygen from the Pitzer correlation.

Ans. 122(°C). An experimental value is 150(°C).

5.36. Show that, in the range of validity of (*5.48*), (*5.49*), and (*5.50*), the following equations hold [see Problem 5.27]:

$$\frac{\Delta U'}{RT} = P_r\left(\frac{dB^0}{dT_r} + \omega\frac{dB^1}{dT_r}\right) \qquad \frac{\Delta A'}{RT} = 0$$

$$\frac{\Delta G'}{RT} = -\frac{P_r}{T_r}(B^0 + \omega B^1)$$

MISCELLANEOUS APPLICATIONS

5.37. Calculate Z and V [in (cm)^3/(g mole)] for isopropanol vapor at 200(°C) and 10(atm) by the following methods:

(*a*) Use the truncated virial equation (*5.21*) with the experimentally determined virial coefficients $B = -388$(cm)^3/(g mole) and $C = -26{,}000$(cm)^6/(g mole)^2.

(*b*) Use the truncated virial equation (*5.22*), employing the generalized correlation of Section 5.5 to estimate B.

(*c*) Use the van der Waals equation (*5.27*), employing (*5.35*) to calculate a and b.

The following physical constants are available for isopropanol: $T_c = 508.2$(K), $P_c = 50.0$(atm), $V_c = 220.4$(cm)^3/(g mole), $\omega = 0.700$.

Ans.

	Z	$V \frac{(\text{cm})^3}{(\text{g mole})}$
Equation (*5.21*)	0.8848	3435
Equation (*5.22*) with correlation	0.8947	3474
Equation (*5.27*) with (*5.35*)	0.9481	3681

5.38. Steam undergoes a change of state from an initial state of 900(°F) and 500(psia) to a final state of 300(°F) and 40(psia). Determine ΔH and ΔS in units of (Btu), (R), and (lb mole) by use of (*a*) the steam tables, (*b*) the generalized correlation of Section 5.5 and the method illustrated in Problem 5.23. The molar heat capacity of steam is given by:

$$C'_P = 7.105 + 0.001467\,T$$

where T is in rankines and C'_P is in (Btu)/(lb mole)(R).

Ans. (*a*) $\Delta H = -5034.6$(Btu)/(lb mole), $\Delta S = 0.0108$(Btu)/(lb mole)(R)

(*b*) $\Delta H = -4971$(Btu)/(lb mole), $\Delta S = 0.080$(Btu)/(lb mole)(R)

Chapter 6

Thermodynamics of Flow Processes

The first law of thermodynamics, given in Chapter 1 by (*1.3*), is rewritten here as

$$\Delta E^t = Q - W$$

This equation is the mathematical formulation of the principle of energy conservation as applied to a process occurring in a closed system, i.e. in a system of constant mass. The total energy of the system E^t includes energy in both its internal and external forms, and thus the total energy change ΔE^t is the sum of several terms:

$$\Delta E^t = \Delta U^t + \Delta E_K^t + \Delta E_P^t$$

where the superscript t signifies that the various energy terms refer to the entire system.

Most of our applications of the first law have been to systems at rest, and we have therefore been able to set $\Delta E^t = \Delta U^t$, giving

$$\Delta U^t = Q - W$$

In differential form this becomes

$$dU^t = \delta Q - \delta W \tag{6.1}$$

For a homogeneous fluid system (*6.1*) may also be written

$$m\,dU = \delta Q - \delta W \tag{6.2}$$

where U is the specific or unit-mass internal energy of the fluid and m is the (constant) mass of the system. If U is the molar internal energy, then m is replaced by n, the number of moles.

Inherent in the formulation of (*6.2*) is the assumption that the system has uniform properties throughout, and hence that each unit mass of fluid in the system experiences the same change of internal energy.

Our initial purpose in this chapter is to develop from (*6.1*) an equation more general than (*6.2*), in particular an equation applicable to closed systems of several parts, each having its own set of uniform properties, but between which fluid may flow. Subsequently we will develop less restricted energy equations, applicable to open systems and to flow processes generally. In addition we will consider flow processes from the standpoint of the second law.

6.1 ENERGY EQUATIONS FOR CLOSED SYSTEMS

A closed system is necessarily one of constant mass. Such systems, considered generally, may consist of several interconnected parts, none of which is itself constrained to constant mass. However, the total mass of all parts taken together must be constant. The kind of system under consideration is illustrated in Fig. 6-1.

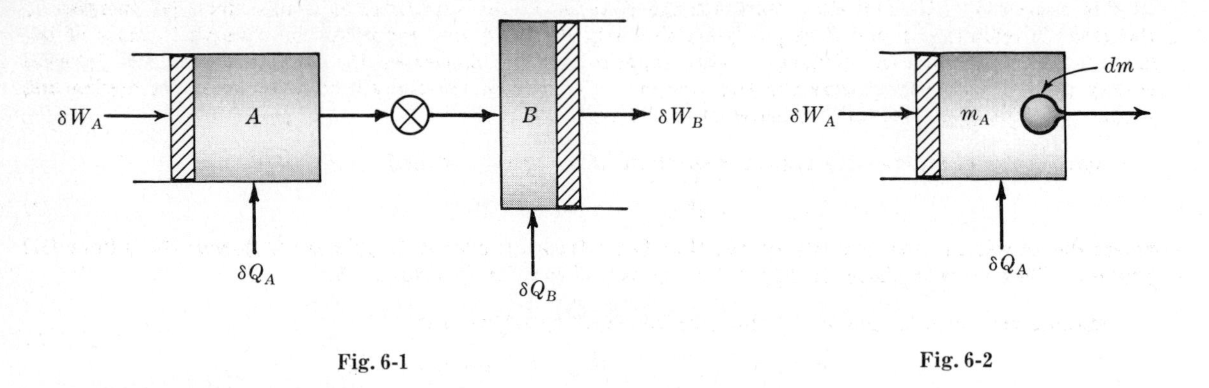

Fig. 6-1

Fig. 6-2

This figure shows two cylinders connected through a partially opened valve, and we may consider that gas is transferred from cylinder A to cylinder B by appropriate motion of the pistons. The two identifiable parts or regions of the system are the two cylinders. We assume that the properties of the gas in each cylinder are uniform throughout (but different in the two cylinders). The total internal energy of the system at any instant is then given by

$$U^t = m_A U_A + m_B U_B$$

and for a differential change in the system

$$dU^t = d(m_A U_A + m_B U_B)$$

The total heat and work quantities are clearly

$$\delta Q = \delta Q_A + \delta Q_B \quad \text{and} \quad \delta W = \delta W_A + \delta W_B$$

Substitution of these last three expressions into (*6.1*) gives

$$d(m_A U_A + m_B U_B) = \delta Q_A + \delta Q_B - (\delta W_A + \delta W_B)$$

This equation applies to a system of two regions as shown in Fig. 6-1. For closed systems having an arbitrary number of interconnected regions, we have the general expression

$$d \sum_R (mU) = \sum \delta Q - \sum \delta W \tag{6.3}$$

where the symbol $\sum_R$ signifies summation over the regions of the system. Since the differential of a sum is equal to the sum of differentials, (*6.3*) may equally well be written

$$\sum_R d(mU) = \sum \delta Q - \sum \delta W \tag{6.3a}$$

The symbol $\sum$ with δQ and δW merely indicates that all terms of the type considered must be included.

Example 6.1. With respect to the process illustrated in Fig. 6-1, expand completely the differential

$$dU^t = d(m_A U_A + m_B U_B)$$

and identify each of the resulting terms with a particular mass of gas in the system.

Expansion of the differential gives

$$dU^t = m_A\, dU_A + U_A\, dm_A + m_B\, dU_B + U_B\, dm_B \tag{1}$$

In this expression dU_A and dU_B represent the changes in the specific or unit-mass internal energies of the gases in cylinders A and B respectively; similarly dm_A and dm_B represent the changes in mass of the gas contained in the two cylinders. This equation merely illustrates the fact that the total internal energy of a region may change for two reasons: a change in the amount of material in the region and a change in the properties of the material in the region.

Since the total mass of the system is constant, $dm_A + dm_B = 0$ and we may write

$$dm_B = -dm_A = dm \tag{2}$$

where dm represents the quantity of gas that flows from cylinder A to cylinder B during the differential process. This mass is shown in Fig. 6-2 at its initial location in cylinder A.

Making the substitutions in (*1*) that are indicated by (*2*), we get

$$dU^t = m_A\,dU_A + (U_B - U_A)\,dm + m_B\,dU_B$$

This equation illustrates an additional method of accounting for the total internal energy change of the system. It indicates the several masses that we have selected as comprising the system; after each such mass is multiplied by its change in specific internal energy, the products are added to give the total internal-energy change. Thus the quantity of gas in cylinder A may be considered in two parts. First is that part which will remain in cylinder A after the process occurs; it has a mass m_A and experiences a change in specific internal energy dU_A. Second is that differential mass of gas dm that will flow out of cylinder A and into cylinder B during the process and which therefore experiences a finite change of internal energy $U_B - U_A$. Finally we identify the mass of gas m_B which is initially in cylinder B and which remains there and experiences an internal-energy change dU_B. It is always possible to determine the total internal-energy change of a closed system by such an accounting scheme.

Integration of (*6.3*) for a finite process gives immediately

$$\Delta \sum_R (mU) = \sum Q - \sum W \tag{6.4}$$

However, such direct integration is not always advantageous, because it may yield an equation in several unknowns. On the other hand, the original differential equation may be reducible to a simpler form, which when integrated can be solved for a single unknown.

It is useful to substitute for U in (*6.3*) so as to obtain an alternative energy equation in terms of the enthalpy H, because the resulting equation is sometimes easier to apply. Since by definition, $H = U + PV$,

$$d\sum_R (mH) = d\sum_R (mU) + d\sum_R (mPV)$$

Combination with (*6.3*) gives

$$\boxed{d\sum_R (mH) = \sum \delta Q - \sum \delta W + d\sum_R (mPV)} \tag{6.5}$$

This equation proves to be particularly useful for application to processes where the pressure is constant in each region of the system (but not necessarily the same in the different regions), because in this case the last two terms can be made to cancel. For this to be true the process must be mechanically reversible (see Sec. 1.5) within each region of the system. We must *assume* this is so if in fact we are to evaluate the work terms. With this assumption, each work term comes from (*1.2*), here rewritten as $\delta W = P\,d(mV)$. If P is constant, then

$$\delta W = d(mPV) \quad \text{and} \quad \sum \delta W = \sum_R d(mPV) = d\sum_R (mPV)$$

Substitution into (*6.5*) reduces it to the particularly simple equation

$$d\sum_R (mH) = \sum \delta Q \tag{6.6}$$

This is a generalization of the result obtained in Example 1.7 for a constant-pressure, mechanically reversible process where the pressure was uniform throughout the system. Equation (*6.6*) shows that the change in total enthalpy for a closed system is equal to the heat transfer, provided the pressure is constant in each region of the system, even though different in different regions, and provided the process is mechanically reversible within each region. Integration of (*6.6*) gives

$$\Delta \sum_R (mH) \quad = \quad \sum Q \tag{6.7}$$

or alternatively

$$\sum_R \Delta(mH) \quad = \quad \sum Q \tag{6.7a}$$

The equations so far developed contain no terms to account for changes in potential or kinetic energy of the fluid that flows from one region of the system to another. The absence of terms to account for changes in gravitational potential energy of fluid moving from one region of the system to another can be justified only if the difference in elevation of the various regions is inconsequential. This is almost always the case. Where it is not, the best course to follow is to employ a more general equation, (*6.12*), developed later in this chapter.

The omission of terms for changes in kinetic energy is based on the presumption that applications of the energy equations developed in this section are to processes which start and end with the fluid at rest. Thus no kinetic energy change occurs, and no term to account for it is necessary. Where this presumption is not fulfilled, one must again turn to a more general energy equation.

Example 6.2. A crude method for determining the enthalpy of steam in a steam line is to bleed steam from the line through a hose into a barrel of water. The mass of the barrel with its contents and the temperature of the water are recorded at the beginning and end of the process. The condensation of steam by the water raises the temperature of the water.

Our energy equations are written for a system of constant mass. It is therefore essential to choose the system so that it includes not only the water originally in the barrel, but also the steam in the line that will be condensed by the water during the process. We may imagine a piston in the steam line that separates the steam in the line into two parts: that which will enter the barrel and that which will remain in the line. This view is represented schematically in Fig. 6-3.

In addition to the piston in the steam line, another piston is imagined to separate the water in the barrel from the atmosphere. Both the atmosphere and the steam in the line exert forces (shown as pressures, i.e. forces distributed over an area) which act on the system. These forces move during the process and therefore do work.

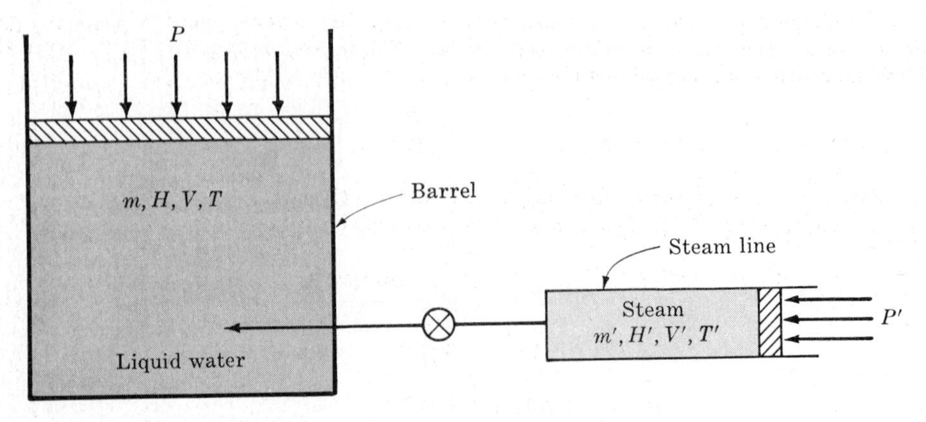

Fig. 6-3

The pressure P exerted by the atmosphere on the water in the barrel is taken as constant, but the other properties of the water change during the process. Thus the mass changes from m_1 to m_2; the specific enthalpy, from H_1 to H_2; the specific volume, from V_1 to V_2; and the temperature, from T_1 to T_2. In the steam line, however, the intensive properties T', P', H', and V' remain constant. The mass of steam considered as part of the system goes from an initial value of m' to zero. From a mass balance,

$$m_2 - m_1 = m'$$

Since the pressure in the barrel P and the pressure in the steam line P' are both constant *(6.7)* is applicable, provided we assume the process to be mechanically reversible within the barrel and the steam line. We really have no choice, because this is the only assumption that allows us to solve the problem. Furthermore, in the absence of any information about heat transfer, we assume the barrel and the steam line to be well insulated, and set $\Sigma Q = 0$. As a result *(6.7)* becomes

$$\Delta \sum_R (mH) = 0$$

The two regions of the system are the barrel and the steam line. In view of this the above equation may be expanded to give

$$\underbrace{m_2H_2}_{\sum_R \text{final}} - \underbrace{(m_1H_1 + m'H')}_{\sum_R \text{initial}} = 0$$

from which

$$H' = \frac{m_2H_2 - m_1H_1}{m'}$$

We have said that the method described is crude. This is so because the assumptions that must be made to solve for an answer are difficult to fulfill. In particular it is quite impossible to prevent heat transfer, and even worse there is no way to determine a value for it.

Example 6.3. A tank of 50(ft)^3 capacity contains 1000(lb$_m$) of liquid water in equilibrium with pure water vapor, which fills the rest of the tank, at a temperature of 212(°F) and 1(atm) absolute pressure. From a water line at slightly above atmospheric pressure 1500(lb$_m$) of water at 160(°F) is to be bled into the tank. How much heat must be transferred to the contents of the tank during this process if the temperature and pressure in the tank are not to change?

The system, which must be chosen to include the initial contents of the tank and the water to be added during the process, is shown in Fig. 6-4.

Since the tank at all times contains liquid and vapor in equilibrium at 212(°F) and 1(atm), the properties H_f and V_f for the liquid and H_g and V_g for the vapor in the tank are the same at the end as at the start of the process. The properties H' and V' of the liquid water in the line are also constant. As liquid is added to the tank, it displaces some of the vapor that is present initially, and this vapor must condense. Thus the tank at the end of the process contains as liquid all the 1000(lb$_m$) of

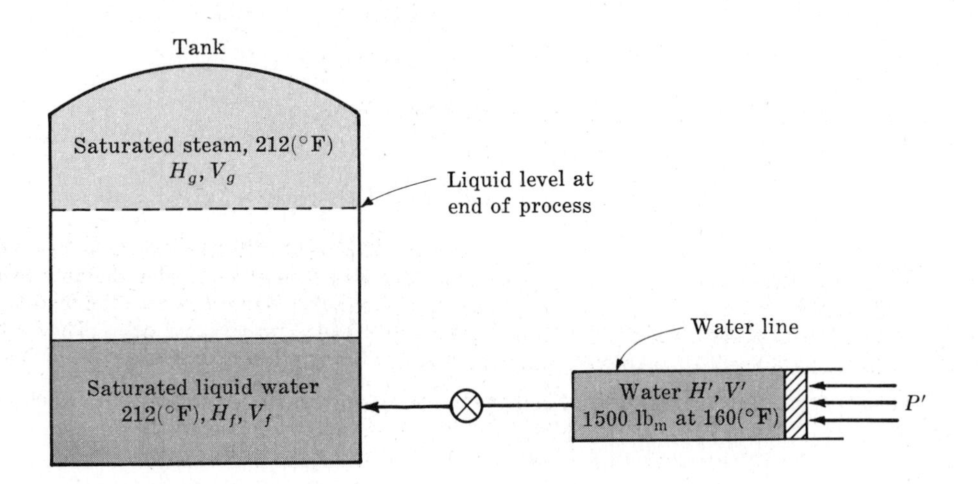

Fig. 6-4

liquid initially present in the tank. This $1000(\text{lb}_m)$ of liquid is present as saturated liquid at $212(°\text{F})$ both at the beginning and end of the process, and hence does not change in properties. In addition, the $1500(\text{lb}_m)$ of liquid added to the tank remains liquid, but it changes from liquid at $160(°\text{F})$ to saturated liquid at $212(°\text{F})$. Also, the vapor which condenses adds to the liquid in the tank at the end of the process. Let the amount of vapor which condenses be $y(\text{lb}_m)$. This mass of material changes from saturated vapor at $212(°\text{F})$ to saturated liquid at $212(°\text{F})$ during the process. The part of the vapor which does not condense remains as saturated vapor at $212(°\text{F})$ and therefore does not change in properties.

Since the pressure is constant in both the tank and the water line, (*6.7*) applies, as does (*6.7a*). Either could be directly used to solve for ΣQ, as both show that ΣQ is equal to the total enthalpy change of the system. However, in the above description of the process we have identified four masses which constitute the system and have described their changes during the process. This suggests that it may be advantageous to determine the total enthalpy change by summation over the changes which occur in these masses, in the manner illustrated in Example 6.1.

Thus we first note that two of the masses – the initial $1000(\text{lb}_m)$ of liquid in the tank and the mass of vapor which does not condense – undergo no change in properties, and may be omitted from consideration. The other two masses are the $1500(\text{lb}_m)$ of water added to the tank from the line and the $y(\text{lb}_m)$ of vapor which condenses. The heat transfer is therefore equal to the enthalpy changes of these two masses:

$$\sum Q = 1500(H_f - H') + y(H_f - H_g)$$

From the steam tables,

$$H_f = 180.16(\text{Btu})/(\text{lb}_m) \qquad H' = 127.96(\text{Btu})/(\text{lb}_m) \qquad H_g = 1150.5(\text{Btu})/(\text{lb}_m)$$

Thus $$\sum Q = 1500(180.16 - 127.96) + y(180.16 - 1150.5) = 78{,}300 - 970.3\,y$$

We must now determine y. This is most easily done by noting that the sum of the volume changes of the masses of the system that we have already identified must be equal to the total volume change of the system. The total volume change of the system is just the volume of the $1500(\text{lb}_m)$ of water at $160(°\text{F})$ that is added to the tank. Since the total volume change is negative,

$$\sum (m\,\Delta V) = -1500V'$$

The same masses that change in enthalpy also change in volume. Thus

$$1500(V_f - V') + y(V_f - V_g) = -1500V'$$

or $$1500V_f + y(V_f - V_g) = 0$$

From the steam tables,

$$V_f = 0.01672(\text{ft})^3/(\text{lb}_m) \qquad V_g = 26.80(\text{ft})^3/(\text{lb}_m)$$

Therefore $$(1500)(0.01672) + y(0.01672 - 26.80) = 0 \qquad \text{or} \qquad y = 0.9364(\text{lb}_m)$$

For the total heat transferred we have

$$\sum Q = 78{,}300 - (970.3)(0.9364) = 77{,}390(\text{Btu})$$

Analysis of this problem through identification of masses has led to a very simple solution. If we had summed over regions, we should eventually have reached the same result, but the calculations would have been much more tedious.

6.2 ENERGY EQUATIONS FOR STEADY-STATE-FLOW PROCESSES

We wish now to consider the kind of process that is referred to as "steady-state flow." Such processes are of primary importance in engineering because the mass production of materials and energy demands continuous operation of processes. The term steady-state-flow process implies the continuous flow of material through an apparatus. The inflow of mass is at all times exactly matched by the outflow of mass, so that there is no accumulation of material within the apparatus. Moreover, conditions at all points within the apparatus are steady or constant with time. Thus, at any point in the apparatus, the thermodynamic properties are constant; although they may vary from point to point, they

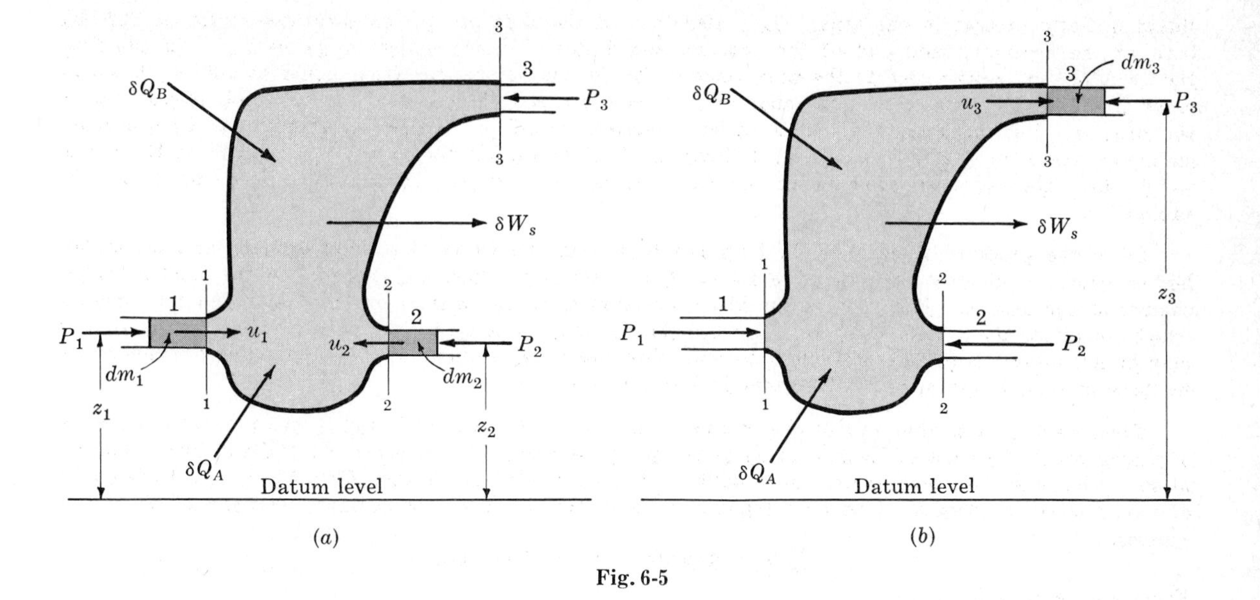

Fig. 6-5

do not change with time at a given point.

Consider the schematic diagram of Fig. 6-5, which shows a region of space bounded by the heavy curves and by the line segments 1-1, 2-2, and 3-3. This region is called the *control volume*. The material contained in the control volume is shown by the light shading. In addition, we recognize three other regions of space which communicate with the control volume: one to the left of 1-1, another to the right of 2-2, and a third to the right of 3-3. These we shall call regions 1, 2, and 3, respectively. For the process considered, regions 1 and 2 contain fluid which flows into the control volume, and region 3 collects fluid which flows from the control volume. There is no limitation on the number of inlets to and outlets from the control volume. We consider here a total of three by way of example.

Our *system* is chosen to include not only the fluid contained initially in the control volume, but in addition all fluid which will enter the control volume during some arbitrarily selected time interval. Thus the initial configuration of the *system* as shown in Fig. 6-5(*a*) includes the masses dm_1 and dm_2 which are to flow into the control volume during a differential time interval. The final configuration of the system after time $d\tau$ is shown in Fig. 6-5(*b*). Here regions 1 and 2 no longer contain the masses dm_1 and dm_2; however, the mass dm_3 has collected in region 3. Since the process considered is one of steady-state flow, the mass contained in the control volume is constant, and

$$dm_3 = dm_1 + dm_2$$

The masses in regions 1, 2, and 3 are considered to have the properties of the fluid as measured at 1-1, 2-2, and 3-3, respectively. These properties include a velocity u and an elevation z above a datum plane, as well as the thermodynamic properties. Thus these masses have kinetic and gravitational potential energy as well as internal energy, and an energy equation for steady-flow processes must contain terms for all these forms of energy.

As we indicated in Sec. 1.6 the kinetic energy of a mass m is equal to

$$E_K = \frac{mu^2}{2g_c}$$

where u is velocity, and g_c is the dimensional constant discussed in Sec. 1.1 and included to allow the use of any self-consistent set of units. The gravitational potential energy of a mass m is

$$E_P = \frac{mzg}{g_c}$$

where z is the height above some arbitrarily chosen datum level, and g is the local acceleration of gravity.

We are now able to write an energy equation for steady-state-flow processes. Note that we have chosen our system to be one of constant mass, i.e. the mass contained in the control volume plus the mass which is to enter during the time interval considered. Thus we may apply the first law in a form already developed, provided that we add terms to account for changes in kinetic and potential energy. Equation (*6.4*) indicates a summation over regions of the system, and the major region is the control volume. However, the mass of fluid in the control volume and the properties of the fluid in the control volume do not change with time. Therefore terms to account for the energy of this region cancel in any energy equation, and may be omitted from the start. The regions that must be considered are those labeled 1, 2, and 3 in Fig. 6-5, and only differential masses leave or enter these regions during the time interval $d\tau$ required for the process that changes the system from the state shown in Fig. 6-5(*a*) to that of Fig. 6-5(*b*). Applied to this process (*6.4*), with the addition of kinetic- and potential-energy terms, appears as

$$\Delta \sum_R (U\,dm) + \Delta \sum_R \left(\frac{u^2\,dm}{2g_c}\right) + \Delta \sum_R \left(\frac{zg\,dm}{g_c}\right) = \sum \delta Q - \sum \delta W$$

Written out, the internal-energy term becomes

$$\Delta \sum_R (U\,dm) = U_3\,dm_3 - U_1\,dm_1 - U_2\,dm_2$$

This result can be expressed more simply by writing

$$\Delta \sum_R (U\,dm) = \sum (U\,dm)_{\text{out}} - \sum (U\,dm)_{\text{in}} = \Delta(U\,dm)_{\substack{\text{flowing}\\\text{streams}}}$$

The kinetic- and potential-energy terms may be expressed similarly. Thus the energy equation may be written

$$\Delta(U\,dm)_{\substack{\text{flowing}\\\text{streams}}} + \Delta\left(\frac{u^2\,dm}{2g_c}\right)_{\substack{\text{flowing}\\\text{streams}}} + \Delta\left(\frac{zg\,dm}{g_c}\right)_{\substack{\text{flowing}\\\text{streams}}} = \sum \delta Q - \sum \delta W$$

where Δ denotes the difference between streams flowing out and streams flowing in.

The $\Sigma\,\delta Q$ term includes all heat transferred to the system. For the example of Fig. 6-5 it would consist of $\delta Q_A + \delta Q_B$. The $\Sigma\,\delta W$ term includes work quantities of two types. The one shown in Fig. 6-5 and designated δW_s represents *shaft work*, i.e. work transmitted across the boundaries of the system by a rotating or reciprocating shaft. The other work quantities are of the kind already considered, which result from the action of external pressures in regions 1, 2, and 3 on the moving boundaries of the system. Thus, for the process of Fig. 6-5, if we assume mechanical reversibility in regions 1, 2, and 3,

$$\sum \delta W = P_3V_3\,dm_3 - P_1V_1\,dm_1 - P_2V_2\,dm_2 + \delta W_s = \Delta(PV\,dm)_{\substack{\text{flowing}\\\text{streams}}} + \delta W_s$$

Substitution in the energy equation and rearrangement gives

$$\Delta(U\,dm)_{\substack{\text{flowing}\\\text{streams}}} + \Delta(PV\,dm)_{\substack{\text{flowing}\\\text{streams}}} + \Delta\left(\frac{u^2\,dm}{2g_c}\right)_{\substack{\text{flowing}\\\text{streams}}} + \Delta\left(\frac{zg\,dm}{g_c}\right)_{\substack{\text{flowing}\\\text{streams}}} = \sum \delta Q - \delta W_s$$

If dm is factored, this equation may be written more simply as

$$\Delta\left[\left(U + PV + \frac{u^2}{2g_c} + \frac{zg}{g_c}\right) dm\right]_{\substack{\text{flowing}\\ \text{streams}}} = \sum \delta Q - \delta W_s$$

By definition, $H = U + PV$, and we have, finally,

$$\Delta\left[\left(H + \frac{u^2}{2g_c} + \frac{zg}{g_c}\right) dm\right]_{\substack{\text{flowing}\\ \text{streams}}} = \sum \delta Q - \delta W_s \qquad (6.8)$$

Equation (*6.8*) is the energy equation for a steady-flow process as written for an infinitesimal time interval $d\tau$. If we divide through by $d\tau$ and denote the various rates by

$$\dot{m} \equiv \frac{dm}{d\tau} \qquad \dot{Q} \equiv \frac{\delta Q}{d\tau} \qquad \dot{W}_s \equiv \frac{\delta W_s}{d\tau}$$

we have

$$\Delta\left[\left(H + \frac{u^2}{2g_c} + \frac{zg}{g_c}\right) \dot{m}\right]_{\substack{\text{flowing}\\ \text{streams}}} = \sum \dot{Q} - \dot{W}_s \qquad (6.9)$$

which expresses the energy equation in terms of rates, all of which are constant in a steady-state-flow process.

One special case of (*6.9*) is very often encountered. If only a single stream enters and a single stream leaves the control volume, then $\dot{m}$ must be the same for both, and (*6.9*) becomes

$$\Delta\left(H + \frac{u^2}{2g_c} + \frac{zg}{g_c}\right) \dot{m} = \sum \dot{Q} - \dot{W}_s \qquad (6.10)$$

Division by $\dot{m}$ gives

$$\Delta\left(H + \frac{u^2}{2g_c} + \frac{zg}{g_c}\right) = \frac{\Sigma \dot{Q}}{\dot{m}} - \frac{\dot{W}_s}{\dot{m}} = \sum Q - W_s$$

or

$$\Delta H + \frac{\Delta u^2}{2g_c} + \Delta z\left(\frac{g}{g_c}\right) = \sum Q - W_s \qquad (6.11)$$

where each term now refers to a *unit mass of fluid* passing through the control volume. All terms in this equation, as in all the energy equations of this chapter, must be expressed in the same energy units.

Example 6.4. Saturated steam at 50(psia) is to be mixed continuously with a stream of water at 60(°F) to produce hot water at 180(°F) at the rate of 500(lb_m)/(min). The inlet and outlet lines to the mixing device all have an internal diameter of 2(in). At what rate must steam be supplied?

Equation (*6.9*) is appropriate for the solution of this example. Evidently, $\dot{W}_s = 0$. We shall assume that the apparatus is insulated so that, to a good approximation, $\Sigma \dot{Q} = 0$. Presumably, also, the device is small enough so that the inlet and outlet elevations are almost the same and the potential-energy terms may be neglected. The velocity of the outlet hot-water stream is given by $u = \dot{m}V/A$, where $\dot{m}$ is 500(lb_m)/(min), V is the specific volume of water at 180(°F), and $A = (\pi/4)D^2$. From the steam tables $V = 0.01651(\text{ft})^3/(\text{lb}_m)$. Thus

$$u = \frac{500(\text{lb}_m)/(\text{min}) \times 0.01651(\text{ft})^3/(\text{lb}_m)}{\frac{\pi}{4} \times \left(\frac{2}{12}\right)^2 (\text{ft})^2 \times 60(\text{s})/(\text{min})} = 6.3(\text{ft})/(\text{s})$$

From this we can calculate the kinetic energy of the outlet water stream:

$$\frac{\dot{m}u^2}{2g_c} = \frac{500(\text{lb}_\text{m})/(\text{min}) \times 6.3^2(\text{ft})^2/(\text{s})^2}{2 \times 32.174(\text{lb}_\text{m})(\text{ft})/(\text{lb}_\text{f})(\text{s})^2} = 308(\text{ft-lb}_\text{f})/(\text{min}) \quad \text{or} \quad 0.396(\text{Btu})/(\text{min})$$

This is an entirely negligible energy quantity compared with the energy changes being considered. [Note that the process raises the temperature of nearly 500(lb_m) of water per minute from 60 to 180(°F). This requires about 60,000(Btu).] We may therefore neglect the kinetic-energy terms for both water streams. Since we do not yet know the flow rate of the steam, we shall for the present neglect this kinetic-energy term also. We can later see if this is justified. Equation (*6.9*) therefore reduces to

$$H_3\dot{m}_3 - H_2\dot{m}_2 - H_1\dot{m}_1 = 0$$

where the subscript 3 refers to the outlet stream; 2, to the inlet water; and 1, to the inlet steam.

From the steam tables:

$$H_1 = 1174.4(\text{Btu})/(\text{lb}_\text{m}) \qquad H_2 = 28.08(\text{Btu})/(\text{lb}_\text{m}) \qquad H_3 = 147.99(\text{Btu})/(\text{lb}_\text{m})$$

Also

$$\dot{m}_3 = 500(\text{lb}_\text{m})/(\text{min}) \qquad \dot{m}_2 = \dot{m}_3 - \dot{m}_1 = 500 - \dot{m}_1$$

Substitution in the energy equation gives

$$(147.99)(500) - (28.08)(500 - \dot{m}_1) - 1174.4\,\dot{m}_1 = 0 \quad \text{or} \quad \dot{m}_1 = 52.30(\text{lb}_\text{m})/(\text{min})$$

The velocity of this stream is

$$u_1 = \frac{\dot{m}_1 V_1}{(\pi/4)D^2} = \frac{52.30(\text{lb}_\text{m})/(\text{min}) \times 8.518(\text{ft})^3/(\text{lb}_\text{m})}{\frac{\pi}{4} \times \left[\frac{2}{12}\right]^2 (\text{ft})^2 \times 60(\text{s})/(\text{min})} = 340(\text{ft})/(\text{s})$$

Its kinetic energy is

$$\frac{\dot{m}_1 u_1^2}{2g_c} = \frac{52.30(\text{lb}_\text{m})/(\text{min}) \times 340^2(\text{ft})^2/(\text{s})^2}{2 \times 32.174(\text{lb}_\text{m})(\text{ft})/(\text{lb}_\text{f})(\text{s})^2} = 93{,}960(\text{ft-lb}_\text{f})/(\text{min})$$

or 120.8(Btu)/(min). We may now rewrite the energy balance to include the kinetic-energy term for the steam flow:

$$(147.99)(500) - (28.08)(500 - \dot{m}_1) - 1174.4\,\dot{m}_1 - 120.8 = 0$$

$$\dot{m}_1 = 52.20(\text{lb}_\text{m})/(\text{min})$$

The inclusion of the kinetic-energy term results in less than a 0.2% change in the answer.

In the preceding example, as in many others, the kinetic-energy term is not significant, even though a stream of rather high velocity is considered. Of course, in cases where kinetic-energy changes are a primary object or result of a process, the kinetic-energy terms are important. For example, the acceleration of an air stream to a high velocity in a wind tunnel requires a considerable work expenditure to produce the kinetic-energy change. For the calculation of this work, one could hardly omit the kinetic-energy terms from the energy equation. The same general comments apply to gravitational potential-energy effects. They are often quite negligible. However, the work generated in a hydroelectric power plant depends directly on the change in elevation of the water flowing through the plant, and the potential-energy terms are of major importance in the energy equation.

6.3 GENERAL ENERGY EQUATIONS

In deriving the energy equation for steady-state-flow processes, we found it useful to introduce the concept of a control volume. The resulting energy equation is seen to connect the properties of the streams flowing into and out of the control volume with the heat and work quantities crossing the boundaries of the control volume. We can extend the use of this concept to include unsteady-state-flow processes. The control volume is still a bounded region of space; however, the boundaries may be flexible to allow for expansion or contraction of the control volume. Furthermore, the mass contained in the control volume need no longer be constant. We deal now with transient conditions, where rates and properties vary with time. During an infinitestimal time interval $d\tau$, the mass entering the control volume may be different from the mass leaving the control volume. The difference must clearly be accounted for by the accumulation or depletion of mass within the control volume. Similarly, the difference between the transport of energy out of the control volume and into it by flowing streams need no longer be accounted for solely by the heat and work terms. Energy may be accumulated or depleted within the control volume. To state the situation precisely, the energy crossing the boundaries of the control volume as heat and work, $\Sigma\,\delta Q - \Sigma\,\delta W$, must equal the change in energy of the material contained within the control volume itself plus the net energy transport of the flowing streams. The energy change of the material in the control volume is $d(mU)_{\text{control volume}}$, and the net energy transport of the flowing streams as shown by (*6.8*) is

$$\Delta\left[\left(H + \frac{u^2}{2g_c} + \frac{zg}{g_c}\right)dm\right]_{\substack{\text{flowing}\\\text{streams}}}$$

The energy equation therefore becomes

$$d(mU)_{\substack{\text{control}\\\text{volume}}} + \Delta\left[\left(H + \frac{u^2}{2g_c} + \frac{zg}{g_c}\right)dm\right]_{\substack{\text{flowing}\\\text{streams}}} = \sum \delta Q - \sum \delta W \tag{6.12}$$

The work term $\Sigma\,\delta W$ may include shaft work, but may also include a term for work resulting from the expansion or contraction of the control volume itself. This equation presumes the control volume to be at rest, and requires that z be measured from a datum level through the center of mass of the control volume.

Equation (*6.12*) is the most general expression for an energy equation that we shall attempt. It has inherent limitations. For example, it assumes no changes in the kinetic and potential energies of the control volume. For the vast majority of applications the control volume may be assumed to change just in internal energy. It is completely impractical to try to write a single energy equation that can be used for all applications. The only suitable guide where complexities arise is strict adherence to the law of conservation of energy, which is the basis for all energy equations. It might be remarked that (*6.12*) reduces to (*6.8*) for steady-state-flow processes. [The first term of (*6.12*) is zero, and $\Sigma\,\delta W$ becomes δW_s.] For a nonflow process, where the control volume contains the entire system, $dm = 0$, and (*6.12*) reduces to

$$d(mU) = \delta Q - \delta W$$

If the system contains several regions, we need to sum the internal-energy term over the regions:

$$\sum_R d(mU) = \sum \delta Q - \sum \delta W$$

which is (*6.3a*).

Example 6.5. Rework Example 6.3 by application of (*6.12*).

We take the tank as our control volume. There is no shaft work, and no expansion work. Therefore, $\Sigma\,\delta W = 0$. There is but one flowing stream, and it flows *into* the control volume. Therefore, the term in (*6.12*) accounting for the energy of the flowing streams reduces to

$$\Delta\left[\left(H + \frac{u^2}{2g_c} + \frac{zg}{g_c}\right)dm\right]_{\substack{\text{flowing}\\\text{streams}}} = -\left(H' + \frac{u'^2}{2g_c} + \frac{z'g}{g_c}\right)dm'$$

However, z' can be taken equal to zero, because the water line and the tank are at essentially the same level. We have no information on u', and can only presume it is small enough so that any contribution from a kinetic-energy term is negligible. This assumption is equivalent to the tacit assumption made in Example 6.3 that the process within the water line is mechanically reversible. This assumption is implicit in the use of (*6.7*) or any equivalent equation. Equation (*6.12*) therefore becomes

$$\sum \delta Q = d(mU)_{\text{tank}} - H'\,dm'$$

We must now integrate this equation over the entire process. Since H' is constant, the result is

$$\sum Q = \Delta(mU)_{\text{tank}} - m'H'$$

By the definition of enthalpy,

$$\Delta(mU)_{\text{tank}} = \Delta(mH)_{\text{tank}} - \Delta(PmV)_{\text{tank}}$$

However, both the volume of the tank $(mV)_{\text{tank}}$ and the pressure in the tank are constant. Therefore $\Delta(PmV)_{\text{tank}} = 0$. As a result,

$$\sum Q = \Delta(mH)_{\text{tank}} - m'H'$$

This equation expresses the fact that for this process the heat transferred is equal to the total enthalpy change caused by the process. This is the same idea upon which we based our earlier solution to this problem.

Clearly, energy equations may be developed from more than one point of view. Different applications yield most readily to different approaches. No formula can substitute for a thorough understanding of the meaning of the law of conservation of energy.

Equation (*6.12*) may be written in terms of rates by dividing through by $d\tau$:

$$\frac{d(mU)_{\substack{\text{control}\\\text{volume}}}}{d\tau} + \Delta\left[\left(H + \frac{u^2}{2g_c} + \frac{zg}{g_c}\right)\dot{m}\right]_{\substack{\text{flowing}\\\text{streams}}} = \sum \dot{Q} - \sum \dot{W}$$

This equation applies at any instant during processes where conditions and flow rates change continuously. Other forms are, of course, also possible. For example, multiplication by $d\tau$ and integration over the time interval from zero to τ gives

$$\Delta(mU)_{\substack{\text{control}\\\text{volume}}} + \int_0^\tau \Delta\left[\left(H + \frac{u^2}{2g_c} + \frac{zg}{g_c}\right)\dot{m}\right]_{\substack{\text{flowing}\\\text{streams}}} d\tau = \sum Q - \sum W$$

The integral can be evaluated only if the various quantities of the integrand are known as functions of time. This requires detailed information describing the process as a function of time.

6.4 APPLICATION OF THE SECOND LAW TO FLOW PROCESSES

In the foregoing development of energy equations, we have implicitly assumed that energy is conserved in all processes, whether flow or nonflow, regardless of whether the system is open or closed. In other words we have extended the validity of the first law of thermodynamics to cover all cases. With respect to the second law, we make exactly the same extension, taking it to be a universally valid law of nature, applicable alike to flow as well as to nonflow processes.

Yet another consideration enters with respect to flow processes, for such processes inevitably result from pressure gradients within the fluid. Moreover, temperature gradients, velocity gradients, and even concentration gradients may exist within the flowing fluid. Thus in contrast to a closed system at uniform conditions throughout, we find in a system where flow occurs a distribution of conditions throughout the system or control volume, and we find it necessary to attribute properties to point masses of fluid. Thus we assume that the intensive properties (density, specific enthalpy, specific entropy, etc.) at a point in a fluid are determined solely by the temperature, pressure, and composition at the point, uninfluenced by the presence of gradients in these conditions at the point. Moreover, we assume that the fluid exhibits a set of intensive properties at the point exactly the same as though the entire system were uniform, and existed at equilibrium at the same temperature, pressure and composition. The implication is that an equation of state applies locally and instantaneously at any point in a fluid system, and that one may employ a concept of *local state*. This concept is independent of the concepts of equilibrium and reversibility, and merely asserts that a mass of fluid of any arbitrary size exists in a definite thermodynamic state to the extent that it exhibits a set of identifiable properties. Experience shows that use of the concept of local state leads for all practical purposes to results in accord with observation. It is therefore universally accepted and employed. At the very worst it represents an acceptable approximation.

Reversible adiabatic flows.

The flow of real fluids is an inherently irreversible process because of the viscous dissipation effects which inevitably accompany flow. Nevertheless, we can imagine the *limiting* process of reversible flow, during which the total entropy is constant. The application of this idea is best illustrated by example.

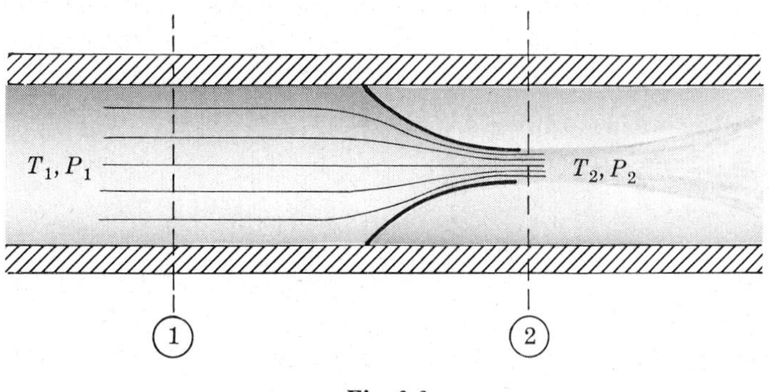

Fig. 6-6

Example 6.6. Consider the steady flow of fluid through a nozzle as illustrated in Fig. 6-6. We wish to consider the change of state which occurs in the fluid as it flows under the influence of the pressure drop $P_1 - P_2$ from station 1 near the nozzle inlet to station 2 at the nozzle exit. The energy equation applicable to this steady-state-flow process is *(6.11)*:

$$\Delta H + \frac{\Delta u^2}{2g_c} + \Delta z\left(\frac{g}{g_c}\right) = \sum Q - W_s$$

In the present application the potential-energy term may be dropped, because the change in elevation Δz from 1 to 2 is zero. Since no shaft work is done by the process, W_s is also zero. Furthermore, we must make an assumption with respect to the heat transfer ΣQ between the control volume (contained within the region between 1 and 2) and the surroundings. Certainly the object of the process is not heat transfer, and therefore steps are usually taken to minimize heat transfer by means of appropriate insulation. Moreover, the amount of heat that could be transferred to each unit mass of fluid flowing through the control volume would normally be quite small, because nozzles accommodate a high rate of flow and the area available for heat transfer is small. We therefore take the process to be adiabatic, and this allows us to write

$$\Delta H + \frac{\Delta u^2}{2g_c} = 0$$

or

$$\frac{u_2^2 - u_1^2}{2g_c} = -(H_2 - H_1) \tag{6.13}$$

This equation expresses the fact that the velocity change caused by flow through an adiabatic nozzle is directly related to the enthalpy change for the process. Presuming we know the upstream conditions P_1, T_1, and u_1, as well as the exit pressure P_2, we would like to determine the conditions T_2 and u_2. In general, however, we need T_2 to find H_2, and H_2 to calculate u_2. Thus the energy equation alone does not suffice for the calculation we wish to make. If on the other hand we consider the limiting case of reversible flow or expansion of fluid through the nozzle, we can write an additional equation, namely $S_2 = S_1$, because for a reversible adiabatic process there can be no change in total entropy. This condition may be imposed on the energy equation by the further qualification

$$\frac{u_2^2 - u_1^2}{2g_c} = -(H_2 - H_1)_S \tag{6.14}$$

which shows that the enthalpy change is that which occurs when the expansion process takes place at constant entropy. This specifies a path for the process, and where data for the fluid are available allows the enthalpy change $(H_2 - H_1)_S$ to be determined. The exit velocity u_2 is then found from (6.14). The velocity given by this equation can very nearly be attained in a properly designed nozzle. However, the design of the nozzle depends on other considerations (fluid mechanics) quite outside the scope of thermodynamics, which provides merely the limit of what can be attained, without prescribing the means of doing it.

Example 6.7. The isentropic expansion of a fluid through a nozzle as discussed in the preceding example produces a fluid stream of increased kinetic energy. This stream can be made to impinge on a turbine blade so as to provide a force to move the blade. The stream thus does work on the turbine blade at the expense of its kinetic energy. This is the principle on which the operation of a turbine depends. A series of nozzles and blades is arranged to expand the fluid in stages and to convert kinetic energy into shaft work. The overall result of the process is the expansion of a fluid from a high pressure to a low pressure with the production of work rather than the production of a high-velocity stream.

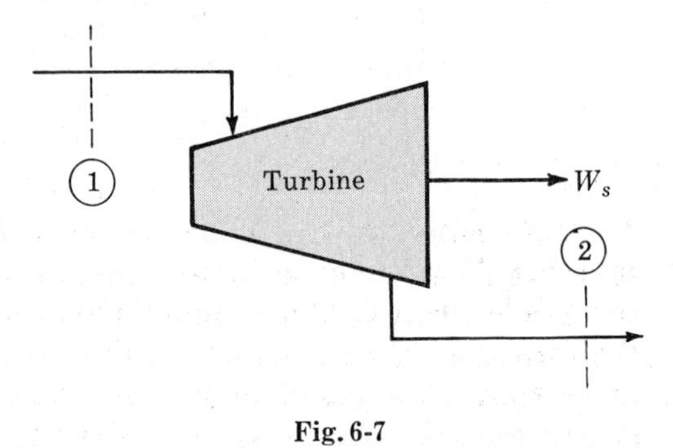

Fig. 6-7

The process is represented in Fig. 6-7. The appropriate energy equation is again (6.11), which applied between stations 1 and 2 reduces to

$$\Delta H = \sum Q - W_s$$

where the potential-energy term is omitted because there is negligible change in elevation, and the kinetic-energy term is omitted because the turbine is normally designed so that both the inlet and exit velocities are relatively low. There is, of course, the possibility of heat exchange between the turbine and the surroundings. However, the object of the process is not heat transfer but the production of work. Therefore in any properly designed turbine $\sum Q$ is entirely negligible, and the term can be omitted, leaving us with the equation

$$W_s = -\Delta H \tag{6.15}$$

Thus the shaft work of a turbine is given directly by the enthalpy change of the expanding fluid, and the only problem is the determination of ΔH. In the design of a turbine one usually knows the intake conditions (and therefore the initial properties at 1) and the discharge pressure P_2. We cannot get H_2 without knowing an additional condition at the turbine discharge. If the turbine operates reversibly as well as adiabatically, i.e. isentropically, then this additional condition is given by $S_2 = S_1$ and in this case (6.15) becomes

$$W_s = -(\Delta H)_S \tag{6.16}$$

The work given by (6.16) is the limiting or *maximum* shaft work that can be produced by adiabatic expansion of a fluid from a given initial state to a given final pressure. Actual machines produce an amount of work equal to 75% or 80% of this. Thus we can define an expansion efficiency as

$$\eta = \frac{W_s\,(\text{actual})}{W_s\,(\text{isentropic})}$$

These two work terms are given by (6.15) and (6.16). That is, $W_s(\text{actual}) = -\Delta H$ and $W_s(\text{isentropic}) = -(\Delta H)_S$. Thus

$$\eta = \frac{\Delta H}{(\Delta H)_S}$$

where ΔH is the actual enthalpy change of the fluid passing through the turbine. These processes, the actual and the reversible, are represented on a Mollier diagram in Fig. 6-8.

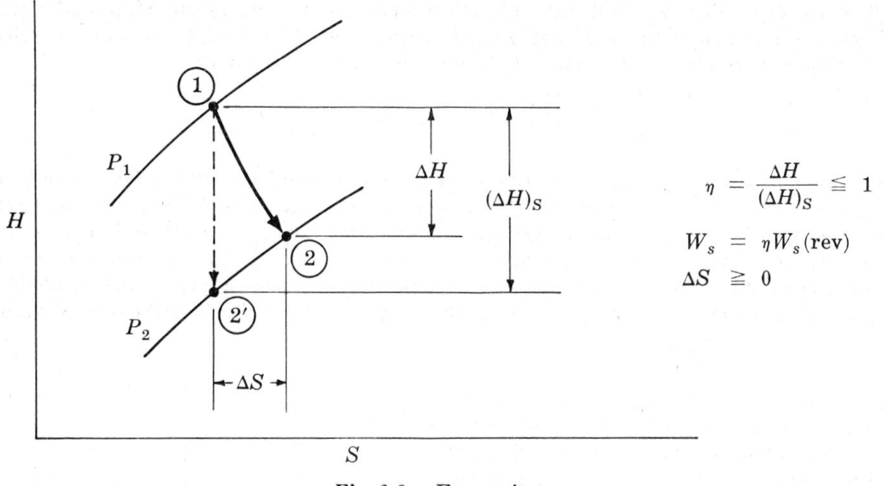

Fig. 6-8. Expansion

The reversible process follows a vertical line of constant entropy from point 1 at the higher pressure P_1 to point 2′ at the discharge pressure P_2. The line representing the actual irreversible process starts again from point 1, but proceeds downward *and to the right,* in the direction of *increasing* entropy. Since the process is adiabatic, the fluid must increase in entropy as a result of any irreversibility. The process terminates at point 2 on the isobar for P_2. The greater the irreversibility, the further this point will lie to the right on the P_2 isobar, giving lower and lower values for the efficiency η of the process.

An adiabatic compression process, represented schematically in Fig. 6-9, is the opposite of an adiabatic expansion process, because work is done on a fluid so as to raise its pressure. In fact, a reversible, adiabatic expansion process can be made to retrace its path in reverse by a reversible, adiabatic compression process. The energy equation is the same for both processes, because the same assumptions of negligible kinetic-energy and potential-energy changes are made. Thus (*6.15*) and (*6.16*) are applicable in either case. There is, however, a difference. In the case of adiabatic compression the reversible work is the *minimum* shaft work required for compression of a fluid from its initial state to a given final pressure. An actual irreversible process requires greater work expenditure. Thus the compression efficiency is defined by

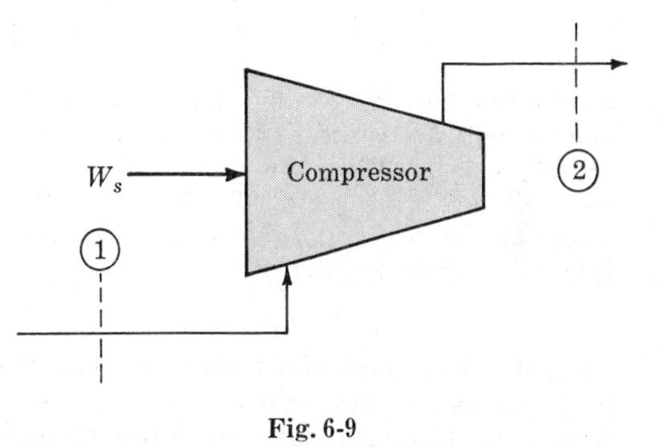

Fig. 6-9

$$\eta = \frac{W_s(\text{isentropic})}{W_s(\text{actual})}$$

or by

$$\eta = \frac{(\Delta H)_S}{\Delta H}$$

The adiabatic compression process is represented on a Mollier diagram in Fig. 6-10. Again the vertical line from 1 to 2′ represents the reversible (isentropic) process. Here P_2 is the higher pressure, and a line representing an actual process rises from point 1, and runs *toward the right* in the direction of *increasing* entropy. The figure clearly shows that in this case $(\Delta H)_S$ is the minimum possible enthalpy increase.

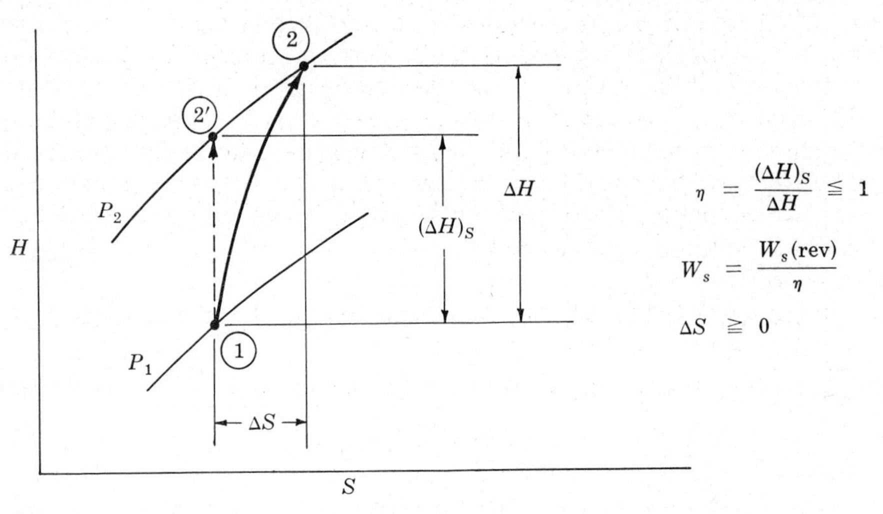

Fig. 6-10. Compression

Throttling processes.

With reference to Fig. 6-8, which depicts adiabatic expansion processes, we see that a possible path for expansion is a horizontal line extending to the right from point 1. Such a line represents an adiabatic expansion process for which there is no change in enthalpy. For this case (*6.15*) becomes

$$W_S = -\Delta H = 0$$

Such a process therefore occurs adiabatically, without the production of shaft work, and with no change in kinetic or potential energy. Thus all terms in (*6.11*), the energy equation for steady flow of a single stream, are zero, and the significant conclusion is that

$$\Delta H = 0$$

The primary result of this kind of process is merely the reduction of pressure, and this result is achieved whenever a fluid flows through a restriction, such as a partially closed valve or a porous plug, without any appreciable gain in kinetic energy. It is known as a *throttling process,* and is inherently irreversible. It cannot even be imagined to be reversible, for as reference to Fig. 6-8 shows, any process represented by a horizontal line inevitably implies an increase in entropy of the fluid.

Figure 6-6 shows a nozzle with reference points taken upstream from the nozzle and at the nozzle exit. We may consider a third station, well downstream from the nozzle, as shown below. The discharge stream from the nozzle, with its high kinetic energy, can be

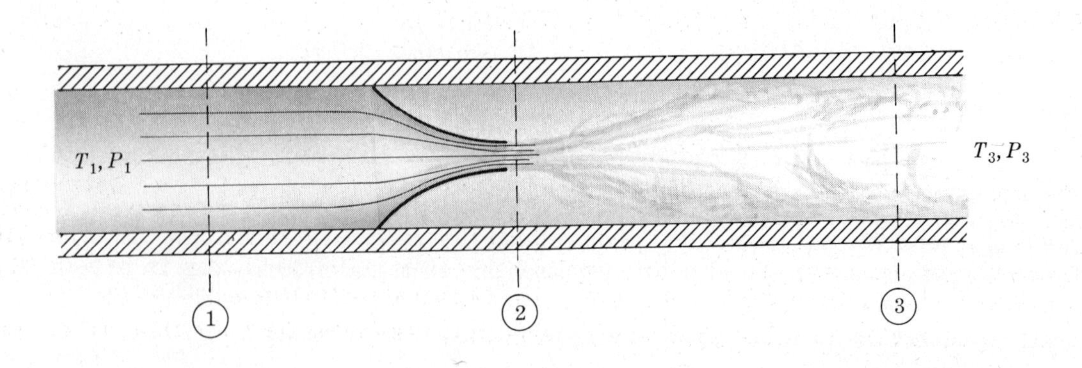

used for the production of work; however, if no work-producing device is present and if the jet from the nozzle merely discharges into the downstream pipe, then it generates turbulence downstream from the nozzle. In this case the flowing fluid swirls about, expanding to fill the pipe, and dissipates its kinetic energy as a result of the damping effects of viscosity. Thus kinetic energy is reconverted into internal energy, and the fluid passes station 3 with a low velocity, and the overall process from 1 to 3 is for all practical purposes a throttling process. Although the change from 1 to 2 can be made nearly reversible, the ensuing process from 2 to 3 cannot, and the overall process must result in an entropy increase.

Example 6.8. What is the temperature change and the entropy change that result when an ideal gas undergoes a throttling process from 2(atm) to 1(atm)?

For a throttling process $\Delta H = 0$, and for an ideal gas (*1.15*) gives $dH = C_P\,dT$. Thus

$$\Delta H = \int_{T_1}^{T_2} C_P\,dT = 0$$

The only way the integral can be zero is for T_2 to be equal to T_1. Thus $\Delta T = 0$, and we conclude that no temperature change occurs when an ideal gas undergoes a throttling process.

The entropy change of an ideal gas is given by (*2.10*), which for $T_2 = T_1$ becomes

$$\Delta S = -R \ln \frac{P_2}{P_1}$$

Thus for $P_2 = 1\text{(atm)}$ and $P_1 = 2\text{(atm)}$

$$\Delta S = -R \ln 0.5 = R \ln 2 = 0.69315\,R$$

Limiting values by the second law.

Use of the second law for numerical calculations requires the assumption of reversibility with respect to flow processes, for otherwise we could make no calculations at all. In this event we use an appropriate energy equation in conjunction with the second law expressed as $\Delta S_{\text{total}} = 0$. The results of any calculations made on this basis are limiting values, representing the best that could be accomplished. Any actual process would be irreversible, accomplishing less than the best. The analysis of irreversible processes will be considered in Chapter 8. For the moment we will assume reversibility, and illustrate the kinds of calculations that this assumption allows us to make.

Example 6.9. A complicated process has been devised to make heat continuously available at a temperature level of 500(°F). The only source of energy is steam, saturated at 250(psia). Cooling water is available in large supply at 70(°F). How much heat can be transferred from the process to a heat reservoir at 500(°F) for every pound of steam condensed in the process?

We do not need to know any details about how the process is accomplished, but we do need to make a couple of basic assumptions. Therefore we assume that steam flows continuously and that it is condensed and subcooled to the cooling-water temperature by the time it emerges. The properties of the inlet steam and outlet condensate are given by the steam tables as:

$$T_1 = 401(°\text{F}) \qquad T_2 = 70(°\text{F})$$

$$H_1 = 1202.1(\text{Btu})/(\text{lb}_\text{m}) \qquad H_2 = 38.1(\text{Btu})/(\text{lb}_\text{m})$$

$$S_1 = 1.5274(\text{Btu})/(\text{lb}_\text{m})(\text{R}) \qquad S_2 = 0.0746(\text{Btu})/(\text{lb}_\text{m})(\text{R})$$

The second law denies the possibility that the *only* effect of this process could be the transfer of heat from the steam at temperatures between 70 and 401(°F) to a heat reservoir at 500(°F). However, in this process we have cooling water available at 70(°F). Thus we may consider that the *two* results of the process are transfer of heat from the steam to a heat reservoir at 500(°F) *and* the transfer of heat to another heat reservoir at 70(°F). A representation of the process is given in Fig. 6-11.

If we neglect potential- and kinetic-energy terms, (*6.11*) as written for the control volume becomes

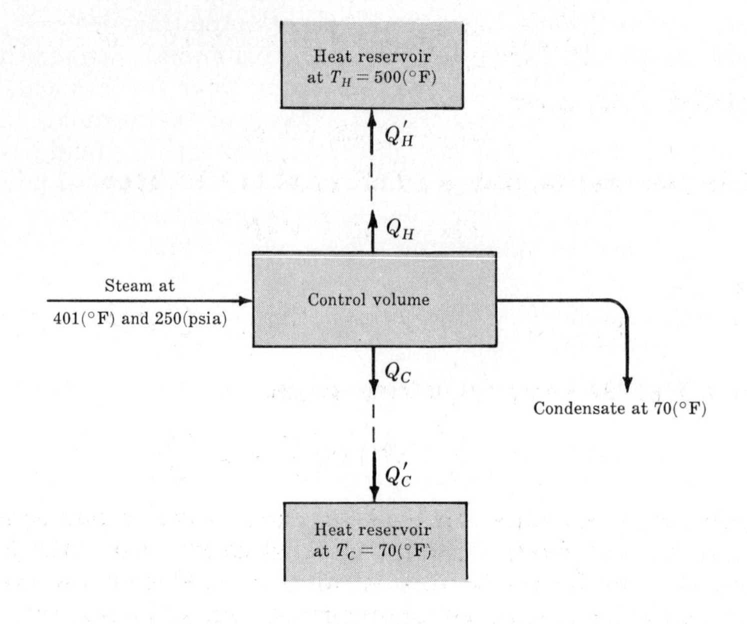

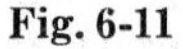
Fig. 6-11

$$\Delta H = \sum Q = Q_H + Q_C$$

Thus $$Q_H + Q_C = 38.1 - 1202.1 = -1164.0(\text{Btu})/(\text{lb}_\text{m})$$

We now apply the second law, $\Delta S_{\text{total}} = 0$. The entropy change of the steam is simply

$$\Delta S = S_2 - S_1 = 0.0746 - 1.5274 = -1.4528(\text{Btu})/(\text{lb}_\text{m})(\text{R})$$

The entropy change of the heat reservoir at 500(°F) is

$$\frac{Q'_H}{500 + 460}\ (\text{Btu})/(\text{lb}_\text{m})(\text{R})$$

and the entropy change of the heat reservoir at 70(°F) is

$$\frac{Q'_C}{70 + 460}\ (\text{Btu})/(\text{lb}_\text{m})(\text{R})$$

The sum of these three entropy changes is zero. However, Q_H and Q_C, taken with reference to the control volume in the energy equation, are opposite in sign to Q'_H and Q'_C, taken with reference to the heat reservoirs in the entropy expressions. Thus, in terms of Q_H and Q_C, the total entropy change is

$$\Delta S_{\text{total}} = -1.4528 - \frac{Q_H}{960} - \frac{Q_C}{530} = 0$$

The energy equation and this entropy equation constitute two equations in two unknowns, and may be solved for Q_H and Q_C. The results for each pound of steam are

$$Q_H = -879.6(\text{Btu})/(\text{lb}_\text{m}) \qquad Q_C = -284.4(\text{Btu})/(\text{lb}_\text{m})$$

The minus signs merely indicate that heat is transferred away from the control volume.

In any actual process more heat would be transferred to the cold reservoir and less to the hot. In order to solve the problem for any case except that of complete reversibility, one would have to know the details of how the process is actually carried out.

6.5 THE MECHANICAL-ENERGY BALANCE

Consider the steady-state flow of fluid through a control volume to which there is but one entrance and one exit. The energy equation which applies is *(6.11)*:

$$\Delta H + \frac{\Delta u^2}{2g_c} + \Delta z\left(\frac{g}{g_c}\right) = \sum Q - W_s$$

The property relation (*3.48*) gives

$$dH = T\,dS + V\,dP$$

and for reversible processes we may set $T\,dS$ equal to δQ. Then

$$dH = \delta Q + V\,dP$$

Integration gives

$$H_2 - H_1 = \Delta H = \sum Q + \int_1^2 V\,dP$$

Substituting for ΔH in the energy equation, we get

$$-W_s = \int_1^2 V\,dP + \frac{\Delta u^2}{2g_c} + \Delta z\left(\frac{g}{g_c}\right)$$

The assumption of reversibility was made in order to derive this equation. However, the viscous nature of real fluids leads to frictional effects that make flow processes inherently irreversible. As it stands, this equation is valid only for an imaginary nonviscous fluid; for real fluids it is at best approximate. An additional term may be incorporated in the equation to account for mechanical energy dissipated through fluid friction. The resulting equation is known as the mechanical-energy balance:

$$-W_s = \int_1^2 V\,dP + \frac{\Delta u^2}{2g_c} + \Delta z\left(\frac{g}{g_c}\right) + \sum F \qquad (6.17)$$

where ΣF is the friction term. The determination of numerical values for ΣF is a problem in fluid mechanics, not thermodynamics, and will not be considered here.

Bernoulli's famous equation is a special case of the mechanical-energy balance applying to nonviscous, incompressible fluids which do not exchange shaft work with the surroundings. For nonviscous fluids, ΣF is zero, and for incompressible fluids,

$$\int_1^2 V\,dP = V\,\Delta P = \frac{\Delta P}{\rho}$$

where ρ is fluid density. Thus Bernoulli's equation becomes

$$\frac{\Delta P}{\rho} + \frac{\Delta u^2}{2g_c} + \Delta z\left(\frac{g}{g_c}\right) = 0 \qquad (6.18)$$

This may also be written

$$\Delta\left[\frac{P}{\rho} + \frac{u^2}{2g_c} + z\left(\frac{g}{g_c}\right)\right] = 0$$

or

$$\frac{P}{\rho} + \frac{u^2}{2g_c} + z\left(\frac{g}{g_c}\right) = \text{const.}$$

The severe limitations on Bernoulli's equation should be carefully noted.

Solved Problems

ENERGY EQUATIONS (Secs. 6.1 through 6.3)

6.1. A 100(ft)³ rigid tank initially contains a mixture of saturated vapor steam and saturated liquid water at 500(psia). Of the total mass, 10% is vapor. Saturated liquid only is bled slowly from the tank until the total mass in the tank drops to half of the initial total mass. During this process the temperature of the contents of the tank is kept constant by the transfer of heat. How much heat is transferred?

We can calculate immediately the initial mass in the tank. The total volume of the tank is the sum of the liquid volume and the vapor volume. Thus

$$V_{\text{tank}} = m_f V_f + m_g V_g = 100(\text{ft})^3$$

Since 10% of the mass is vapor and 90% is liquid then $m_f/m_g = 9$, and

$$V_{\text{tank}} = 9m_g V_f + m_g V_g = m_g(9V_f + V_g) = 100(\text{ft})^3$$

The steam tables provide the values

$$V_f = 0.01975(\text{ft})^3/(\text{lb}_\text{m}) \qquad V_g = 0.9283(\text{ft})^3/(\text{lb}_\text{m})$$

Thus $$m_g[(9)(0.01975) + 0.9283] = 100 \quad \text{or} \quad m_g = 90.41(\text{lb}_\text{m})$$

Since this represents 10% of the total mass, we have $m_t = 904.1(\text{lb}_\text{m})$, and the mass removed from the tank during the process as saturated liquid is given by

$$m' = m_t/2 = 452.1(\text{lb}_\text{m})$$

Since the tank is kept at constant temperature, and at all times contains saturated liquid and vapor in equilibrium, the pressure will also remain constant at 500(psia), the saturation pressure, and the specific properties of the phases will not change. Only the amounts of the two phases will change.

If we take the tank as the control volume, and apply *(6.12)*, we get

$$d(mU)_{\text{tank}} + H_f\,dm' = \sum \delta Q$$

The kinetic- and potential-energy terms have been assumed negligible, and since no work is involved, δW is set equal to zero. In addition, there is but a single flowing stream, which flows *out* of the control volume, and it has the enthalpy of the saturated liquid H_f. Since this enthalpy remains constant during the process, integration of the energy equation gives

$$\Delta(mU)_{\text{tank}} + H_f m' = Q$$

It is convenient to replace U in this equation in terms of H. Since $U = H - PV$,

$$\Delta(mU)_{\text{tank}} = \Delta(mH)_{\text{tank}} - \Delta(mPV)_{\text{tank}}$$

However, P is constant and therefore

$$\Delta(mPV)_{\text{tank}} = P\,\Delta(mV) = P\,\Delta V_{\text{tank}} = 0$$

Thus $$\Delta(mU)_{\text{tank}} = \Delta(mH)_{\text{tank}}$$

and the energy equation can be written

$$\Delta(mH)_{\text{tank}} + H_f m' = Q$$

Since the tank contains both saturated liquid and saturated vapor at the start and end of the process, we determine the total enthalpy change of the tank contents by

$$\Delta(mH)_{\text{tank}} = (m_{f_2}H_f - m_{f_1}H_f) + (m_{g_2}H_g - m_{g_1}H_g)$$

The energy equation now becomes

$$H_g(m_{g_2} - m_{g_1}) - H_f(m_{f_1} - m_{f_2} - m') = Q$$

By a mass balance

$$m_{g_2} + m_{f_2} + m' = m_{g_1} + m_{f_1}$$

or

$$m_{g_2} - m_{g_1} = m_{f_1} - m_{f_2} - m' \equiv y$$

The energy equation now simplifies to

$$Q = y(H_g - H_f) = yH_{fg}$$

We have merely to determine y, the mass of liquid vaporized in the tank, for H_{fg} is given directly in the steam tables for saturated steam at 500(psia). This is done by noting that the volume of the tank is constant. Thus

$$V_{\text{tank}} = m_{f_1}V_f + m_{g_1}V_g = m_{f_2}V_f + m_{g_2}V_g$$

Therefore

$$V_g(m_{g_1} - m_{g_2}) = V_f(m_{f_2} - m_{f_1})$$

or

$$V_g y = V_f(y + m')$$

Solution for y gives

$$y = \frac{m'V_f}{V_g - V_f} = \frac{452.1 \times 0.01975}{0.9283 - 0.01975} = 9.828(\text{lb}_\text{m})$$

From the steam tables $H_{fg} = 755.8(\text{Btu})/(\text{lb}_\text{m})$ and the energy equation gives finally

$$Q = yH_{fg} = 9.828(\text{lb}_\text{m}) \times 755.8(\text{Btu})/(\text{lb}_\text{m}) = 7428(\text{Btu})$$

In this rather labored solution we have applied a very general energy equation (*6.12*) to a simple process. It is possible to reach the same result more readily if we take a different point of view at the start. We may imagine the saturated liquid that flows from the tank to be collected in a piston-cylinder device as indicated in the figure just below. Since what happens to the

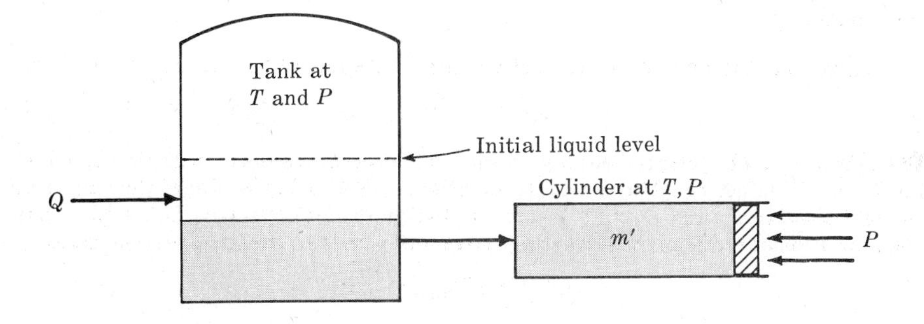

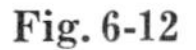
Fig. 6-12

liquid taken from the tank is not specified and does not affect the remaining contents of the tank, we are free to assume that it is collected reversibly in the cylinder as saturated liquid at the T and P of the tank. Then the process is one of constant pressure in a closed system for which the heat transfer is equal to the total enthalpy change: $Q = \Delta H^t$. The only enthalpy change that occurs is the change of $y(\text{lb}_\text{m})$ of liquid into vapor. Thus

$$Q = yH_{fg}$$

Similarly the only volume change that occurs in the system is the change of $y(\text{lb}_\text{m})$ of liquid into vapor. Thus $\Delta V^t = yV_{fg}$. However, the total volume change is clearly also given by $\Delta V^t = m'V_f$. Thus

$$y = \frac{m'V_f}{V_{fg}}$$

These equations are the same ones derived earlier.

6.2. A standard problem solved by the methods of this chapter involves the flow of a gas or vapor from a source at constant temperature and pressure into a closed tank of known volume initially at a lower pressure than the source. For example, a tank containing air at pressure P_1 and temperature T_1 could be connected to a compressed-air line which supplies air at the steady conditions P' and T'. The tank is initially isolated from the line by a closed valve, and the problem is to determine the amount of gas that flows into the tank when the valve is opened long enough for a finite change to occur. Develop the general energy equation which applies to this process.

We will take the tank as our control volume and apply (*6.12*). The process is shown in Fig. 6-13. Since there is but one flowing stream, and since it *enters* the tank, (*6.12*) becomes

$$d(mU)_{\text{tank}} - \left(H' + \frac{u'^2}{2g_c} + \frac{z'g}{g_c}\right)dm' = \sum \delta Q - \sum \delta W$$

However, there is no work involved in the process; there is but one heat term; and we must assume in the absence of detailed information that the kinetic- and potential-energy terms are negligible. The energy equation therefore becomes

$$d(mU)_{\text{tank}} - H'\,dm' = \delta Q$$

A mass balance requires that $dm' = dm$. Therefore we have

$$d(mU)_{\text{tank}} - H'\,dm = \delta Q$$

This may be integrated immediately, because H' is constant, to give

$$m_2U_2 - m_1U_1 - H'(m_2 - m_1) = Q$$

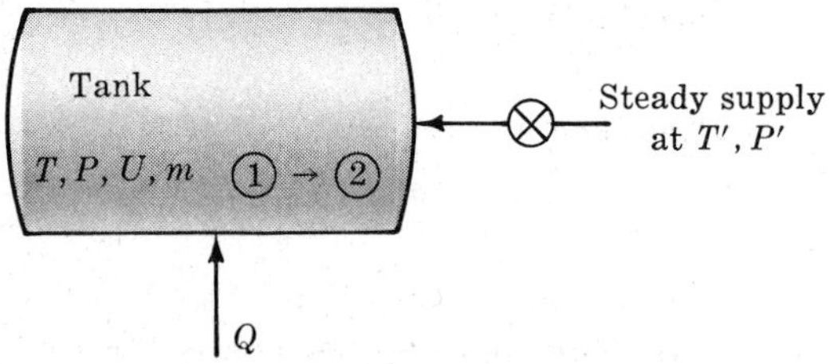

Fig. 6-13

Upon rearrangement we have

$$m_2(U_2 - H') - m_1(U_1 - H') = Q \tag{1}$$

This energy equation is generally applicable to the process described. The properties U_1, U_2, and H' are specific properties, i.e. they are based on a unit mass of material. It is sometimes advantageous to express (*1*) in terms of numbers of moles. Then we have

$$n_2(U_2 - H') - n_1(U_1 - H') = Q \tag{2}$$

where U_1, U_2, and H' are now molar properties.

If we take the gas to be ideal and to have constant heat capacities, then (*2*) can be further developed as follows. Since

$$H' = U' + P'V' = U' + RT'$$

we can substitute for H' in (*2*):

$$n_2(U_2 - U' - RT') - n_1(U_1 - U' - RT') = Q$$

However $\Delta U = C_V\,\Delta T$ and $R = C_P - C_V$. Substituting, we obtain

$$n_2(C_VT_2 - C_PT') - n_1(C_VT_1 - C_PT') = Q \tag{3}$$

For an ideal gas bled into an *evacuated* tank in an *adiabatic* process, n_1 and Q are zero, and (*3*) gives

$$C_VT_2 = C_PT'$$

or

$$T_2 = \gamma T' \tag{4}$$

where $\gamma = C_P/C_V$. This equation shows that for the very special case considered the temperature in the tank is independent of the amount of gas bled into the tank. Thus if air at 27(°C) is bled *adiabatically* into an evacuated tank, then (taking $\gamma = 1.4$)

$$T_2 = 1.4(27 + 273) = 420(\text{K}) \quad \text{or} \quad 147(°\text{C})$$

regardless of the size of the tank and regardless of how much air enters the tank.

6.3. A well-insulated steel tank having a volume of $13(\text{m})^3$ contains air at a pressure of 1(bar) and a temperature of 22(°C). This tank is attached to a compressed-air line, which supplies air at the steady conditions of 5(bar) and 30(°C). Initially the tank is shut off from the air line by a valve. If the valve is opened long enough so that air flows into the tank until the pressure in the tank reaches 3(bar), calculate how much air the tank will contain and what its temperature will be, (*a*) if the process is adiabatic (no heat exchange with the gas); (*b*) if the tank wall exchanges heat with the air so rapidly that the wall and the air are always at the same temperature.

Data: Air may be considered an ideal gas for which $C_P = 7$ and $C_V = 5$(cal)/(g mole)(K). The steel tank has a mass of 1200(kg), and steel has a heat capacity of 0.107(cal)/(g)(K).

The energy equation that applies is (*3*) of Problem 6.2:

$$n_2(C_V T_2 - C_P T') - n_1(C_V T_1 - C_P T') = Q$$

The required values that are provided by the problem statement include C_V, C_P, and the temperatures

$$T_1 = 22 + 273 = 295(\text{K}) \qquad T' = 30 + 273 = 303(\text{K})$$

The initial number of moles in the tank is given by the ideal-gas law:

$$n_1 = \frac{P_1 V_{\text{tank}}}{RT_1} = \frac{1(\text{bar}) \times 13(\text{m})^3 \times [100(\text{cm})/(\text{m})]^3}{83.14(\text{cm}^3\text{-bar})/(\text{g mole})(\text{K}) \times 295(\text{K})} = 530.0(\text{g mole})$$

(*a*) If the process is adiabatic, then $Q = 0$, and the energy equation becomes:

$$n_2 = n_1 \frac{C_V T_1 - C_P T'}{C_V T_2 - C_P T'}$$

Substitution of known numerical values gives

$$n_2 = \frac{-342{,}400}{5T_2 - 2120}$$

Clearly an additional relation between n_2 and T_2 is needed, and it is provided by the ideal-gas law:

$$n_2 = \frac{P_2 V_{\text{tank}}}{RT_2} = \frac{3 \times 13 \times 100^3}{83.14\, T_2} = \frac{469{,}100}{T_2}$$

Equating the two expressions for n_2 and solving for T_2, we obtain

$$T_2 = 370(\text{K}) \quad \text{or} \quad 97(°\text{C})$$

The number of moles n_2 is now given by

$$n_2 = 469{,}100/370 = 1268(\text{g mole})$$

(*b*) If the tank wall exchanges heat with the air so that it is always at the air temperature, then Q is given by

$$-Q = m_{\text{wall}} C_{\text{steel}} (T_2 - T_1)$$

or

$$-Q = 1200(\text{kg}) \times 1000(\text{g})/(\text{kg}) \times 0.107(\text{cal})/(\text{g})(\text{K}) \times [T_2 - 295](\text{K})$$

This expression gives $-Q$ because all terms on the right refer to the tank wall rather than to the system. Thus the quantity on the right is the heat transfer with respect to the tank wall, and this is the negative of Q, the heat transfer with respect to the air. Thus $Q = -128{,}400[T_2 - 295]$(cal) and the energy equation now becomes

$$n_2(5T_2 - 7 \times 303) - 530.0(5 \times 295 - 7 \times 303) = -128{,}400(T_2 - 295)$$

or

$$n_2 = \frac{-128{,}400(T_2 - 295) - 342{,}400}{5T_2 - 2120}$$

The expression for n_2 as given by the ideal-gas law is exactly the same as in part (*a*):

$$n_2 = 469{,}100/T_2$$

Again, we set these expressions equal, and solve for T_2. The final results are

$$T_2 = 300(\text{K}) = 27(^\circ\text{C}) \qquad n_2 = 1564(\text{g mole})$$

The great difference between the results of (*a*) and of (*b*) illustrates the importance of the assumptions made in the initial formulation of the problem. We have worked here two limiting cases. In practice (*b*) would be much more nearly right.

6.4. A donkey engine for service in a chemical plant is powered by a steam engine which draws its steam from a large insulated tank containing mostly high-pressure saturated liquid water. The water is in equilibrium with saturated vapor steam, which occupies a relatively small vapor space above the saturated liquid. As the engine operates it draws off saturated vapor, causing liquid to evaporate, and this in turn lowers the temperature and pressure in the tank. Eventually, the engine must visit the plant power station to recharge its tank. A particular engine carries a tank having a volume of 600(ft)³. For efficient operation it requires steam at pressures ranging from 150 to 100(psia). If the engine visits the power station when the tank pressure has fallen to 100(psia) and the saturated liquid occupies 90% of the tank volume, how much saturated steam at 160(psia) from the power station must be charged to the tank to raise its pressure to 150(psia)? Neglect all heat transfer.

The energy equation applicable here is (*1*) of Problem 6.2, with Q set equal to zero:

$$\frac{m_2}{m_1} = \frac{U_1 - H'}{U_2 - H'} = \frac{H' - U_1}{H' - U_2}$$

Here m_1 and m_2 represent the total mass in the tank at the start and end of the charging process, U_1 and U_2 represent the specific internal energy of the total tank contents at start and end of the process, and H' is the specific enthalpy of the steady supply of saturated steam from the power station at 160(psia). To find m_1, we note that

$$0.9\, V_{\text{tank}} = m_{f_1} V_{f_1} \qquad \text{and} \qquad 0.1\, V_{\text{tank}} = m_{g_1} V_{g_1}$$

Thus

$$m_{f_1} = \frac{0.9 \times 600(\text{ft})^3}{0.017736(\text{ft})^3/(\text{lb}_\text{m})} = 30{,}446(\text{lb}_\text{m})$$

$$m_{g_1} = \frac{0.1 \times 600(\text{ft})^3}{4.434(\text{ft})^3/(\text{lb}_\text{m})} = 13.53(\text{lb}_\text{m})$$

where values of V_{f_1} and V_{g_1} were taken from the steam tables at $P = 100(\text{psia})$. Therefore $m_1 = m_{f_1} + m_{g_1} = 30{,}460(\text{lb}_\text{m})$. Since the tank contains saturated liquid and saturated vapor, it may be considered to have the quality

$$x_1 = m_{g_1}/m_1 = 13.53/30{,}460 = 0.000444$$

The initial internal energy U_1 is therefore given by

$$U_1 = U_{f_1} + x_1 U_{fg_1} = 298.28 + (0.000444)(807.5) = 298.64(\text{Btu})/(\text{lb}_\text{m})$$

From the steam tables we have for saturated vapor at 160(psia) that $H' = 1196.0(\text{Btu})/(\text{lb}_\text{m})$. The energy equation now becomes

$$\frac{m_2}{m_1} = \frac{1196.0 - 298.64}{1196.0 - U_2} = \frac{897.4}{1196.0 - U_2}$$

Both m_2 and U_2 are unknown, and we must make use of another relationship connecting the initial and final states of the system. It is provided by

$$V_{\text{tank}} = m_1 V_1 = m_2 V_2 \qquad \text{or} \qquad m_2/m_1 = V_1/V_2$$

Since V_1 is given by $V_1 = V_{f_1} + x_1 V_{fg_1}$, we have

$$V_1 = 0.017736 + (0.000444)(4.416) = 0.01970(\text{ft})^3/(\text{lb}_m)$$

Combination of the two equations for m_2/m_1 and substitution for V_1 gives

$$\frac{0.01970}{V_2} = \frac{897.4}{1196.0 - U_2}$$

However
$$V_2 = V_{f_2} + x_2 V_{fg_2} = 0.018089 + 2.998\,x_2$$

$$U_2 = U_{f_2} + x_2 U_{fg_2} = 330.24 + 781.0\,x_2$$

where data have been taken from the saturation tables at $P = 150$(psia). As a result we have the equation

$$\frac{0.01970}{0.018089 + 2.998\,x_2} = \frac{897.4}{865.8 - 781.0\,x_2}$$

which may be solved for x_2, giving

$$x_2 = 0.0003043$$

Now we can calculate the final volume:

$$V_2 = 0.018089 + (2.998)(0.0003043) = 0.01900$$

Therefore
$$\frac{m_2}{m_1} = \frac{V_1}{V_2} = \frac{0.01970}{0.01900}$$

which together with $m_1 = 30{,}460(\text{lb}_m)$ yields

$$m_2 - m_1 = 1122(\text{lb}_m)$$

This is the amount of saturated steam at 160(psia) which must be charged to the tank from the power station.

6.5. Another standard problem in unsteady flow involves the flow of gas or vapor *out* of a tank. One usually wishes to determine the conditions in the gas remaining in a tank after sufficient gas has been bled out to cause a finite change in state. Develop the energy equation which applies.

The kind of process considered is shown in Fig. 6-14. We take the tank as the control volume and apply (*6.12*). There is no work involved in the process, and we will take the kinetic- and potential-energy terms to be negligible. The only flowing stream flows *out* of the control volume; so (*6.12*) becomes

$$d(mU)_{\text{tank}} + H'\,dm' = \delta Q$$

However, the stream leaving the tank at any instant must have the properties of the gas in the tank at that instant. Thus $H' = H$. Furthermore $dm' = -dm$ and the energy equation may be written

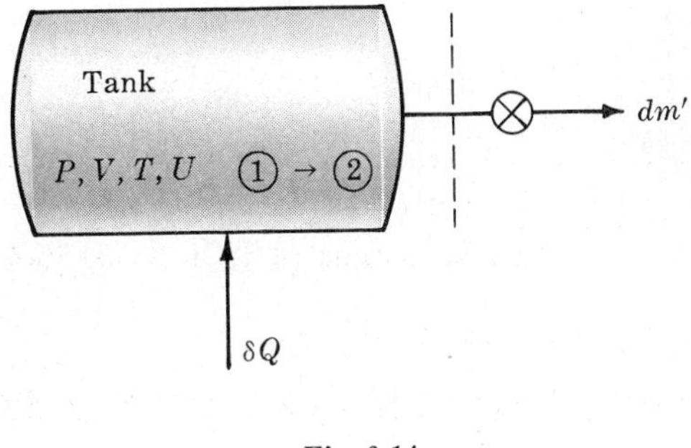

Fig. 6-14

$$d(mU)_{\text{tank}} - H\,dm = \delta Q \tag{1}$$

Integration of this equation over a finite process presents a problem, because the properties of the gas in the tank continually change. Thus a relationship is needed between the enthalpy H of the gas in the tank and the mass m of gas in the tank. Since this depends on the details of the process, no general relationship is known. An alternative form of (*1*) can be developed by expansion of the first term:

$$m\,dU + U\,dm - H\,dm = \delta Q \qquad \text{or} \qquad m\,dU - (H - U)\,dm = \delta Q$$

However $H - U = PV$, so that

$$m\,dU - PV\,dm = \delta Q \tag{2}$$

The properties U and V are the unit-mass or specific properties of the gas in the tank, and m is the mass of gas in the tank. This equation can equally well be written in terms of n, the number of moles in the tank:

$$n\,dU - PV\,dn = \delta Q \tag{3}$$

where now U and V are molar properties.

If the gas in the tank is taken to be ideal, then

$$PV = RT \qquad dU = C_V\,dT$$

and (3) becomes

$$nC_V\,dT - RT\,dn = \delta Q \tag{4}$$

When C_V is constant, this equation can be solved for the two limiting cases of (a) an adiabatic process where no heat is exchanged with the gas and (b) a process where heat is exchanged between the tank walls and the gas so rapidly that the gas and the tank are always at the same temperature.

(a) If the process is adiabatic, $\delta Q = 0$, and

$$nC_V\,dT = RT\,dn \qquad \text{or} \qquad \frac{dT}{T} = \frac{R}{C_V}\frac{dn}{n}$$

This equation can be integrated directly; however, we find it convenient to change variables through use of the relationship

$$V_{\text{tank}} = nV$$

Since the volume of the tank V_{tank} is constant, differentiation gives

$$\frac{dn}{n} = -\frac{dV}{V}$$

Therefore we can write

$$\frac{dT}{T} = -\frac{R}{C_V}\frac{dV}{V}$$

Integration for constant C_V gives

$$\ln\frac{T_2}{T_1} = \frac{R}{C_V}\ln\frac{V_1}{V_2}$$

However, $R/C_V = \gamma - 1$ and therefore

$$\frac{T_2}{T_1} = \left(\frac{V_1}{V_2}\right)^{\gamma-1}$$

This is exactly the expression one gets for the reversible, adiabatic expansion of an ideal gas with constant heat capacities (see Examples 1.10 and 2.5). We have, of course, assumed the process to be adiabatic and the gas to be ideal with constant heat capacities, but we have not specifically assumed reversibility. However, in the development of the energy equation, our assumptions that the stream leaving the tank has negligible velocity and has properties identical with those of the gas in the tank are equivalent to the assumption of reversibility. Thus we could have started with the assumption that for an adiabatic process the gas that remains in the tank at any instant has undergone a reversible, adiabatic expansion from its initial state. We can then write immediately any equation which applies to such a process. Figure 6-15 allows one to picture the expansion process. The shaded portion of the tank volume represents the gas that will leave the tank during a particular process. The remainder of the gas merely expands during the process until it fills the entire volume. If this is done slowly and adiabatically then the expansion should be reversible and adiabatic, i.e. isentropic.

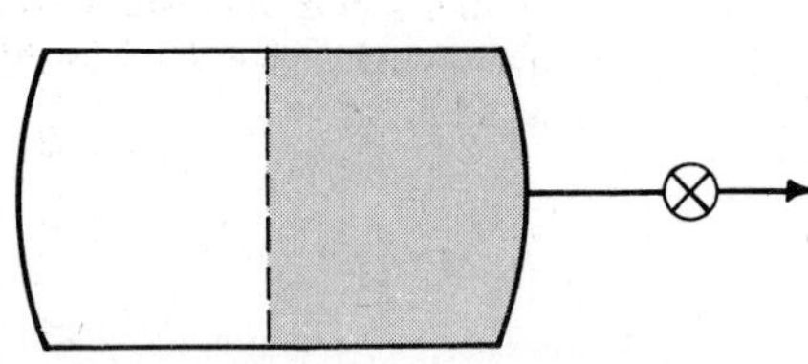

Fig. 6-15

(b) If the gas does not expand adiabatically, but exchanges heat with the tank walls at a sufficient rate to maintain the wall temperature equal to the gas temperature, then the heat transfer *with respect to the wall* is

$$\delta Q_{\text{wall}} = (mC)_{\text{wall}}\, dT$$

where m is the mass of the tank itself and C is its specific heat capacity. In our energy equation δQ is heat transfer with respect to the gas, and therefore has the opposite sign of δQ_{wall}. Thus

$$\delta Q = -(mC)_{\text{wall}}\, dT$$

and substitution in (*4*) yields

$$\frac{dT}{T} = \frac{R\, dn}{nC_V + mC} = \frac{R}{C_V}\frac{d(nC_V)}{nC_V + mC} = (\gamma - 1)\frac{d(nC_V + mC)}{nC_V + mC}$$

where we have for simplicity omitted the subscript "wall" on mC. Integration gives

$$\frac{T_2}{T_1} = \left(\frac{n_2 C_V + mC}{n_1 C_V + mC}\right)^{\gamma - 1}$$

We have the additional equations

$$P_1 V_{\text{tank}} = n_1 R T_1 \qquad P_2 V_{\text{tank}} = n_2 R T_2$$

Knowing P_1, P_2, T_1 and V_{tank} (as well as C_V, m, and C) these equations may be solved by trial for T_2 and n_2. Note that if the total heat capacity of the walls, mC, is much larger than nC_V, the result will be $T_2 \cong T_1$.

6.6. Consider a well-insulated tank of 100(ft)³ capacity containing superheated steam at 400(°F) and 50(psia). A valve is opened, and steam is bled out until the pressure in the tank is reduced to 1(atm). Calculate the final temperature of the steam in the tank and the amount of steam bled out if (*a*) the process occurs rapidly enough so that no heat is transferred between the tank wall and the steam in the tank, and (*b*) the process occurs slowly enough so that heat transfer between the tank wall and the steam keeps them always at the same temperature. The mass of the tank is 1400(lb_m), and its specific heat capacity is 0.107(Btu)/(lb_m)(R).

(*a*) If there is no heat transfer to the steam in the tank then, as discussed in Problem 6.5, we can assume that the steam left in the tank at any instant has undergone a reversible, adiabatic (isentropic) expansion from its initial conditions. These conditions are

$$T_1 = 400(°\text{F}) \qquad P_1 = 50(\text{psia})$$

and from the tables for superheated steam we have

$$S_1 = 1.7348(\text{Btu})/(\text{lb}_m)(\text{R}) \qquad V_1 = 10.061(\text{ft})^3/(\text{lb}_m)$$

Since $S_2 = S_1 = 1.7348(\text{Btu})/(\text{lb}_m)(\text{R})$, we must find from the steam tables the conditions at which steam has this entropy at 1(atm) or 14.696(psia). Examination of the tables shows that the steam must be wet, and is therefore at its saturation temperature of 212(°F). At these conditions

$$S_{f_2} = 0.3121 \qquad S_{fg_2} = 1.4446(\text{Btu})/(\text{lb}_m)(\text{R})$$

The quality x_2 is

$$x_2 = \frac{S_2 - S_{f_2}}{S_{fg_2}} = \frac{1.7348 - 0.3121}{1.4446} = 0.9848$$

and the specific volume is given by

$$V_2 = V_{f_2} + x_2 V_{fg_2} = 0.016715 + (0.9848 \times 26.78) = 26.39(\text{ft})^3/(\text{lb}_m)$$

The initial and final masses of steam in the tank are given by:

$$m_1 = \frac{V_{\text{tank}}}{V_1} = \frac{100(\text{ft})^3}{10.061(\text{ft})^3/(\text{lb}_m)} = 9.939(\text{lb}_m)$$

$$m_2 = \frac{V_{\text{tank}}}{V_2} = \frac{100}{26.39} = 3.789(\text{lb}_m)$$

The amount of steam bled from the tank is therefore $9.939 - 3.789 = 6.150(\text{lb}_m)$.

(*b*) When the tank wall exchanges heat with the expanding steam, the appropriate energy equation is (*2*) of Problem 6.5:

$$m\,dU - PV\,dm = \delta Q$$

The following development of this equation is intended to put it into a form most convenient for integration. Since $V_{\text{tank}} = mV = \text{constant}$, differentiation gives $-V\,dm = m\,dV$ and the energy equation becomes

$$m(dU + P\,dV) = \delta Q$$

However, by (*2.6*), $dU + P\,dV = T\,dS$; hence

$$mT\,dS = \delta Q$$

In this equation δQ represents heat transfer with respect to the steam in the tank. This heat comes from the tank wall and with respect to the wall is given by

$$\delta Q' = (mC)_{\text{tank}}\,dT$$

Since $\delta Q = -\delta Q'$:

$$mT\,dS = -(mC)_{\text{tank}}\,dT$$

In addition $m = V_{\text{tank}}/V$, so that we have finally

$$\frac{V_{\text{tank}}}{V}T\,dS = -(mC)_{\text{tank}}\,dT$$

or

$$\frac{dS}{dT} = -\left(\frac{mC}{V}\right)_{\text{tank}}\frac{V}{T}$$

From the data:

$$\left(\frac{mC}{V}\right)_{\text{tank}} = \frac{1400(\text{lb}_\text{m}) \times 0.107(\text{Btu})/(\text{lb}_\text{m})(\text{R})}{100(\text{ft})^3} = 1.498\frac{(\text{Btu})}{(\text{R})(\text{ft})^3}$$

Therefore

$$\frac{dS}{dT} = -1.498\frac{V}{T} \tag{1}$$

where V is in $(\text{ft})^3/(\text{lb}_\text{m})$, T is in (R), and S is in $(\text{Btu})/(\text{lb}_\text{m})(\text{R})$. It is this equation that must be integrated. Since S, T, and V are related for any given state of superheated steam by numerical data in the steam tables, we are able to carry out a numerical integration as follows.

The initial conditions of T_1 and V_1 for the steam in the tank are known. From these (*1*) allows calculation of an initial value for dS/dT. Using this and an incremental temperature change ΔT, we can calculate the change in S:

$$\Delta S = \frac{dS}{dT}\Delta T$$

The new values $S = S_1 + \Delta S$ and $T = T_1 + \Delta T$ identify a new state of superheated steam in the steam tables, for which there is a new value for V (and also P). The new values for T and V yield a new value for dS/dT by (*1*), and the process may now be repeated for another increment ΔT. Thus we calculate succesive values of S and V (and the corresponding P) for successive increments ΔT. The process continues until the given final pressure of 14.696(psia) is reached. The calculations are made increasingly accurate by decreasing the increment ΔT. More sophisticated numerical integration procedures can also be employed, but are not necessary here. The results obtained for the present problem at $P_2 = 14.696(\text{psia})$ are

$$T_2 = 395.5(^\circ\text{F}) \qquad S_2 = 1.8716(\text{Btu})/(\text{lb}_\text{m})(\text{R}) \qquad V_2 = 34.49(\text{ft})^3/(\text{lb}_\text{m})$$

Thus

$$m_2 = \frac{V_{\text{tank}}}{V_2} = \frac{100}{34.49} = 2.899(\text{lb}_\text{m})$$

and the amount of steam bled from the tank is $9.939 - 2.899 = 7.040(\text{lb}_\text{m})$.

Comparison of the results from parts (*a*) and (*b*) of this problem reveals the considerable effect of the initial assumptions made in setting up the solution:

	(*a*)	(*b*)
Final state of the steam	wet	superheated
Final temperature of the steam	$212(^\circ\text{F})$	$395.5(^\circ\text{F})$
Mass of steam bled from the tank	$6.15(\text{lb}_\text{m})$	$7.04(\text{lb}_\text{m})$

6.7. A steam boiler with a capacity of $1000(\text{ft})^3$ contains saturated liquid water and saturated steam in equilibrium at 100(psia). Initially the liquid and vapor occupy equal volumes. During a certain time interval saturated vapor is withdrawn from the boiler while simultaneously $31{,}160(\text{lb}_\text{m})$ of liquid water at 100(°F) is added to the boiler. During the process the boiler is maintained at the constant pressure of 100(psia) by the addition of heat. At the end of the process the saturated liquid remaining in the boiler occupies one-quarter of the boiler volume, and saturated vapor fills the remainder. How much heat was added to the boiler during the process?

First we determine the masses of material contained in the boiler and flowing from the boiler. Within the boiler the properties of saturated liquid and saturated vapor remain constant throughout the process. From the tables for saturated steam, we have at 100(psia)

$$V_f = 0.017736 \qquad V_g = 4.434(\text{ft})^3/(\text{lb}_\text{m})$$

Thus the initial and final masses of liquid and vapor in the boiler are given by

$$m_{f_1} = \frac{500(\text{ft})^3}{0.017736(\text{ft})^3/(\text{lb}_\text{m})} = 28{,}192(\text{lb}_\text{m})$$

$$m_{g_1} = \frac{500(\text{ft})^3}{4.434(\text{ft})^3/(\text{lb}_\text{m})} = 113(\text{lb}_\text{m})$$

Therefore $m_1 = m_{f_1} + m_{g_1} = 28{,}192 + 113 = 28{,}305(\text{lb}_\text{m})$. Similarly

$$m_{f_2} = \frac{250}{0.017736} = 14{,}096(\text{lb}_\text{m}) \qquad m_{g_2} = \frac{750}{4.436} = 169(\text{lb}_\text{m})$$

and $m_2 = 14{,}265(\text{lb}_\text{m})$. Thus the depletion of mass in the boiler amounts to

$$m_1 - m_2 = 28{,}305 - 14{,}265 = 14{,}040(\text{lb}_\text{m})$$

The mass of vapor withdrawn from the boiler must be the sum of this mass and the mass of liquid added. Thus

$$m_g' = 14{,}040 + 31{,}160 = 45{,}200(\text{lb}_\text{m})$$

We now apply (*6.12*) to the boiler taken as the control volume. There is no work associated with the process, and we will assume the kinetic- and potential-energy terms to be negligible. There is a saturated vapor stream flowing out, having an enthalpy H_g', and a liquid stream flowing in, with an enthalpy H_f'. Equation (*6.12*) is therefore written

$$d(mU)_\text{boiler} + H_g'\,dm_g' - H_f'\,dm_f' = \delta Q$$

Since H_g' and H_f' are constant, integration gives

$$m_2U_2 - m_1U_1 + H_g'm_g' - H_f'm_f' = Q$$

The first two terms represent the total internal energy of the contents of the tank at the end and start of the process. These terms are most readily calculated as follows

$$m_2U_2 = m_{f_2}U_f + m_{g_2}U_g = (14{,}096)(298.28) + (169)(1105.8) = 4{,}391{,}400(\text{Btu})$$

$$m_1U_1 = m_{f_1}U_f + m_{g_1}U_g = (28{,}192)(298.28) + (113)(1105.8) = 8{,}534{,}100(\text{Btu})$$

where the values of U_f and U_g were taken from the tables for saturated steam at 100(psia). We may now solve for Q. The value for H_g' is taken from the tables for saturated steam at 100(psia) and H_f' is taken as the enthalpy of saturated liquid at 100(°F). Thus

$$Q = 4{,}391{,}400 - 8{,}534{,}100 + (45{,}200)(1187.8) - (31{,}160)(68.05) = 47{,}425{,}500(\text{Btu})$$

6.8. Wet steam at 15(bar) is throttled adiabatically in a continuous-flow process to 2(bar). The resulting stream has a temperature of 130(°C). What are the temperature and quality of the wet steam? Calculate ΔS of the steam as a result of the process.

For a throttling process $\Delta H = 0$ or $H_1 = H_2$. At the final conditions of 130(°C) and 2(bar) we find from the tables for superheated steam: $H_2 = 2727.3$(J)/(g). Since the steam in its initial condition is wet, it must be at its saturation temperature, and this is found from the tables for saturated steam to be $T_1 = 198.3$(°C). In addition, $H_{f_1} = 844.9$(J)/(g) and $H_{fg_1} = 1947.3$(J)/(g). Therefore

$$H_1 = H_{f_1} + x_1 H_{fg_1} = 844.9 + x_1(1947.3) = H_2 = 2727.3\text{(J)/(g)}$$

Solution for x_1 gives

$$x_1 = 0.9667$$

Thus the wet steam is 96.67% vapor on a mass basis.

The initial entropy of the steam is determined by

$$S_1 = S_{f_1} + x_1 S_{fg_1} = 2.3150 + 0.9667 \times 4.1298 = 6.3073\text{(J)/(g)(K)}$$

For the final state $S_2 = 7.1789$ and

$$\Delta S = S_2 - S_1 = 7.1789 - 6.3073 = 0.8716\text{(J)/(g)(K)}$$

6.9. A thermometer is inserted in a pipe through which is flowing a steady stream of gas at temperature T_1 and velocity u_1. Will the indicated temperature be higher than, lower than, or the same as T_1?

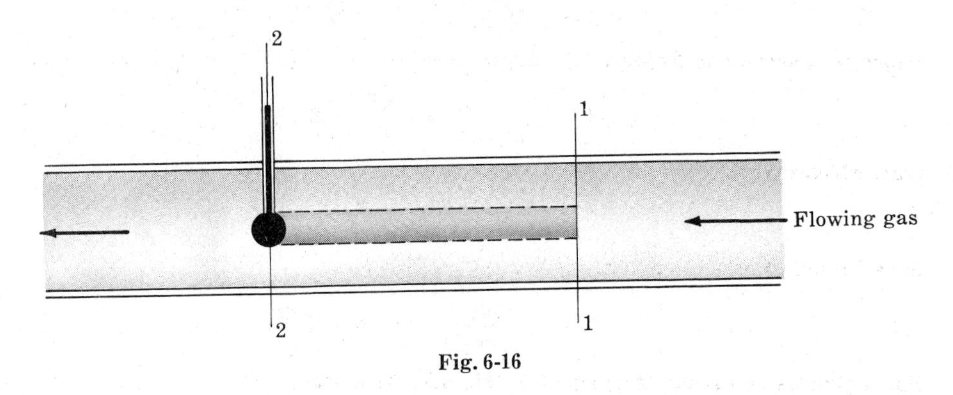

Fig. 6-16

The experimental arrangement is shown in Fig. 6-16. We choose as a control volume an imaginary "tube" (shown by the dashed lines) whose cross-section is the projected area of the thermometer bulb. The bounding planes 2-2 and 1-1 are drawn, respectively, through the centerline of the thermometer, and at a position far enough upstream so that conditions are unaffected by the presence of the thermometer. It is assumed that gas reaching 2-2 in the control volume is decelerated to a low velocity and that it enters the control volume at 1-1. The flow is steady, and we consider only the single stream leaving the control volume at 2-2 and entering at 1-1. Clearly, $W_s = \Delta z = 0$, and we also assume that no heat is transferred to or from the gas. Thus (*6.11*) reduces to (*6.13*), which was previously derived for flow through an adiabatic nozzle:

$$H_2 - H_1 = -\frac{u_2^2 - u_1^2}{2g_c} \tag{6.13}$$

For an ideal gas with constant heat capacity $H_2 - H_1 = C_P(T_2 - T_1)$. Substitution of this expression and $u_2 = 0$ into (*6.13*) gives

$$C_P(T_2 - T_1) = \frac{u_1^2}{2g_c}$$

from which we obtain for the thermometer temperature

$$T_2 = T_1 + \frac{u_1^2}{2g_c C_P} \tag{1}$$

The measured temperature is therefore *higher than* the gas temperature T_1; the temperature T_2 defined by (*1*) is sometimes called the *stagnation temperature.* Many of the fluid-mechanical assumptions implicit in the solution of this problem are of limited validity, however, so actual thermometer readings for such arrangements are only approximately equal to the stagnation temperature.

6.10. Propane at 10(atm) and 320(K) expands in a steady-flow process through an orifice to a pressure of 1(atm). Heat is added so as to maintain the downstream temperature also at 320(K). Upstream from the orifice the propane flows with a velocity of 2(m)/(s), and downstream from the orifice its velocity is approximately 20(m)/(s). How much heat must be added per kilogram of propane flowing through the orifice? What percentage error would be introduced if the kinetic-energy change of the propane were neglected? The volumetric properties of propane at these conditions are well represented by (*5.22*), the simplest form of the virial equation, and the second virial coefficient may be determined from the generalized correlation of Sec. 5.5.

The appropriate energy equation is (*6.11*), which reduces to

$$\Delta H + \frac{\Delta u^2}{2g_c} = Q \tag{1}$$

because there is no work, and the change in potential energy of the propane is evidently negligible. To determine ΔH, we make use of the equation

$$\left(\frac{\partial H}{\partial P}\right)_T = -T\left(\frac{\partial V}{\partial T}\right)_P + V \tag{2}$$

which was derived in Problem 3.3. Substitution of $Z = PV/RT$ in (*5.22*) and solution for V gives

$$V = \frac{RT}{P} + B$$

from which we get

$$\left(\frac{\partial V}{\partial T}\right)_P = \frac{R}{P} + \frac{dB}{dT}$$

Substitution of these two equations in (*2*) yields

$$\left(\frac{\partial H}{\partial P}\right)_T = B - T\frac{dB}{dT}$$

For a process occurring at constant T this may be written

$$dH = \left(B - T\frac{dB}{dT}\right)dP \qquad \text{(const. } T\text{)}$$

Since $B = B(T)$, integration at constant T gives

$$\Delta H = \left(B - T\frac{dB}{dT}\right)\Delta P \tag{3}$$

The generalized correlation of Sec. 5.5 provides the equations

$$\frac{BP_c}{RT_c} = B^0 + \omega B^1 \tag{5.39}$$

$$\frac{P_c}{R}\frac{dB}{dT} = \frac{dB^0}{dT_r} + \omega\frac{dB^1}{dT_r} \tag{5.40}$$

and Appendix 3 provides the values

$$T_c = 369.9\text{(K)} \qquad P_c = 42\text{(atm)} \qquad \omega = 0.152$$

Thus $T_r = 320/369.9 = 0.865$, and from Figs. 5-6 and 5-7 we read the values

$$B^0 = -0.44 \qquad B^1 = -0.24$$

$$dB^0/dT_r = 0.94 \qquad dB^1/dT_r = 1.75$$

Thus from (*5.39*) and (*5.40*):

$$\frac{BP_c}{RT_c} = -0.44 - 0.152 \times 0.24 = -0.48 \qquad \text{and} \qquad \frac{P_c}{R}\frac{dB}{dT} = 0.94 + 0.152 \times 1.75 = 1.21$$

Therefore

$$B = -0.48\frac{RT_c}{P_c} \qquad \text{and} \qquad \frac{dB}{dT} = 1.21\frac{R}{P_c}$$

and substitution in (*3*) gives

$$\Delta H = \left(-0.48\frac{RT_c}{P_c} - 1.21\frac{RT}{P_c}\right)\Delta P = -(0.48\,T_c + 1.21\,T)R\frac{\Delta P}{P_c}$$

$$= -[0.48 \times 369.9 + 1.21 \times 320](\text{K}) \times 8314(\text{J})/(\text{kg mole})(\text{K}) \times \frac{-9}{42}$$

$$= 1.01 \times 10^6(\text{J})/(\text{kg mole})$$

Since the molecular weight of propane is 44,

$$\Delta H = \frac{1.01 \times 10^6}{44} = 22{,}950(\text{J})/(\text{kg})$$

We now evaluate the kinetic-energy term:

$$\frac{\Delta u^2}{2g_c} = \frac{u_2^2 - u_1^2}{2g_c} = \frac{[20^2 - 2^2](\text{m})^2/(\text{s})^2}{2 \times 1(\text{kg})(\text{m})/(\text{N})(\text{s})^2} = 198(\text{N-m})/(\text{kg}) \quad \text{or} \quad 198(\text{J})/(\text{kg})$$

Equation (*1*) now allows calculation of Q:

$$Q = 22{,}950 + 198 = 23{,}150(\text{J})/(\text{kg})$$

The % error that would be caused by neglect of the kinetic-energy term is

$$\frac{198}{23{,}150} \times 100 = 0.86\%$$

6.11. Methane gas undergoes a continuous throttling process from upstream conditions of 40(°C) and 20(atm) to a downstream pressure of 5(atm). What is the gas temperature on the downstream side of the throttling device? For the conditions of this problem, the volumetric properties of methane are adequately described by the truncated virial equation (*5.22*), used in conjunction with the generalized correlation of Sec. 5.5. An expression for the molar heat capacity of methane in the ideal-gas state is

$$C_P' = 3.381 + 18.044 \times 10^{-3}T - 4.300 \times 10^{-6}T^2 \qquad (1)$$

where T is in (K) and C_P' is in (cal)/(g mole)(K).

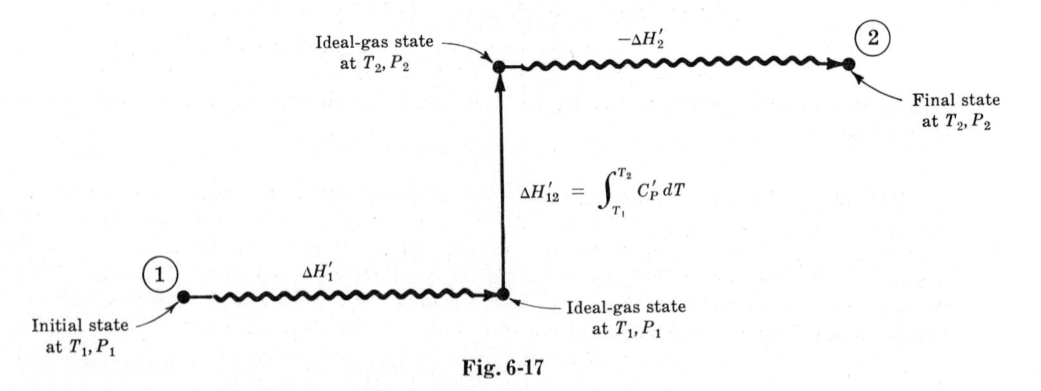

Fig. 6-17

As shown in Sec. 6.4, (*6.11*) applied to a throttling process gives $\Delta H = 0$. Thus the downstream gas temperature must be such that there is no enthalpy change across the throttling device.

In order to calculate the final temperature, we set up a calculational path similar to that of

Problem 5.23. By inspection of Fig. 6-17, we see that the requirement that $\Delta H = 0$ is equivalent to

$$\Delta H_1' + \Delta H_{12}' - \Delta H_2' = 0$$

or

$$\Delta H_2' - \int_{T_1}^{T_2} C_P' \, dT = \Delta H_1' \tag{2}$$

The given upstream conditions determine $\Delta H_1'$; however, the terms on the left-hand side of (*2*) depend on the final temperature, so T_2 must be found by trial.

For methane, from Appendix 3,

$$T_c = 190.7(\text{K}) \qquad P_c = 45.8(\text{atm}) \qquad \omega = 0.013$$

For the initial state

$$T_{r_1} = 313.2/190.7 = 1.64 \qquad P_{r_1} = 20/45.8 = 0.437$$

and from Figs. 5-6 and 5-7 for $T_{r_1} = 1.64$

$$B^0 = -0.11 \qquad B^1 = 0.14$$

$$dB^0/dT_r = 0.185 \qquad dB^1/dT_r = 0.082$$

We have:

$$\frac{\Delta H'}{RT} = P_r\left[\left(\frac{dB^0}{dT_r} - \frac{B^0}{T_r}\right) + \omega\left(\frac{dB^1}{dT_r} - \frac{B^1}{T_r}\right)\right] \tag{5.49}$$

Therefore

$$\frac{\Delta H_1'}{RT_1} = 0.437\left[\left(0.185 - \frac{-0.11}{1.64}\right) + 0.013\left(0.082 - \frac{0.14}{1.64}\right)\right] = 0.110$$

from which

$$\Delta H_1' = 0.110 \times 1.987(\text{cal})/(\text{g mole})(\text{K}) \times 313.2(\text{K}) = 68.5(\text{cal})/(\text{g mole})$$

Substitution into (*2*) of this value of $\Delta H_1'$ together with the heat-capacity expression (*1*) gives the equation to be solved for T_2:

$$\Delta H_2' + 3.381(313.2 - T_2) + 9.022 \times 10^{-3}[(313.2)^2 - T_2^2]$$
$$- 1.433 \times 10^{-6}[(313.2)^3 - T_2^3] = 68.5 \tag{3}$$

We proceed by assuming a trial value for T_2. Next, $\Delta H_2'$ is calculated by (*5.49*). Finally, $\Delta H_2'$ and T_2 are substituted into (*3*). The correct T_2 will give a value of 68.5 for the left-hand side of (*3*). As a first trial, we assume a temperature drop of 5(°C), that is, $T_2 = 308.2(\text{K})$. The trial values for T_{r_2} and P_{r_2} are then

$$T_{r_2} = 308.2/190.7 = 1.62 \qquad P_{r_2} = 5/45.8 = 0.109$$

At these conditions, (*5.49*) and Figs. 5-6 and 5-7 give $\Delta H_2' = 17.8(\text{cal})/(\text{g mole})$, and the left-hand side of (*3*) becomes

$$17.8 + 3.381(313.2 - 308.2) + 9.022 \times 10^{-3}[(313.2)^2 - (308.2)^2]$$
$$- 1.433 \times 10^{-6}[(313.2)^3 - (308.2)^3] = 60.7 < 68.5$$

Evidently the assumed T_2 is too high. Repeating the above calculations, we obtain finally as the solution to (*3*):

$$T_2 = 307.3(\text{K}) = 34.1(°\text{C})$$

The effect of the throttling process of this problem was to *decrease* the temperature of the flowing gas stream by about 6(°C), and in fact a usual purpose of a throttling process (e.g. in refrigeration engineering) is to achieve a temperature drop. However it is also possible to obtain a temperature *increase* under certain conditions. The thermodynamic property which determines the sign and magnitude of the temperature change is the *Joule-Thomson coefficient* $(\partial T/\partial P)_H$ [see Problems 3.4, 5.4, 5.8, and 5.33]. Thus for a change of state at constant H,

$$\Delta T = \int_{P_1}^{P_2} \left(\frac{\partial T}{\partial P}\right)_H dP \qquad (\text{const. } H)$$

Clearly, if $P_2 < P_1$, as in a throttling process, a temperature drop occurs if $(\partial T/\partial P)_H$ is positive. Conversely, if $(\partial T/\partial P)_H$ is negative, the temperature will increase on throttling. Certain gases, notably hydrogen and helium, have negative Joule-Thomson coefficients at normal conditions.

SECOND-LAW APPLICATIONS (Sec. 6.4)

6.12. A test made on a stand-by turbine power unit produced the following results. With steam supplied to the turbine at 200(psia) and 700(°F) the discharge from the turbine at 1(psia) was saturated vapor only. What is the efficiency of the turbine?

Assuming the turbine to be properly designed, we can neglect heat transfer and kinetic- and potential-energy terms. Under these circumstances (see the discussion following Example 6.7), $\eta = \Delta H/(\Delta H)_S$. Thus we need to determine from the steam tables the actual enthalpy change of the steam and the enthalpy change that would be obtained were expansion carried out isentropically to 1(psia).

At the initial conditions of the steam

$$P_1 = 200(\text{psia}) \qquad T_1 = 700(°\text{F})$$

From the table for superheated vapor we find

$$H_1 = 1373.8(\text{Btu})/(\text{lb}_m) \qquad S_1 = 1.7234(\text{Btu})/(\text{lb}_m)(\text{R})$$

From the table for saturated steam at 1(psia) we find

$$H_2 = 1105.8(\text{Btu})/(\text{lb}_m)$$

Thus $$\Delta H = H_2 - H_1 = 1105.8 - 1373.8 = -268.0(\text{Btu})/(\text{lb}_m)$$

We must now find the enthalpy of steam at 1(psia) for which the entropy $S_2' = S_1 = 1.7234$. (See Fig. 6-8.) We note that the entropy of saturated vapor at 1(psia) is 1.9779(Btu)/(lb_m)(R), which is higher than the required S_2'. Thus the required state must be wet steam, and we must determine its quality. We have at 1(psia):

$$S_2' = 1.7234 \qquad S_f = 0.1327 \qquad S_{fg} = 1.8453$$

Since $S_2' = S_f + xS_{fg}$

$$1.7234 = 0.1327 + 1.8453\,x \qquad \text{or} \qquad x = 0.862$$

The required enthalpy H_2' is now given by

$$H_2' = H_f + xH_{fg}$$

At 1(psia) we have

$$H_f = 69.7 \qquad H_{fg} = 1036.0$$

Therefore $$H_2' = 69.7 + 0.862 \times 1036.0 = 962.7(\text{Btu})/(\text{lb}_m)$$

and $$(\Delta H)_S = H_2' - H_1 = 962.7 - 1373.8 = -411.1(\text{Btu})/(\text{lb}_m)$$

Finally $$\eta = \frac{\Delta H}{(\Delta H)_S} = \frac{-268.0}{-411.1} = 0.652 \quad \text{or} \quad 65.2\%$$

6.13. Helium gas is to be compressed continuously in an adiabatic compressor from initial conditions of 1(bar) and 20(°C) to a final pressure of 5(bar). If it is assumed that the compressor will operate with an efficiency of 75% compared with isentropic compression, what will be the final temperature of the helium? Helium may be considered an ideal gas for which $C_P = 5(\text{cal})/(\text{g mole})(\text{K})$.

For isentropic compression of an ideal gas with constant heat capacities we have from Example 2.5 that

$$T_2 = T_1\left(\frac{P_2}{P_1}\right)^{(\gamma-1)/\gamma} \qquad \text{(isentropic)}$$

With $T_1 = 20 + 273 = 293(\text{K})$ this becomes

$$T_2 = (293)(5)^{0.4} = 557.8(\text{K})$$

Thus $$(\Delta T)_S = 557.8 - 293 = 264.8(\text{K})$$

From *(6.16)* and *(1.15)* we have

$$W_s = -(\Delta H)_S = -C_P(\Delta T)_S \qquad \text{(isentropic)}$$

and
$$W_s = -\Delta H = -C_P(\Delta T) \quad \text{(actual)}$$

By the definition of efficiency for a compression process:

$$\eta = \frac{W_s\ \text{(isentropic)}}{W_s\ \text{(actual)}} = \frac{(\Delta H)_S}{\Delta H} = \frac{C_P(\Delta T)_S}{C_P(\Delta T)} = \frac{(\Delta T)_S}{\Delta T}$$

Thus
$$\Delta T = (\Delta T)_S/\eta = 264.8/0.75 = 353(\text{K})$$

and
$$T_2 = T_1 + \Delta T = 293 + 353 = 646(\text{K})$$

6.14. Methane gas is to be compressed at the rate of 1,000,000(ft)^3/(hr) [as measured at 1(atm) and 60(°F)] from 100(psia) and 80(°F) to 500(psia). The compressor is to operate adiabatically and is expected to have an efficiency of 80% compared with isentropic compression. After compression the methane is to be cooled at a constant pressure of 500(psia) to a temperature of 100(°F). What will be the required horsepower of the compressor and what will be the rate of heat removal in the cooler? Data for methane:

At 100(psia) and 80(°F): $H = 407.0$(Btu)/(lb$_m$), $S = 1.450$(Btu)/(lb$_m$)(R)

At 500(psia):

T(°F)	H(Btu)/(lb$_m$)	S(Btu)/(lb$_m$)(R)
100	408.0	1.257
300	530.4	1.441
310	536.9	1.450
320	543.4	1.459
330	549.9	1.467
340	556.4	1.475
350	562.9	1.483
360	569.4	1.490

At 1(atm) and 60(°F) methane is essentially an ideal gas, and the molar rate of compression is calculated as

$$\dot{n} = \frac{P\dot{V}^t}{RT} = \frac{1(\text{atm}) \times 10^6(\text{ft})^3/(\text{hr})}{0.7302(\text{atm})(\text{ft})^3/(\text{lb mole})(\text{R}) \times 520(\text{R})} = 2634(\text{lb mole})/(\text{hr})$$

Since the molecular weight of methane is 16, the mass flow rate is

$$\dot{m} = 16 \times 2634 = 42{,}140(\text{lb}_m)/(\text{hr})$$

For reversible, adiabatic compression the process would occur at constant entropy. Therefore from the data given (see Fig. 6-10)

$$S_2' = S_1 = 1.450(\text{Btu})/(\text{lb}_m)(\text{R})$$

At 500(psia) the state having this entropy occurs at a temperature of 310(°F) and an enthalpy $H_2' = 536.9$(Btu)/(lb$_m$). Therefore

$$(\Delta H)_S = 536.9 - 407.0 = 129.9(\text{Btu})/(\text{lb}_m)$$

The efficiency of a compression process is given by

$$\eta = \frac{(\Delta H)_S}{\Delta H}$$

Therefore the actual enthalpy change is given by

$$\Delta H = \frac{(\Delta H)_S}{\eta} = \frac{129.9}{0.8} = 162.4(\text{Btu})/(\text{lb}_m)$$

and since $W_s = -\Delta H$, the work input to the compressor is 162.4(Btu)/(lb$_m$). The power requirement is calculated as follows:

$$\text{Power} = \frac{162.4(\text{Btu})/(\text{lb}_m) \times 42{,}140(\text{lb}_m)/(\text{hr})}{60(\text{min})/(\text{hr}) \times 42.408\dfrac{(\text{Btu})/(\text{min})}{(\text{HP})}} = 2690(\text{HP})$$

The actual final enthalpy of the compressed methane is

$$H_2 = H_1 + \Delta H = 407.0 + 162.4 = 569.4(\text{Btu})/(\text{lb}_m)$$

and this value is obtained at a temperature of 360(°F).

In the cooler no work is done, and the kinetic- and potential-energy terms should be negligible. Therefore (*6.10*) becomes

$$\dot{Q} = \dot{m}\,\Delta H = 42{,}140(\text{lb}_m)/(\text{hr}) \times [408.0 - 569.4](\text{Btu})/(\text{lb}_m)$$

$$= -6.801 \times 10^6(\text{Btu})/(\text{hr})$$

6.15. Ethylene gas is to be continuously compressed from an initial state of 1(atm) and 21(°C) to a final pressure of 18(atm) in an adiabatic compressor. If compression is 75% efficient compared with an isentropic process, what will be the work requirement and what will be the final temperature of the ethylene? The virial equation of state (*5.22*) adequately describes the volumetric behavior of ethylene, and the second virial coefficient is given by the generalized correlation of Sec. 5.5. The molar heat capacity of ethylene in the ideal-gas state is represented as a function of temperature by

$$C_P' = 2.83 + 28.60 \times 10^{-3}T - 8.73 \times 10^{-6}T^2 \qquad (1)$$

where T is in (K) and C_P' is in (cal)/(g mole)(K).

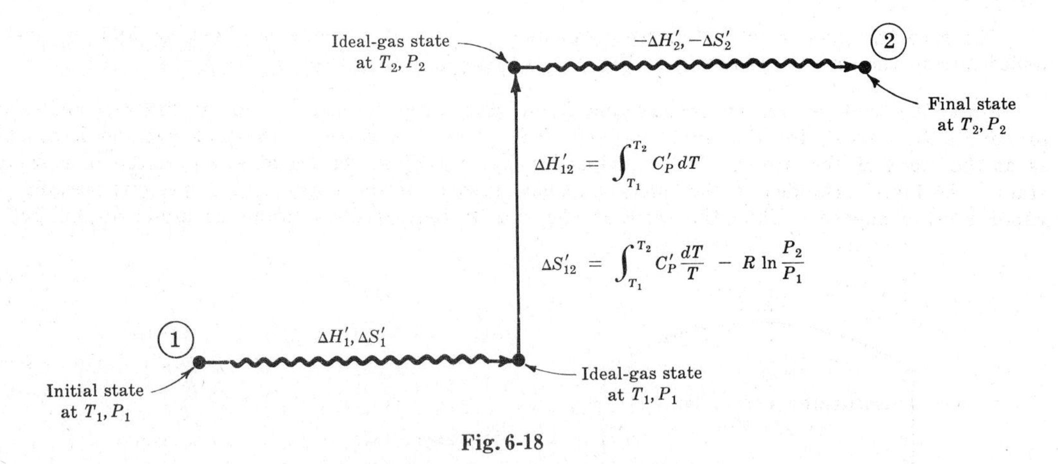

Fig. 6-18

This problem is similar to Problem 5.23 in that a calculational path must be devised in order to determine the property changes between the initial and final states. For the data available the most convenient path is that shown in Fig. 6-18. The difficulty here is that the final temperature T_2 is not known, and without it one cannot determine values for $\Delta H_2'$ and $\Delta S_2'$. Thus a trial value for T_2 must be assumed for purposes of getting started toward a solution to the problem.

If we consider the *isentropic* compression of ethylene, then the condition imposed on the process is that $S_2 = S_1$. In this event,

$$\Delta S_1' + \Delta S_{12}' - \Delta S_2' = 0$$

or

$$\Delta S_1' + \int_{T_1}^{T_2} C_P'\frac{dT}{T} - R\ln\frac{P_2}{P_1} - \Delta S_2' = 0 \qquad (2)$$

The only way to find the value of T_2 for which this equation is satisfied is by trial. To get a first estimate for T_2 we use the expression valid for an ideal gas with constant heat capacities:

$$T_2 = T_1\left(\frac{P_2}{P_1}\right)^{(\gamma-1)/\gamma}$$

A value for γ can be estimated by $\gamma = C'_P/C'_V = C'_P/(C'_P - R)$, where C'_P is evaluated from (*1*) at $T_1 = 21 + 273 = 294(\text{K})$. Thus

$$C'_P = 2.83 + 28.60\times10^{-3}\times294 - 8.73\times10^{-6}\times(294)^2$$

$$= 10.5(\text{cal})/(\text{g mole})(\text{K})$$

and
$$\gamma = \frac{10.5}{10.5-2.0} = 1.235 \quad \text{or} \quad \frac{\gamma-1}{\gamma} = \frac{0.235}{1.235} = 0.19$$

Our first estimate of T_2 is then

$$T_2 = 294\times(18)^{0.19} = 509(\text{K})$$

Making use of this value, we can now evaluate the various terms of (*2*). For ethylene, from Appendix 3:

$$T_c = 283.1(\text{K}) \qquad P_c = 50.5(\text{atm}) \qquad \omega = 0.085$$

Thus for the initial conditions, which are fixed,

$$T_{r_1} = 294/283.1 = 1.04 \qquad P_{r_1} = 1/50.5 = 0.02$$

and from Figs. 5-6 and 5-7 for this value of T_{r_1}

$$B^0 = -0.31 \qquad B^1 = -0.05$$

$$dB^0/dT_r = 0.58 \qquad dB^1/dT_r = 0.76$$

We have:

$$\frac{\Delta H'}{RT} = P_r\left[\left(\frac{dB^0}{dT_r} - \frac{B^0}{T_r}\right) + \omega\left(\frac{dB^1}{dT_r} - \frac{B^1}{T_r}\right)\right] \qquad (5.49)$$

Thus

$$\frac{\Delta H'_1}{RT_1} = 0.02\left[\left(0.58 - \frac{-0.31}{1.04}\right) + 0.085\left(0.76 - \frac{-0.05}{1.04}\right)\right] = 0.019$$

$$\Delta H'_1 = 0.019\times1.987(\text{cal})/(\text{g mole})(\text{K})\times294(\text{K}) = 11(\text{cal})/(\text{g mole})$$

From

$$\frac{\Delta S'}{R} = P_r\left(\frac{dB^0}{dT_r} + \omega\frac{dB^1}{dT_r}\right) \qquad (5.50)$$

we find

$$\frac{\Delta S'_1}{R} = 0.02(0.58 + 0.085\times0.76) = 0.013$$

$$\Delta S'_1 = 0.013\times1.987 = 0.03(\text{cal})/(\text{g mole})(\text{K})$$

For the change from T_1, P_1 to T_2, P_2 in the ideal-gas state:

$$\Delta S'_{12} = \int_{T_1}^{T_2} C'_P\frac{dT}{T} - R\ln\frac{P_2}{P_1}$$

$$= \int_{294}^{509}\left(\frac{2.83}{T} + 28.60\times10^{-3} - 8.73\times10^{-6}T\right)dT - R\ln\frac{P_2}{P_1}$$

$$= 2.83\ln\frac{509}{294} + 28.60\times10^{-3}(509-294) - \frac{8.73\times10^{-6}}{2}[(509)^2-(294)^2] - 1.987\ln18$$

$$= 1.20(\text{cal})/(\text{g mole})(\text{K})$$

For a final pressure of 18(atm) and a tentative final temperature of 509(K) we have

$$T_{r_2} = \frac{509}{283.1} = 1.80 \qquad P_{r_2} = \frac{18}{50.5} = 0.356$$

and from Fig. 5-7

$$dB^0/dT_r = 0.15 \qquad dB^1/dT_r = 0.05$$

Use of these values in (*5.50*) gives

$$\Delta S_2' = 0.11\text{(cal)/(g mole)(K)}$$

As a result of these calculations we find

$$\Delta S = \Delta S_1' + \Delta S_{12}' - \Delta S_2' = 0.03 + 1.20 - 0.11 = 1.12 \neq 0$$

Thus the condition that $S_2 = S_1$ is not satisfied. The trial temperature $T_2 = 509\text{(K)}$ was evidently too high; so we choose a lower value and continue the trial-and-error process until $\Delta S = S_2 - S_1 = 0$. The value for T_2 found in this way is

$$T_2 = 473\text{(K)}$$

We can now calculate the enthalpy change for isentropic compression:

$$(\Delta H)_S = \Delta H_1' + \Delta H_{12}' - \Delta H_2'$$

The value for $\Delta H_1'$ has already been determined. $\Delta H_{12}'$ is found from

$$\Delta H_{12}' = \int_{T_1}^{T_2} C_P' \, dT = \int_{294}^{473} (2.83 + 28.60 \times 10^{-3}T - 8.73 \times 10^{-6}T^2)\, dT$$

$$= 2.83(473 - 294) + \frac{0.0286}{2}[(473)^2 - (294)^2] - \frac{8.73 \times 10^{-6}}{3}[(473)^3 - (294)^3]$$

$$= 2235\text{(cal)/(g mole)}$$

For $T_2 = 473\text{(K)}$ and $P_2 = 18\text{(atm)}$, we have

$$T_{r_2} = \frac{473}{283.1} = 1.67 \qquad P_{r_2} = 0.356$$

and from Figs. 5-6 and 5-7

$$B^0 = -0.11 \qquad B_1 = 0.14$$

$$dB^0/dT_r = 0.185 \qquad dB^1/dT_r = 0.078$$

By (*5.49*)

$$\frac{\Delta H_2'}{RT_2} = 0.356\left[\left(0.185 - \frac{-0.11}{1.67}\right) + 0.085\left(0.078 - \frac{0.14}{1.67}\right)\right] = 0.0895$$

$$\Delta H_2' = 0.0895 \times 1.987 \times 473 = 84\text{(cal)/(g mole)}$$

Thus

$$(\Delta H)_S = 11 + 2235 - 84 = 2160\text{(cal)/(g mole)}$$

For a compressor efficiency of 75% we have

$$W_s = -\Delta H = \frac{-(\Delta H)_S}{0.75} = \frac{-2160}{0.75} = -2880\text{(cal)/(g mole)}$$

The actual final temperature will therefore be that temperature for which

$$\Delta H = \Delta H_1' + \Delta H_{12}' - \Delta H_2' = 2880\text{(cal)/(g mole)}$$

Again this must be determined by trial. If we take $T_2 = 521\text{(K)}$, then we find

$$\Delta H_{12}' = 2950\text{(cal)/(g mole)}$$

For this T_2, $T_{r_2} = 521/283.1 = 1.84$ and from Figs. 5-6 and 5-7

$$B^0 = -0.08 \qquad B^1 = 0.16$$

$$dB^0/dT_r = 0.145 \qquad dB^1/dT_r = 0.045$$

Application of (*5.49*) gives

$$\Delta H_2' = 71\text{(cal)/(g mole)}$$

Thus

$$\Delta H = 11 + 2950 - 71 = 2890\text{(cal)/(g mole)}$$

which is close enough to the required value of 2880 so that we may take the actual final temperature T_2 of the ethylene to be very close to 521(K) or 248(°C).

6.16. (*a*) For steady flow in an adiabatic nozzle of cross-sectional area A, show that

$$u^2 = -g_c V \frac{dH}{dV}\left[\frac{d(\ln A)}{d(\ln u)} + 1\right]$$

(*b*) If in addition the flow is reversible, show that

(i) $$u^2 = -g_c V^2 \left(\frac{\partial P}{\partial V}\right)_S \left[\frac{d(\ln A)}{d(\ln u)} + 1\right]$$

(ii) $$\mathbf{M}^2 - 1 = \frac{d(\ln A)}{d(\ln u)}$$

where $\mathbf{M}$ is the *Mach number,* defined as the ratio of the velocity u to the sonic velocity c, i.e. $\mathbf{M} = u/c$.

(*a*) The energy equation for an adiabatic nozzle is given by (*6.13*), which in differential form becomes

$$\frac{dH}{du} = -\frac{u}{g_c}$$

from which we see that dH/du is *always negative*. We may also write

$$dH = -\frac{u\,du}{g_c} = -\frac{u^2}{g_c} d(\ln u) = -\frac{u^2}{g_c}\frac{d(\ln u)}{d(\ln V)}\frac{dV}{V}$$

Solving for u^2, we have

$$u^2 = -g_c V \frac{dH}{dV}\frac{d(\ln V)}{d(\ln u)} \tag{1}$$

For steady flow the mass flow rate $\dot{m}$ must be constant for the full length of the nozzle. Thus conservation of mass is expressed by

$$\frac{uA}{V} = \dot{m} = \text{constant}$$

where A is the cross-sectional area of the nozzle. By logarithmic differentiation we get

$$d(\ln u) + d(\ln A) - d(\ln V) = 0$$

or

$$\frac{d(\ln V)}{d(\ln u)} = \frac{d(\ln A)}{d(\ln u)} + 1$$

Substitution of this relation into (*1*) gives the required result:

$$u^2 = -g_c V \frac{dH}{dV}\left[\frac{d(\ln A)}{d(\ln u)} + 1\right] \tag{2}$$

(*b*) If we now impose the condition that the flow be reversible as well as adiabatic, i.e. if it is isentropic, then (*2*) may be written

$$u^2 = -g_c V \left(\frac{\partial H}{\partial V}\right)_S \left[\frac{d(\ln A)}{d(\ln u)} + 1\right] \tag{3}$$

and $(\partial H/\partial V)_S$ can be determined from the general property relation

$$dH = T\,dS + V\,dP \tag{3.48}$$

Division of this equation by dV and restriction to constant S provides

$$\left(\frac{\partial H}{\partial V}\right)_S = V\left(\frac{\partial P}{\partial V}\right)_S$$

Substitution into (*3*) yields the required result:

$$u^2 = -g_c V^2 \left(\frac{\partial P}{\partial V}\right)_S \left[\frac{d(\ln A)}{d(\ln u)} + 1\right] \tag{4}$$

We also note that the property relation (*3.48*) gives $(\partial H/\partial P)_S = V$. Thus for isentropic flow dH/dP is *always positive*.

From physics we have the following equation for the sonic velocity c:

$$c^2 = -g_c V^2 \left(\frac{\partial P}{\partial V}\right)_S$$

Combining this equation with (4) we get

$$\frac{u^2}{c^2} = \frac{d(\ln A)}{d(\ln u)} + 1$$

Since the ratio u/c is defined as the Mach number $\mathbf{M}$, we have finally

$$\mathbf{M}^2 - 1 = \frac{d(\ln A)}{d(\ln u)} \tag{5}$$

A number of qualitative observations can be made as a result of this equation. Clearly there are three cases to consider, corresponding to

$\mathbf{M} < 1$ (subsonic flow)

$\mathbf{M} = 1$ (sonic flow)

$\mathbf{M} > 1$ (supersonic flow)

We also keep in mind the signs of the two derivatives noted earlier:

$$dH/du = \ominus \qquad \text{and} \qquad dH/dP = \oplus$$

When $\mathbf{M} < 1$, and the flow is subsonic, (5) shows that $d \ln A/d \ln u$ is negative. There are two cases to be considered: (1) A decreases in the direction of flow while u increases; (2) A increases in the direction of flow while u decreases. If $\mathbf{M} = 1$, the flow is sonic, and by (5) $d \ln A/d \ln u = 0$. There is but one case to be treated: (3) A is constant. When $\mathbf{M} > 1$, the flow is supersonic, and from (5) we have that $d \ln A/d \ln u$ is positive. Again there are two cases: (4) A increases as u increases; (5) A decreases as u decreases. The qualitative conclusions which follow from the equations given are displayed for each case below.

(1)	(2)	(3)	(4)	(5)
$\mathbf{M} < 1$		$\mathbf{M} = 1$	$\mathbf{M} > 1$	
$\frac{d(\ln A)}{d(\ln u)} = \ominus$		$\frac{d(\ln A)}{d(\ln u)} = 0$	$\frac{d(\ln A)}{d(\ln u)} = \oplus$	
Flow is subsonic		Flow is sonic	Flow is supersonic	
A decreases (converging)	A increases (diverging)	A is constant	A increases (diverging)	A decreases (converging)
u increases	u decreases	u may increase or decrease	u increases	u decreases
H decreases	H increases	H and P may decrease or increase	H decreases	H increases
P decreases	P increases		P decreases	P increases
Converging Nozzle	***Diverging Diffuser***	***Point Condition at an End or Transition between Subsonic & Supersonic***	***Diverging Nozzle***	***Converging Diffuser***

Case (1) is the familiar converging nozzle through which a fluid flows with steadily increasing velocity as the pressure drops (see Fig. 6-19). The limiting velocity is the sonic velocity, and for insentropic flow this can be attained only at the exit of the nozzle where the area has become constant.

Case (2) applies to a device that receives a high-velocity (but subsonic) stream and decreases its velocity, converting kinetic energy into internal energy, thus causing an increase

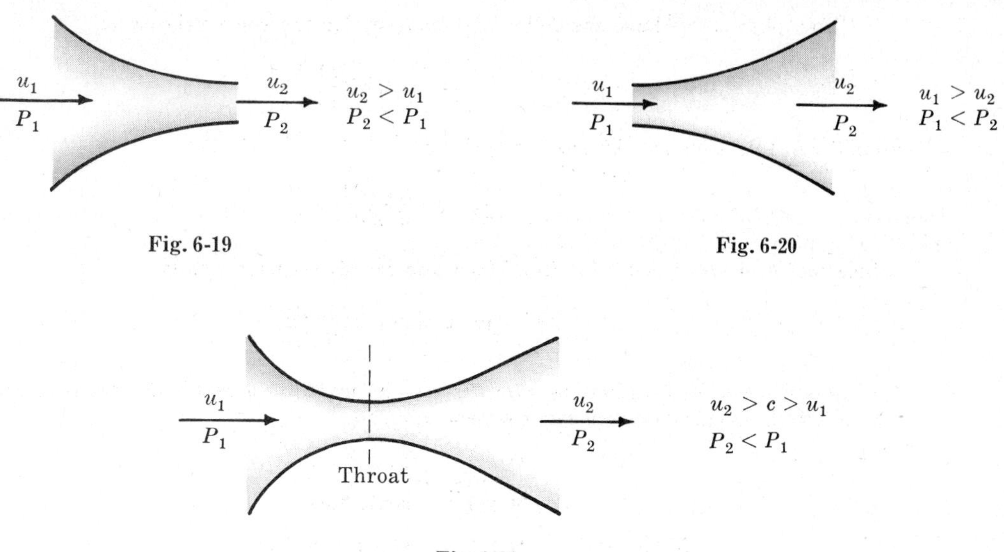

Fig. 6-19

Fig. 6-20

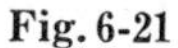

Fig. 6-21

in enthalpy and pressure (see Fig. 6-20). Such a device is called a diffuser. A common use of the diffuser is for the compression of the intake air for a jet airplane engine.

Case (4) is again a nozzle, but a *diverging* nozzle. Once flow has reached the sonic velocity, further increases in velocity can be attained only if the nozzle area increases. Thus the diverging nozzle is almost always found as part of a converging-diverging nozzle, a combination of case (1) and case (4) (see Fig. 6-21). At the throat where the converging and diverging sections join, the flow is sonic and we have a transition from subsonic to supersonic flow, represented by case (3).

Case (5) is again a diffuser, one that *converges* and operates at supersonic velocities. Just as a diverging cross-section is required to increase the velocity of a supersonic flow, so a converging cross-section is required to decrease the velocity of a supersonic flow in an isentropic process. Such a diffuser is used on a supersonic jet aircraft for the compression of the intake air. To bring the air to subsonic velocities (relative to the aircraft) case (5) is combined with case (2) to make a converging-diverging diffuser. The throat of such a device is again represented by case (3), where the transition is from supersonic to subsonic flow.

6.17. Steam flows through an adiabatic, reversible nozzle at the rate of $4(\text{lb}_\text{m})/(\text{s})$. Inlet conditions are $940(°\text{F})$ and $400(\text{psia})$, and the entrance velocity is $30(\text{ft})/(\text{s})$. The discharge pressure is $20(\text{psia})$. Plot V, u, and A against P for the full pressure range of the nozzle from $400(\text{psia})$ down to $20(\text{psia})$.

The appropriate expression of the first law is (*6.14*):

$$\frac{u^2 - u_1^2}{2g_c} = -(H - H_1)_S$$

where the unsubscripted symbols u and H apply to any point downstream from the nozzle inlet. Solution for u gives:

$$u = \sqrt{u_1^2 + 2g_c(H_1 - H)_S} \qquad (1)$$

In addition we have the continuity equation $\dot{m} = uA/V$ or

$$A = \frac{\dot{m}V}{u} \qquad (2)$$

For the initial state we find the following properties from the tables for superheated steam:

$$V_1 = 2.041(\text{ft})^3/(\text{lb}_\text{m}) \qquad H_1 = 1491.5(\text{Btu})/(\text{lb}_\text{m}) \qquad S_1 = 1.7407(\text{Btu})/(\text{lb}_\text{m})(\text{R})$$

Application of (*2*) at the nozzle entrance gives

$$A_1 = \frac{\dot{m}V_1}{u_1} = \frac{4(\text{lb}_\text{m})/(\text{s}) \times 2.041(\text{ft})^3/(\text{lb}_\text{m})}{30(\text{ft})/(\text{s})} \times 144(\text{in})^2/(\text{ft})^2 = 39.19(\text{in})^2$$

Consider now a point in the nozzle where $P = 395$(psia). Since the expansion process in the nozzle is considered to be reversible and adiabatic, it occurs at constant entropy. Thus

$$S = S_1 = 1.7407(\text{Btu})/(\text{lb}_\text{m})(\text{R})$$

and this is the value of S throughout the nozzle. We now locate the state in the tables for superheated steam for which $P = 395$(psia) and $S = 1.7407$(Btu)/(lb$_\text{m}$)(R). This state occurs at a temperature just above 936(°F), and by interpolation we find

$$H = 1489.1 \qquad V = 2.060$$

The velocity at this point is given by (*1*):

$$u = \sqrt{30^2(\text{ft})^2/(\text{s})^2 - 2 \times 32.174(\text{lb}_\text{m})(\text{ft})/(\text{lb}_\text{f})(\text{s})^2 \times [1491.5 - 1489.1](\text{Btu})/(\text{lb}_\text{m}) \times 778(\text{ft-lb}_\text{f})/(\text{Btu})}$$

$$= 347.9(\text{ft})/(\text{s})$$

From (*2*) the corresponding value of the area is

$$A = \frac{4 \times 2.060}{347.9} \times 144 = 3.410(\text{in})^2$$

We continue in like fashion to successively lower pressures. The results are given in the table below and are displayed in Fig. 6-22.

P(psia)	H(Btu)/(lb$_\text{m}$)	V(ft)^3/(lb$_\text{m}$)	u(ft)/(s)	A(in)^2
400	1491.5	2.041	30.0	39.19
395	1489.1	2.060	347.9	3.410
390	1487.2	2.081	464.9	2.578
380	1483.2	2.124	645.3	1.896
350	1471.5	2.266	1000.1	1.305
300	1449.3	2.556	1453.8	1.013
250	1424.0	2.947	1838.5	0.923
200	1394.3	3.504	2206.1	0.915
150	1358.1	4.380	2584.4	0.976
100	1311.1	5.988	3005.4	1.148
80	1287.1	7.111	3199.0	1.280
60	1257.7	8.865	3421.3	1.492
40	1219.4	12.078	3690.9	1.885
30	1194.7	15.050	3854.8	2.249
20	1162.4	20.484	4059.1	2.907

It is not possible from thermodynamics to determine the relationship of cross-sectional area of the nozzle to longitudinal distance. That is a problem in fluid mechanics. Thus the proper shape of a nozzle so that it may approach reversible operation depends on design principles that are not part of thermodynamics.

6.18. A stream of air at atmospheric pressure is to be cooled continuously from 38(°C) to 15(°C) for purposes of air conditioning a building. The amount of air required is 30(m)^3/(min) [as measured at 1(atm) and 25(°C)]. The temperature of the ambient air to which heat is discarded is 38(°C). What is the minimum power requirement of a mechanical refrigeration system designed for this purpose?

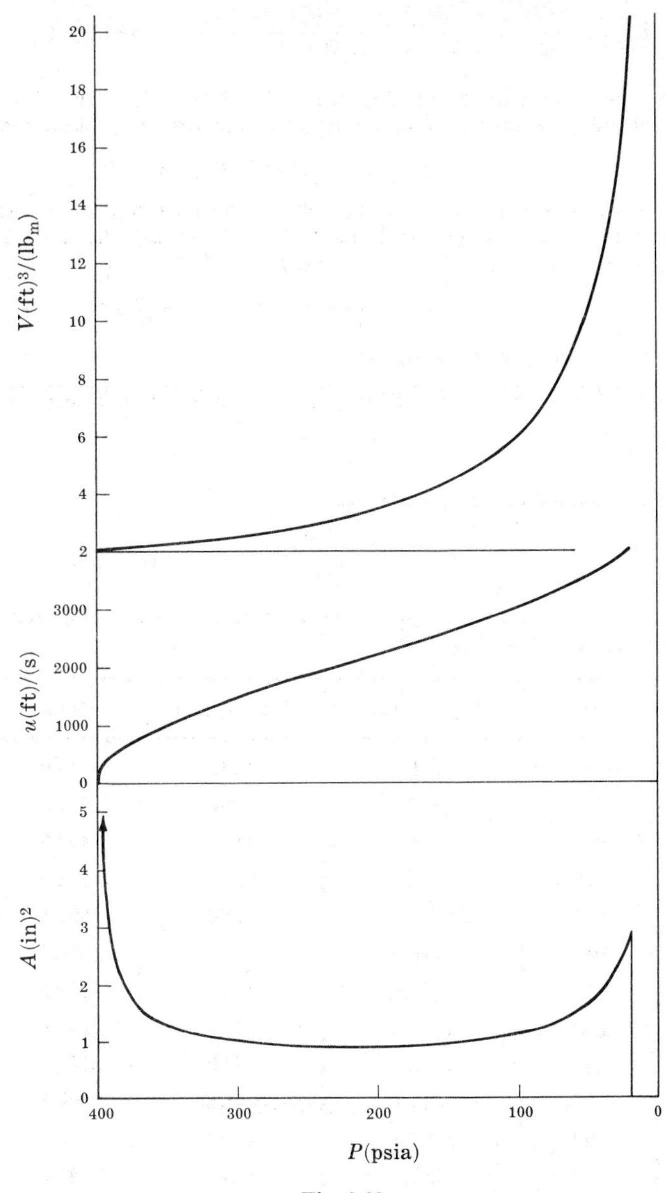

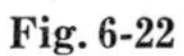

Fig. 6-22

We will consider air to be an ideal gas for which $C_P = 7\text{(cal)/(g mole)(K)}$. The amount of air cooled is

$$\dot{n} = \frac{P\dot{V}^t}{RT} = \frac{1\text{(atm)} \times 30\text{(m)}^3\text{/(min)} \times [100\text{(cm)/(m)}]^3}{82.05\text{(cm}^3\text{-atm)/(g mole)(K)} \times 298\text{(K)}} = 1227\text{(g mole)/(min)}$$

The process that occurs is represented in Fig. 6-23.

The energy equation *(6.10)* can be written for this process as

$$(\Delta H)\dot{n} = \dot{Q} - \dot{W}_s \qquad (1)$$

where the potential- and kinetic-energy terms are omitted as negligible. If the minimum work $\dot{W}_s$ is to be required, then the process must be reversible, and

$$\Delta S_{\text{total}} = 0$$

There are two entropy changes associated with the process. First, the air stream changes in entropy by the amount $(\Delta S)\dot{n}$, and, second, the surroundings, acting as a heat reservoir, change in

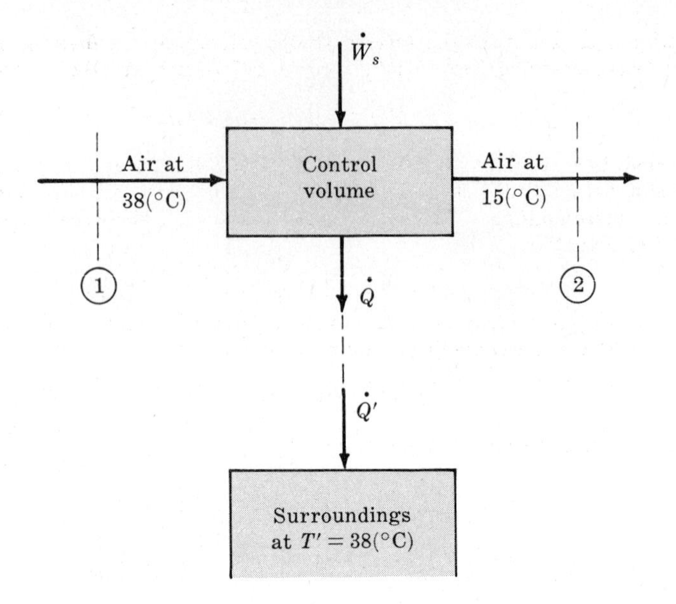

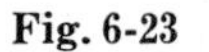

Fig. 6-23

entropy by $\dot{Q}'/T' = -\dot{Q}/T'$. The sum of these two entropy changes must be zero:

$$(\Delta S)\dot{n} - \dot{Q}/T' \;=\; 0 \tag{2}$$

Elimination of $\dot{Q}$ between (1) and (2) gives

$$\dot{W}_s \;=\; (T'\Delta S - \Delta H)\dot{n}$$

Since air is taken to be an ideal gas

$$\Delta H \;=\; C_P(T_2 - T_1) \;=\; 7(15 - 38) \;=\; -161(\text{cal})/(\text{g mole})$$

$$\Delta S \;=\; C_P \ln \frac{T_2}{T_1} \;=\; 7 \ln \frac{273 + 15}{273 + 38} \;=\; -0.5378(\text{cal})/(\text{g mole})(\text{K})$$

Therefore

$$\dot{W}_s \;=\; [(273 + 38)(-0.5378) + 161](1227) \;=\; -7680(\text{cal})/(\text{min}) \quad \text{or} \quad -536(\text{J})/(\text{s})$$

Thus the power required is 536(W) or 0.536(kW).

6.19. Show whether or not the following continuous process is possible: Nitrogen gas at 6(atm) and 21(°C) enters a device which has no moving parts and which is thoroughly insulated from its surroundings. Half of the nitrogen flows out of the device at 82(°C) and the other half at −40(°C), both halves at 1(atm). Assume nitrogen at these conditions behaves as an ideal gas for which $C_P = 7(\text{cal})/(\text{g mole})(\text{K})$.

If the first and second laws of thermodynamics are not violated by this process, then it is possible. The first law is expressed by an energy equation, which for this steady-flow process is given by (*6.9*). Since the device is well insulated, we may take the heat transfer to be zero. Since the device has no moving parts, the shaft work must be zero. If we neglect the kinetic- and potential-energy terms (*6.9*) becomes

$$\Delta(\dot{m}H)_{\substack{\text{flowing}\\ \text{streams}}} \;=\; 0$$

If we let $\dot{m}'$ represent the mass flow rate of the colder exit stream and $\dot{m}''$ the flow rate of the warmer exit stream, then we may write

$$\dot{m}'H_2' + \dot{m}''H_2'' - (\dot{m}' + \dot{m}'')H_1 \;=\; 0$$

or

$$\dot{m}'(H_2' - H_1) + \dot{m}''(H_2'' - H_1) \;=\; 0$$

All this equation says is that the total enthalpy change of the flowing streams is zero. For the particular process described, $\dot{m}' = \dot{m}''$, and therefore the first law requires that

$$(H_2' - H_1) + (H_2'' - H_1) = 0$$

The second law requires that the total entropy change as a result of the process be positive, or in the limit, zero. Since there is no heat exchange with the surroundings, there is no entropy change of the surroundings. The total entropy change is then the entropy change of the flowing streams, and is given by

$$\dot{m}'(S_2' - S_1) + \dot{m}''(S_2'' - S_1)$$

It is this quantity that must be equal to or greater than zero. Since the two mass flow rates are equal and positive, an equivalent requirement is that

$$(S_2' - S_1) + (S_2'' - S_1) \geqq 0$$

For an ideal gas we have the equations

$$H_2 - H_1 = C_P(T_2 - T_1)$$

$$S_2 - S_1 = C_P \ln \frac{T_2}{T_1} - R \ln \frac{P_2}{P_1}$$

From the problem statement

$$P_1 = 6(\text{atm}) \qquad P_2 = 1(\text{atm})$$

$$T_1 = 21(°\text{C}) \quad \text{or} \quad 294(\text{K})$$

$$T_2' = -40(°\text{C}) \quad \text{or} \quad 233(\text{K}) \qquad T_2'' = 82(°\text{C}) \quad \text{or} \quad 355(\text{K})$$

Therefore $\quad H_2' - H_1 = 7(\text{cal})/(\text{g mole})(\text{K}) \times [233 - 294](\text{K}) = -427(\text{cal})/(\text{g mole})$

$$S_2' - S_1 = 7 \ln \frac{233}{294} - 1.987 \ln \frac{1}{6} = 1.93(\text{cal})/(\text{g mole})(\text{K})$$

Similarly $\quad H_2'' - H_1 = 7 \times [355 - 294] = 427(\text{cal})/(\text{g mole})$

$$S_2'' - S_1 = 7 \ln \frac{355}{294} - 1.987 \ln \frac{1}{6} = 4.88(\text{cal})/(\text{g mole})(\text{K})$$

We therefore have

$$(H_2' - H_1) + (H_2'' - H_1) = -427 + 427 = 0$$

$$(S_2' - S_1) + (S_2'' - S_1) = 1.93 + 4.88 = 6.81(\text{cal})/(\text{K}) > 0$$

Thus the process as described satisfies the requirements of the laws of thermodynamics, and is therefore *possible*. Whether any particular device can be made to cause changes determined to be possible depends on the ingenuity of its design. Thermodynamics puts a limit on what can be done, but it still allows wide latitude for the exercise of ingenuity or genius.

MECHANICAL-ENERGY BALANCE (Sec. 6.5)

6.20. Water is drained from a 10(ft) diameter tank through a 2(in) hole in the side located 1(ft) above the bottom of the tank (see Fig. 6-24). The top of the tank is open to the atmosphere. What is the velocity of the exiting stream when the water level in the tank is 8(ft)?

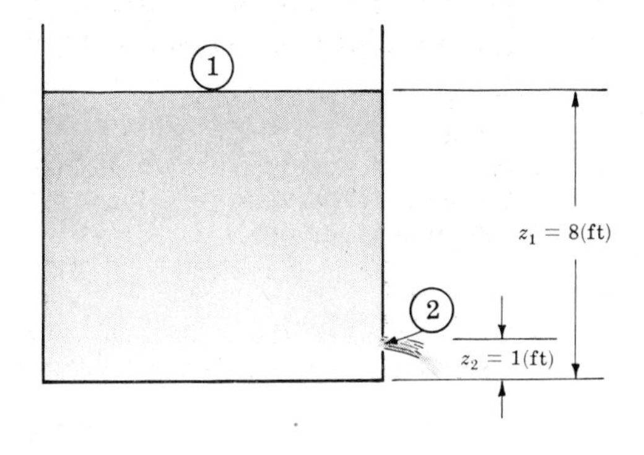

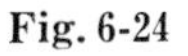

Fig. 6-24

Although the liquid level in the tank slowly recedes, the ratio of the tank to hole diameters is so large that the velocity of the water surface in the tank is essentially zero. We can therefore treat the drainage process as a "quasi-steady" flow for which the control volume, taken as the total volume of water in the tank, remains virtually constant. There is no shaft work done on or by the fluid; so, if we assume the water to be nonviscous and incompressible, we can apply the Bernoulli equation (*6.18*) between locations 1 (the liquid surface in the tank) and 2 (the exiting jet of liquid). Thus we write

$$\frac{P_2 - P_1}{\rho} + \frac{u_2^2 - u_1^2}{2g_c} + (z_2 - z_1)\left(\frac{g}{g_c}\right) = 0$$

Since $u_1 \cong 0$ by assumption, and $P_1 = P_2 =$ atmospheric pressure, solution of the above equation for u_2 gives

$$u_2 = \sqrt{2g(z_1 - z_2)}$$

By the statement of the problem, $z_2 = 1\text{(ft)}$ and $z_1 = 8\text{(ft)}$. Therefore

$$u_2 = \sqrt{2 \times 32.174\text{(ft)/(s)}^2 \times 7\text{(ft)}} = \sqrt{450.4} = 21.2\text{(ft)/(s)}$$

The actual surface velocity u_1 can be estimated from the continuity equation applied between the drainage hole and a plane parallel to and just below the liquid surface. The mass flow rate $\dot{m}_1$ into this plane will be equal to the mass flow rate $\dot{m}_2$ at 2:

$$\dot{m}_1 = \dot{m}_2$$

or

$$\frac{u_1 A_1}{V_1} = \frac{u_2 A_2}{V_2}$$

But $V_1 = V_2$ by the incompressibility assumption, so

$$u_1 = u_2\left(\frac{A_2}{A_1}\right) = 21.2\text{(ft)/(s)} \times \left[\frac{\frac{2}{12}\text{(ft)}}{10\text{(ft)}}\right]^2 = 5.9 \times 10^{-3}\text{(ft)/(s)}$$

and the surface velocity is indeed negligible with respect to u_2.

6.21. **Air at a pressure of 100(psia) and a temperature of 70(°F) enters a horizontal steam-jacketed pipe with a velocity of 100(ft)/(s). Measurements at the exit end of the pipe show that the air is heated to 390(°F) and that the pressure is 97(psia). What percentage of the pressure drop in the pipe can be attributed to fluid friction? Assume that the air behaves as an ideal gas.**

The mechanical-energy balance (*6.17*) may be written in differential form as

$$-\delta W_s = V\,dP + \frac{u\,du}{g_c} + \frac{g}{g_c}dz + \delta F$$

Since there is no shaft work and no change in elevation, this reduces to

$$dP = \frac{-u}{Vg_c}du - \frac{\delta F}{V}$$

The mass flow rate of air is the same at the pipe entrance as at the exit. Thus by the continuity equation

$$\dot{m}_1 = \frac{u_1 A_1}{V_1} = \dot{m}_2 = \frac{u_2 A_2}{V_2}$$

Since $A_1 = A_2$, we have

$$\frac{u_1}{V_1} = \frac{u_2}{V_2} = \frac{u}{V} = \text{constant}$$

The expression for dP may therefore be written

$$dP = -\left(\frac{u_1}{V_1 g_c}\right)du - \frac{\delta F}{V}$$

and integration gives

$$\Delta P = -\left(\frac{u_1}{V_1 g_c}\right)\Delta u - \int_1^2 \frac{\delta F}{V}$$

The total pressure drop is given as the sum of two terms, and we may associate the first of these terms with the pressure drop that results in a velocity increase and the second term with the pressure drop attributed to fluid friction. To determine this last term, we rearrange the equation:

$$-\int_1^2 \frac{\delta F}{V} = \Delta P + \left(\frac{u_1}{V_1 g_c}\right)\Delta u$$

The initial specific volume of the air is given by the ideal-gas law:

$$V_1 = \frac{m}{M}\left(\frac{RT}{P}\right) = \frac{1(\text{lb}_\text{m})}{29(\text{lb}_\text{m})/(\text{lb mole})}\left[\frac{10.73(\text{psia})(\text{ft})^3/(\text{lb mole})(\text{R}) \times 530(\text{R})}{100(\text{psia})}\right]$$

$$= 1.961(\text{ft})^3/(\text{lb}_\text{m})$$

Since $$u_2 = u_1\frac{V_2}{V_1} = u_1\left(\frac{T_2}{T_1}\right)\left(\frac{P_1}{P_2}\right) = 100 \times \frac{850}{530} \times \frac{100}{97} = 165(\text{ft})/(\text{s})$$

then $\Delta u = u_2 - u_1 = 165 - 100 = 65(\text{ft})/(\text{s})$ and

$$-\int_1^2 \frac{\delta F}{V} = -3(\text{lb}_\text{f})/(\text{in})^2 \times 144(\text{in})^2/(\text{ft})^2 + \frac{100(\text{ft})/(\text{s}) \times 65(\text{ft})/(\text{s})}{1.961(\text{ft})^3/(\text{lb}_\text{m}) \times 32.174(\text{lb}_\text{m})(\text{ft})/(\text{lb}_\text{f})(\text{s})^2}$$

$$= -432 + 103 = -329(\text{lb}_\text{f})/(\text{ft})^2$$

The percentage of δP attributed to fluid friction is therefore $(329/432) \times 100 = 76.1\%$.

MISCELLANEOUS APPLICATIONS

6.22. Water at 100(°F) and 1(psia) is pumped from the condenser of a power plant to the boiler pressure of 1000(psia). If the pump operates reversibly and adiabatically, how much work is required? What is the temperature rise of the water?

This is a steady-flow process to which (*6.11*) is applicable. It reduces in this case to $W_s = -\Delta H$ because there is no heat transfer and the kinetic- and potential-energy terms are negligible for any properly designed pump. One way to evaluate ΔH is by (*3.48*): $dH = T\,dS + V\,dP$. Since the process is reversible and adiabatic, $dS = 0$, and

$$\Delta H = \int_1^2 V\,dP$$

Therefore the work is given by

$$W_s = -\int_1^2 V\,dP$$

[This same result is obtained from the mechanical-energy balance (*6.17*), because for a reversible process $\Sigma F = 0$.] Water at 100(°F) is only slightly compressible, and we may therefore evaluate the integral to a good approximation by taking V constant at its initial value:

$$W_s \cong -V_1(P_2 - P_1)$$

From the steam tables we find the specific volume of saturated liquid water at 100(°F) to be

$$V_1 = 0.01613(\text{ft})^3/(\text{lb}_\text{m})$$

Therefore $$W_s \cong -0.01613(\text{ft})^3/(\text{lb}_\text{m}) \times [1000 - 1](\text{lb}_\text{f})/(\text{in})^2 \times 144(\text{in})^2/(\text{ft})^2$$

$$\cong -2320(\text{ft-lb}_\text{f})/(\text{lb}_\text{m})$$

We now need an expression for the temperature change of the water in an isentropic process, and this can be found from (*3.55*) by setting $dS = 0$:

$$dT = \frac{T}{C_P}\left(\frac{\partial V}{\partial T}\right)_P dP \qquad \text{(constant } S\text{)}$$

The volume expansivity is defined by (*3.17*) as

$$\beta = \frac{1}{V}\left(\frac{\partial V}{\partial T}\right)_P$$

Thus
$$dT = \frac{TV\beta}{C_P}\,dP \qquad \text{(constant } S\text{)}$$

Since T, V, β, and C_P will change very little with pressure, we may to a good approximation consider them constant at their initial values. Integration then gives

$$\Delta T \cong \frac{T_1 V_1 \beta_1}{C_{P_1}}\,\Delta P$$

For water at 100(°F)

$$V_1 = 0.01613(\text{ft})^3/(\text{lb}_m) \qquad \beta_1 = 210 \times 10^{-6}(\text{R})^{-1} \qquad C_{P_1} = 1.0(\text{Btu})/(\text{lb}_m)(°\text{F})$$

Thus
$$\Delta T \cong \left[\frac{560(\text{R}) \times 0.01613(\text{ft})^3/(\text{lb}_m) \times 210 \times 10^{-6}(\text{R})^{-1}}{1(\text{Btu})/(\text{lb}_m)(°\text{F}) \times 778(\text{ft-lb}_f)/(\text{Btu})}\right] \times [999(\text{lb}_f)/(\text{in})^2 \times 144(\text{in})^2/(\text{ft})^2]$$

$$\cong 0.35(°\text{F})$$

This problem may also be solved by direct use of values from the steam tables. For saturated liquid at 100(°F)

$$H_1 = 68.05(\text{Btu})/(\text{lb}_m) \qquad S_1 = 0.12963(\text{Btu})/(\text{lb}_m)(\text{R})$$

[The liquid at 100(°F) and 1(psia) is not actually saturated, because the saturation pressure is slightly less than 1(psia); however, the effect of this small pressure difference on H and S is entirely negligible.] Since the pumping operation is isentropic, we must find in the tables for compressed liquid the state for which

$$S_2 = S_1 = 0.12963$$

at 1000(psia). By interpolation, this state is found to have the properties:

$$T_2 = 100.36(°\text{F}) \qquad H_2 = 71.05(\text{Btu})/(\text{lb}_m)$$

Thus
$$W_s = -\Delta H = -(H_2 - H_1) = -(71.05 - 68.05) = -3(\text{Btu})/(\text{lb}_m)$$

or $-2330(\text{ft-lb}_f)/(\text{lb}_m)$ and

$$\Delta T = 100.36 - 100.00 = 0.36(°\text{F})$$

Clearly the two methods used give equivalent results.

Supplementary Problems

Many of the following problems require data from the steam tables for solution.

ENERGY EQUATIONS (Secs. 6.1 through 6.3)

6.23. A steam turbine operates adiabatically and produces 4000(HP). Steam is fed to the turbine at 300(psia) and 900(°F). The exhaust from the turbine is saturated steam at 1.5(psia), and it enters a condenser where it is condensed and cooled to 90(°F). What is the steam rate for the turbine, and at what rate must cooling water be supplied to the condenser if the water enters at 65(°F) and is heated to 85(°F)?

Ans. $\dot{m}_{\text{steam}} = 468.7(\text{lb}_m)/(\text{min})$; $\dot{m}_{\text{water}} = 24{,}705(\text{lb}_m)/(\text{min})$

6.24. Saturated steam at 1(bar) is taken continuously into a compressor at low velocity and compressed to 3(bar). It then enters a nozzle where it expands back to a pressure of 1(bar) and its initial condition of saturated steam. However, it now has a velocity of 600(m)/(s). The steam rate is 2.5(kg)/(s). It is found necessary to cool the compressor at the rate of 150(kJ)/(s). What is the power requirement of the compressor? *Ans.* Power = 600(kW)

6.25. A well-insulated closed tank has a volume of 2500(ft)^3. Initially, it contains 50,000(lb$_m$) of water distributed between liquid and vapor phases at 80(°F). Saturated steam at 160(psia) is admitted to the tank until the pressure reaches 100(psia). How many pounds mass of steam are added?

Ans. 14,210(lb$_m$)

6.26. A well-insulated tank of 3(m)^3 capacity contains 1400(kg) of liquid water in equilibrium with its vapor, which fills the remainder of the tank. The initial temperature is 280(°C). Liquid water at 70(°C) in the amount of 900(kg) flows into the tank, and nothing is removed. How much heat must be added during the process if the temperature in the tank is to be unchanged?

Ans. 784.54×10^6(J)

6.27. A tank with a volume of 1(m)^3 is initially evacuated. Atmospheric air leaks into the tank through a weld imperfection. The process is slow, and heat transfer with the surroundings keeps the tank and its contents at the ambient temperature of 27(°C). Calculate the amount of heat exchanged with the surroundings during the time it takes for the pressure in the tank to reach 1(atm).

Ans. $Q = -101.3$(kJ)

6.28. A continuous process for liquefying nitrogen includes a heat-exchanger system and throttle valve with no moving parts. The nitrogen enters the system at 100(atm) and 80(°F). Two streams leave the system: first, unliquefied nitrogen gas at 70(°F) and 1(atm), and second, saturated liquid nitrogen at 1(atm). The apparatus is well insulated, and it is estimated that only 25(Btu) of heat will leak into the system for each pound mole of entering nitrogen. What fraction of the entering nitrogen is liquefied under these conditions?

Available data: When nitrogen at 100(atm) and 80(°F) expands adiabatically through a throttle to 1(atm), the temperature drops to 45(°F). The latent heat of vaporization of nitrogen at its normal boiling point of −320.4(°F) is 2400(Btu)/(lb mole). The heat capacity of nitrogen gas may be taken as constant at 6.72(Btu)/(lb mole)(°F).

Ans. 2.84%

6.29. A frictionless piston, having a cross-sectional area of 0.8(ft)^2, slides in a cylinder as shown in Fig. 6-25. The piston works to compress a spring, which exerts a force on the piston $F = 30{,}000x$, where x is in (ft) and F is in (lb$_f$). Steam from a line at 600(psia) and 800(°F) is admitted slowly to the cylinder through the valve, and is allowed to flow until the pressure in the cylinder is 200(psia). Assuming that no heat is exchanged with the steam and that $x_1 = 0$, determine: (*a*) an equation giving $H_2 - H_1 = f(P_2, V_2)$, where subscript 2 denotes steam in the cylinder, and subscript 1 refers to steam in the line; (*b*) the enthalpy and mass of the steam in the cylinder at the end of the process.

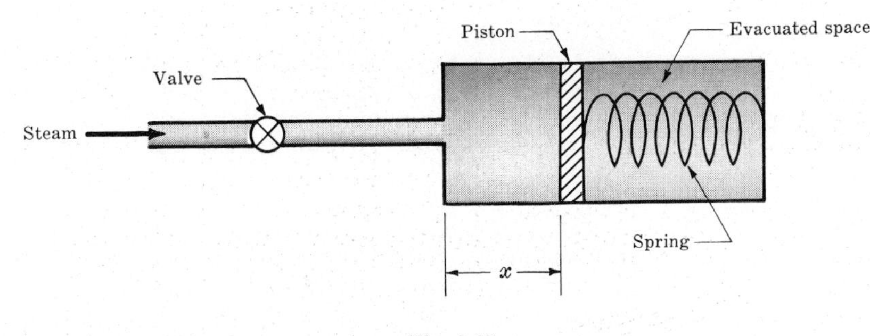

Fig. 6-25

Ans. (*a*) $H_2 - H_1 = P_2 V_2/2$. Note that this equation must be solved in conjunction with the steam tables, which provide a second relation connecting P_2, V_2, and H_2.

(*b*) $H_2 = 1482.3$(Btu)/(lb$_m$), $m_2 = 0.1523$(lb$_m$)

6.30. A portable power-supply system consists of a 1(ft)^3 bottle of compressed helium, charged to 2000(psia) at 80(°F), connected to a small turbine. During operation the helium drives the turbine continuously until the pressure in the bottle drops to 100(psia). For a particular use at high altitude the turbine exhausts at 5(psia). Neglecting all heat transfer with the gas, calculate the maximum possible shaft work obtainable from the device. Assume helium to be an ideal gas with constant heat capacities, $C_V = 3$, $C_P = 5$(Btu)/(lb mole)(R). [R may therefore be taken as 2(Btu)/(lb mole)(R).] *Ans.* $W_s = 461$(Btu)

SECOND-LAW APPLICATIONS (Sec. 6.4)

6.31. Two perfectly insulated tanks A and B of equal volume contain equal quantities of the same ideal gas at the same pressure and temperature. Tank A is connected to the inlet of a small reversible adiabatic turbine, which drives an electric generator, and tank B discharges through an insulated valve. Both the turbine and the valve discharge gas to the atmosphere. Both devices are allowed to operate until gas discharge ceases. Regarding these operations, there are three alternative choices for completion of each of the following statements, namely: *equal to*, *greater than*, or *less than*. Indicate in each case which is correct. Assume no heat transfer to or from the gas.

(*a*) When discharge ceases in each system, the temperature in tank A is ____________ the temperature in B.

(*b*) When the pressures in both tanks have fallen to one-half the initial pressure, the temperature of the gas discharged from the *turbine* will be ____________ the temperature of the gas issuing from the *valve*.

(*c*) During the discharge process, the temperature of the gas leaving the *turbine* will be ____________ the temperature of the gas leaving *tank A* at the same instant.

(*d*) During the discharge process, the temperature of the gas leaving the *valve* will be ____________ the temperature of the gas leaving *tank B* at the same instant.

(*e*) When the tank pressures have reached atmospheric, the quantity of gas in *tank A* is ____________ the quantity in *tank B*.

Ans. (*a*) equal to, (*b*) less than, (*c*) less than, (*d*) equal to, (*e*) equal to

6.32. Determine the efficiency (with respect to isentropic operation) of the turbine of Problem 6.23.

Ans. 77.26%

6.33. For an ideal gas with constant heat capacities flowing reversibly and adiabatically through a nozzle from an initial pressure P_1 to a final pressure P_2, show that

$$u_2 = \sqrt{2g_c C_P T_1\left[1 - \left(\frac{P_2}{P_1}\right)^{(\gamma-1)/\gamma}\right]}$$

where u_1 has been taken to be negligible, and C_P is the *specific* heat capacity of the fluid.

6.34. Air expands through a nozzle from a negligible initial velocity to a final velocity of 1100(ft)/(s). Calculate the temperature drop of the air, assuming that air is an ideal gas for which $C_P = 7$(Btu)/(lb mole)(°F). The molecular weight of air is 29.

Ans. Temperature drop = 100(°F)

6.35. (*a*) Saturated steam at 7(bar) expands reversibly and adiabatically through a turbine to a pressure of 0.34(bar). What is the work of the turbine? (*b*) If for the same initial conditions and the same final pressure as in (*a*) the steam expands adiabatically but irreversibly through a turbine and produces 80% of the work of (*a*), what is the final state of the steam?

Ans. (*a*) $W_s = 484.6$(J)/(g); (*b*) wet steam for which $x = 0.8908$

6.36. Consideration is being given to the use of a steady-flow expander (or gas turbine) powered by a stream of hot compressed gases. It is required that the gases should discharge from the turbine at atmospheric pressure and 80(°F). It is also required that the turbine produce 1250(HP) for a flow rate of 50(lb mole)/(min) of gas. Estimate the initial temperature and pressure required for the gas stream. For the purpose of an estimate, assume that the turbine will operate isentropically and that the gases are ideal with C_P constant at 10.5(Btu)/(lb mole)(°F).

Ans. $T_1 = 181(°F)$, $P_1 = 2.46(atm)$

6.37. Saturated steam is compressed continuously from 15(psia) to 70(psia) in a centrifugal compressor which operates adiabatically. For an efficiency of 75% compared with isentropic compression, determine the required work and the final state of the steam.

Ans. $W_s = -181.3(Btu)/(lb_m)$; superheated steam at 70(psia) and 601.4(°F)

6.38. A power plant employs two adiabatic steam turbines in series. Steam enters the first turbine at 1100(°F) and 950(psia), and discharges from the second turbine at 2(psia). The system was designed so that equal work would be done by the two turbines, and the design was based on an efficiency of 80% compared with isentropic operation *in each turbine separately*. If the turbines perform according to these design conditions, what should be the temperature and pressure of the steam between the turbines? What is the overall efficiency of the two turbines considered together, compared with isentropic expansion from the initial state to the final pressure?

Ans. $T = 586.5(°F)$, $P = 94.6(psia)$, $\eta = 83.4\%$. Note that solution for T and P is by trial.

6.39. A tank contains $1(lb_m)$ of steam at a pressure of 300(psia) and a temperature of 700(°F). It is connected through a valve to a vertical cylinder which contains a piston, as shown in Fig. 6-26. The piston has a mass such that a pressure of 100(psia) is required to support it. Initially, the piston rests on the bottom of the cylinder. The valve is cracked so as to allow steam to flow slowly into the cylinder until the pressure is uniform throughout the system. Assuming that no heat is transferred from the steam to the surroundings and that no heat is exchanged between the two parts of the system, determine the final temperatures in the tank and in the cylinder.

Ans. $T_{tank} = 439.3(°F)$, $T_{cyl} = 569.4(°F)$

6.40. Flake ice at 32(°F) is to be produced continuously from liquid water at 70(°F) at the rate of $1000(lb_m)/(hr)$. The process is to employ exhaust steam available saturated at 25(psia) as the sole source of energy. Heat rejection is to be to the atmosphere. On a day when the ambient temperature is 80(°F) what is the absolute minimum steam rate required?

Ans. $69(lb_m)/(hr)$

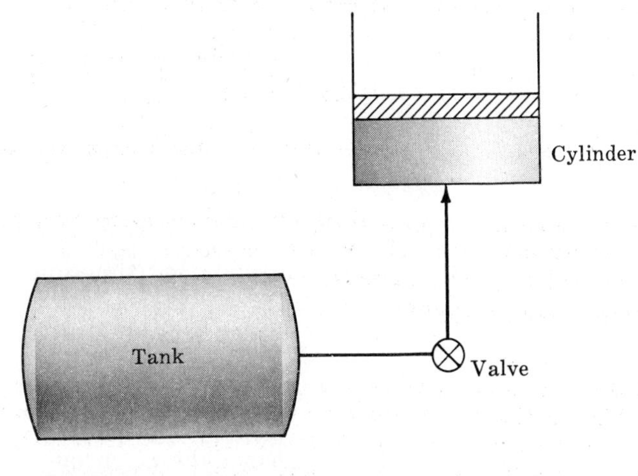

Fig. 6-26

Chapter 7

Chemical Thermodynamics

In preceding chapters we have dealt primarily with systems containing a single chemical species. However, many systems of practical interest contain two or more distinct chemical species. In addition, a system may be made up of several phases, and the distribution of species is rarely uniform among the phases. Moreover, some of the species of a system may participate in chemical reactions, which alter the composition of the system. In this chapter we treat such systems with respect to phase and chemical equilibria.

Any quantitative treatment of these topics requires methods for the description of the thermodynamic properties of mixtures. Thus Secs. 7.1 through 7.5 constitute an introduction to the thermodynamics of solutions. The remaining sections deal with phase equilibrium (Secs. 7.6 and 7.7) and chemical equilibrium (Secs. 7.8 and 7.9). Because of the introductory nature of this outline, the examples and problems of these last four sections are limited to the simpler applications of theory; however, sufficient material is included to serve as a basis for solving more complex problems.

7.1 PARTIAL MOLAR PROPERTIES

We found in Sec. 3.6 that the chemical potential μ_i plays a central role in the description of phase equilibrium. According to (*3.40*), μ_i can be defined as a mole-number derivative of the Gibbs function:

$$\mu_i = \left[\frac{\partial(nG)}{\partial n_i}\right]_{T,P,n_j} \tag{3.40}$$

For a constant-composition mixture, the temperature and pressure derivatives of μ_i are given by

$$\left(\frac{\partial \mu_i}{\partial T}\right)_{P,n} = -\left[\frac{\partial(nS)}{\partial n_i}\right]_{T,P,n_j} \tag{3.45}$$

$$\left(\frac{\partial \mu_i}{\partial P}\right)_{T,n} = \left[\frac{\partial(nV)}{\partial n_i}\right]_{T,P,n_j} \tag{3.46}$$

It is anticipated from (*3.40*), (*3.45*), and (*3.46*), then, that functions of the form

$$\boxed{\bar{M}_i = \left[\frac{\partial(nM)}{\partial n_i}\right]_{T,P,n_j}} \tag{7.1}$$

will be of importance in the treatment of thermodynamic properties of solutions, particularly as regards phase equilibrium considerations. The functions $\bar{M}_i$, called *partial molar properties,* in fact have a general significance to solution thermodynamics, whether one is concerned with phase equilibrium calculations or not.

It is observed experimentally that for single-phase systems at constant T and P certain thermodynamic functions, such as the total volume V^t, the total enthalpy H^t, etc., are homogeneous in the first degree in the mole numbers n_i (or masses m_i). That is, given a system containing mole numbers $n_1, n_2, \ldots, n_m$ and having a total property M^t, it is found that increasing each mole number by the same factor α at constant T and P results in a new value αM^t of the property. Properties M^t conforming to this behavior are *extensive* properties. As shown in Problem 7.1, extensive total properties M^t $(= nM)$ are related to the corresponding partial molar properties $\bar{M}_i$ by

$$\boxed{nM = \sum n_i \bar{M}_i} \tag{7.2}$$

Division of (*7.2*) by the total number of moles n yields an expression for the molar property M in terms of the $\bar{M}_i$ and the mole fractions x_i:

$$\boxed{M = \sum x_i \bar{M}_i} \tag{7.3}$$

It is apparent from (*7.3*) that the $\bar{M}_i$, like M, are *intensive* properties. That is, although they may in general be functions of T, P, and the x_i, they do not depend on the extent of a system. For a pure material, $\bar{M}_i$ is equal to M_i, the molar property of pure i (see for example Problems 3.11 and 3.32).

If one chooses to deal with the *masses* (rather than with the mole numbers) of the components of a system, then equations analogous to (*7.1*) through (*7.3*) follow with n_i replaced by m_i and n by m. In this case, x_i is the weight fraction of i, and the $\bar{M}_i$ are called partial *specific* properties. Most of the equations of this chapter will be developed on a molar basis, but conversion to a specific basis is simply a matter of making the above formal substitutions.

Example 7.1. A group of students came across an unsuspected supply of laboratory alcohol, containing 96 weight-percent ethanol and 4 weight-percent water. As an experiment they decided to convert 2(liter) of this material into vodka, having a composition of 56 weight-percent ethanol and 44 weight-percent water. Wishing to perform the experiment carefully, they searched the literature and found the following partial-specific-volume data for ethanol-water mixtures at 25(°C) and 1(atm):

	In 96% ethanol	In vodka
$\bar{V}_{H_2O}$	0.816(cm)^3/(g)	0.953(cm)^3/(g)
$\bar{V}_{EtOH}$	1.273(cm)^3/(g)	1.243(cm)^3/(g)

The specific volume of water at 25(°C) is 1.003(cm)^3/(g). How many liters of water should be added to the 2(liter) of laboratory alcohol, and how many liters of vodka result?

Let m_w be the mass of water added to mass m_a of laboratory alcohol to produce a mass m_v of vodka. The overall material balance is

$$m_a + m_w = m_v$$

and a material balance on the water gives

$$0.04\, m_a + m_w = 0.44\, m_v$$

Solution of these equations for m_w and m_v in terms of m_a yields

$$m_w = 0.7143\, m_a \qquad m_v = 1.7143\, m_a$$

The *total* volumes V_w^t, V_a^t, and V_v^t are related to the corresponding masses through the specific volumes:

$$V_w^t = m_w V_w \qquad V_a^t = m_a V_a \qquad V_v^t = m_v V_v$$

and thus the equations for calculation of V_w^t and V_v^t are

$$V_w^t = 0.7143 \frac{V_w V_a^t}{V_a} \qquad V_v^t = 1.7143 \frac{V_v V_a^t}{V_a}$$

According to (*7.3*), the specific volume of a binary solution is given in terms of the partial specific volumes of the components as

$$V = x_1\bar{V}_1 + x_2\bar{V}_2$$

For the laboratory alcohol and the vodka, then,

$$V_a = (0.04)(0.816) + (0.96)(1.273) = 1.255(\text{cm})^3/(\text{g})$$

$$V_v = (0.44)(0.953) + (0.56)(1.243) = 1.115(\text{cm})^3/(\text{g})$$

and by the statement of the problem

$$V_w = 1.003(\text{cm})^3/(\text{g}) \qquad V_a^t = 2000(\text{cm})^3$$

Substitution of these values into the equations for V_w^t and V_v^t gives

$$V_w^t = (0.7143)\frac{(1.003)(2000)}{(1.255)} = 1142(\text{cm})^3 = 1.142(\text{liter})$$

$$V_v^t = (1.7143)\frac{(1.115)(2000)}{(1.255)} = 3046(\text{cm})^3 = 3.046(\text{liter})$$

Note that the total volume of material mixed is $2.000 + 1.142 = 3.142$(liter), while, alas, only 3.046(liter) of vodka is obtained.

It is convenient to rewrite the definition (*7.1*) in terms of the intensive variables M and x_i. First, we expand the derivative:

$$\left[\frac{\partial(nM)}{\partial n_i}\right]_{T,P,n_j} = M\left(\frac{\partial n}{\partial n_i}\right)_{T,P,n_j} + n\left(\frac{\partial M}{\partial n_i}\right)_{T,P,n_j}$$

But $\left(\frac{\partial n}{\partial n_i}\right)_{T,P,n_j} = 1$ and (*7.1*) becomes

$$\bar{M}_i = M + n\left(\frac{\partial M}{\partial n_i}\right)_{T,P,n_j} \tag{7.4}$$

Now the intensive property M of an m-component mixture is a function of T, P, and $m-1$ independent mole fractions. For convenience, we consider these mole fractions to be $x_1, x_2, \ldots, x_{i-1}, x_{i+1}, \ldots, x_m$; that is, we eliminate the mole fraction x_i of the component of interest. We can then write, at constant T and P,

$$dM = \sum_k \left(\frac{\partial M}{\partial x_k}\right)_{T,P,x_l} dx_k \qquad (\text{const. } T, P)$$

where the summation over k excludes component i, and the subscript x_l indicates that all mole fractions other than x_i and x_k are held constant. Division of this equation by dn_i and restriction to constant n_j gives

$$\left(\frac{\partial M}{\partial n_i}\right)_{T,P,n_j} = \sum_k \left(\frac{\partial M}{\partial x_k}\right)_{T,P,x_l} \left(\frac{\partial x_k}{\partial n_i}\right)_{n_j} \tag{7.5}$$

We must now find an expression for $(\partial x_k/\partial n_i)_{n_j}$. By definition, $x_k = n_k/n$, from which

$$\left(\frac{\partial x_k}{\partial n_i}\right)_{n_j} = -\frac{n_k}{n^2} = -\frac{x_k}{n} \quad \text{for } k \neq i \tag{7.6}$$

Combination of (*7.4*), (*7.5*), and (*7.6*) gives the desired result:

$$\boxed{\bar{M}_i = M - \sum_{k\neq i} x_k \left(\frac{\partial M}{\partial x_k}\right)_{T,P,x_{l\neq k,i}}} \tag{7.7}$$

where the restrictions on the indices k and l are shown explicitly. Equation (*7.7*) is of course merely an alternate form of (*7.1*), from which it was derived. However, it finds greater use in the treatment of experimental data, because compositions are generally given as x_i, rather than as n_i, and because the molar properties M, rather than M^t, are usually of interest.

Example 7.2. Write (7.7) for the components of a binary mixture and show how the $\bar{M}_i$ for a binary system can be found from a plot of M versus x_1 at constant T and P.

From (7.7),

$$\bar{M}_1 = M - x_2 \frac{dM}{dx_2}$$

or

$$\bar{M}_1 = M + (1 - x_1) \frac{dM}{dx_1} \tag{7.8}$$

where we have used the fact that $x_1 + x_2 = 1$, and that $dx_2 = -dx_1$. Since conditions of constant T and P are understood, there is only one independent variable, which we have chosen as x_1; the derivative dM/dx_1 is therefore written as a total derivative. Similarly,

$$\bar{M}_2 = M - x_1 \frac{dM}{dx_1} \tag{7.9}$$

A representative plot of M against x_1 is shown in Fig. 7-1. At any composition x_1 the derivative dM/dx_1 can be found by constructing a tangent to the curve. Call the intersections of this tangent with the M-axes I_2 (at $x_1 = 0$) and I_1 (at $x_1 = 1$). From the geometry of the figure, we can write two expressions for dM/dx_1:

$$\frac{dM}{dx_1} = \frac{M - I_2}{x_1} \qquad \text{and} \qquad \frac{dM}{dx_1} = I_1 - I_2$$

Solving for I_2 and I_1, we obtain

$$I_1 = M + (1 - x_1) \frac{dM}{dx_1} \qquad I_2 = M - x_1 \frac{dM}{dx_1}$$

Comparison of these two equations with (7.8) and (7.9) gives

$$I_1 = \bar{M}_1 \qquad I_2 = \bar{M}_2$$

Thus the $\bar{M}_i$ values for the two components of a binary solution are equal to the M-intercepts of the tangent drawn to the M versus x_1 curve at the composition of interest, as shown by Fig. 7-1. It is clear from this construction that a tangent drawn at $x_1 = 1$ gives $\bar{M}_1 = M_1$ and one drawn at $x_1 = 0$ ($x_2 = 1$) gives $\bar{M}_2 = M_2$. This is seen from Fig. 7-2, and is in accord with the requirement that for a pure material

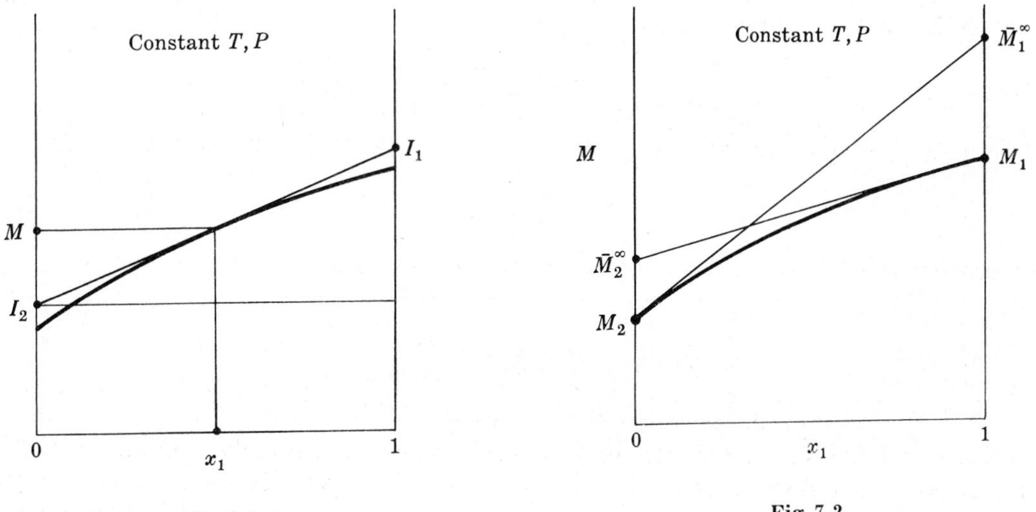

Fig. 7-1

Fig. 7-2

$\bar{M}_i = M_i$. The other ends of the tangents shown in Fig. 7-2 intercept the opposite axes, and give the partial molar property of the component present at *infinite dilution*, designated $\bar{M}_i^{\infty}$. Thus, for $x_1 = 1$ and $x_2 = 0$, $\bar{M}_2 = \bar{M}_2^{\infty}$, and when $x_1 = 0$ and $x_2 = 1$, $\bar{M}_1 = \bar{M}_1^{\infty}$.

A numerical example of the procedure just described is shown in Fig. 7-3, which shows a plot of the specific volume of ethanol-water mixtures versus weight fraction of ethanol at 25(°C) and 1(atm). The tangent drawn at $x_{EtOH} = 0.5$ gives intercepts which provide values as shown of $\bar{V}_{H_2O}$ and $\bar{V}_{EtOH}$. Thus we find for a 50-weight-percent mixture of ethanol in water that $\bar{V}_{H_2O} = 0.963(cm)^3/(g)$ and $\bar{V}_{EtOH} = 1.235(cm)^3/(g)$. The $\bar{V}_i$ values obtained from the same curve for the *entire* composition range are shown in Fig. 7-4.

7.2 FUGACITY. FUGACITY COEFFICIENT

Two auxiliary functions which are of particular use in the thermodynamic treatment of solutions are the *fugacity* f and the *fugacity coefficient* ϕ. Their definitions for the three cases of interest are:

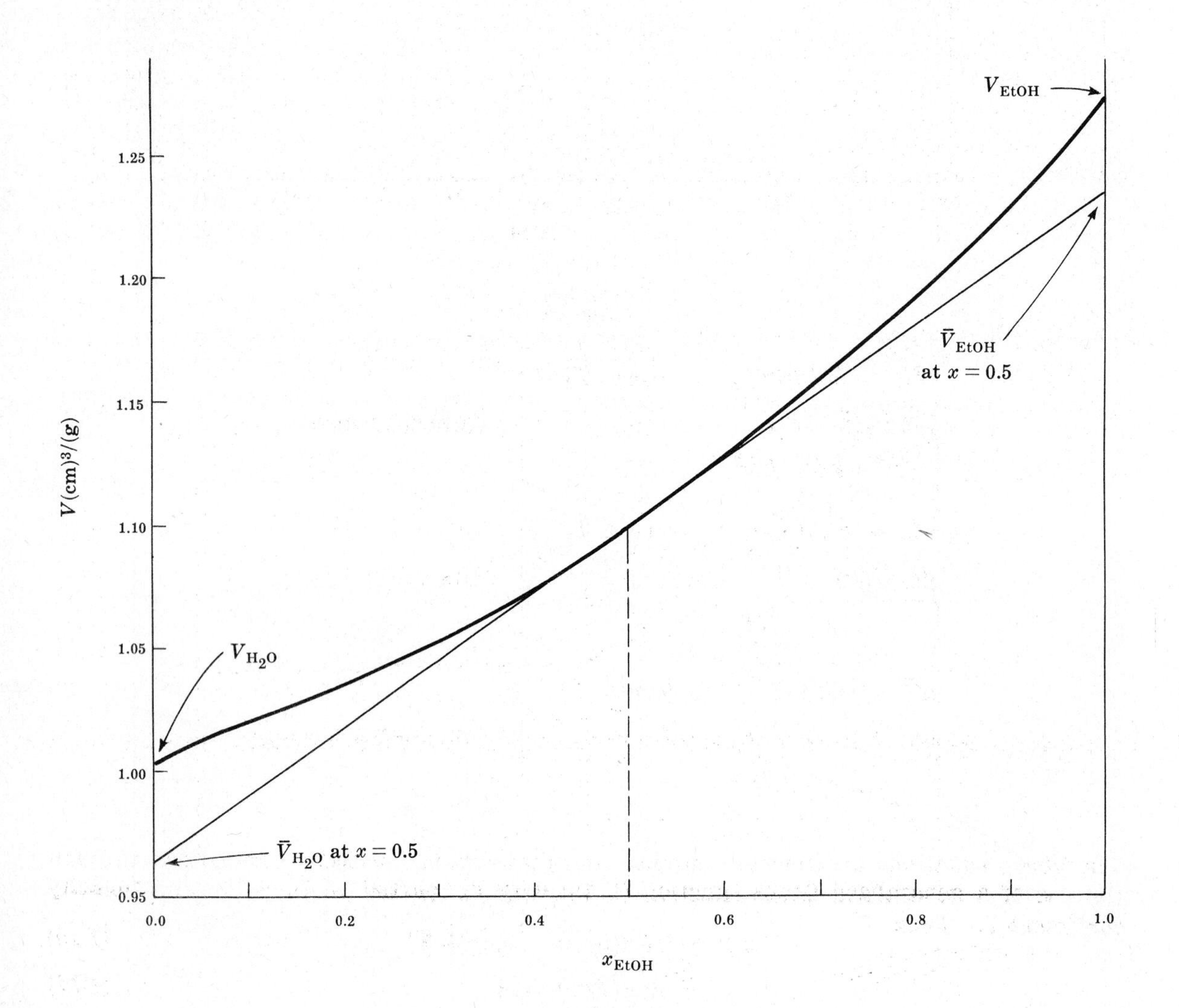

Fig. 7-3

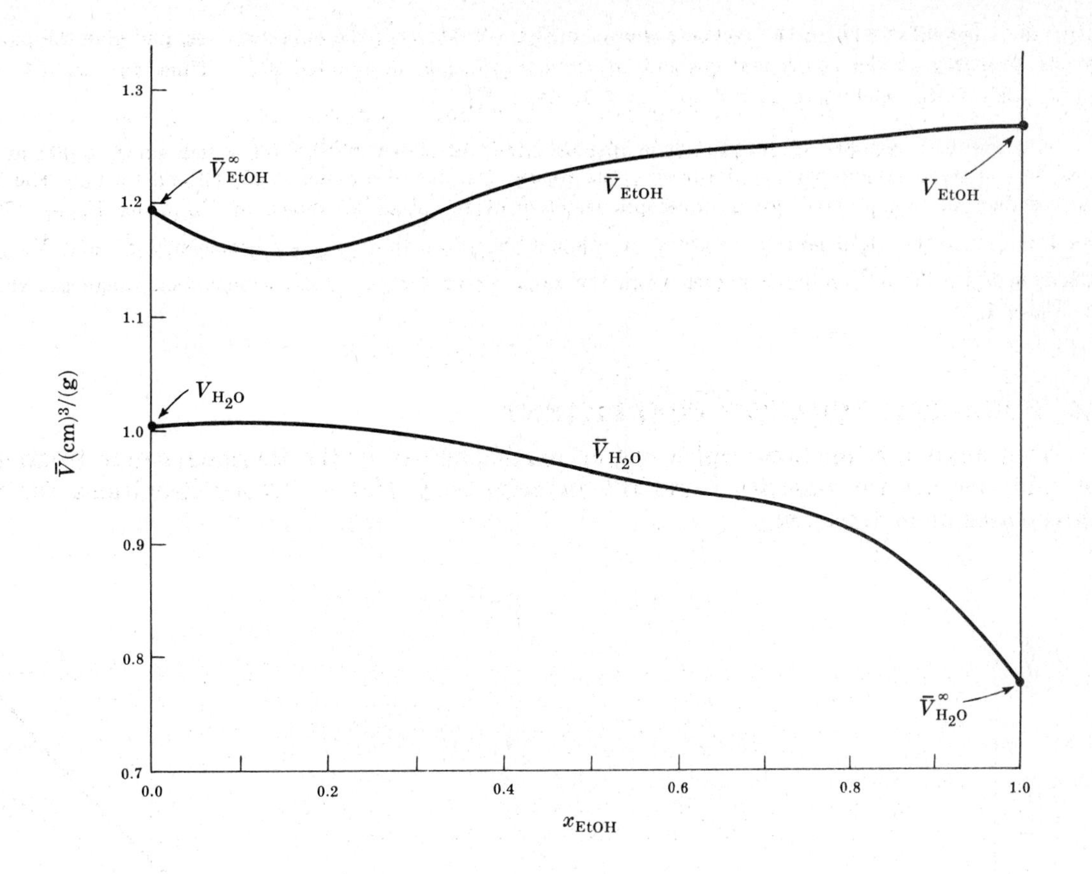

Fig. 7-4

$$\left.\begin{aligned} dG_i &= RT\,d(\ln f_i) \qquad \text{(const. } T\text{)} \\ \lim_{P\to 0}\,(f_i/P) &= 1 \\ \phi_i &= f_i/P \end{aligned}\right\} \quad \text{Pure component}$$

$$\left.\begin{aligned} dG &= RT\,d(\ln f) \qquad \text{(const. } T\text{)} \\ \lim_{P\to 0}\,(f/P) &= 1 \\ \phi &= f/P \end{aligned}\right\} \quad \text{Mixture}$$

$$\left.\begin{aligned} d\bar{G}_i &= RT\,d(\ln \hat{f}_i) \qquad \text{(const. } T\text{)} \\ \lim_{P\to 0}\,(\hat{f}_i/x_iP) &= 1 \\ \hat{\phi}_i &= \hat{f}_i/x_iP \end{aligned}\right\} \quad \text{Component in solution}$$

The above equations are formally similar and they can be concisely summarized through the use of a generalized Gibbs function $\mathcal{G}$, fugacity $\mathcal{f}$, "partial pressure" $\mathcal{P}$, and fugacity coefficient φ. Thus,

$$d\mathcal{G} = RT\,d(\ln \mathcal{f}) \qquad \text{(const. } T\text{)} \tag{7.10}$$

$$\lim_{P\to 0}\,(\mathcal{f}/\mathcal{P}) = 1 \tag{7.11}$$

$$\varphi = \mathcal{f}/\mathcal{P} \tag{7.12}$$

Combination of (*7.11*) and (*7.12*) gives an equation for the zero-pressure limit of φ:

$$\lim_{P\to 0} \varphi = 1 \tag{7.13}$$

The appropriate forms of (*7.10*) through (*7.13*) can be recovered for the three special cases by selecting the specializations of the generalized variables from the following table:

	$\mathcal{G}$	f	$\mathcal{P}$	φ
Pure component i	G_i	f_i	P	ϕ_i
Mixture	G	f	P	ϕ
Component i in solution	$\bar{G}_i\ (=\mu_i)$	$\hat{f}_i$	x_iP	$\hat{\phi}_i$

With the restriction to constant T understood, (*7.10*) merely provides a change in variable between $\mathcal{G}$ and f. Integration of (*7.10*) at constant T gives

$$\Delta\mathcal{G} = RT \ln\left(\frac{f_{\text{final}}}{f_{\text{initial}}}\right) \tag{7.14}$$

where $\Delta\mathcal{G} = \mathcal{G}_{\text{final}} - \mathcal{G}_{\text{initial}}$. Equation (*7.14*) is of considerable utility, for it allows for both pressure and composition changes, and for changes in physical state. According to (*7.11*), f has units of pressure. The fugacity coefficient φ is dimensionless and, by (*7.13*), it is defined in such a way as to approach the convenient value of unity as P approaches zero.

Both f and ϕ are intensive properties; because of the way they are defined, however, the appropriate extensive variables are $n \ln f$ and $n \ln \phi$, rather than nf and $n\phi$. Since $\hat{f}_i$ and $\hat{\phi}_i$ are *not* related to f and ϕ as partial molar properties, we use a carat (^) rather than an overbar on $\hat{f}_i$ and $\hat{\phi}_i$. However, as with partial molar properties, $\hat{f}_i = f_i$ and $\hat{\phi}_i = \phi_i$ for $x_i = 1$.

We shall now develop equations for calculation of f or ϕ for a pure component or for a constant-composition mixture from PVT data or from a thermal equation of state. For conciseness, the subscript i will be deleted; however, the resultant equations are valid for both cases. We start with the specialization of (*7.10*):

$$dG = RT\, d(\ln f) \qquad \text{(const. } T) \tag{7.15}$$

At constant T and composition, (*3.35*) becomes for one mole of solution

$$dG = V\, dP \qquad \text{(const. } T, x) \tag{7.16}$$

Combination of (*7.15*) and (*7.16*) gives

$$RT\, d(\ln f) = V\, dP \qquad \text{(const. } T, x) \tag{7.17}$$

From the specialization of (*7.12*), we obtain $d(\ln f) = d(\ln \phi) + d(\ln P)$ or

$$d(\ln f) = d(\ln \phi) + \frac{dP}{P} \tag{7.18}$$

Substitution of (*7.18*) in (*7.17*), rearrangement, and use of the definition of Z gives

$$d(\ln \phi) = (Z-1)\frac{dP}{P} \qquad \text{(const. } T, x) \tag{7.19}$$

Equation (*7.19*) can now be integrated from the zero-pressure state [where, according to (*7.13*), $\phi = 1$] to pressure P, giving

$$\boxed{\ln \phi = \int_0^P (Z-1)\frac{dP}{P}} \qquad \text{(const. } T, x) \qquad (7.20)$$

Since $f = \phi P$, the corresponding expression for $\ln f$ is

$$\boxed{\ln f = \ln P + \int_0^P (Z-1)\frac{dP}{P}} \qquad \text{(const. } T, x) \qquad (7.21)$$

Equations analogous to (7.20) and (7.21) for which V, rather than P, is the variable of integration, are easily derived. They are

$$\ln \phi = (Z-1) - \ln Z - \int_\infty^V (Z-1)\frac{dV}{V} \qquad \text{(const. } T, x) \qquad (7.22)$$

$$\ln f = \ln\frac{RT}{V} + (Z-1) - \int_\infty^V (Z-1)\frac{dV}{V} \qquad \text{(const. } T, x) \qquad (7.23)$$

Important special cases of (7.20) through (7.23) result for ideal gases, for which $Z = 1$. Thus

$$\phi_i = \phi = 1 \qquad \text{(ideal gas)} \qquad (7.24)$$

$$f_i = f = P \qquad \text{(ideal gas)} \qquad (7.25)$$

The analogous equations for a component of an ideal-gas mixture are

$$\hat{\phi}_i = 1 \qquad \text{(ideal gas)} \qquad (7.26)$$

$$\hat{f}_i = y_i P \qquad \text{(ideal gas)} \qquad (7.27)$$

where y_i is the mole fraction of i in a gas phase.

Example 7.3. Derive an expression giving $\ln \phi$ for a gas described by (5.22) and by the generalized correlation of Sec. 5.5. Estimate ϕ and f for carbon dioxide gas at -40(°C) and 5(atm).

Equation (7.20) is used to find ϕ from (5.22):

$$\ln \phi = \int_0^P \left(\frac{BP}{RT}\right)\frac{dP}{P}$$

or

$$\ln \phi = \frac{BP}{RT} \qquad (7.28)$$

But, by (5.39),

$$B = \frac{RT_c}{P_c}(B^0 + \omega B^1)$$

Substituting the last equation into (7.28) and noting the definitions of P_r and T_r, we obtain

$$\ln \phi = \frac{P_r}{T_r}(B^0 + \omega B^1) \qquad (7.29)$$

The ϕ for carbon dioxide can be estimated from (7.29). From Appendix 3 we find $T_c = 304.2$(K), $P_c = 72.9$(atm), and $\omega = 0.225$. Then for the stated conditions

$$T_r = \frac{273.2 - 40}{304.2} = 0.767 \qquad P_r = \frac{5}{72.9} = 0.0686$$

and from Fig. 5-6

$$B^0 = -0.55 \qquad B^1 = -0.45$$

Thus

$$\ln \phi = \frac{0.0686}{0.767}[-0.55 + 0.225(-0.45)] = -0.0582$$

or
$$\phi = 0.943$$

Since $f = \phi P$, the corresponding value of the fugacity is $f = 0.943 \times 5 = 4.72(\text{atm})$.

The function related to $\ln f$ as a partial molar property can now be determined. First, we apply (*7.14*) to a hypothetical change of state from an ideal-gas mixture to a real solution in any physical state at the same T, P, and composition. Thus, in (*7.14*), $\mathcal{G}_{\text{final}} = G$, $\mathcal{G}_{\text{initial}} = G'$, $\mathcal{f}_{\text{final}} = f$, and $\mathcal{f}_{\text{initial}} = f'$. But by (*7.25*) $f' = P$, and (*7.14*) therefore becomes

$$G - G' = RT \ln (f/P) \tag{7.30}$$

Multiplication of (*7.30*) by the total number of moles n gives

$$nG - nG' = RT(n \ln f) - nRT \ln P \tag{7.31}$$

Differentiating (*7.31*) with respect to n_i at constant T, P, and n_j, we obtain

$$\bar{G}_i - \bar{G}_i' = RT\left[\frac{\partial(n \ln f)}{\partial n_i}\right]_{T,P,n_j} - RT \ln P \tag{7.32}$$

where (*7.1*) has been used to identify the first two terms as partial molar properties.

An alternative expression for $\bar{G}_i - \bar{G}_i'$ can be derived by applying (*7.14*) *directly* to component i in solution, again considering a change from the ideal-gas state to the real state at the same T, P, and composition. The result is

$$\bar{G}_i - \bar{G}_i' = RT \ln (\hat{f}_i/x_i P) \tag{7.33}$$

Combination of (*7.32*) and (*7.33*) gives

$$\boxed{\ln (\hat{f}_i/x_i) = \left[\frac{\partial(n \ln f)}{\partial n_i}\right]_{T,P,n_j}} \tag{7.34}$$

Comparison of (*7.34*) with (*7.1*) shows that the function $\ln (\hat{f}_i/x_i)$, rather than $\ln \hat{f}_i$, is related to $\ln f$ as a partial molar property. It can be similarly shown that $\ln \hat{\phi}_i$ is the partial molar property for $\ln \phi$:

$$\boxed{\ln \hat{\phi}_i = \left[\frac{\partial(n \ln \phi)}{\partial n_i}\right]_{T,P,n_j}} \tag{7.35}$$

Equations (*7.34*) and (*7.35*), when used in conjunction with (*7.20*) through (*7.23*), provide the means for obtaining $\hat{f}_i$ and $\hat{\phi}_i$ from equations of state. Such calculations require mixing rules for the material parameters in the equation of state; we illustrate in Example 7.4 how the mixing rules enter into the derivations.

The T- and P-derivatives of $\ln f$ are readily obtained from (*7.30*), rewritten as

$$\ln f = \frac{G}{RT} - \frac{G'}{RT} + \ln P \tag{7.36}$$

Differentiation of (*7.36*) with respect to T yields

$$\left(\frac{\partial \ln f}{\partial T}\right)_{P,x} = \left[\frac{\partial(G/RT)}{\partial T}\right]_{P,x} - \left[\frac{\partial(G'/RT)}{\partial T}\right]_{P,x}$$

where the subscript x indicates constant composition. But the Gibbs-Helmholtz equation (Problem 3.9) gives

$$\left[\frac{\partial(G/RT)}{\partial T}\right]_{P,x} = -\frac{H}{RT^2} \quad \text{and} \quad \left[\frac{\partial(G'/RT)}{\partial T}\right]_{P,x} = -\frac{H'}{RT^2}$$

Noting the definition (*4.13*) of the residual enthalpy $\Delta H'$, we obtain finally on combination of the last three equations:

$$\left(\frac{\partial \ln f}{\partial T}\right)_{P,x} = \frac{\Delta H'}{RT^2} \tag{7.37}$$

The pressure derivative is found in a similar manner. Differentiation of (*7.36*) with respect to P yields

$$\left(\frac{\partial \ln f}{\partial P}\right)_{T,x} = \left[\frac{\partial(G/RT)}{\partial P}\right]_{T,x} - \left[\frac{\partial(G'/RT)}{\partial P}\right]_{T,x} + \frac{1}{P}$$

But, from (*3.38*)

$$\left[\frac{\partial(G/RT)}{\partial P}\right]_{T,x} = \frac{V}{RT}$$

$$\left[\frac{\partial(G'/RT)}{\partial P}\right]_{T,x} = \frac{V'}{RT} = \frac{(RT/P)}{RT} = \frac{1}{P}$$

Combination of the last three equations gives

$$\left(\frac{\partial \ln f}{\partial P}\right)_{T,x} = \frac{V}{RT} \tag{7.38}$$

Example 7.4. Find expressions for $\ln \hat{\phi}_i$ and $\ln \hat{f}_i$ of the components in a gas mixture described by the truncated virial equation (*5.22*). Calculate and plot as a function of composition $\hat{\phi}_i$ and $\hat{f}_i$ for methane and n-hexane in a binary vapor mixture of methane and n-hexane at 75(°C) and 1(atm).

Equation (*7.28*) gives $\ln \phi$ for the mixture. Substituting (*7.28*) into (*7.35*), we obtain

$$\ln \hat{\phi}_i = \left[\frac{\partial(nBP/RT)}{\partial n_i}\right]_{T,P,n_l}$$

or

$$\ln \hat{\phi}_i = \frac{P}{RT}\left[\frac{\partial(nB)}{\partial n_i}\right]_{T,P,n_l} \tag{7.39}$$

where subscript n_l here indicates that all mole numbers except n_i are constant. We now need the mixing rule for B in order to evaluate the derivative in (*7.39*). Rather than use the original rule (*5.14*), we employ the equivalent expression written in terms of δ functions [Problem 5.10, (*5*) and (*6*)]. Thus,

$$B = \sum_j y_j B_{jj} + \frac{1}{2}\sum_j \sum_k y_j y_k \delta_{jk} \tag{7.40}$$

where $\delta_{jk} = 2B_{jk} - B_{jj} - B_{kk}$. Note that we use dummy summation indices j and k, rather than i and j, to avoid possible confusion with the particular subscript i. Multiplication of (*7.40*) by n gives

$$nB = \sum_j n_j B_{jj} + \frac{1}{2n}\sum_j \sum_k n_j n_k \delta_{jk}$$

from which we obtain

$$\left[\frac{\partial(nB)}{\partial n_i}\right]_{T,P,n_l} = B_{ii} - \frac{1}{2n^2}\sum_j \sum_k n_j n_k \delta_{jk} + \frac{1}{n}\sum_j n_j \delta_{ji}$$

$$= B_{ii} - \frac{1}{2}\sum_j \sum_k y_j y_k \delta_{jk} + \sum_j y_j \delta_{ji}$$

The last term is unaffected by multiplication by $\sum_k y_k$, because the mole fractions sum to unity. Doing this and rearranging, we obtain finally

$$\left[\frac{\partial(nB)}{\partial n_i}\right]_{T,P,n_l} = B_{ii} + \frac{1}{2}\sum_j \sum_k y_j y_k (2\delta_{ji} - \delta_{jk})$$

and (*7.39*) becomes

$$\ln \hat{\phi}_i = \frac{P}{RT}\left[B_{ii} + \frac{1}{2}\sum_j \sum_k y_j y_k (2\delta_{ji} - \delta_{jk})\right] \tag{7.41}$$

The corresponding expression for $\ln \hat{f}_i$ is now easily found from (*7.41*) by use of the definition (*7.12*). Thus,

$$\ln \hat{f}_i = \ln y_i P + \frac{P}{RT}\left[B_{ii} + \frac{1}{2}\sum_j \sum_k y_j y_k (2\delta_{ji} - \delta_{jk})\right] \tag{7.42}$$

Although (*7.41*) and (*7.42*) appear complex, they are the simplest realistic relationships available for the composition dependence of $\ln \hat{\phi}_i$ and $\ln \hat{f}_i$ for the components in a real gas mixture. Because $\delta_{jj} = 0$ and $\delta_{jk} = \delta_{kj}$, the summation reduces to relatively compact expressions even for mixtures containing large numbers of components. Thus, for species 1 and 2 in a binary mixture, (*7.41*) and (*7.42*) give

$$\ln \hat{\phi}_1 = \frac{P}{RT}[B_{11} + y_2^2 \delta_{12}]$$
$$\ln \hat{\phi}_2 = \frac{P}{RT}[B_{22} + y_1^2 \delta_{12}] \tag{7.43}$$

and

$$\ln \hat{f}_1 = \ln y_1 P + \frac{P}{RT}[B_{11} + y_2^2 \delta_{12}]$$
$$\ln \hat{f}_2 = \ln y_2 P + \frac{P}{RT}[B_{22} + y_1^2 \delta_{12}] \tag{7.44}$$

For the methane(1)–n-hexane(2) system at 75(°C):

$$B_{11} = -26(\text{cm})^3/(\text{g mole}), \quad B_{22} = -1239(\text{cm})^3/(\text{g mole}), \quad \delta_{12} = 905(\text{cm})^3/(\text{g mole})$$

The $\hat{\phi}_i$ and $\hat{f}_i$ curves at 1(atm) calculated from (*7.43*) and (*7.44*) with these values are shown by Fig. 7-5. On the scale of this figure the $\hat{f}_i$ curves appear to be straight lines; actually, they deviate slightly from linearity because of the δ_{12} terms in (*7.44*). When $\delta_{12} = 0$, then (*7.44*) gives the linear equation

$$\hat{f}_i = y_i P \exp\left(\frac{PB_{ii}}{RT}\right) = y_i f_i \qquad (i = 1, 2)$$

where the last equality follows from (*7.28*). Similarly, the departure of the $\hat{\phi}_i$ curves from straight, horizontal lines, quite evident in Fig. 7-5, results from the δ_{12} terms in (*7.43*). When $\delta_{12} = 0$ then (*7.43*) gives *constant* values for $\hat{\phi}_1$ and $\hat{\phi}_2$:

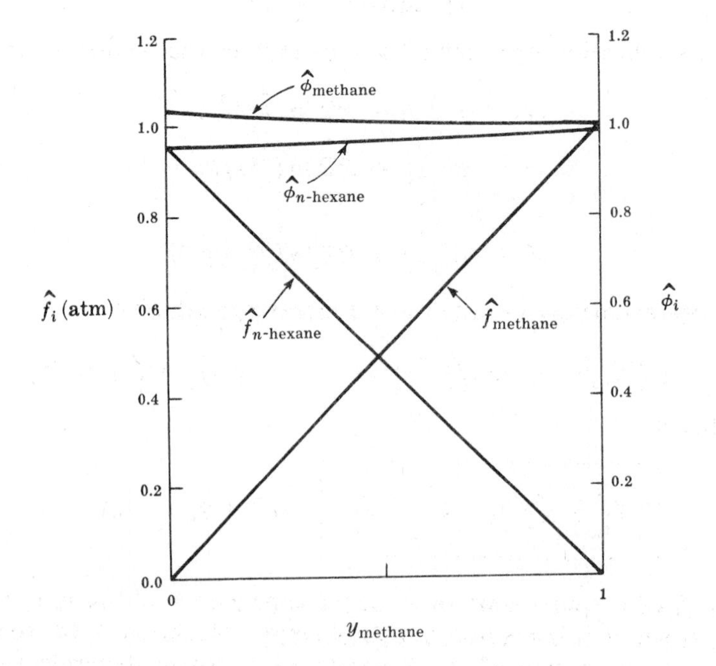

Fig. 7-5

$$\hat{\phi}_i = \exp\left(\frac{PB_{ii}}{RT}\right) = \phi_i \qquad (i = 1, 2)$$

Thus mixtures of real gases for which $\delta_{ij} = 0$ exhibit particularly simple $\hat{\phi}_i$ and $\hat{f}_i$ behavior at low pressures; we show in Problem 7.3 that such behavior is a characteristic of *ideal solutions*.

7.3 EQUALITY OF FUGACITIES AS A CRITERION OF PHASE EQUILIBRIUM

In Sec. 3.6 we found that the criterion for phase equilibrium for PVT systems of uniform T and P could be concisely stated by use of the chemical potential:

$$\mu_i^\alpha = \mu_i^\beta = \cdots = \mu_i^\pi \qquad (i = 1, 2, \ldots, m) \tag{3.76}$$

Equation (*3.76*) holds for each component i in an m-component system containing π phases at equilibrium. The relationship between the fugacity and the Gibbs function expressed in the definition (*7.10*) allows us to derive an alternative criterion for phase equilibrium which is often of greater practical utility than (*3.76*).

We proceed by applying (*7.14*) to each component i in each phase p, letting $\mathcal{G} = \bar{G}_i = \mu_i$ and $f = \hat{f}_i$:

$$\mu_i^p\,(\text{final}) - \mu_i^p\,(\text{initial}) = RT \ln\left[\frac{\hat{f}_i^p\,(\text{final})}{\hat{f}_i^p\,(\text{initial})}\right] \tag{7.45}$$

Since (*7.45*) is valid both for changes in composition and in physical state, we can without loss of generality let the "initial" state be the same for all phases, namely that of component i as it exists in the πth phase at equilibrium conditions. Thus, for $i = 1, 2, \ldots, m$,

$$\mu_i^p\,(\text{initial}) = \mu_i^\pi$$

$$\hat{f}_i^p\,(\text{initial}) = \hat{f}_i^\pi$$

and (*7.45*) yields the following equations for μ_i in each of the phases α through $\pi - 1$:

$$\begin{aligned}
\mu_i^\alpha &= \mu_i^\pi + RT\ln(\hat{f}_i^\alpha/\hat{f}_i^\pi)\\
\mu_i^\beta &= \mu_i^\pi + RT\ln(\hat{f}_i^\beta/\hat{f}_i^\pi)\\
&\vdots\\
\mu_i^{\pi-1} &= \mu_i^\pi + RT\ln(\hat{f}_i^{\pi-1}/\hat{f}_i^\pi)
\end{aligned}$$

Substitution of these equations in (*3.76*) and rearrangement gives

$$\ln(\hat{f}_i^\alpha/\hat{f}_i^\pi) = \ln(\hat{f}_i^\beta/\hat{f}_i^\pi) = \cdots = \ln(\hat{f}_i^{\pi-1}/\hat{f}_i^\pi) = 0$$

from which we obtain

$$\boxed{\hat{f}_i^\alpha = \hat{f}_i^\beta = \cdots = \hat{f}_i^\pi} \qquad (i = 1, 2, \ldots, m) \tag{7.46}$$

Thus the fugacity $\hat{f}_i$ of a component in a multicomponent multiphase system must be the same in all phases in which it is present at equilibrium. Equation (*7.46*) constitutes a major justification for the introduction of the fugacity as a thermodynamic variable, and will be the starting point for many of the applications considered in later sections.

As an example of a specific application of (7.46), consider the vapor(v)-liquid(l) equilibrium of a single pure component i. Since for a pure component $\hat{f}_i = f_i$, (7.46) becomes

$$f_i^l = f_i^v \tag{7.47}$$

Similarly, for the triple point we have solid(s)-liquid-vapor equilibrium of pure i, and

$$f_i^s = f_i^l = f_i^v$$

Example 7.5. Using volumetric data from the steam tables, calculate and plot ϕ and f versus P for water at 260(°C) and at pressures from 0 to 100(bar).

At 260(°C), the saturation pressure of water is 46.88(bar). Therefore the required calculations refer to both the gas phase [from 0 to 46.88(bar)] and the liquid phase [from 46.88 to 100(bar)]. In Example 5.4 we found a four-term virial expansion of Z in $1/V$ which described the gas-phase data up to 46.88(bar), and we employ these results in the calculation of ϕ and f for the gas phase:

$$Z = 1 + \frac{B}{V} + \frac{C}{V^2} + \frac{D}{V^3}$$

Equation (7.22) is convenient for use with equations of state that give Z as an explicit function of V, and from it we obtain:

$$\ln \phi^v = \frac{2B}{V} + \frac{3C}{2V^2} + \frac{4D}{3V^3} - \ln Z \tag{7.48}$$

Similarly, (7.23) gives

$$\ln f^v = \ln\frac{RT}{V} + \frac{2B}{V} + \frac{3C}{2V^2} + \frac{4D}{3V^3} \tag{7.49}$$

It was shown in Example 5.4 that vapor volumes at 260(°C) calculated from the four-term virial equation were virtually identical with those given in the steam tables. Thus in applying (7.48) and (7.49) we need only find V and Z from data in the steam tables at each P, and use these values together with the derived virial coefficients [$B = -142.2(\text{cm})^3/(\text{g mole})$, $C = -7140(\text{cm})^6/(\text{g mole})^2$, and $D = 1.51 \times 10^6(\text{cm})^9/(\text{g mole})^3$] to calculate ϕ and f.

The above procedure takes us to $P^{\text{sat}} = 46.88$(bar), above which pressure water is a liquid at 260(°C). According to (7.47), at the saturation conditions of 260(°C) and 46.88(bar),

$$f^l(\text{sat}) = f^v(\text{sat}) = f^{\text{sat}}$$

and similarly

$$\phi^l(\text{sat}) = \phi^v(\text{sat}) = \phi^{\text{sat}}$$

Thus the vapor and liquid branches of the ϕ versus P and f versus P curves are continuous at P^{sat}, and f^{sat} and ϕ^{sat} calculated for the gas phase can be used as reference values for determining the liquid-phase quantities. Equation (7.19) may be rewritten as:

$$d(\ln \phi^l) = \left(\frac{V}{RT} - \frac{1}{P}\right) dP \qquad \text{(const. } T\text{)}$$

Integration from P^{sat} to P gives, on rearrangement,

$$\ln \phi^l = \ln \phi^{\text{sat}} - \ln\frac{P}{P^{\text{sat}}} + \frac{1}{RT}\int_{P^{\text{sat}}}^{P} V\,dP \qquad \text{(const. } T\text{)} \tag{7.50}$$

Evaluation of the integral in (7.50) requires an expression for the isothermal dependence of V on P. For liquids, one can employ the Tait equation of state (Problem 5.14), choosing the reference state for the Tait equation as the saturated liquid:

$$V = V^{\text{sat}} - D\ln\left(\frac{P+E}{P^{\text{sat}}+E}\right)$$

Substituting this equation into (7.50) and integrating, we obtain

$$\ln \phi^l = \ln \phi^{\text{sat}} + \ln\frac{P^{\text{sat}}}{P} + \frac{(V^{\text{sat}}+D)(P-P^{\text{sat}})}{RT} - \frac{D(P+E)}{RT}\ln\left(\frac{P+E}{P^{\text{sat}}+E}\right) \tag{7.51}$$

The corresponding expression for $\ln f^l$ is

$$\ln f^l = \ln f^{\text{sat}} + \frac{(V^{\text{sat}} + D)(P - P^{\text{sat}})}{RT} - \frac{D(P+E)}{RT} \ln\left(\frac{P+E}{P^{\text{sat}} + E}\right) \tag{7.52}$$

For liquid water at 260(°C), $V^{\text{sat}} = 22.985(\text{cm})^3/(\text{g mole})$, and for pressures up to 1400(bar), Tait parameters D and E at 260(°C), derived from the volumetric data of the steam tables, are

$$D = 2.260(\text{cm})^3/(\text{g mole}) \qquad E = 534.4(\text{bar})$$

The values calculated for f and ϕ from (*7.48*), (*7.49*), (*7.51*), and (*7.52*) are plotted as functions of P in Fig. 7-6. For comparison three common approximations are indicated by dashed lines. The first, labeled $\phi^v = 1$ and $f^v = P$, represents the assumption of an ideal gas. For water vapor at 260(°C) this leads to increasing error with increasing pressure up to the saturation pressure, where it is about 18%.

The second approximation, which is a natural extension of the first to the liquid state, is based on the assumption that $f^l = P^{\text{sat}}$ at the saturation pressure and at all higher pressures. In the present case this leads to a nearly uniform error of 18% in both f^l and ϕ^l.

The third common approximation rests on the assumption that $f^l = f^{\text{sat}}$ at the saturation pressure and at all higher pressures. This assumption leads to much smaller errors than the second, giving in this case a maximum error of 2.5% at 100(bar). With respect to (*7.51*) and (*7.52*) this asumption is realized by setting $V^{\text{sat}} = D = 0$, and reference to the Tait equation shows this to be equivalent to the assumption that the liquid volume is zero. A much better assumption is that the liquid volume is equal to V^{sat}, not only at P^{sat}, but at all higher pressures. With respect to (*7.51*) and (*7.52*) this assumption amounts to setting $D = 0$. This yields the equations

$$\ln \phi^l = \ln \phi^{\text{sat}} - \ln \frac{P}{P^{\text{sat}}} + \frac{V^{\text{sat}}(P - P^{\text{sat}})}{RT} \tag{7.53}$$

$$\ln f^l = \ln f^{\text{sat}} + \frac{V^{\text{sat}}(P - P^{\text{sat}})}{RT} \tag{7.54}$$

Values of ϕ^l and f^l calculated from these equations are not shown on Fig. 7-6, as they lead to very small errors, only 0.01% at 100(bar) and 0.7% at 500(bar) for water at 260(°C).

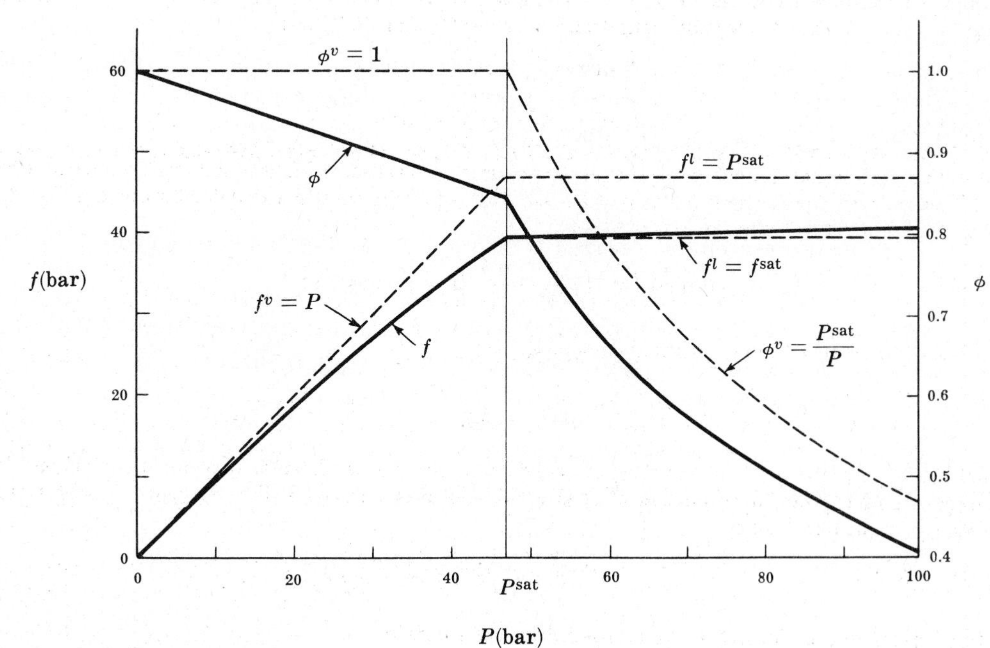

Fig. 7-6

7.4 IDEAL SOLUTIONS. STANDARD STATES. PROPERTY CHANGES OF MIXING. ACTIVITY

The correlation and prediction of the properties of real materials is greatly facilitated when one has a standard or model of "ideal" behavior to which real behavior can be compared. Such models should be simple, and they should conform to the behavior of real materials at least in some limiting condition. Thus, as we have seen, the ideal gas serves as a useful model of the behavior of gases, and it represents real-gas behavior in the limit as P approaches zero.

Ideal solutions.

Another model of behavior, appropriate for nonelectrolyte solutions, is provided by the *ideal solution.* To motivate the definition of this model, we present in Fig. 7-7 a typical plot of $\hat{f}_i$ versus x_i for one of the components of a binary solution at constant T and P. Two features of this plot are of particular significance. First, we see that as x_i approaches unity, the curve representing $\hat{f}_i$ becomes tangent to the straight line given by the equation:

$$\hat{f}_i = x_i f_i$$

where f_i is the fugacity of pure i, the fugacity at $x_i = 1$. This observation is given mathematically by

$$\lim_{x_i \to 1} (\hat{f}_i / x_i) = f_i \qquad (7.55)$$

Equation (7.55) is the *Lewis and Randall rule.*

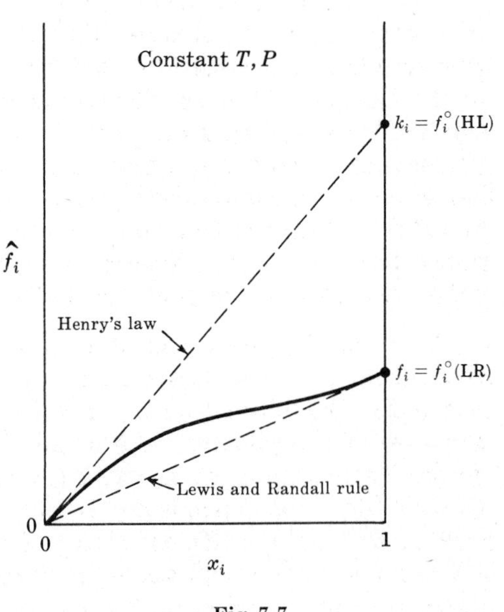

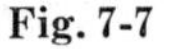
Fig. 7-7

The second important characteristic of the $\hat{f}_i$ versus x_i plot obtains at the low-concentration end of the curve, where x_i approaches zero. Here it is observed that the curve can be approximated by a straight line of finite slope. This is indicated in Fig. 7-7 by the tangency of the solid curve with the upper dashed line, and is expressed mathematically by

$$\lim_{x_i \to 0} (\hat{f}_i / x_i) = k_i \qquad (7.56)$$

Equation (7.56) is the statement of *Henry's law,* and k_i, the slope of the limiting tangent, is called *Henry's constant.*

The simplest possible $\hat{f}_i$ versus x_i relationship is a direct proportionality, and Fig. 7-7 and (7.55) and (7.56) assert that such a relationship is indeed obtained for the components of real solutions at the extremes of the composition range. An ideal solution conforms to an extension of this limiting behavior; it is defined as a solution for which the fugacity of every component is directly proportional to its mole fraction over the *entire* composition range at any fixed conditions of temperature and pressure:

$$\boxed{\hat{f}_i^{\text{id}} = x_i f_i^\circ} \quad \text{(const. } T \text{ and } P) \qquad (7.57)$$

Standard states.

The proportionality factor f_i° is called the *standard-state fugacity* of component i. We see from (*7.57*) that when $x_i = 1$, $\hat{f}_i^{id} = f_i^\circ$, and therefore f_i° is interpreted as the fugacity of component i as a pure material at the temperature and pressure of interest. Thus the *standard-state* for component i in solution is a state of pure i, either the real state of pure i or a hypothetical one (as described below), at the T and P of the solution.

Equation (*7.57*) is used in two different ways: first, for approximating the fugacities of components in real solutions, and second, for generating values of $\hat{f}_i^{id}$ to serve as references to which actual values of $\hat{f}_i$ may be compared. In either case only two choices of the standard state have proved to be generally useful: one is based on the Lewis and Randall rule, and the other on Henry's law.

The lower dashed line of Fig. 7-7 represents the idealized behavior $\hat{f}_i^{id}$ versus x_i expected of a component i in solution that is ideal in the sense of the Lewis and Randall rule. In this case the standard-state fugacity f_i° (LR) is the fugacity f_i of pure i as it actually exists at the T and P of the solution and in the same physical state (e.g. liquid or gas) as the solution. This standard state is invariably used for components that are stable in the same physical state as the solution at the given T and P; for example, for mixtures of superheated gases or sub-cooled liquids. However, there is always a range of conditions of T and P for which the full curve of Fig. 7-7 for a given phase (liquid or gas) cannot be determined because the phase becomes unstable at higher values of x_i. In this event one must either extrapolate the curve to find f_i° (LR) or use a standard state based on Henry's law.

The upper dashed line of Fig. 7-7 represents ideal-solution behavior in the sense of Henry's law. The standard-state fugacity f_i° (HL) is equal to k_i, the Henry's law constant, and is the fugacity of pure i in a fictitious or hypothetical state at the temperature and pressure of the solution. Although f_i° (HL) is mathematically well-defined [see (*7.56*)], its determination requires actual data for the specific solution to which it applies. Unlike f_i° (LR), its value depends on the nature of the solution of which i is a component. Standard states based on Henry's law are used for species in liquid solutions that are of limited solubility and which do not exist pure in the same physical state as the solution, e.g. for gases or solids dissolved in liquids. However, in these cases standard states for the *solvent* are based on the Lewis and Randall rule.

A standard state for component i of a solution, as we have defined it, is a state of pure i, either real or hypothetical, at the temperature and pressure of the solution. Thus as temperature and pressure change, the standard state changes also, and so do the properties of the standard state. The only property of the standard state so far discussed is the fugacity f_i°, and this property is, of course, a function of T and P. Once a relationship between f_i°, T, and P is established, then the other thermodynamic properties of the standard state can be derived from the property relations for pure materials. This is possible because f is defined by (*7.10*) in direct relation to the Gibbs function G, and because T and P are the canonical variables for G (see Sec. 3.3). Thus, given f_i° as a function of T and P, one can calculate standard-state values for the volume V_i°, the enthalpy H_i°, the entropy S_i°, etc. (See Problem 7.5.)

Property changes of mixing.

In describing the properties of solutions, one often deals with *property changes of mixing*. These functions, designated collectively by ΔM, are defined as the difference between an actual molar (or specific) property of a solution M and the mole-fraction (or mass-fraction) average of the standard-state molar (or specific) properties M_i° of the components of the solution:

$$\Delta M = M - \sum x_i M_i^\circ \tag{7.58}$$

Thus ΔV is the *volume change of mixing*:

$$\Delta V = V - \sum x_i V_i^\circ$$

Similarly, ΔH is the *enthalpy change of mixing*:

$$\Delta H = H - \sum x_i H_i^\circ$$

Other property changes of mixing are defined analogously.

Example 7.6. A mass m_1 of pure component 1 is mixed with a mass m_2 of pure component 2 under conditions of constant pressure. Find an equation for the heat which must be added to maintain the solution at constant temperature. Assume that negligible stirring work is required to effect the mixing, and that the pure components are initially at the same temperature as that of the final solution.

If we take m_1 and m_2 together to be the system, then, by the statement of the problem, (6.7a) applies:

$$\sum_R \Delta(mH) = Q \tag{6.7a}$$

The process is depicted in Fig. 7-8.

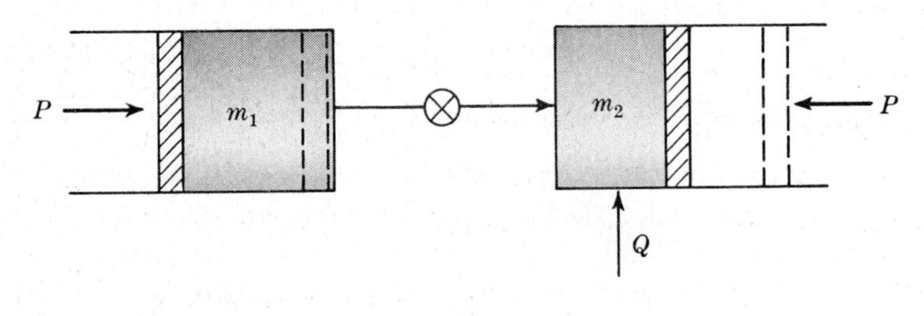

Fig. 7-8

For the left-hand cylinder,

$$\Delta(mH) = 0 - m_1 H_1$$

and for the right-hand cylinder

$$\Delta(mH) = (m_1 + m_2)H - m_2 H_2$$

where H is the specific enthalpy of the final mixture. But, by (7.58)

$$H = x_1 H_1 + x_2 H_2 + \Delta H$$

where ΔH is the specific enthalpy change of mixing based on standard states taken as the real pure components (in conformance with the Lewis and Randall rule), and x_1 and x_2 are the mass (or weight) fractions of 1 and 2 in the solution. Substitution of the last three equations into (6.7a) yields the simple result

$$Q = (m_1 + m_2)\,\Delta H = \Delta H^t \tag{7.59}$$

Thus the total heat added in the prescribed process is equal to the total enthalpy change of mixing. It is for this reason that ΔH is often called the *heat of mixing*. The type of process just described is realizable in the laboratory, and affords a sensitive method for the determination of the thermal properties of solutions. Negative values for ΔH indicate that the mixing process is *exothermic*, i.e. that heat is liberated; positive values represent *endothermic* mixing, for which heat is absorbed.

The *instantaneous* heat effect resulting from the addition of a component i to a solution is found from the differential form of (7.59):

$$\delta Q = d\,\Delta H^t = d(m\,\Delta H)$$

where $m = m_1 + m_2$. Since T and P are constant, the change in ΔH^t results solely from the addition of component i. Thus we can write for this process

$$d\,\Delta H^t = d(m\,\Delta H) \equiv \left[\frac{\partial(m\,\Delta H)}{\partial m_i}\right]_{T,P,m_j} dm_i$$

Since ΔH is a proper thermodynamic function, the partial derivative can be identified as $\overline{\Delta H}_i$, the partial specific enthalpy change of mixing of component i in solution. The equation for δQ then becomes

$$\delta Q = \overline{\Delta H}_i \, dm_i \tag{7.60}$$

from which

$$\overline{\Delta H}_i = \frac{\delta Q}{dm_i} \tag{7.61}$$

Equation (7.61) shows that $\overline{\Delta H}_i$ can be interpreted as the instantaneous heat effect produced by the addition of a unit mass of i to an infinite amount of solution. These quantities depend on the composition of the solution, as shown for example by Fig. 7-9 for the sulfuric acid–water system at 25(°C). The values were calculated from data for ΔH by the methods of Example 7.2. The two curves for $\overline{\Delta H}_{H_2SO_4}$ and $\overline{\Delta H}_{H_2O}$ are highly asymmetric, and therefore the instantaneous heat effects are quite different depending on the concentration of the solution and on which component is added. In particular, the value of $\overline{\Delta H}^{\infty}_{H_2O}$ is more than triple the value of $\overline{\Delta H}^{\infty}_{H_2SO_4}$. Addition of a small amount of water t a large amount of concentrated sulfuric acid therefore results in the liberation of substantially more hea than the addition of sulfuric acid to water. The water-to-acid heat effect is large enough to cause loc vaporization, with attendant sputtering. Addition of concentrated acid to water, on the other hand, accompanied by a much smaller heat effect which does not cause vaporization.

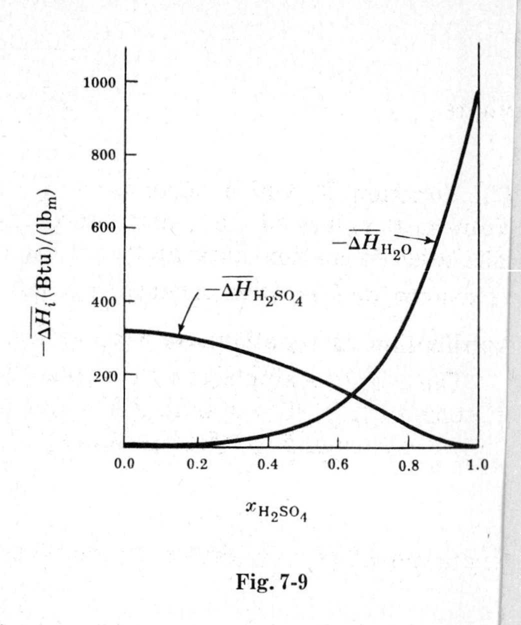

Fig. 7-9

Activity.

The choice of standard state for a component in solution is arbitrary (within the r striction to a state of pure i at the T and P of the solution) and numerical values for t M_i° and hence for ΔM, depend on the choices made. For any given choice, all proper changes of mixing are related to $\hat{f}_i$ and f_i°. We start the development of these relatio ships by using (7.3) to change (7.58) into an alternate form:

$$\Delta M = \sum x_i(\bar{M}_i - M_i^\circ) \tag{7.6}$$

Application of (7.62) to the Gibbs function gives:

$$\Delta G = \sum x_i(\bar{G}_i - G_i^\circ) \tag{7.62}$$

The terms $\bar{G}_i - G_i^\circ$ are replaced by application of (7.14) where

$$\Delta \mathcal{G} = \bar{G}_i - G_i^\circ \qquad f_{\text{final}} = \hat{f}_i \qquad f_{\text{initial}} = f_i^\circ$$

giving

$$\bar{G}_i - G_i^\circ = RT \ln(\hat{f}_i/f_i^\circ) \tag{7.}$$

Substituting (7.63) into (7.62a) and rearranging, we get an expression in dimensionl form for the Gibbs function change of mixing:

$$\frac{\Delta G}{RT} = \sum x_i \ln(\hat{f}_i/f_i^\circ) \tag{7.}$$

Similar expressions can be derived for other property changes of mixing. The m useful of these properties are the dimensionless functions $P\,\Delta V/RT$, $\Delta H/RT$ and $\Delta S/R$, a they are given by (see Problem 7.6)

$$\frac{P\,\Delta V}{RT} = \sum x_i \left[\frac{\partial \ln (\hat{f}_i/f_i^\circ)}{\partial \ln P}\right]_{T,x} \tag{7.65}$$

$$\frac{\Delta H}{RT} = -\sum x_i \left[\frac{\partial \ln (\hat{f}_i/f_i^\circ)}{\partial \ln T}\right]_{P,x} \tag{7.66}$$

$$\frac{\Delta S}{R} = -\sum x_i \ln (\hat{f}_i/f_i^\circ) - \sum x_i \left[\frac{\partial \ln (\hat{f}_i/f_i^\circ)}{\partial \ln T}\right]_{P,x} \tag{7.67}$$

Equations (*7.64*) through (*7.67*) illustrate explicitly how property changes of mixing are related to the standard-state fugacities and their temperature and pressure derivatives. It is apparent from these equations that the dimensionless ratio $\hat{f}_i/f_i^\circ$ plays an important part in solution thermodynamics; for this reason it is given a special name, the *activity*, and symbol $\hat{a}_i$:

$$\boxed{\hat{a}_i = \hat{f}_i/f_i^\circ} \tag{7.68}$$

If a component is in its standard state, $\hat{a}_i = 1$, because $\hat{a}_i^\circ = f_i^\circ/f_i^\circ = 1$. If a component i in solution behaves *ideally*, $\hat{a}_i = x_i$, because, from (*7.57*) and (*7.68*),

$$\hat{a}_i^{\text{id}} = \frac{\hat{f}_i^{\text{id}}}{f_i^\circ} = \frac{x_i f_i^\circ}{f_i^\circ} = x_i$$

As with M_i° and ΔM, the numerical value of an activity depends on the choice made for a standard state.

Example 7.7. Write (*7.64*) through (*7.67*) in terms of activities and find the corresponding expressions for ideal solutions.

Substitution of $\hat{a}_i$ for the ratio $\hat{f}_i/f_i^\circ$ gives directly

$$\frac{\Delta G}{RT} = \sum x_i \ln \hat{a}_i \tag{7.69}$$

$$\frac{P\Delta V}{RT} = \sum x_i \left[\frac{\partial(\ln \hat{a}_i)}{\partial(\ln P)}\right]_{T,x} \tag{7.70}$$

$$\frac{\Delta H}{RT} = -\sum x_i \left[\frac{\partial(\ln \hat{a}_i)}{\partial(\ln T)}\right]_{P,x} \tag{7.71}$$

$$\frac{\Delta S}{R} = -\sum x_i \ln \hat{a}_i - \sum x_i \left[\frac{\partial(\ln \hat{a}_i)}{\partial(\ln T)}\right]_{P,x} \tag{7.72}$$

For an ideal solution, $\hat{a}_i = x_i$ for all components at all T and P. The above equations thus become

$$\frac{\Delta G^{\text{id}}}{RT} = \sum x_i \ln x_i \tag{7.73}$$

$$\frac{P\,\Delta V^{\text{id}}}{RT} = 0 \tag{7.74}$$

$$\frac{\Delta H^{\text{id}}}{RT} = 0 \tag{7.75}$$

$$\frac{\Delta S^{\text{id}}}{R} = -\sum x_i \ln x_i \tag{7.76}$$

The second and third relationships follow from the fact that the derivatives are evaluated at constant composition; they state that the volume change of mixing and the enthalpy change of mixing are zero for an ideal solution. The functions ΔG^{id} and ΔS^{id} are *not* zero, however. Figure 7-10 is a graph of $\Delta G^{\text{id}}/RT$ and $\Delta S^{\text{id}}/R$ against x_1 for a binary solution. We note that $\Delta G^{\text{id}}/RT$, which is just the negative of $\Delta S^{\text{id}}/R$, is symmetrical about $x_1 = 0.5$. Moreover, ΔG^{id} is negative (and ΔS^{id} positive) for all compositions; this is true for an ideal solution containing any number of components.

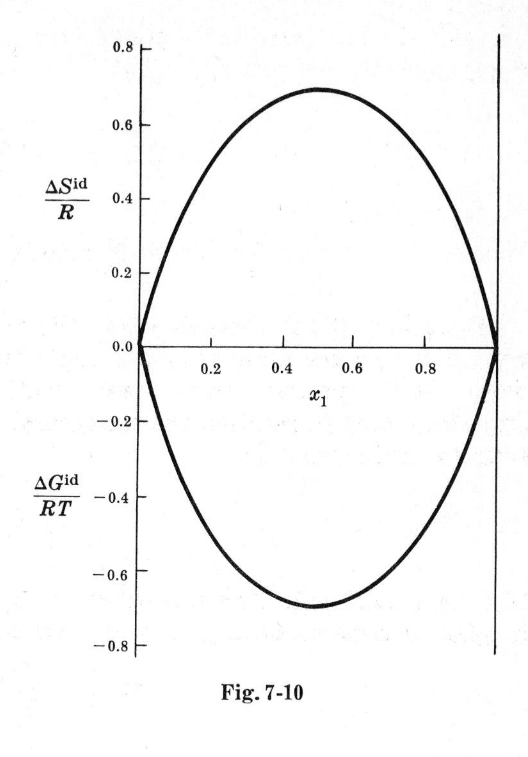

Fig. 7-10

7.5 EXCESS PROPERTIES. ACTIVITY COEFFICIENT

An *excess property* of a solution is defined as the difference between the actual mixture property and that which would obtain for an ideal solution at the same temperature, pressure, and composition. Thus we define

$$M^E = M - M^{\text{id}} \tag{7.77}$$

$$\Delta M^E = \Delta M - \Delta M^{\text{id}} \tag{7.78}$$

where M^E is the excess property and ΔM^E is the *excess property change of mixing*. Actually, M^E and ΔM^E are identical, as can be seen by substitution of (*7.58*), and its specialization for an ideal solution, into (*7.78*). For conciseness in notation, therefore, we shall usually employ the symbol M^E, rather than ΔM^E.

The numerical value of an excess property depends on the choices of standard states for all of the components. Like the corresponding ΔM, the M^E are proper thermodynamic functions, and a number of useful thermodynamic formulas relating various excess functions can be derived which are completely analogous to those presented earlier for properties of pure components or constant-composition solutions. We list a few of the more important equations here:

$$\frac{G^E}{RT} = \frac{H^E}{RT} - \frac{S^E}{R} \tag{7.79}$$

$$C_P^E = \left(\frac{\partial H^E}{\partial T}\right)_{P,x} = T\left(\frac{\partial S^E}{\partial T}\right)_{P,x} \tag{7.80}$$

$$V^E = \left(\frac{\partial G^E}{\partial P}\right)_{T,x} \tag{7.81}$$

$$S^E = -\left(\frac{\partial G^E}{\partial T}\right)_{P,x} \tag{7.82}$$

$$\frac{H^E}{RT} = -T\left[\frac{\partial(G^E/RT)}{\partial T}\right]_{P,x} \tag{7.83}$$

According to (*7.74*) and (*7.75*), $\Delta V^{\text{id}} = \Delta H^{\text{id}} = 0$, and so for these properties (and also for certain others) $M^E(= \Delta M^E) = \Delta M$. Thus, the excess volume is equal to the volume change of mixing, etc. However, by (*7.76*), $\Delta S^{\text{id}} \neq 0$; so for the entropy and for entropy-related functions (such as G and A) the excess property is *not* in general equal to the property change on mixing.

When M is taken to be G/RT, (7.78) becomes

$$\frac{\Delta G^E}{RT} = \frac{\Delta G}{RT} - \frac{\Delta G^{id}}{RT}$$

Combination of this equation with (7.64) and (7.73) gives:

$$\frac{G^E}{RT} = \frac{\Delta G^E}{RT} = \sum x_i \ln(\hat{f}_i/f_i^\circ) - \sum x_i \ln x_i$$

or

$$\frac{G^E}{RT} = \sum x_i \ln(\hat{f}_i/x_i f_i^\circ) \tag{7.84}$$

We find it convenient to employ a new function γ_i, called the *activity coefficient*, and defined as

$$\boxed{\gamma_i = \frac{\hat{a}_i}{x_i} = \frac{\hat{f}_i}{x_i f_i^\circ}} \tag{7.85}$$

Equation (7.84) can then be written

$$\frac{G^E}{RT} = \sum x_i \ln \gamma_i \tag{7.86}$$

Comparison of this equation with (7.69) suggests that similar expressions for the other excess functions analogous to (7.70), (7.71), and (7.72) can be obtained by replacing ΔM by M^E and $\hat{a}_i$ by γ_i. These are in fact rigorous, and are given as

$$\frac{PV^E}{RT} = \sum x_i \left[\frac{\partial(\ln \gamma_i)}{\partial(\ln P)}\right]_{T,x} \tag{7.87}$$

$$\frac{H^E}{RT} = -\sum x_i \left[\frac{\partial(\ln \gamma_i)}{\partial(\ln T)}\right]_{P,x} \tag{7.88}$$

$$\frac{S^E}{R} = -\sum x_i \ln \gamma_i - \sum x_i \left[\frac{\partial(\ln \gamma_i)}{\partial(\ln T)}\right]_{P,x} \tag{7.89}$$

For an ideal solution, by (7.57) and (7.85),

$$\gamma_i^{id} = \frac{\hat{f}_i^{id}}{x_i f_i^\circ} = \frac{x_i f_i^\circ}{x_i f_i^\circ} = 1$$

for every component at all T and P. Thus, by (7.86) through (7.89) all of the excess functions are zero for ideal solutions. This last conclusion of course also follows trivially from the definition (7.77) of an excess function.

Equation (7.86) suggests that $\ln \gamma_i$ is related to G^E/RT as a partial molar property, and this is readily demonstrated. Therefore

$$\ln \gamma_i = \left[\frac{\partial(nG^E/RT)}{\partial n_i}\right]_{T,P,n_j} \tag{7.90}$$

Excess functions provide the most sensitive numerical measure of solution properties in relation to the properties of the pure components in their standard states, and analytical representations of these functions find widespread use. Since temperature, pressure, and composition are the favored experimental variables, the most concise and convenient representation is through G^E as a function of T, P, and composition. Such a representation is *complete*, because T, P, and composition are the canonical variables for the Gibbs function. Of course, any use of excess functions requires knowledge of the standard-state properties, and these are readily determined once the f_i°-T-P relationships are established (see Problem 7.5).

Although classical thermodynamics imposes certain general requirements on equations used to represent G^E as a function of T, P, and x, it offers no clues as to the functional forms that might be reasonable for such equations, and they must therefore be established either empirically or on the basis of some molecular theory. Thus equations for G^E play a role in solution thermodynamics analogous to that of thermal equations of state in the thermodynamics of single-component PVT systems. If it happens that an equation of state with accurate mixing rules is known for a solution, then G^E can be calculated from the equation of state. As a matter of fact, in this event one might not deal with excess functions at all, because it is just as convenient to calculate directly the solution properties of interest. This is the approach usually adopted for mixtures of gases up to moderate pressures, where use of a truncated virial equation with its exact mixing rules is appropriate. For condensed phases equations of state with accurate mixing rules are rarely known, and recourse must be made to empirical equations for the excess functions.

Many expressions have been proposed to give the composition dependence of G^E. For binary solutions and standard states based on the Lewis and Randall rule, the simpler equations are usually special cases of one of the following:

$$\frac{G^E}{x_1x_2RT} = B + C(x_1-x_2) + D(x_1-x_2)^2 + \cdots \tag{7.91}$$

$$\frac{x_1x_2RT}{G^E} = B' + C'(x_1-x_2) + D'(x_1-x_2)^2 + \cdots \tag{7.92}$$

where the coefficients B, B', C, C', etc. (not to be confused with virial coefficients) are functions of T and P, but not of x. Equations (*7.91*) and (*7.92*) are merely power series expansions in either mole fraction x_1 or x_2. The particular symmetrical form in which the composition variable is displayed is chosen for convenience, and these equations may be put in other forms by use of the identities

$$x_1 - x_2 = 2x_1 - 1 = 1 - 2x_2$$

The choice between (*7.91*) and (*7.92*) and the degree of truncation depends on an empirical determination of which choice best represents real behavior in any particular case.

Example 7.8. The dimensionless excess Gibbs function for liquid mixtures of benzene and cyclohexane is well represented by (*7.91*) with $C = D = \cdots = 0$. Let us calculate and plot as a function of composition G^E/RT, H^E/RT, S^E/R, and the activity coefficients for this system at 40(°C) and 1(atm). Experimental values for B at 1(atm) are

T(°C)	B
35	0.479
40	0.458
45	0.439

Starting from

$$\frac{G^E}{RT} = Bx_1x_2 \tag{7.93}$$

we obtain by application of (7.83) to (7.93):

$$\frac{H^E}{RT} = -T\left[\frac{\partial(G^E/RT)}{\partial T}\right]_{P,x} = -Tx_1x_2\left(\frac{\partial B}{\partial T}\right)_P$$

or

$$\frac{H^E}{RT} = -x_1x_2\left[\frac{\partial B}{\partial(\ln T)}\right]_P \tag{7.94}$$

The corresponding equation for S^E/R is obtained by substitution of (7.93) and (7.94) into (7.79) and rearrangement:

$$\frac{S^E}{R} = -x_1x_2\left\{B + \left[\frac{\partial B}{\partial(\ln T)}\right]_P\right\} \tag{7.95}$$

Finally, application of (7.90) to (7.93) yields expressions for $\ln\gamma_1$ and $\ln\gamma_2$:

$$\ln\gamma_1 = \left[\frac{\partial(nG^E/RT)}{\partial n_1}\right]_{T,P,n_2} = n_2B\left[\frac{\partial(n_1/n)}{\partial n_1}\right]_{n_2} = n_2B\left(\frac{1}{n} - \frac{n_1}{n^2}\right)$$

or

$$\ln\gamma_1 = Bx_2^2 \tag{7.96a}$$

and similarly

$$\ln\gamma_2 = Bx_1^2 \tag{7.96b}$$

From the data, $B = 0.458$ at 40(°C). The temperature derivative at 40(°C) can be estimated from the values of B at 35(°C) and 45(°C):

$$\left[\frac{\partial B}{\partial(\ln T)}\right]_P \cong \frac{0.439 - 0.479}{\ln 318.2 - \ln 308.2} = -1.25$$

Substitution of these numbers into (7.93) through (7.96b) gives the following expressions

$$\frac{G^E}{RT} = 0.458\,x_1x_2 \qquad \frac{H^E}{RT} = 1.25\,x_1x_2 \qquad \frac{S^E}{R} = 0.792\,x_1x_2$$

$$\ln\gamma_1 = 0.458\,x_2^2 \qquad \ln\gamma_2 = 0.458\,x_1^2$$

which are plotted as functions of composition on Fig. 7-11.

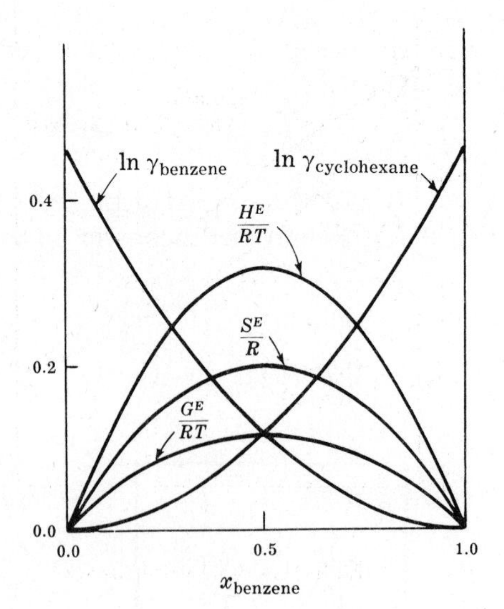

Fig. 7-11

7.6 PHASE DIAGRAMS FOR BINARY SYSTEMS

Quantitative methods for the calculation of phase equilibrium conditions can be formulated from concepts presented in Secs. 7.1 through 7.5. First, however, we show how the phase behavior of binary systems may be represented by diagrams.

For a nonreacting PVT system containing two chemical species ($m = 2$), the phase rule (*3.77*) becomes $F = 4 - \pi$, and the maximum number of independent intensive variables required to specify the thermodynamic state of a stable system is therefore *three*, corresponding to the case of a single equilibrium phase ($\pi = 1$). If the three intensive variables are chosen as P, T, and one of the mole fractions (or a weight fraction), then all equilibrium states of the system can be represented in three-dimensional P-T-composition space. Within this space, the states of *pairs* of phases, coexisting at equilibrium ($F = 4 - 2 = 2$), define *surfaces*; similarly, the states of three phases in equilibrium ($F = 4 - 3 = 1$) are represented as *space curves*.

Two-dimensional *phase diagrams* are obtained from intersections of the three-dimensional surfaces and curves with planes of constant pressure or constant temperature. By the former construction, one obtains a diagram having temperature and composition as its coordinates; this is called a Tx diagram. The second construction yields a diagram with the coordinates pressure and composition, called a Px diagram. The features of these diagrams are indicated by Figs. 7-12 through 7-20.

Vapor-liquid systems.

Figure 7-12 is a Tx diagram for the vapor-liquid equilibrium (VLE) of the cyclohexane-toluene system at 1(atm) total pressure, obtained from the intersection of the VLE surface with the $P = 1$(atm) plane. Figure 7-13 is a Px diagram for the same system at 90(°C), obtained from the intersection of the VLE surface with the $T = 90$(°C) plane. The "lens" shape of these figures is typical for systems containing components of similar chemical nature but having dissimilar vapor pressures.

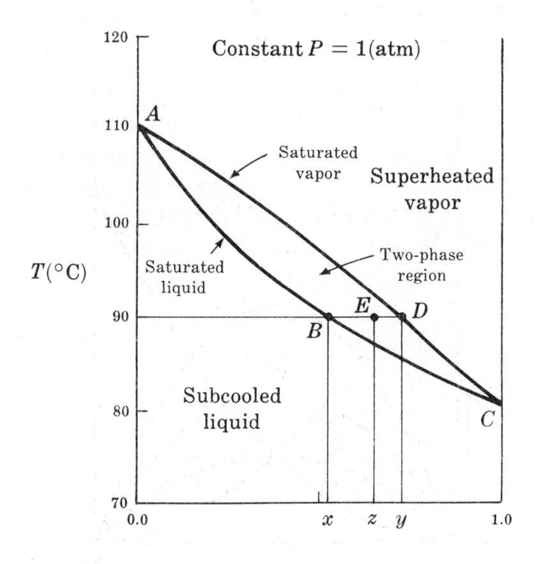

Fig. 7-12

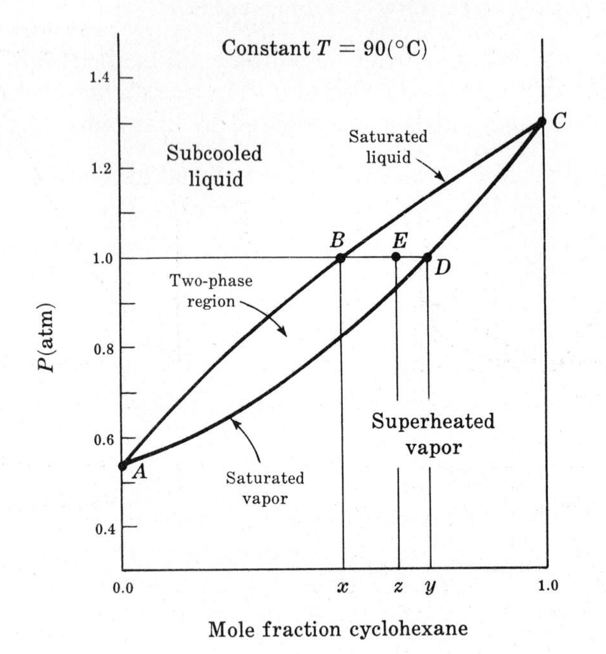

Fig. 7-13

Curve ABC in Figs. 7-12 and 7-13 represents the states of saturated liquid mixtures; it is called the *bubble-point* curve. Curve ADC, the *dew-point* curve, represents states of saturated vapor. The bubble- and dew-point curves converge to the pure-component saturation values on the T- and P-axes at the composition extremes $x = 0$ and $x = 1$. Thus curves ABC and ADC in Fig. 7-12 intersect the T-axes at 110.6(°C) and 80.7(°C), which are the boiling points at 1(atm) of toluene and cyclohexane. Similarly, curves ABC and ADC in Fig. 7-13 intersect the P-axes at 0.535(atm) and 1.310(atm), the vapor pressures of toluene and cyclohexane at 90(°C).

The region below ABC in Fig. 7-12, and above ABC in Fig. 7-13, corresponds to states of subcooled liquid; the region above ADC in Fig. 7-12, and below ADC in Fig. 7-13, is for superheated vapor. The area between ABC and ADC in both figures is the vapor-liquid two-phase region. Mixtures whose (T, x) or (P, x) coordinates fall within this area will spontaneously split into a liquid and a vapor phase. The equilibrium compositions of the phases formed in such a separation are determined from the phase diagrams by the intersections with the dew- and bubble-point curves of a horizontal straight line drawn through the point representing the overall state of the mixture; this construction derives from the requirement that T and P of coexisting phases be the same. For example, a mixture of mole fraction z of cyclohexane, when brought to a temperature of 90(°C) at 1(atm) total pressure (point E in Figs. 7-12 and 7-13), forms a liquid containing mole fraction x (point B) and a vapor of mole fraction y (point D). Straight lines such as BD which connect states in phase equilibrium are called *tie lines.*

Tie lines have a useful stoichiometric property, derived as follows. Let n be the total number of moles of a mixture having a mole fraction z_1 of component 1, which separates into n^l moles of liquid with mole fraction x_1 and n^v moles of vapor of mole fraction y_1. A mole number balance on component 1 gives

$$x_1 n^l + y_1 n^v = z_1 n$$

while an overall mole number balance yields

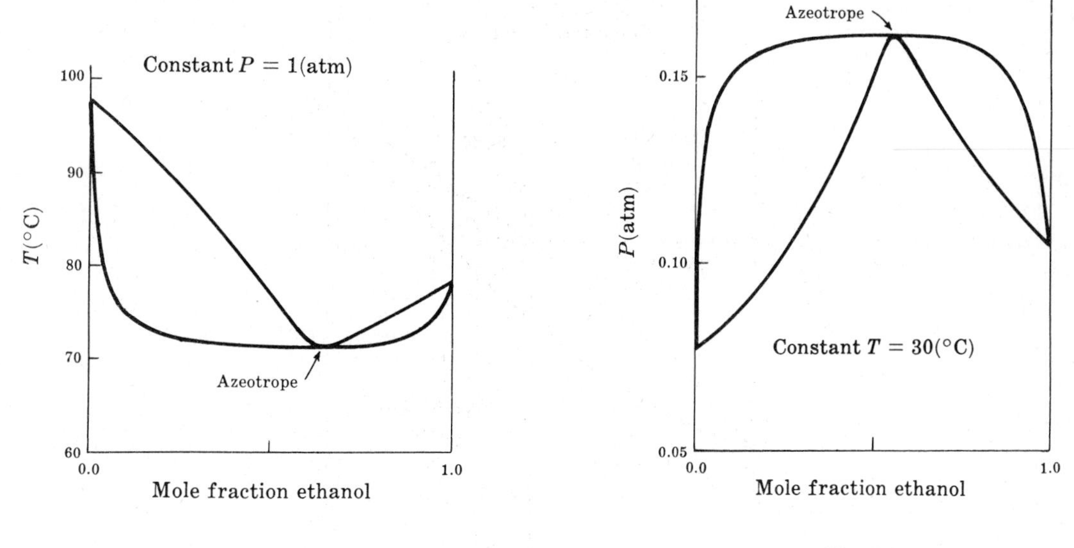

Fig. 7-14

Fig. 7-15

$$n^l + n^v = n$$

Solving these equations for the mole number ratio n^l/n^v, we obtain

$$\frac{n^l}{n^v} = \frac{y_1 - z_1}{z_1 - x_1}$$

But we see from Figs. 7-12 and 7-13 (letting component 1 be cyclohexane) that $y_1 - z_1$ and $z_1 - x_1$ are the lengths of segments ED and BE of the tie line BD. Thus the last equation admits the geometrical interpretation

$$\frac{n^l}{n^v} = \frac{ED}{BE}$$

This is a statement of the *lever principle,* which asserts that the ratio of the numbers of moles of the equilibrium phases is inversely proportional to the ratio of the lengths of the corresponding segments of the tie line. The lever principle also holds for mass units if the analogous quantities are substituted for mole numbers and mole fractions in the above equations.

A common variation on the type of VLE behavior shown in Figs. 7-12 and 7-13 is exemplified by the ethanol–n-heptane system. Figure 7-14 is a Tx diagram for this system at 1(atm), and Fig. 7-15 is a Px diagram for $T = 30(°C)$. The significant feature of these figures is the occurrence of a state of intermediate composition for which the equilibrium liquid and vapor compositions are identical. Such a state is called an *azeotrope.*

The azeotropic state of a binary system is special in that it possesses only a single degree of freedom, rather than two as required for normal two-component, two-phase equilibrium. Thus specification of any one of the coordinates T, P, or x_1 for a binary azeotrope fixes the other two, provided that the azeotrope actually exists. Binary azeotropes are therefore similar to the saturation states of pure components.

An important characteristic of the azeotropic state, apparent in Figs. 7-14 and 7-15, is the occurrence of a minimum or a maximum on the Tx and Px diagrams at the azeotropic composition. The maximum or minimum occurs on both the bubble-point and dew-point curves, and satisfies the appropriate pair of equations:

$$\left(\frac{\partial T}{\partial x_1}\right)_{P,\,\mathrm{az}} = \left(\frac{\partial T}{\partial y_1}\right)_{P,\,\mathrm{az}} = 0$$

or

$$\left(\frac{\partial P}{\partial x_1}\right)_{T,\,\mathrm{az}} = \left(\frac{\partial P}{\partial y_1}\right)_{T,\,\mathrm{az}} = 0$$

Although the ethanol–n-heptane system has a maximum-pressure (minimum-temperature) azeotrope, minimum-pressure (maximum-temperature) azeotropes are common. Azeotropes also may occur in systems containing more than two components.

Example 7.9. A 50 mole % liquid mixture of cyclohexane in toluene is confined in a vertical piston-and-cylinder apparatus at 85(°C) and 1(atm) pressure. The temperature of the mixture is increased to 105(°C) by addition of heat. Show on the appropriate phase diagram the physical states assumed by the system during the process.

If the piston can be considered frictionless, and if the heating is done slowly enough, then the process occurs at a constant uniform pressure of 1(atm), and all states assumed by the system are equilibrium states. The appropriate phase diagram is the Tx diagram of Fig. 7-12, redrawn for this example as Fig. 7-16.

At 85(°C) and 1(atm), the state of the mixture is subcooled liquid, shown as point a on the diagram. The system is closed, and the overall composition remains constant during the process. Therefore the states of the system *as a whole* fall on a vertical line passing through a. When the temperature reaches the bubble point of 90.6(°C) [point b], the first bubble of vapor appears. This vapor, represented by point b' with mole fraction $y_{b'} = 0.70$, is richer in cyclohexane than the original mixture. As heating is continued, the amount of vapor increases and the amount of liquid decreases, with the states of the two phases following the paths $b'c'$ and bc, respectively. Finally, at the dew point of 97.1(°C) [point c'], the last drop (dew) of liquid with composition $y_c = 0.29$ disappears. The system now contains only vapor of the original 50 mole % composition, and further heating to the final state of 105(°C) and 1(atm) [point d] is through the superheated vapor region.

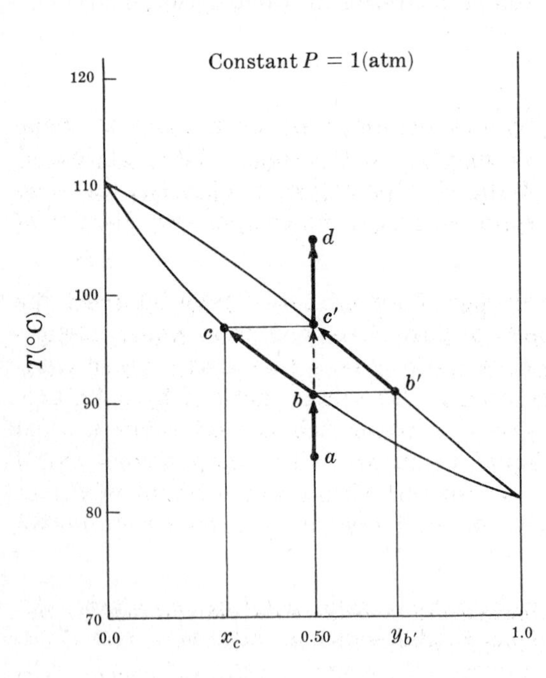

Fig. 7-16

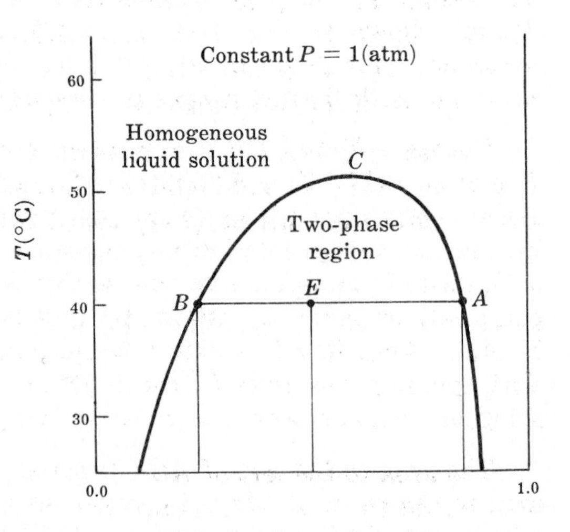

Fig. 7-17

Liquid-liquid systems.

Pairs of components are often incompletely miscible in the liquid state over certain ranges of P, T, and composition. Since the behavior of condensed phases is often quite insensitive to pressure, the important variables are T and composition. A Tx diagram of the liquid-liquid equilibria for the partially miscible system methanol–n-heptane is shown in Fig. 7-17. The data were taken at atmospheric pressure.

The liquid-liquid two-phase region is delineated by the dome-shaped curve, which extends to lower temperatures than are shown in the figure. Mixtures whose (T, x) coordinates fall outside of the dome (but not inside any other two-phase region) exist as single homogeneous liquid phases at 1(atm). A mixture whose (T, x) coordinates are inside the dome cannot exist as a single stable phase, but will separate into two liquid phases of different compositions. Thus, as shown on Fig. 7-17, a 50 mole % mixture of methanol in n-heptane, when brought to 40(°C) at 1(atm) [point E], will split into a n-heptane-rich phase (point B) containing 24 mole % methanol, and a methanol-rich phase (point A) containing 85 mole % methanol. The ratio of the total numbers of moles in the two phases at these conditions is then given by the lever principle as

$$\frac{n(\text{methanol-rich})}{n(\text{heptane-rich})} = \frac{0.50 - 0.24}{0.85 - 0.50} = 0.743$$

Liquid-liquid Tx diagrams are sometimes called *mutual solubility curves*, and the compositions of the coexisting phases are commonly reported as *solubilities*. Thus, at 40(°C) and 1(atm), the solubility of methanol in n-heptane is 24 mole % and the solubility of n-heptane in methanol is $100 - 85 = 15$ mole %.

Point C on Fig. 7-17 represents the highest temperature [51.2(°C)] for which the two liquid phases can coexist in equilibrium at 1(atm) pressure. It is called an upper *consolute temperature*, or upper *critical-solution temperature*. Liquid systems may have an upper consolute temperature, a lower consolute temperature, or in some cases both.

Liquid-solid and solid-solid systems.

The greatest variety of binary phase behavior is exhibited by such systems. The liquid-solid Tx diagram (composition in *weight %* copper) for the copper-silver system at 1(atm), shown in Fig. 7-18, page 262, is one of the simpler diagrams obtained for such systems. This diagram is typical for binary systems in which the components form *solid solutions* with limited ranges of composition.

Curves AE and EB are *freezing curves* for copper-silver mixtures; they intersect the T-axes at 960.5(°C) and 1083(°C), the melting points of pure silver and pure copper, respectively. Mixtures whose (T, x) coordinates lie above these curves (but below other equilibrium curves) are stable homogeneous liquid solutions. The area enclosed by $AECA$ is a liquid-solid two-phase region, within which crystals of silver-rich α solid solution with compositions given by curve AC coexist with liquid mixtures with compositions given by AE. Area $BEDB$ is also a liquid-solid two-phase region, within which liquid mixtures with compositions given by curve BE are in equilibrium with crystals of copper-rich β solid solution with compositions given by BD.

The area to the left of ACF is a region of homogeneous α solid solution; similarly, the area to the right of BDG is a region of homogeneous β solid solution. Mixtures with (T, x) coordinates falling in the area $FCEDGF$ will separate into α and β solid solutions, with compositions given by the curves CF and DG, respectively.

There is a single temperature at which a liquid solution is in equilibrium with both α and β solid solutions; this temperature is defined by the intersection of curves AE and EB.

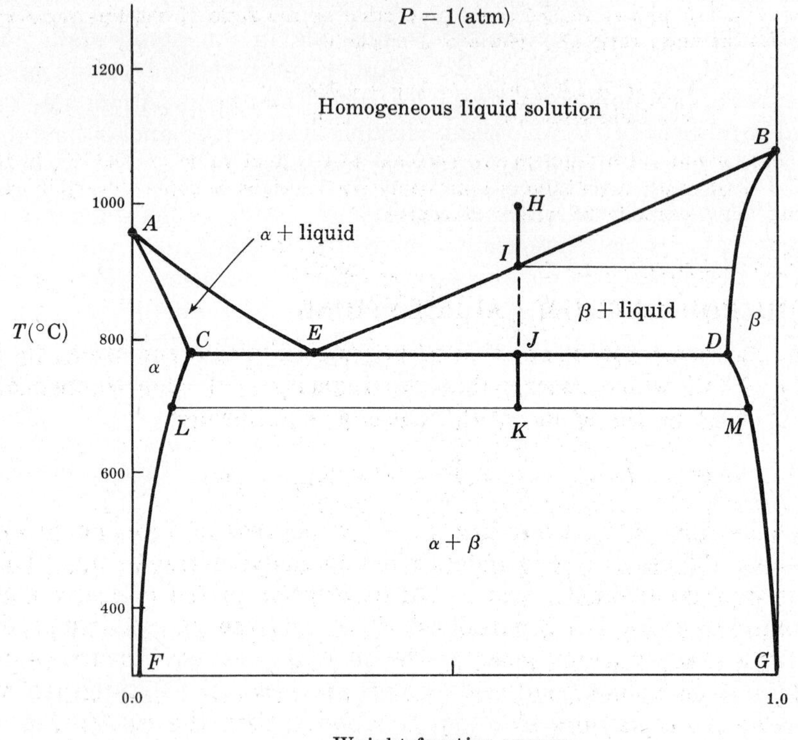

Fig. 7-18

The point of intersection at E is called the *eutectic* point. For this system, the eutectic occurs at 779(°C) and 28.5 weight % copper; at the eutectic, the equilibrium α and β solid solutions contain 8.8 and 92.0 weight % copper, respectively. Curves ACE and BDE are the *melting curves,* and thus the eutectic temperature is the lowest melting point for systems of this type. Except for mixtures of eutectic composition, melting points and freezing points are in general different for binary systems. Significantly, binary systems can have mixture freezing points and melting points which are considerably lower than those of the constituent pure components. Note that points A and B represent *both* the freezing points and the melting points of the pure constituents.

Example 7.10. A 60 weight % mixture of copper in silver is cooled from 1000(°C) to 700(°C). Describe the physical states assumed by the mixture.

The path of the cooling process is shown by the vertical line $HIJK$ in Fig. 7-18. The initial state of the mixture (point H) is homogeneous liquid solution. When the temperature reaches approximately 910(°C) [point I], crystals of β solid solution containing about 93 weight % copper freeze out. On further cooling, the amount of solid phase increases relative to that of liquid, while the equilibrium liquid and β solid solution both become richer in silver. By the time the eutectic temperature of 779(°C) is reached, the silver composition in the liquid (point E) has increased to $100 - 28.5 = 71.5$ weight %, and the relative mass of solid solution to liquid solution has become

$$\frac{m(\beta \text{ solid solution})}{m(\text{eutectic solution})} = \frac{0.60 - 0.285}{0.92 - 0.60} = 0.98$$

At exactly 779(°C) crystals of α solid solution intermixed with crystals of β solid solution begin to freeze out, and further extraction of heat from the system results in no temperature change. During this isothermal process the remaining liquid solution (point E), having the eutectic composition of 28.5 weight % copper, freezes into α and β phases in the same proportion as the ratio of the line segments ED/CE. Once freezing is complete the mass ratio of β phase to α phase is

$$\frac{m(\beta \text{ solid solution})}{m(\alpha \text{ solid solution})} = \frac{0.60 - 0.088}{0.92 - 0.60} = 1.6$$

Further extraction of heat results in a temperature decrease to the final value of 700(°C), in which state the system consists of α solid solution containing approximately 6 weight % copper in equilibrium with β solid solution containing approximately 95 weight % copper.

7.7 VAPOR-LIQUID EQUILIBRIUM CALCULATIONS

The thermodynamic basis for calculation of phase equilibrium conditions in PVT systems is provided by (*7.46*), which asserts that the fugacity $\hat{f}_i$ of each component i in an m-component system must be the same in all phases at equilibrium:

$$\hat{f}_i^\alpha = \hat{f}_i^\beta = \cdots = \hat{f}_i^\pi \qquad (i = 1, 2, \ldots, m) \tag{7.46}$$

According to the phase rule (*3.77*), there are $2 - \pi + m$ degrees of freedom in an m-component, π-phase system. Thus, $2 - \pi + m$ independent intensive variables must be specified just to render the system determinate. An actual *description* of the intensive state of the system, however, requires $\pi(m-1)+2$ variables: T, P, and the $(m-1)$ independent mole fractions in each of the π equilibrium phases. The general phase equilibrium problem may be posed as follows: Given values for $2 - \pi + m$ of the $\pi(m-1)+2$ intensive variables, what values (if any) of the remaining $m(\pi - 1)$ variables satisfy the equilibrium criterion (*7.46*)?

VLE equations. Raoult's law.

If equations of state are available which accurately describe the volumetric properties of each equilibrium phase for every anticipated T, P, and composition, then solution of the problem is possible by direct application of (*7.46*). Unfortunately, the required equations of state are *not* usually available, particularly for condensed phases (liquids, solids), and this "direct" method finds only limited use. Vapor-liquid equilibrium (VLE) calculations at low and moderate pressures are usually done by use of an alternative approach, described below.

For equilibrium between a single vapor phase and a single liquid phase, (*7.46*) gives

$$\hat{f}_i^v = \hat{f}_i^l \qquad (i = 1, 2, \ldots, m) \tag{7.97}$$

The specialization of (*7.12*) for a component in a vapor mixture yields $\hat{f}_i^v = y_i \hat{\phi}_i P$, and an expression for $\hat{f}_i^l$ is found from application of (*7.85*) to component i in the liquid phase: $\hat{f}_i^l = x_i \gamma_i f_i^\circ$. Substitution in (*7.97*) then gives

$$\boxed{y_i \hat{\phi}_i P = x_i \gamma_i f_i^\circ} \qquad (i = 1, 2, \ldots, m) \tag{7.98}$$

Equation (*7.98*) is a perfectly rigorous alternative equilibrium criterion to (*7.97*). In applying it, we assume the availability of a vapor-phase equation of state for calculation of $\hat{\phi}_i$ (see, e.g., Example 7.4), and a G^E expression for the liquid phase for calculation of γ_i (see, e.g., Example 7.8). In addition, we require values for the standard-state fugacities f_i° upon which G^E is based. Note that we have not affixed v and l superscripts to the quantities in (*7.98*), for it will be understood that $\hat{\phi}_i$ refers to the vapor phase and that γ_i and f_i° are liquid-phase properties.

If the G^E expression is based on Lewis and Randall standard states, it proves convenient to deal with a more explicit expression for f_i°, derived as follows. Since f_i° is the fugacity of pure i at the system T and P, we can write

$$f_i^\circ = f_i(P) \equiv P_i^{\text{sat}} \times \frac{f_i^{\text{sat}}}{P_i^{\text{sat}}} \times \frac{f_i(P)}{f_i^{\text{sat}}}$$

where all quantities are evaluated at the system temperature T. Here P_i^{sat} is the saturation (vapor) pressure of pure i, and f_i^{sat} is the fugacity of pure i at P_i^{sat}. As demonstrated in Example 7.5, the first ratio on the right-hand side of the above equation is calculable from the vapor-phase equation of state for pure i; similarly, the second ratio may be evaluated from knowledge of the volumetric behavior of pure liquid i. Thus we obtain from (*7.21*)

$$\frac{f_i^{\text{sat}}}{P_i^{\text{sat}}} = \exp\left[\int_0^{P_i^{\text{sat}}} (Z_i - 1)\frac{dP}{P}\right] \qquad \text{(const. } T\text{)}$$

and from (*7.50*)

$$\frac{f_i(P)}{f_i^{\text{sat}}} = \exp\left[\frac{1}{RT}\int_{P_i^{\text{sat}}}^{P} V_i\,dP\right] \qquad \text{(const. } T\text{)}$$

where Z_i is the vapor-phase compressibility factor and V_i is the liquid-phase molar volume. Substituting the resulting expression for f_i° into (*7.98*), we obtain

$$y_i\hat{\phi}_i P = x_i\gamma_i P_i^{\text{sat}} \exp\left[\int_0^{P_i^{\text{sat}}} (Z_i - 1)\frac{dP}{P} + \frac{1}{RT}\int_{P_i^{\text{sat}}}^{P} V_i\,dP\right] \qquad (i = 1, 2, \ldots, m) \tag{7.99}$$

Apart from the restriction to standard states based on the Lewis and Randall rule, (*7.99*) is as valid as the more abstract equilibrium criterion (equality of the chemical potentials μ_i) upon which it is ultimately based. The great advantage of the present equation is that it displays some of the thermodynamic variables in a more transparent (although still partially implicit) form, and serves as a suitable point of departure from which one can make reasonable simplifications. The most common of these simplifications are listed below.

(*a*) The liquid molar volume is independent of pressure, and equal to the saturated liquid volume V_i^{sat}:

$$\frac{1}{RT}\int_{P_i^{\text{sat}}}^{P} V_i\,dP \cong \frac{V_i^{\text{sat}}}{RT}(P - P_i^{\text{sat}}) \tag{7.100a}$$

(*b*) The liquid molar volume is negligible, or $P \cong P_i^{\text{sat}}$:

$$\frac{1}{RT}\int_{P_i^{\text{sat}}}^{P} V_i\,dP \cong 0 \tag{7.100b}$$

(*c*) The vapor phase is an ideal solution:

$$\hat{\phi}_i = \phi_i \tag{7.100c}$$

(*d*) The vapor phase is a mixture of ideal gases ($Z_i = 1$):

$$\left.\begin{aligned} \int_0^{P_i^{\text{sat}}} (Z_i - 1)\frac{dP}{P} &= 0 \\ \hat{\phi}_i &= 1 \end{aligned}\right\} \tag{7.100d}$$

(*e*) The liquid phase is an ideal solution:

$$\gamma_i = 1 \tag{7.100e}$$

Depending on the chemical nature of the system, and on the temperature and pressure levels, one or more of the above approximations may be applicable to a given problem. Equations (*7.100b*) and (*7.100d*) provide useful first approximations when the pressure is low, and (*7.100a*) and (*7.100c*) are often suitable to moderate pressure levels. The fifth approximation, (*7.100e*), is not generally satisfactory, because the majority of mixtures of real liquids do not behave as ideal solutions. The most widely used simplifications to (*7.99*), resulting from use of (*7.100b*) and (*7.100d*), and of (*7.100b*), (*7.100d*), and (*7.100e*), respectively, are

$$\boxed{y_i P = x_i \gamma_i P_i^{\text{sat}}} \qquad (i = 1, 2, \ldots, m) \qquad (7.101)$$

$$\boxed{y_i P = x_i P_i^{\text{sat}}} \qquad (i = 1, 2, \ldots, m) \qquad (7.102)$$

Equation (*7.102*) is the statement of *Raoult's law*; it represents the simplest possible vapor-liquid equilibrium behavior.

Example 7.11. VLE in the benzene-toluene system is well represented by Raoult's law at low and moderate pressures. Construct the Px diagram at 90(°C) and the Tx diagram at 1(atm) total pressure for this system. Vapor-pressure data [in (atm)] for benzene(1) and toluene(2) are:

T(°C)	P_1^{sat}	P_2^{sat}	T(°C)	P_1^{sat}	P_2^{sat}
80.1	1.000	0.384	98	1.683	0.689
84	1.126	0.439	100	1.777	0.732
88	1.268	0.501	104	1.978	0.825
90	1.343	0.535	108	2.196	0.928
94	1.506	0.608	110.6	2.347	1.000

Application of Raoult's law (*7.102*) to each component gives

$$y_1 P = x_1 P_1^{\text{sat}} \qquad (7.103a)$$

$$y_2 P = x_2 P_2^{\text{sat}} \qquad (7.103b)$$

The mole fractions in each phase sum to unity; the vapor-phase mole fractions may therefore be eliminated by addition of the two equations, giving

$$P = P_2^{\text{sat}} + x_1(P_1^{\text{sat}} - P_2^{\text{sat}}) \qquad (7.104)$$

Substitution of (*7.104*) in (*7.103a*) and solution for y_1 yields

$$y_1 = \frac{x_1 P_1^{\text{sat}}}{P_2^{\text{sat}} + x_1(P_1^{\text{sat}} - P_2^{\text{sat}})} \qquad (7.105)$$

Equations (*7.104*) and (*7.105*) are equations for the bubble-point and dew-point curves, respectively, written in forms convenient for construction of the Px diagram. Since the temperature is fixed for this type of phase diagram, the pure-component vapor pressures are constant over the entire composition range. Thus the bubble-point pressure P and the corresponding equilibrium value of y_1 at the dew point are given directly as functions of x_1.

Pressure, not temperature, is the fixed coordinate for a Tx diagram, and solution of (*7.104*) for the equilibrium temperatures at specified x_1 is inconvenient. The bubble-point curve is most readily constructed by solving (*7.104*) for the bubble-point compositions x_1 at representative values of T between the saturation temperatures of the pure components:

$$x_1 = \frac{P - P_2^{\text{sat}}}{P_1^{\text{sat}} - P_2^{\text{sat}}} \qquad (7.106)$$

The corresponding equilibrium compositions on the dew-point curve are obtained by substitution of (*7.106*) in (*7.103a*) and solution for y_1:

$$y_1 = \left(\frac{P_1^{\text{sat}}}{P}\right)\left(\frac{P - P_2^{\text{sat}}}{P_1^{\text{sat}} - P_2^{\text{sat}}}\right) \qquad (7.107)$$

We now illustrate the application of the above equations for the benzene(1)-toluene(2) system. For the Px diagram, we calculate the P and equilibrium y_1 for $x_1 = 0.20$ at 90(°C). For 90(°C), we are given

$$P_1^{\text{sat}} = 1.343(\text{atm}) \qquad P_2^{\text{sat}} = 0.535(\text{atm})$$

Then, by (*7.104*) and (*7.105*),

$$P = 0.535 + 0.20(1.343 - 0.535) = 0.697(\text{atm})$$

$$y_1 = \frac{0.20 \times 1.343}{0.697} = 0.385$$

Thus, at 90(°C) and 0.697(atm), liquid containing 20 mole % benzene is in equilibrium with vapor containing 38.5 mole % benzene. These compositions are shown as points B and D on Fig. 7-19, which is the complete Px diagram constructed by application of (*7.104*) and (*7.105*) to the entire liquid composition range ($x_1 = 0$ to $x_1 = 1$).

For the Tx diagram, we calculate the equilibrium values of x_1 and y_1 for $T = 100$(°C) at 1(atm) total pressure. At 100(°C), from the data,

$$P_1^{\text{sat}} = 1.777(\text{atm}) \qquad P_2^{\text{sat}} = 0.732(\text{atm})$$

Equations (*7.106*) and (*7.107*) then yield

$$x_1 = \frac{1 - 0.732}{1.777 - 0.732} = 0.256$$

$$y_1 = \left(\frac{1.777}{1}\right) \times 0.256 = 0.455$$

These compositions are shown as points B' and D' on the complete Tx diagram of Fig. 7-20, obtained by application of (*7.106*) and (*7.107*) to the entire range of saturation temperatures [80.1(°C) to 110.6(°C)].

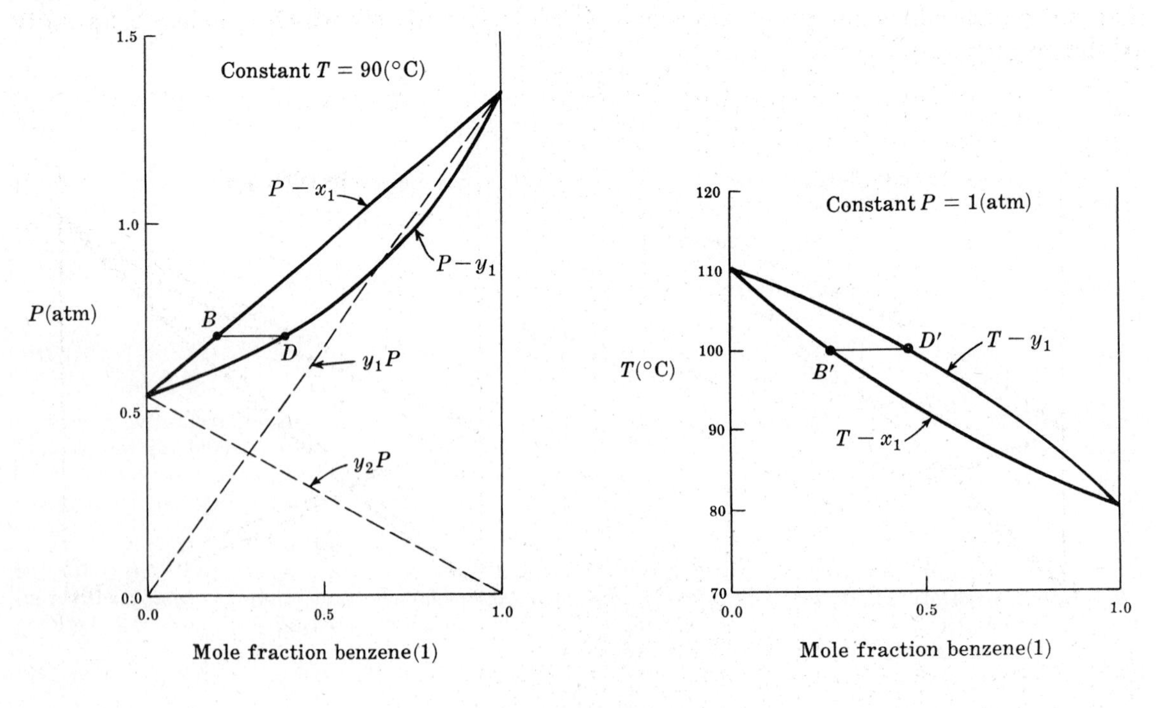

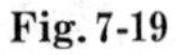
Fig. 7-19

Fig. 7-20

Deviations from Raoult's law.

A significant result of Example 7.11 is that the bubble-point curve of a system which obeys Raoult's law is linear on the Px diagram (Fig. 7-19). Similarly, the *partial pressures* y_1P and y_2P of the two components are proportional to x_1 and x_2, respectively, as shown by the two dashed lines in Fig. 7-19. These linear relationships resulting from Raoult's law provide a convenient basis for classification of VLE behavior. Systems for which the bubble-point and partial-pressure curves lie above the Raoult's-law lines, as in Fig. 7-21, are said to exhibit *positive deviations* from Raoult's law. Similarly, if these curves fall below the Raoult's-law lines, as in Fig. 7-22, the systems show *negative deviations* from Raoult's law.

The types of effects which promote deviations from Raoult's law can be determined by comparison of a rigorous expression for partial pressure with the partial-pressure equation of Raoult's law. Designating the partial pressure y_iP by the symbol P_i we can write Raoult's law (*7.102*) as

$$P_i\,(\text{RL}) = x_i P_i^{\text{sat}}$$

where (RL) signifies Raoult's law. Combination of this equation with the exact expression (*7.99*) then gives, on rearrangement,

$$\frac{P_i - P_i\,(\text{RL})}{P_i\,(\text{RL})} = \left\{\frac{\gamma_i}{\hat{\phi}_i}\exp\left[\int_0^{P_i^{\text{sat}}}(Z_i - 1)\frac{dP}{P} + \frac{1}{RT}\int_{P_i^{\text{sat}}}^{P} V_i\,dP\right]\right\} - 1$$

The sign of the deviation $P_i - P_i\,(\text{RL})$ is thus determined by the magnitude of the term in braces; if this term is greater than unity, positive deviations result, while negative deviations obtain if the term is less than unity.

It is clear from the last equation that both liquid-phase and vapor-phase behavior contribute to deviations from Raoult's law, the former through γ_i and the integral of V_i, the latter through $\hat{\phi}_i$ and the integral containing Z_i. At low pressures, however, the dominant effect is usually that due to γ_i. Thus, systems with liquid-phase activity coefficients greater than unity generally show positive deviations, while systems for which γ_i is less than unity exhibit negative deviations.

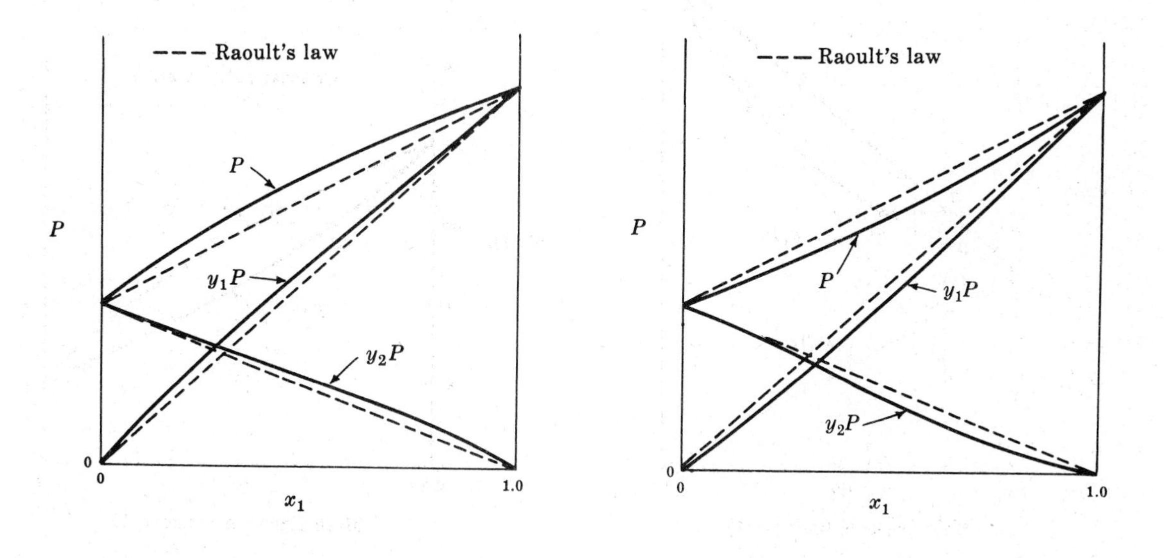

Fig. 7-21 Fig. 7-22

Significant deviations from Raoult's law are often manifested by azeotrope formation. For systems in which the pure components have significantly different vapor pressures (or boiling points), the occurrence of azeotropes generally implies very large (for maximum-pressure azeotropes) or very small (for minimum-pressure azeotropes) values of the liquid-phase activity coefficients. Systems with similar pure-component vapor pressures, however, can exhibit azeotropes with only moderately nonideal liquid-solution behavior. This is illustrated in Example 7.12.

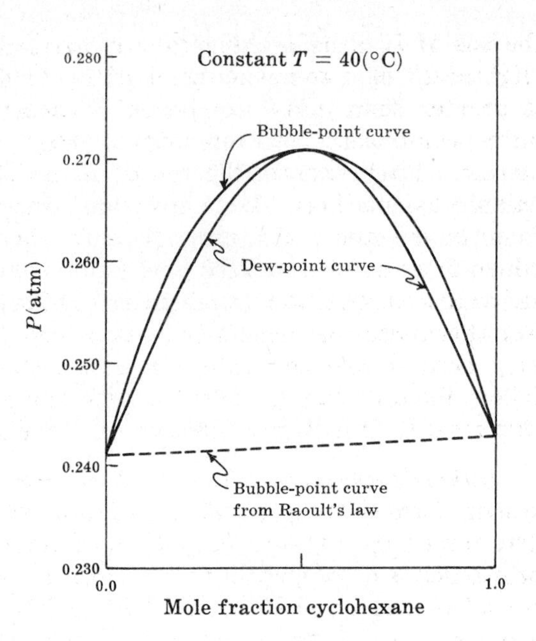

Fig. 7-23

Example 7.12. Construct the Px diagram for the cyclohexane(1)-benzene(2) system at 40(°C). Use (*7.101*) and the activity coefficients derived in Example 7.8. At 40(°C), $P_1^{sat} = 0.243$(atm) and $P_2^{sat} = 0.241$(atm).

Application of (*7.101*) to each component gives

$$y_1 P = x_1 \gamma_1 P_1^{sat} \qquad (7.108a)$$

$$y_2 P = x_2 \gamma_2 P_2^{sat} \qquad (7.108b)$$

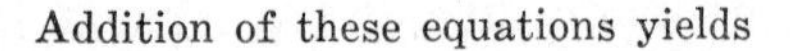

Addition of these equations yields

$$P = \gamma_2 P_2^{sat} + x_1(\gamma_1 P_1^{sat} - \gamma_2 P_2^{sat}) \qquad (7.109)$$

The activity coefficients for this system are given by the symmetrical expressions (*7.96a*) and (*7.96b*). Substitution of these equations into (*7.109*) gives

$$P = P_2^{sat} \exp(Bx_1^2) + x_1\{P_1^{sat} \exp[B(1-x_1)^2] - P_2^{sat} \exp(Bx_1^2)\} \qquad (7.110)$$

Equation (*7.110*) is the equation of the bubble-point curve. Corresponding equilibrium points on the dew-point curve are found by substituting (*7.110*) into (*7.108a*) and solving for y_1, noting that

$$\gamma_1 = \exp[B(1-x_1)^2]$$

The Px diagram, calculated from these equations with the given vapor pressures and with $B = 0.458$, is shown on Fig. 7-23. It is assumed in these calculations that B is independent of pressure, at least over the small range of pressures encountered. Because the benzene and cyclohexane vapor pressures are nearly identical, the effect of the activity coefficients is to produce a maximum-pressure azeotrope; if the liquid solutions were ideal, the bubble-point curve would follow Raoult's law as indicated by the dashed line.

K-values and flash calculations.

The primary application of phase equilibrium relationships is in the design of separation processes which depend on the tendency of given chemical species to distribute themselves preferentially in one or another equilibrium phase. A convenient measure of this tendency with respect to VLE is the equilibrium ratio y_i/x_i, which can be expressed as a function of thermodynamic variables as a result of (*7.98*):

$$K_i = \frac{\gamma_i f_i^\circ}{\hat{\phi}_i P} \qquad (7.111)$$

where

$$K_i \equiv \frac{y_i}{x_i} \qquad (7.112)$$

The use of K_i to represent the ratio y_i/x_i is so common that this quantity is usually called an equilibrium *K-value*. Another name is the vapor-liquid *distribution coefficient*. Although the use of K_i adds nothing to our thermodynamic knowledge of VLE, it is a measure of "lightness" of a component, i.e. of its tendency to concentrate in the vapor phase. If K_i is greater than unity, component i concentrates in the vapor phase. If K_i is less than unity, component i concentrates in the liquid phase, and is regarded as a "heavy" component. Furthermore, the use of K_i provides for computational convenience in material balance calculations, allowing formal elimination of one of the sets of variables x_i or y_i in favor of the other. However, K_i as a thermodynamic variable is related by (*7.111*) to γ_i, which is a function of T, P, and liquid composition, and to $\hat{\phi}_i$, which is a function of T, P, and vapor composition, and is therefore itself a complex function of T, P, the x_i, and the y_i. For this reason no generally satisfactory direct correlation of K_i with these variables has ever been developed. The simplest behavior displayed by K-values is for systems that follow Raoult's law (*7.102*); for such systems (*7.111*) becomes $K_i(\text{RL}) = P_i^{\text{sat}}/P$. Even for this case, K_i is a strong function of T and P.

Duhem's theorem (Problem 3.18) asserts that the equilibrium state of a closed PVT system formed from specified initial amounts of prescribed chemical species is completely determined by any *two* properties of the system taken as a whole, provided that these two properties are independently variable at equilibrium. Since T and P are uniform throughout all phases at equilibrium, they qualify as such properties for any system of more than one component. According to Duhem's theorem, then, one can in principle calculate the compositions of equilibrium phases at a specified T and P if one knows the overall compositions $z_1, z_2, \ldots, z_m$ of the m components. This type of computation when done for a VLE problem is called a *flash calculation*.

The term derives from the fact that a liquid mixture at a pressure above its bubble-point pressure will "flash" or partially evaporate if the pressure is lowered to a value between the bubble-point and dew-point pressures. Such a process can be carried out continuously if a liquid is throttled through an orifice into a tank maintained at an appropriate pressure. The liquid and vapor phases formed in the flash tank are equilibrium phases existing at a particular T and P, depending on the process.

The x_i and y_i values which result from a flash calculation must of course satisfy the equilibrium criterion as expressed by (*7.111*) and (*7.112*); in addition, they must satisfy certain material balance requirements, derived as follows. At the specified T and P, one mole of mixture of composition $z_1, z_2, \ldots, z_m$ will separate into L moles of liquid of composition $x_1, x_2, \ldots, x_m$ and V moles of vapor of composition $y_1, y_2, \ldots, y_m$. An overall mole balance requires that

$$1 = L + V$$

and the component mole balances are

$$z_i = x_i L + y_i V \qquad (i = 1, 2, \ldots, m)$$

Elimination of V from the second equation yields

$$z_i = x_i L + y_i(1 - L) \qquad (i = 1, 2, \ldots, m)$$

Now eliminating y_i in favor of x_i by application of (*7.112*), and solving for x_i, we obtain

$$x_i = \frac{z_i}{L + K_i(1 - L)} \qquad (i = 1, 2, \ldots, m) \tag{7.113}$$

But the x_i must sum to unity; hence

$$\sum \frac{z_i}{L + K_i(1 - L)} = 1 \tag{7.114}$$

where the summation is taken over all components i. A different pair of equations, corresponding to (*7.113*) and (*7.114*), results if y_i and V, rather than x_i and L, are retained in the derivation; they are

$$y_i = \frac{K_i z_i}{(1-V) + K_i V} \tag{7.115}$$

$$\sum \frac{K_i z_i}{(1-V) + K_i V} = 1 \tag{7.116}$$

Solution of (*7.113*) and (*7.114*) for L and the x_i or of (*7.115*) and (*7.116*) for V and the y_i constitutes the material-balance portion of a flash calculation. The thermodynamic requirement is expressed by (*7.111*), and it is here that the difficulty lies. The K-values used in (*7.113*) through (*7.116*) should be those which satisfy (*7.111*). However, γ_i is a function of the x_i, and $\hat{\phi}_i$ is a function of the y_i. Thus K-values depend on the values of x_i and y_i to be determined, and iterative calculations are necessary. K-values are first estimated on any reasonable basis available (perhaps the assumption of Raoult's law), and values of x_i and y_i are determined from the material balance equations. These, together with the specified T and P, allow (if sufficient data are available) the calculation of K-values by (*7.111*). With these new K-values the calculations are repeated, and the process is continued to convergence.

The quantities L and V are numerically equal to the fractions that the liquid and vapor phases represent of the total moles of the system, and must therefore satisfy the inequalities

$$0 \leqq L \leqq 1 \qquad 0 \leqq V \leqq 1$$

These inequalities will in fact be satisfied (assuming the use of correct K-values) only if the specified T and P are within the range for which a two-phase system can exist. It is therefore important to establish this first, before one makes a flash calculation.

The limiting case $L = 0$, $V = 1$ represents the dew point, for which (*7.114*) becomes

$$\sum \frac{z_i}{K_i} = 1 \tag{7.117}$$

where the z_i ($\equiv y_i$) are the compositions of the saturated vapor. The temperature and pressure at the dew point for a vapor of fixed composition are not independently variable, and specification of one automatically fixes the other. Thus the dew-point temperature (or pressure) at a specified P (or T) is the value for which the equilibrium K-values satisfy (*7.117*). The corresponding equilibrium liquid compositions are found from the defining equation for K_i, (*7.112*). As with a flash calculation, the equilibrium values of x_i and y_i ($\equiv z_i$) are those consistent with (*7.111*). Since the K_i are in general functions of composition (both liquid and vapor) as well as of T and P, the calculation of equilibrium values of either x_i or y_i usually requires an iterative procedure.

At the bubble point, $L = 1$, $V = 0$, and (*7.116*) becomes

$$\sum K_i z_i = 1 \tag{7.118}$$

where the z_i ($\equiv x_i$) are the compositions of saturated liquid. The bubble-point conditions are determined by calculations similar to those for the dew point. Again, an iterative procedure is generally necessary to establish values of P (or T), x_i ($\equiv z_i$), and y_i which satisfy (*7.111*).

Example 7.13. A liquid mixture of methane(1), ethane(2), and propane(3) is fed to an equilibrium flash vaporizer controlled at 50(°F) and 200(psia). Calculate the fraction of the original feed stream which leaves the vaporizer as liquid and the compositions of the equilibrium phases. The process is continuous, and the feed composition (mole fractions) is

$$z_1 = 0.10 \qquad z_2 = 0.20 \qquad z_3 = 0.70$$

and the K-values at 50(°F) and 200(psia) are

$$K_1 = 10.2 \qquad K_2 = 1.76 \qquad K_3 = 0.543$$

The flash calculation will make use of (*7.113*) and (*7.114*). Direct solution of (*7.114*) for L is cumbersome, and hand calculations are often performed by an alternative but equivalent procedure. A value of L is assumed, and the x_i are calculated from (*7.113*). If the calculated x_i sum to unity, then the assumed value of L is consistent with the given K-values and feed composition; if not, then a new value of L must be chosen and the calculations repeated until $\Sigma x_i = 1$.

The calculations for this problem are summarized in the table.

Component	K_i	x_i for $L = 0.65$	x_i for $L = 0.75$	x_i for $L = 0.703$	$y_i = K_i x_i$
methane	10.2	0.024	0.030	0.027	0.275
ethane	1.76	0.158	0.168	0.162	0.285
propane	0.543	0.833	0.790	0.811	0.440
Σx_i		1.015	0.988	1.000	$\Sigma y_i = 1.000$

For the first trial, we let $L = 0.65$. Equation (*7.113*) applied to each component gives

$$x_1 = \frac{0.10}{0.65 + 10.2(1-0.65)} = 0.024$$

$$x_2 = \frac{0.20}{0.65 + 1.76(1-0.65)} = 0.158$$

$$x_3 = \frac{0.70}{0.65 + 0.543(1-0.65)} = 0.833$$

These x_i values sum to 1.015, and a new value of L must be tried. For $L = 0.75$, as shown in the table, the calculated x_i sum to 0.988, so the true L is between 0.65 and 0.75. Subsequent calculations give $L = 0.703$; the corresponding equilibrium x_i and y_i values are shown in the last two columns of the table.

The K-values given in the statement of this problem were calculated directly from *experimental* values of x_i and y_i by the defining equation, $K_i = y_i/x_i$. Thus the flash calculation performed merely regenerates values of x_i and y_i already known. Were this a "real" flash calculation, the answers would not be known, and correct K-values could not be computed from the defining equation. Rather, an iterative process would be required. The calculation given here simulates the last iteration of this process.

7.8 CHEMICAL-REACTION STOICHIOMETRY. PROPERTY CHANGES OF REACTION

Up to this point, interphase transport of matter has been the only mechanism considered as causing changes in the chemical composition of a phase in a closed system. Under suitable conditions, however, composition changes may occur as a result of depletion or creation of chemical species due to *chemical reaction*. The calculation of property changes associated with chemical reactions, and of equilibrium conditions in reacting systems, are the subjects of the remaining sections of this chapter.

Stoichiometric coefficients; reaction coordinates.

Most scientists and engineers are familiar with the usual "chemical" notation for single chemical reactions. Thus, the combination of gaseous hydrogen, $H_2(g)$, with gaseous oxygen, $O_2(g)$, to form liquid water, $H_2O(l)$, is written as

$$H_2(g) + \tfrac{1}{2}O_2(g) \longrightarrow H_2O(l)$$

The use of chemical notation serves adequately as an aid in material balance or thermodynamic calculations for simple systems in which only a few chemical reactions occur. However, the systematic development of the methods of chemical-reaction thermodynamics, and the application of the results to complex reacting systems, is facilitated by use of an equivalent but slightly more abstract "algebraic" notation. We shall find it convenient to use both notations; the algebraic notation is described briefly below.

A single chemical reaction in a system containing m chemical species is written algebraically as

$$0 = \sum_i \nu_i X_i(\mathrm{p}) \qquad (i = 1, 2, \ldots, m) \tag{7.119}$$

where $X_i(\mathrm{p})$ represents the chemical formula for species i and includes a designation (p) of the physical state of i, solid (p) = (s), liquid (p) = (l), or gas (p) = (g). The ν_i are *stoichiometric coefficients* or *stoichiometric numbers* determined from the "chemical" notation for a reaction. By convention, ν_i is negative for a *reactant* and positive for a *product*. If a particular species k in a system does not participate in the reaction (if it is "inert"), then $\nu_k = 0$. For the reaction $H_2(g) + \frac{1}{2}O_2(g) \rightarrow H_2O(l)$, (*7.119*) becomes

$$0 = \nu_1 X_1(g) + \nu_2 X_2(g) + \nu_3 X_3(l)$$

where

$$X_1(g) \equiv H_2(g) \qquad \nu_1 = -1$$

$$X_2(g) \equiv O_2(g) \qquad \nu_2 = -\tfrac{1}{2}$$

$$X_3(l) \equiv H_2O(l) \qquad \nu_3 = 1$$

Chemical reactions can be "combined" in the algebraic sense. Consider q chemical reactions for which the stoichiometric coefficients of the species $X_i(\mathrm{p})$ are $\nu_{i,j}$, where $j = 1, 2, \ldots, q$ identifies the particular reaction. We write (*7.119*) for each reaction:

$$\begin{aligned} 0 &= \sum_i \nu_{i,1} X_i(\mathrm{p}) \\ 0 &= \sum_i \nu_{i,2} X_i(\mathrm{p}) \\ &\;\vdots \\ 0 &= \sum_i \nu_{i,q} X_i(\mathrm{p}) \end{aligned} \tag{7.120}$$

where each set of species $\{i\}$ is specific to the particular reaction and need not include all species of the system. Summation of these q equations gives on rearrangement

$$0 = \sum_i \left(\sum_j \nu_{i,j} \right) X_i(\mathrm{p}) \tag{7.121}$$

If on comparing (*7.121*) with (*7.119*) we find that for each species i

$$\nu_i = \sum_j \nu_{i,j} \tag{7.122}$$

then the single chemical reaction (*7.119*) may formally be considered the "sum" of the q reactions (*7.120*).

Example 7.14. Consider the water-gas synthesis reaction

$$C(s) + H_2O(g) \rightarrow CO(g) + H_2(g)$$

In the algebraic form (*7.119*), this reaction is written as

$$0 = \nu_1 X_1(s) + \nu_2 X_2(g) + \nu_3 X_3(g) + \nu_4 X_4(g) \tag{1}$$

where
$$\begin{aligned} X_1(s) &\equiv C(s) & \nu_1 &= -1 \\ X_2(g) &\equiv H_2O(g) & \nu_2 &= -1 \\ X_3(g) &\equiv CO(g) & \nu_3 &= 1 \\ X_4(g) &\equiv H_2(g) & \nu_4 &= 1 \end{aligned} \tag{2}$$

We wish to show that reaction (*1*) may be considered the sum of the two reactions

$$C(s) + \tfrac{1}{2}O_2(g) \longrightarrow CO(g)$$
$$H_2O(g) \longrightarrow H_2(g) + \tfrac{1}{2}O_2(g)$$

In algebraic notation these become

$$0 = \nu_{1,1}X_1(s) + \nu_{5,1}X_5(g) + \nu_{3,1}X_3(g) \tag{3}$$
$$0 = \nu_{2,2}X_2(g) + \nu_{4,2}X_4(g) + \nu_{5,2}X_5(g) \tag{4}$$

where
$$\begin{aligned} X_1(s) &\equiv C(s) & \nu_{1,1} &= -1 & & \\ X_2(g) &\equiv H_2O(g) & & & \nu_{2,2} &= -1 \\ X_3(g) &\equiv CO(g) & \nu_{3,1} &= 1 & & \\ X_4(g) &\equiv H_2(g) & & & \nu_{4,2} &= 1 \\ X_5(g) &\equiv O_2(g) & \nu_{5,1} &= -\tfrac{1}{2} & \nu_{5,2} &= \tfrac{1}{2} \end{aligned}$$

It is now necessary to show that (*7.122*) is satisfied for each stoichiometric coefficient ν_i of reaction (*1*). Thus

$$\begin{aligned} \nu_1 &= \nu_{1,1} + \nu_{1,2} = -1 + 0 = -1 \\ \nu_2 &= \nu_{2,1} + \nu_{2,2} = 0 - 1 = -1 \\ \nu_3 &= \nu_{3,1} + \nu_{3,2} = 1 + 0 = 1 \\ \nu_4 &= \nu_{4,1} + \nu_{4,2} = 0 + 1 = 1 \\ \nu_5 &= \nu_{5,1} + \nu_{5,2} = -\tfrac{1}{2} + \tfrac{1}{2} = 0 \end{aligned}$$

Since these results are in agreement with (*2*), reaction (*1*) is indeed the sum of reactions (*3*) and (*4*).

Given a system containing many chemical species, one could postulate the occurrence of a great number of chemical reactions. Similarly, a single known overall reaction could conceivably be considered, in the *algebraic* sense, the net result of combinations of many simpler reactions. Application of thermodynamic methods to prediction of equilibrium states of such systems requires prior knowledge of the possible *independent* chemical reactions consistent with the particular species assumed present in a system. The number r and the nature of these reactions are determined as follows:

(*a*) Write reaction equations for the formation of each chemical species from its elements. (Formation reactions.)

(*b*) Combine these equations so as to eliminate from the set of equations all elements not considered to be present in the system.

The resulting r equations, or any r equations which are linear combinations of them, represent the desired independent reactions.

Example 7.15. Determine the independent chemical reactions in a system containing $C(s)$, $H_2O(g)$, $CO(g)$, $CO_2(g)$ and $H_2(g)$.

We can write three formation reactions:

$$H_2(g) + \tfrac{1}{2}O_2(g) \longrightarrow H_2O(g) \tag{1}$$
$$C(s) + \tfrac{1}{2}O_2(g) \longrightarrow CO(g) \tag{2}$$
$$C(s) + O_2(g) \longrightarrow CO_2(g) \tag{3}$$

The element $O_2(g)$ is assumed not present in the system, and it must be eliminated from (*1*), (*2*), and (*3*). Subtraction of (*1*) from (*2*) gives

$$C(s) + H_2O(g) \longrightarrow CO(g) + H_2(g) \qquad (4)$$

Similarly, multiplication of (*1*) by 2 and subtraction of the result from (*3*) yields

$$C(s) + 2H_2O(g) \longrightarrow CO_2(g) + 2H_2(g) \qquad (5)$$

Thus there are *two* possible independent reactions ($r = 2$) for the given system. The two independent reactions need not necessarily be taken as (*4*) and (*5*); multiplication of (*2*) by 2 and subtraction of (*3*) from the result gives

$$C(s) + CO_2(g) \longrightarrow 2CO(g) \qquad (6)$$

Either (*4*) or (*5*) in combination with (*6*) would also constitute a pair of independent reactions.

If $CO_2(g)$ were assumed *not* present in the system, then the formation equation (*3*) would not be written. Subtraction of (*1*) from (*2*) to eliminate $O_2(g)$ would then yield the single independent ($r = 1$) reaction (*4*), which is the water-gas reaction of Example 7.14.

In a closed system undergoing chemical reaction, the changes in the numbers of moles of the chemical species present are not independent of one another. For the general case of r independent reactions, they are related through the stoichiometric coefficients to r independent variables called *reaction coordinates* ϵ_j. In the remainder of this section we will restrict treatment to the case of a single independent reaction for which the reaction coordinate is ϵ.

We represent this single reaction as $0 = \nu_1 X_1 + \nu_2 X_2 + \nu_3 X_3 + \nu_4 X_4$; as the reaction proceeds changes in the numbers of moles of the species X_1, X_2, X_3, and X_4 are directly related to the stoichiometric coefficients ν_1, ν_2, ν_3, and ν_4 according to

$$\frac{dn_2}{dn_1} = \frac{\nu_2}{\nu_1} \quad \text{or} \quad \frac{dn_2}{\nu_2} = \frac{dn_1}{\nu_1}$$

$$\frac{dn_3}{dn_1} = \frac{\nu_3}{\nu_1} \quad \text{or} \quad \frac{dn_3}{\nu_3} = \frac{dn_1}{\nu_1}$$

$$\frac{dn_4}{dn_1} = \frac{\nu_4}{\nu_1} \quad \text{or} \quad \frac{dn_4}{\nu_4} = \frac{dn_1}{\nu_1}$$

The above relations may be combined to give

$$\frac{dn_1}{\nu_1} = \frac{dn_2}{\nu_2} = \frac{dn_3}{\nu_3} = \frac{dn_4}{\nu_4} \equiv d\epsilon$$

or, more generally,

$$\frac{dn_1}{\nu_1} = \frac{dn_2}{\nu_2} = \cdots = \frac{dn_m}{\nu_m} \equiv d\epsilon \qquad (7.123)$$

Thus the relationship between a differential change dn_i in the number of moles of a species and the differential change $d\epsilon$ in the reaction coordinate is *by definition* a simple proportionality, given from (*7.123*) as

$$\boxed{dn_i = \nu_i \, d\epsilon} \qquad (i = 1, 2, \ldots, m) \qquad (7.124)$$

The definition of ϵ itself is completed by the specification that it be *zero* for some particular condition of the system, usually the initial unreacted state. Designating the mole numbers for this initial state as n_{i_0}, we have

$$\epsilon = 0 \quad \text{for} \quad n_i = n_{i_0} \qquad (i = 1, 2, \ldots, m)$$

Integration of (*7.124*) for each species i then gives

$$\int_{n_{i_0}}^{n_i} dn_i = n_i - n_{i_0} = \nu_i \int_0^{\epsilon} d\epsilon = \nu_i \epsilon$$

from which

$$\boxed{n_i = n_{i_0} + \nu_i \epsilon} \qquad (i = 1, 2, \ldots, m) \tag{7.125}$$

Summation of (*7.125*) over the m species gives

$$\boxed{n = n_0 + \nu \epsilon} \tag{7.126}$$

where $$n = \sum n_i \qquad n_0 = \sum n_{i_0} \qquad \nu = \sum \nu_i$$

Division of (*7.125*) by (*7.126*) gives the *overall* mole fraction z_i of species i in the system:

$$z_i = \frac{n_i}{n} = \frac{n_{i_0} + \nu_i \epsilon}{n_0 + \nu \epsilon}$$

or $$z_i = \frac{z_{i_0} + \nu_i \xi}{1 + \nu \xi} \qquad (i = 1, 2, \ldots, m) \tag{7.127}$$

where $z_{i_0} = n_{i_0}/n_0$ and $\xi = \epsilon/n_0$.

Example 7.16. In a bomb calorimeter 0.02(g mole) of propane (C_3H_8) is completely oxidized to carbon dioxide (CO_2) and water vapor (H_2O) on burning with 0.6(g mole) of air. What is the final composition of the system? Take air to contain 21 mole % oxygen (O_2) and 79 mole % nitrogen (N_2).

The oxidation reaction is written in conventional chemical notation as

$$C_3H_8(g) + 5O_2(g) \longrightarrow 3CO_2(g) + 4H_2O(g) \tag{1}$$

However, the system contains *five* components – C_3H_8(1), O_2(2), N_2(3), CO_2(4), and H_2O(5) – and thus reaction (*1*) becomes in algebraic notation

$$0 = \sum_{i=1}^{5} \nu_i X_i(p)$$

with

$$\begin{aligned} X_1(g) &\equiv C_3H_8(g) & \nu_1 &= -1 \\ X_2(g) &\equiv O_2(g) & \nu_2 &= -5 \\ X_3(g) &\equiv N_2(g) & \nu_3 &= 0 \\ X_4(g) &\equiv CO_2(g) & \nu_4 &= 3 \\ X_5(g) &\equiv H_2O(g) & \nu_5 &= 4 \\ & & \nu = \sum \nu_i &= 1 \end{aligned}$$

By the statement of the problem, the initial mole numbers n_{i_0} are

$$n_{1_0} = 0.02\text{(g mole)}$$

$$n_{2_0} = (0.21)(0.6) = 0.126\text{(g mole)}$$

$$n_{3_0} = (0.79)(0.6) = 0.474\text{(g mole)}$$

$$n_{4_0} = n_{5_0} = 0$$

and the final number of moles of propane is zero ($n_1 = 0$). The corresponding value of the reaction coordinate is found by solution of (*7.125*) for ϵ with $i = 1$:

$$\epsilon = \frac{n_1 - n_{1_0}}{\nu_1} = \frac{0 - 0.02}{-1} = 0.02\text{(g mole)}$$

The remaining final mole numbers n_i ($i = 2, 3, 4, 5$) are given by (*7.125*) when the above values of n_{i_0} and ϵ are inserted. For example, for oxygen

$$n_2 = n_{2_0} + \nu_2\epsilon = 0.126 + (-5)(0.02) = 0.026\text{(g mole)}$$

The results are summarized in the table as mole numbers n_i and mole fractions y_i.

Species	Initial		Final	
	n_{i_0}	y_{i_0}	n_i	y_i
C_3H_8	0.02	0.0323	0.0	0.0
O_2	0.126	0.2032	0.026	0.0406
N_2	0.474	0.7645	0.474	0.7406
CO_2	0.0	0.0	0.06	0.0938
H_2O	0.0	0.0	0.08	0.1250
	$n_0 = \Sigma n_{i_0} = 0.62$		$n = \Sigma n_i = 0.64$	

Property changes of reaction.

For each phase of a π-phase, m-component system (*3.35*) provides an expression for the total differential of the Gibbs function expressed as a function of its canonical variables, and it is rewritten here as

$$d(nG)^p = -(nS)^p\,dT^p + (nV)^p\,dP^p + \sum_i \mu_i^p\,dn_i^p \tag{7.128}$$

where μ_i^p ($\equiv \bar{G}_i^p$) is the chemical potential of species i in phase p. If the system is in phase equilibrium, then the temperature, pressure, and chemical potential of each species are uniform throughout all phases, and we may sum the p equations represented by (*7.128*) to get an equation for the differential of the total Gibbs function G^t for the entire system of p phases:

$$dG^t = -S^t\,dT + V^t\,dP + \sum_i \mu_i\,dn_i \tag{7.129}$$

Although this is a very general equation, its usual application is to reacting systems for which any particular species appears in just one phase. In this event, the identification of a species also identifies its phase. If changes in the mole numbers n_i occur as the result of a single chemical reaction in a closed system, then by (*7.124*) each dn_i may be eliminated in favor of the product $\nu_i\,d\epsilon$, and (*7.129*) becomes

$$dG^t = -S^t\,dT + V^t\,dP + \Delta G_{T,P}\,d\epsilon \tag{7.130}$$

where by definition

$$\Delta G_{T,P} = \sum_i \nu_i\mu_i \tag{7.131}$$

The function $\Delta G_{T,P}$ is the *Gibbs function change of reaction*. From (*7.130*) it is evidently equal to $(\partial G^t/\partial\epsilon)_{T,P}$, and therefore provides a measure of the variation of the total system property G^t at constant T and P with the extent of reaction as represented by the reaction coordinate ϵ. Moreover, the units of $\Delta G_{T,P}$ must be the units of G^t divided by the units of ϵ. Thus when G^t is expressed in (cal) and ϵ is expressed in (g mole), then $\Delta G_{T,P}$ has the units of (cal)/(g mole), and is therefore regarded as an intensive property of the system for a given reaction with a specific set of stoichiometric coefficients. (Note that these coefficients are regarded as pure numbers.)

It is convenient to write $\Delta G_{T,P}$ as the sum of two terms, the first of which refers to the chemical species in *standard states*. The basis for this is the application of (*7.14*) to species i, where

$$\mathcal{G} = \mu_i, \quad \mathcal{f}_{\text{final}} = \hat{f}_i, \quad \text{and} \quad \mathcal{f}_{\text{initial}} = f_i^\circ$$

This gives

$$\mu_i = \mu_i^\circ + RT \ln (\hat{f}_i / f_i^\circ) \tag{7.132}$$

The standard states are denoted by the superscript (°), and they are states of the *pure* chemical species at the temperature of the system but at *fixed* pressure. (This contrasts with the standard states of Section 7.4, where the pressure was taken as that of the system.) Substitution of (*7.132*) into (*7.131*) gives

$$\boxed{\Delta G_{T,P} = \Delta G^\circ + RT \ln \prod (\hat{f}_i / f_i^\circ)^{\nu_i}} \tag{7.133}$$

where

$$\boxed{\Delta G^\circ = \sum_i \nu_i \mu_i^\circ} \tag{7.134}$$

and $\prod$ signifies the product over all species i. The function ΔG° is the *standard* (state) *Gibbs function change of reaction,* and depends on the temperature only. The fugacity ratio $\hat{f}_i / f_i^\circ$ is defined by (*7.68*) as the activity $\hat{a}_i$, and (*7.133*) may be rewritten

$$\Delta G_{T,P} = \Delta G^\circ + RT \ln \prod \hat{a}_i^{\nu_i} \tag{7.133a}$$

Other *standard* (state) *property changes of reaction* are defined similarly:

$$\boxed{\Delta M^\circ = \sum_i \nu_i M_i^\circ} \tag{7.135}$$

and are related to ΔG° by equations analogous to property relations already derived. Thus we have a Gibbs-Helmholtz equation (Problem 3.9) which gives the *standard* (state) *enthalpy change of reaction* ΔH°:

$$\boxed{\Delta H^\circ = -RT^2 \frac{d(\Delta G^\circ / RT)}{dT}} \tag{7.136}$$

This quantity is often called a *standard heat of reaction.* In addition, the *standard heat-capacity change of reaction* ΔC_P° and the *standard entropy change of reaction* ΔS° are given by:

$$\boxed{\Delta C_P^\circ = \frac{d\,\Delta H^\circ}{dT}} \tag{7.137}$$

$$\boxed{\Delta S^\circ = \frac{\Delta H^\circ - \Delta G^\circ}{T}} \tag{7.138}$$

Ordinary derivatives are used in (*7.136*) and (*7.137*) because the standard property changes of reaction are functions of temperature only.

Two important stoichiometric properties of the ΔM° follow directly from the definition (*7.135*). First, if the stoichiometric coefficients for a given reaction are all multiplied by the same factor α, then the value of ΔM° for the new reaction is just α times that for the original reaction. For example, the value of ΔM° for the reaction

$$2C(s) + O_2(g) \longrightarrow 2CO(g)$$

is exactly *twice* the value for the reaction

$$C(s) + \tfrac{1}{2}O_2(g) \longrightarrow CO(g)$$

Similarly, the value of ΔM° for the reaction

$$H_2O(g) \longrightarrow H_2(g) + \tfrac{1}{2}O_2(g)$$

is the *negative* of the value of ΔM° for the reaction

$$H_2(g) + \tfrac{1}{2}O_2(g) \longrightarrow H_2O(g)$$

A second important attribute of these functions is an additivity property. If we can consider a particular overall chemical reaction to be the algebraic sum of q subsidiary reactions, then the stoichiometric coefficients of the overall and subsidiary reactions are related by (*7.122*). Substitution of this equation into (*7.135*) yields

$$\Delta M^\circ = \sum_i \Big(\sum_j \nu_{i,j}\Big) M_i^\circ = \sum_j \Big(\sum_i \nu_{i,j} M_i^\circ\Big)$$

or

$$\Delta M^\circ = \sum_j \Delta M_j^\circ \tag{7.139}$$

where

$$\Delta M_j^\circ = \sum_i \nu_{i,j} M_i^\circ$$

Thus the value of ΔM° for an overall reaction is the sum of the values ΔM_j° for any set of subsidiary reactions that sum to the overall reaction.

Example 7.17. Derive equations for the calculation of ΔH° and ΔG° at any temperature T from given values ΔH_0° and ΔG_0° at some reference temperature T_0. Assume that ΔC_P° is known as a function of T.

Integration of (*7.137*) from T_0 to T gives the desired expression for ΔH°:

$$\boxed{\Delta H^\circ = \Delta H_0^\circ + \int_{T_0}^{T} \Delta C_P^\circ \, dT} \tag{7.140}$$

Similarly, integration of (*7.136*) from T_0 to T yields

$$\frac{\Delta G^\circ}{RT} = \frac{\Delta G_0^\circ}{RT_0} - \int_{T_0}^{T} \frac{\Delta H^\circ}{RT^2}\, dT \tag{7.141}$$

Substituting (*7.140*) for ΔH° and integrating, we obtain

$$\boxed{\frac{\Delta G^\circ}{RT} = \frac{\Delta G_0^\circ - \Delta H_0^\circ}{RT_0} + \frac{\Delta H_0^\circ}{RT} + \frac{1}{RT}\int_{T_0}^{T} \Delta C_P^\circ \, dT - \frac{1}{R}\int_{T_0}^{T} \frac{\Delta C_P^\circ}{T}\, dT} \tag{7.142}$$

Compilation of ΔM° values as a function of T for all reactions of practical interest is neither feasible nor necessary. Equations (*7.140*) and (*7.142*) show that ΔH° and ΔG° for *any* T can be calculated from values of ΔH° and ΔG° at a single reference temperature T_0, provided one knows ΔC_P° over the temperature range from T_0 to T. Equation (*7.139*) shows further that it is unnecessary to tabulate ΔH° and ΔG° even at T_0 for *every* reaction of interest; one needs no more than entries for a representative set of subsidiary reactions which can be combined algebraically to yield other reactions. These subsidiary reactions are most conveniently taken as compound *formation reactions*. (See Example 7.15.)

Standard property changes of formation are designated ΔM_f°, and modern compilations of data are for a temperature of 25(°C). The standard state for a gas is taken as the hypothetical ideal-gas state of the pure gas at 1(atm) pressure, and the standard state for a liquid or solid is taken as the real state of the pure liquid or solid at 1(atm) pressure. It follows from (*7.25*) that the standard state for a gas is a state of *unit fugacity* when the fugacity is expressed in atmospheres.

The temperature dependence of ΔH° and ΔG° is established as follows. According to (*7.135*), ΔC_P° is given by

$$\Delta C_P^\circ = \sum_i \nu_i C_{P_i}^\circ \tag{7.143}$$

An expression for the temperature dependence of ΔC_P° is obtained by substitution of appropriate equations for $C_{P_i}^\circ$ into (*7.143*). For most purposes the following empirical equation for $C_{P_i}^\circ$ (see Secs. 4.6 and 4.7) is sufficiently general:

$$C_{P_i}^\circ = a_i + b_i T + c_i T^2 + d_i/T^2 \tag{7.144}$$

where a_i, b_i, c_i, and d_i are constants for species i. Substitution in (*7.143*) yields

$$\Delta C_P^\circ = \Delta a + (\Delta b)T + (\Delta c)T^2 + (\Delta d)/T^2 \tag{7.145}$$

where

$$\Delta a = \sum_i \nu_i a_i, \quad \ldots, \quad \Delta d = \sum_i \nu_i d_i \tag{7.146}$$

Combination of (*7.145*) with (*7.140*) and (*7.142*) then gives, on integration and rearrangement,

$$\Delta H^\circ = \Delta H_0^\circ + T_0(\Delta a)(\tau - 1) + \frac{T_0^2 \Delta b}{2}(\tau^2 - 1) + \frac{T_0^3 \Delta c}{3}(\tau^3 - 1) + \frac{\Delta d}{T_0}\left(\frac{\tau - 1}{\tau}\right) \tag{7.147}$$

$$\Delta G^\circ = \Delta H_0^\circ + (\Delta G_0^\circ - \Delta H_0^\circ)\tau + T_0(\Delta a)(\tau - 1 - \tau \ln \tau) - \frac{T_0^2 \Delta b}{2}(\tau^2 - 2\tau + 1) - \frac{T_0^3 \Delta c}{6}(\tau^3 - 3\tau + 2) - \frac{\Delta d}{2T_0}\left(\frac{\tau^2 - 2\tau + 1}{\tau}\right) \tag{7.148}$$

where $\tau = T/T_0$ is a dimensionless absolute temperature.

Example 7.18. Calculate ΔH° and ΔG° at 1000(°C) for the water-gas synthesis reaction

$$C(s) + H_2O(g) \longrightarrow CO(g) + H_2(g) \tag{1}$$

Standard formation data at 25(°C) are, for $H_2O(g)$,

$$\Delta H_f^\circ = -57{,}798\text{(cal)/(g mole)} \qquad \Delta G_f^\circ = -54{,}635\text{(cal)/(g mole)}$$

and, for $CO(g)$,

$$\Delta H_f^\circ = -26{,}416\text{(cal)/(g mole)} \qquad \Delta G_f^\circ = -32{,}808\text{(cal)/(g mole)}$$

Constants in the heat-capacity equation (*7.144*) for the pure chemical species in their standard states are [units of (cal)/(g mole)(K) for C_P°]

Compound	a	$10^3 b$	$10^7 c$	$10^{-5} d$
$C(s)$	4.10	1.02	0.0	−2.10
$H_2O(g)$	7.256	2.298	2.83	0.0
$CO(g)$	6.420	1.665	−1.96	0.0
$H_2(g)$	6.947	−0.200	4.81	0.0

The required values of ΔH° and ΔG° will be calculated from (*7.147*) and (*7.148*). As shown in Example 7.14, reaction (*1*) can be considered the sum of the two reactions

$$C(s) + \tfrac{1}{2}O_2(g) \longrightarrow CO(g) \tag{2}$$

$$H_2O(g) \longrightarrow H_2(g) + \tfrac{1}{2}O_2(g) \tag{3}$$

But (*2*) is the formation reaction for $CO(g)$, and (*3*) is just the reverse of the formation reaction for $H_2O(g)$. The functions ΔH_0° and ΔG_0° for reaction (*1*) at $T_0 = 25(°C) = 298(K)$ are then given by

$$\Delta H_0^\circ = \Delta H_f^\circ[CO(g)] - \Delta H_f^\circ[H_2O(g)] = -26{,}416 - (-57{,}798) = 31{,}382\text{(cal)/(g mole)}$$

$$\Delta G_0^\circ = \Delta G_f^\circ[\mathrm{CO(g)}] - \Delta G_f^\circ[\mathrm{H_2O(g)}] = -32{,}808 - (-54{,}635) = 21{,}827\mathrm{(cal)/(g\ mole)}$$

Values for Δa, Δb, Δc, and Δd are found by application of (*7.146*) to the data given in the problem statement. Thus,

$$\Delta a = (-1)(4.10) - (1)(7.256) + (1)(6.420) + (1)(6.947) = 2.011\mathrm{(cal)/(g\ mole)(K)}$$

Similarly, we find

$$\Delta b = -1.853 \times 10^{-3}\mathrm{(cal)/(g\ mole)(K)^2}$$

$$\Delta c = 2 \times 10^{-9}\mathrm{(cal)/(g\ mole)(K)^3}$$

$$\Delta d = 2.10 \times 10^5\mathrm{(cal)(K)/(g\ mole)}$$

We can now calculate ΔH° and ΔG°. For $T = 1000(^\circ\mathrm{C})$ and $T_0 = 25(^\circ\mathrm{C})$, the dimensionless temperature τ is

$$\tau = \frac{T}{T_0} = \frac{1000 + 273}{25 + 273} = 4.272$$

Substitution of numerical values into (*7.147*) gives

$$\begin{aligned}\Delta H^\circ = {} & 31{,}382 + (298)(2.011)(4.272 - 1) \\ & + (\tfrac{1}{2})(298)^2(-1.853 \times 10^{-3})[(4.272)^2 - 1] \\ & + (\tfrac{1}{3})(298)^3(2 \times 10^{-9})[(4.272)^3 - 1] \\ & + (\tfrac{1}{298})(2.10 \times 10^5)[(4.272 - 1)/4.272] \\ = {} & 32{,}465\mathrm{(cal)/(g\ mole)}\end{aligned}$$

Similarly, solution of (*7.148*) for ΔG° yields

$$\begin{aligned}\Delta G^\circ = {} & 31{,}382 + (21{,}827 - 31{,}382)(4.272) \\ & + (298)(2.011)(4.272 - 1 - 4.272 \ln 4.272) \\ & - (\tfrac{1}{2})(298)^2(-1.853 \times 10^{-3})[(4.272)^2 - (2)(4.272) + 1] \\ & - (\tfrac{1}{6})(298)^3(2 \times 10^{-9})[(4.272)^3 - (3)(4.272) + 2] \\ & - (\tfrac{1}{2})(\tfrac{1}{298})(2.10 \times 10^5)\left[\frac{(4.272)^2 - (2)(4.272) + 1}{4.272}\right] \\ = {} & -11{,}197\mathrm{(cal)/(g\ mole)}\end{aligned}$$

The values calculated for ΔH° and ΔG° at $1000(^\circ\mathrm{C})$ are for the reaction (*1*) with the stoichiometric coefficients as written, and for the conversion of C(s) and H_2O(g) in their standard states at $1000(^\circ\mathrm{C})$ into CO(g) and H_2(g) in their standard states at $1000(^\circ\mathrm{C})$.

7.9 CHEMICAL-REACTION-EQUILIBRIUM CALCULATIONS

A mixture of chemical species, when brought in contact with a catalyst in a well-stirred vessel, will react to a greater or lesser extent to form new chemical species at the expense of all or some of the original compounds present in the system. After the passage of sufficient time (perhaps forever, if the catalyst is poor) the composition of the system will attain a true steady-state value, different in general from the original composition. This section deals with the application of thermodynamic methods to the prediction of such states of *chemical equilibrium.* The equilibrium states, while possible in the thermodynamic sense, may or may not be realizable on a realistic time scale in the plant or the laboratory, and questions relating to reaction *rates* cannot be answered by thermodynamics. Nonetheless, knowledge of the equilibrium states of a reacting system is important, because the *equilibrium* conversion of reactant to product species (as measured by, e.g., the reaction coordinate ϵ) is the *maximum* conversion possible under specified conditions of T and P.

A criterion for chemical equilibrium.

The condition for equilibrium with respect to variations in mole numbers in a closed, heterogeneous system is given by

$$(dG^t)_{T,P} = 0 \qquad (3.69)$$

If the changes in mole numbers occur as the result of a single chemical reaction, then (*7.130*) becomes $(dG^t)_{T,P} = \Delta G_{T,P}\, d\epsilon$, and by (*3.69*) we have $\Delta G_{T,P}\, d\epsilon = 0$. Since this last equation must hold fcr arbitrary $d\epsilon$, it follows that

$$\boxed{\Delta G_{T,P} = 0} \qquad (7.149)$$

is a criterion for chemical equilibrium in a system in which there occurs a single chemical reaction. This result is readily generalized to systems in which there occur r *independent* chemical reactions. Thus we obtain

$$(\Delta G_{T,P})_j = 0 \qquad (j = 1, 2, \ldots, r) \qquad (7.150)$$

where $(\Delta G_{T,P})_j$ is the Gibbs function change of reaction for the jth independent chemical reaction.

Since $(\Delta G_{T,P})_j = \sum_i \nu_{i,j}\mu_i$ by definition (*7.131*), and since each μ_i is in general a function of T, P, and the mole fractions of the various species, (*7.150*) constitutes r implicit relationships that must hold among the intensive variables at chemical equilibrium. Attainment of a state of complete equilibrium in a multiphase reacting system requires in addition that the condition (*3.76*) for phase equilibrium be satisified. The generalization of the phase rule (*3.77*) to include the additional r constraints of (*7.150*) is obtained by a simple extension of the arguments of Example 3.14. The result is

$$\boxed{F = 2 - \pi + m - r} \qquad (7.151)$$

Thus the effect of the occurrence of chemical reactions is to *decrease* the number of degrees of freedom of a PVT system at equilibrium. As noted in Problem 3.18, however, Duhem's theorem remains unchanged.

Example 7.19. Determine the number of degrees of freedom at equilibrium of a chemically reactive system containing solid sulfur S and the three gases O_2, SO_2, and SO_3.

There are four chemical species ($m = 4$) and two phases ($\pi = 2$) in the system. The number of independent reactions is *two* ($r = 2$), because the elements S(s) and O_2(g), which appear in the formation reactions

$$S(s) + O_2(g) \longrightarrow SO_2(g) \qquad S(s) + \tfrac{3}{2}O_2(g) \longrightarrow SO_3(g)$$

are both assumed present at equilibrium. The extended phase rule (*7.151*) then gives $F = 2 - 2 + 4 - 2 = 2$.

Thus there are *two* degrees of freedom for the system, and specification of any two of the intensive variables P, T, and the equilibrium mole fractions of the gaseous species (the mole fraction of sulfur in the solid phase is always unity) fixes the equilibrium state of the system.

Practical application of the chemical equilibrium criterion (*7.149*) is facilitated if the thermodynamic variables are displayed more explicitly. This is done by use of relationships presented in Sec. 7.8. Substitution of (*7.133*) into (*7.149*) gives, on rearrangement,

$$\prod (\hat{f}_i/f_i^\circ)^{\nu_i} = \exp\left(\frac{-\Delta G^\circ}{RT}\right)$$

or

$$\boxed{\prod (\hat{f}_i/f_i^\circ)^{\nu_i} = K} \qquad (7.152)$$

where
$$K = \exp\left(\frac{-\Delta G^\circ}{RT}\right) \tag{7.153}$$

The function K, which depends on T only, is called a chemical *equilibrium "constant"*. Numerical values of ΔG°, and hence of K, can be determined at any temperature from tabulated formation data and heat-capacity values, as illustrated in Example 7.18. Expressions for $\hat{f}_i$ must in general be found from an equation of state or from a G^E expression.

Application to reactions between gases.

The case of a single gas-phase reaction is of special importance, and will be considered in some detail. The standard state of a gaseous species is a state of unit fugacity if the fugacities are expressed in (atm); thus $f_i^\circ = 1\text{(atm)}$ and (*7.152*) becomes

$$\prod \hat{f}_i^{\nu_i} = K$$

Replacement of $\hat{f}_i$ by the equivalent expression $\hat{f}_i = y_i \hat{\phi}_i P$ gives, on rearrangement,

$$\prod (y_i \hat{\phi}_i)^{\nu_i} = P^{-\nu} K \tag{7.154}$$

where $\nu = \sum \nu_i$ and P must be in (atm).

The usual application of (*7.154*) is to the determination of equilibrium conditions for a gas mixture of known initial composition. This amounts to an application of Duhem's theorem, which requires that two independently variable equilibrium conditions be specified in order to make the problem determinate. For the present case, the component mole fractions are eliminated in favor of the initial compositions and the dimensionless reaction coordinate ξ by use of (*7.127*), written in terms of y in place of z:

$$y_i = \frac{y_{i_0} + \nu_i \xi}{1 + \nu \xi} \qquad (i = 1, 2, \ldots, m) \tag{7.155}$$

With the y_i replaced by (*7.155*), (*7.154*) becomes a single equation relating the three variables T, P, and ξ. Specification of any two of them, in conformance with Duhem's theorem, allows determination of the remaining one. However, the solution of the problem may be complicated because of the dependence of $\hat{\phi}_i$ on T, P, and ξ.

Ideal-gas reactions.

The simplest case of (*7.154*) obtains for a reacting mixture of ideal gases, for which $\hat{\phi}_i = 1$:

$$\prod y_i^{\nu_i} = P^{-\nu} K \tag{7.156}$$

For this special case, the T-, P-, and composition-dependent portions of the equilibrium equation are distinct and separable, and solution of (*7.156*) for ξ, P, or T is straightforward.

Example 7.20. Ethanol (C_2H_5OH) can be manufactured by the vapor-phase hydration of ethylene (C_2H_4) according to the reaction

$$C_2H_4(g) + H_2O(g) \longrightarrow C_2H_5OH(g)$$

The feed to a reactor in which the above reaction takes place is a gas mixture containing 25 mole % ethylene and 75 mole % steam. Estimate the product composition if the reaction occurs at 125(°C) and 1(atm). A value for ΔG° at 125(°C), calculated by the method of Example 7.18, is 1082(cal)/(g mole).

The best estimate we can make of the product composition is the *equilibrium* composition at the given T and P. It will be assumed that the gaseous mixtures behave as mixtures of ideal gases; therefore

(*7.156*) is applicable. For $C_2H_4 \equiv (1)$, $H_2O \equiv (2)$, and $C_2H_5OH \equiv (3)$, the stoichiometric coefficients are $\nu_1 = -1$, $\nu_2 = -1$, and $\nu_3 = 1$. Then $\nu = \Sigma \nu_i = -1$, and the material balance equation (*7.155*) gives

$$\begin{aligned} y_1 &= \frac{y_{1_0} + \nu_1\xi}{1 + \nu\xi} = \frac{0.25 - \xi}{1 - \xi} \\ y_2 &= \frac{y_{2_0} + v_2\xi}{1 + \nu\xi} = \frac{0.75 - \xi}{1 - \xi} \\ y_3 &= \frac{y_{3_0} + \nu_3\xi}{1 + \nu\xi} = \frac{\xi}{1 - \xi} \end{aligned} \tag{1}$$

The equilibrium constant K is calculated from the definition (*7.153*), with $\Delta G^\circ = 1082(\text{cal})/(\text{g mole})$ and $T = 273 + 125 = 398(\text{K})$:

$$K = \exp\left(\frac{-1082}{1.987 \times 398}\right) = 0.254$$

Equation (*7.156*) then gives

$$\left(\frac{0.25 - \xi}{1 - \xi}\right)^{-1}\left(\frac{0.75 - 1}{1 - \xi}\right)^{-1}\left(\frac{\xi}{1 - \xi}\right)^{+1} = (1)^{+1} \times 0.254$$

or

$$\frac{\xi(1 - \xi)}{(0.25 - \xi)(0.75 - \xi)} = 0.254$$

Solution for ξ yields $\xi = 0.0395$ and substitution of this value into (*1*) gives the equilibrium composition of the product:

$$y_1 = 0.219 \qquad y_2 = 0.740 \qquad y_3 = 0.041$$

By definition, the reaction coordinate ξ (or ϵ) provides a measure of the extent of reaction or of the "yield" of a reaction. Thus, an increase of ξ (or ϵ) for a given reaction indicates a "shift to the right," or the conversion of more reactant species to product species. For the particular case of an ideal-gas reaction, one can make some important generalizations regarding the effects of thermodynamic properties and process variables on the equilibrium value of ξ (or ϵ); these generalizations, although strictly valid only for the ideal-gas case, often may be applied to systems of real gases as well.

We start with (*7.156*), written in the equivalent form

$$K_y = P^{-\nu}K \tag{7.157}$$

where

$$K_y \equiv \prod y_i^{\nu_i} \tag{7.158}$$

According to (*7.157*), K_y is a function of T and P only, because K is a function of T only. Alternatively, by (*7.158*) and (*7.155*), K_y can be considered a function of the single variable ξ for fixed values of the initial compositions y_{i_0}. It can be shown moreover (the proof is lengthy and we shall not reproduce it here) that

$$\frac{d\xi}{dK_y} > 0 \tag{7.159}$$

and (*7.159*) lets us draw conclusions about trends in ξ from corresponding trends in K_y.

The first generalization has to do with the relative magnitude of ξ. According to the definition (*7.153*) of K, a large negative value of ΔG° will give a large value of K, and hence, by (*7.157*), of K_y. This implies, through (*7.159*), a large equilibrium ξ relative to the corresponding ξ for a reaction of identical stoichiometry but smaller negative ΔG°. Conversely, a reaction with a large positive ΔG° will yield a small equilibrium ξ relative to that for a similar reaction with smaller positive ΔG°. The sign and magnitude of ΔG° thus offer clues as to the yield one can expect from a given reaction; large negative values of ΔG°

imply high equilibrium conversions to the product species, whereas large positive values of ΔG° imply small equilibrium conversions to the product species.

Two other generalizations are concerned with the effects of changes in temperature and pressure on the equilibrium value of ξ. The temperature derivative of ξ is given by

$$\left(\frac{\partial \xi}{\partial T}\right)_P = \left(\frac{d\xi}{dK_y}\right)\left(\frac{\partial K_y}{\partial T}\right)_P$$

But from (*7.157*)

$$\left(\frac{\partial K_y}{\partial T}\right)_P = P^{-\nu}\frac{dK}{dT}$$

and from (*7.153*)

$$\frac{dK}{dT} = \left[\exp\left(\frac{-\Delta G^\circ}{RT}\right)\right]\frac{d(-\Delta G^\circ/RT)}{dT}$$

Combining this with (*7.153*) and (*7.136*) we get

$$\frac{dK}{dT} = \frac{K\,\Delta H^\circ}{RT^2}$$

The equation for $(\partial\xi/\partial T)_P$ then becomes

$$\left(\frac{\partial \xi}{\partial T}\right)_P = \frac{d\xi}{dK_y}P^{-\nu}\frac{K\,\Delta H^\circ}{RT^2}$$

or

$$\left(\frac{\partial \xi}{\partial T}\right)_P = \left[\frac{K_y}{RT^2}\frac{d\xi}{dK_y}\right]\Delta H^\circ \qquad (7.160)$$

Similarly, we find for the pressure derivative of ξ

$$\left(\frac{\partial \xi}{\partial P}\right)_T = \left[\frac{K_y}{P}\frac{d\xi}{dK_y}\right](-\nu) \qquad (7.161)$$

The bracketed terms in (*7.160*) and (*7.161*) are always positive, and therefore the signs of $(\partial\xi/\partial T)_P$ and $(\partial\xi/\partial P)_T$ are determined solely by the signs of ΔH° and $\nu(=\sum \nu_i)$, respectively. We may therefore conclude the following:

(*a*) If at a given temperature ΔH° is positive, i.e. if the standard reaction is *endothermic,* then an increase in T at constant P will cause an increase in the equilibrium ξ. If ΔH° is negative, i.e. if the standard reaction is *exothermic,* then an increase in T at constant P causes a decrease in the equilibrium ξ.

(*b*) If the total stoichiometric coefficient $\nu(=\sum \nu_i)$ is *negative,* then an increase in P at constant T results in an increase in the equilibrium ξ. If ν is *positive,* then an increase in P at constant T causes a decrease in the equilibrium ξ.

Example 7.21. State the probable effects of increasing P and T on the yield of ammonia (NH_3) from the reaction

$$\tfrac{1}{2}N_2(g) + \tfrac{3}{2}H_2(g) \longrightarrow NH_3(g) \qquad (1)$$

For this reaction, $\Delta H^\circ = -11{,}040(\text{cal})/(\text{g mole})$ at 25(°C).

In the absence of data on which to base a rigorous calculation, we assume that the reacting system is a mixture of ideal gases. The total stoichiometric coefficient ν is

$$\nu = \sum \nu_i = -\tfrac{1}{2} - \tfrac{3}{2} + 1 = -1$$

According to (*7.161*), then, an increase in pressure will favor the production of *more* ammonia by reaction (*1*). The standard reaction is exothermic at 25(°C), and therefore by (*7.160*) increasing T at this temperature level will cause a *decrease* in the conversion to ammonia. Actually, ΔH° for reaction (*1*) is negative for *all* temperatures of practical interest, so this last conclusion holds generally for the ideal-gas ammonia synthesis reaction.

Both of the above effects are put to use in the commercial production of ammonia via reaction (*1*). Most modern ammonia plants employ very high pressures, in conjunction with the lowest temperatures consistent with a reasonable rate of reaction.

Multireaction equilibria.

So far we have considered the prediction of equilibrium states only for systems in which a single independent chemical reaction occurs. The approach adopted for this special case was to develop an expression for $(dG^t)_{T,P}$ as a function of the composition variables and to set it equal to zero in accord with the equilibrium condition (*3.69*). This method is readily generalized to apply when there are r independent chemical reactions. However, this method leads to r complicated equations which require simultaneous solution, and hand calculations are usually extremely tedious. Moreover, these equations are not well suited to solution by automatic machine computation. The modern procedure for solving multireaction equilibrium problems is therefore based on the alternative and equivalent method noted in Sec. 3.6.

This method requires an expression for G^t in terms of the composition variables, and the equilibrium compositions of the species are then determined directly as the set of composition values which *minimizes* G^t at a given T and P, as required by (*3.68*). Solution of the equations developed would again be most tedious by hand calculation, but solution by automatic machine computation is facilitated. In the present treatment, we consider only the calculation of equilibrium compositions for gas-phase reactions at specified T, P, and initial compositions.

The total Gibbs function G^t of the system is given by (*7.2*):

$$G^t = \sum_i n_i \bar{G}_i = \sum_i n_i \mu_i$$

This is the function to be minimized with respect to the mole numbers n_i at constant T and P, subject to restraints imposed by material balances for a closed system. The μ_i are given by (*7.132*) as

$$\mu_i = \mu_i^\circ + RT \ln (\hat{f}_i / f_i^\circ)$$

and $\hat{f}_i$ is related to $\hat{\phi}_i$ by

$$\hat{f}_i = y_i \hat{\phi}_i P$$

The standard states are those adopted previously for chemical reaction calculations, and μ_i° is taken as $\Delta G_{f_i}^\circ$, the Gibbs function change for the standard formation reaction of species i. We shall work with pressure units of (atm), and therefore $f_i^\circ = 1$. Combination of the last three equations, with $f_i^\circ = 1$ and $\mu_i^\circ = \Delta G_{f_i}^\circ$, gives on rearrangement

$$G^t = \sum_i n_i \Delta G_{f_i}^\circ + \left(\sum_i n_i\right) RT \ln P + RT \sum_i n_i \ln n_i - RT \sum_i n_i \ln\left(\sum_i n_i\right) + RT \sum_i n_i \ln \hat{\phi}_i \qquad (7.162)$$

where the y_i have been eliminated in favor of the mole numbers n_i.

We must now find the set $\{n_i\}$ which minimizes G^t [as given by (*7.162*)] at constant T and P, subject to the restraints of the material balances. The standard solution to this type of problem is through the method of Lagrange's undetermined multipliers. This requires that the restraints imposed by the material balances be incorporated in the expression for G^t. The material balance equations are developed as follows.

Although molecular species are not conserved in a chemically reacting system, mass is. Let A_k be the total number of atomic weights of the kth element present in the system, as determined by the initial constitution of the system. Let a_{ik} be the number of atoms of the kth element present in each molecule of chemical species i. Then for each element k, $\sum_i n_i a_{ik} = A_k$ or

$$\sum_i (n_i a_{ik}) - A_k = 0$$

This is multiplied by an undetermined constant λ_k, and the expression is then summed over all k, giving

$$\sum_k \left\{ \lambda_k \left[\sum_i (n_i a_{ik}) - A_k \right] \right\} = 0$$

Since this quantity is zero, it may be added to the right-hand side of (*7.162*):

$$G^t = \sum_i n_i \Delta G^\circ_{f_i} + \left(\sum_i n_i \right) RT \ln P + RT \sum_i n_i \ln n_i - RT \sum_i n_i \ln \left(\sum_i n_i \right)$$

$$+ RT \sum_i n_i \ln \hat{\phi}_i + \sum_k \left\{ \lambda_k \left[\sum_i (n_i a_{ik}) - A_k \right] \right\} \qquad (7.163)$$

The final step is to differentiate (*7.163*) with respect to each mole number n_i to form the derivatives $[\partial G^t / \partial n_i]_{T,P,n_j}$, which are then set equal to zero. This leads to

$$\Delta G^\circ_{f_i} + RT \ln P + RT \ln y_i + RT \ln \hat{\phi}_i + \sum_k \lambda_k a_{ik} = 0 \qquad (i = 1, 2, \ldots, m) \qquad (7.164)$$

There are m such equations. In addition, there there are w material balance equations of the form

$$\sum_i y_i a_{ik} = A_k \Big/ \sum_i n_i \qquad (k = 1, 2, \ldots, w) \qquad (7.165)$$

where w is the number of elements present in the compounds which make up the system. Also, the y_i must satisfy

$$\sum_i y_i = 1 \qquad (7.166)$$

This provides a total of $m + w + 1$ equations.

The unknowns in these equations are the y_i, of which there are m, the λ_k, of which there are w, and the term $\sum_i n_i$, giving a total of $m + w + 1$ unknowns. Thus the set of equations (*7.164*), (*7.165*), and (*7.166*) can be solved for all unknowns.

However, (*7.164*) was derived on the presumption that all the $\hat{\phi}_i$ are known. If the phase is an ideal gas, then each $\hat{\phi}_i$ is indeed known, and is unity, but for real gases each $\hat{\phi}_i$ is a function of the various y_i, which are to be determined. Thus an iterative procedure is indicated, which is initiated by setting the $\hat{\phi}_i$ equal to unity. Solution of the equations then provides a preliminary set of y_i. For low pressures or high temperatures this result is usually quite adequate. Where it is not, an equation of state is used together with the calculated y_i to give a new and more nearly correct set of $\hat{\phi}_i$ for use in (*7.164*), and a new set of y_i is determined. The process is repeated until successive iterations produce no significant change in the y_i. All calculations are done by computer, including calculation of the $\hat{\phi}_i$ by equations such as (*7.41*).

It is important to note that in this procedure the question of what chemical reactions are involved never enters directly into any of the equations. However, the choice of a set of species is entirely equivalent to the choice of a set of independent reactions among the species. In any event a set of species or a set of independent reactions must be assumed, and different assumptions will, in general, produce different results.

Example 7.22. Calculate the equilibrium composition at 1000(K) and 1(atm) of a system containing the gaseous species CH_4, H_2O, CO, CO_2, and H_2, if in the initial unreacted state there are present 2(g mole) of

CH_4 and 3(g mole) of H_2O. Values of ΔG_f° at 1000(K) are

$$\Delta G^\circ_{f_{CH_4}} = 4610\text{(cal)/(g mole)} \qquad \Delta G^\circ_{f_{H_2O}} = -46{,}030\text{(cal)/(g mole)}$$

$$\Delta G^\circ_{f_{CO}} = -47{,}940\text{(cal)/(g mole)} \qquad \Delta G^\circ_{CO_2} = -94{,}610\text{(cal)/(g mole)} \qquad \Delta G^\circ_{f_{H_2}} = 0$$

The required values of A_k are determined from the initial numbers of moles, and the values of a_{ik} come directly from the chemical formulas of the chemical species. These are shown in the table.

	Element k		
	Carbon	Oxygen	Hydrogen
	A_k = no. of atomic wts. of k in system		
	$A_C = 2$	$A_O = 3$	$A_H = 14$
Species i	a_{ik} = no. of atoms of k per molecule of i		
CH_4	$a_{CH_4,\,C} = 1$	$a_{CH_4,\,O} = 0$	$a_{CH_4,\,H} = 4$
H_2O	$a_{H_2O,\,C} = 0$	$a_{H_2O,\,O} = 1$	$a_{H_2O,\,H} = 2$
CO	$a_{CO,\,C} = 1$	$a_{CO,\,O} = 1$	$a_{CO,\,H} = 0$
CO_2	$a_{CO_2,\,C} = 1$	$a_{CO_2,\,O} = 2$	$a_{CO_2,\,H} = 0$
H_2	$a_{H_2,\,C} = 0$	$a_{H_2,\,O} = 0$	$a_{H_2,\,H} = 2$

At 1(atm) and 1000(K) the assumption of ideal gases should be fully justified, and the $\ln\hat{\phi}_i$ term of (*7.164*) can be omitted. In addition $\ln P$ is zero and (*7.164*) becomes

$$\Delta G^\circ_{f_i} + RT\ln y_i + \sum_k \lambda_k a_{ik} = 0$$

The five equations resulting from (*7.164*) are [with $RT = 1987$(cal)/(g mole)]:

$$\begin{aligned}
CH_4&: & 4610 + RT\ln y_{CH_4} + \lambda_C + 4\lambda_H &= 0\\
H_2O&: & -46{,}030 + RT\ln y_{H_2O} + 2\lambda_H + \lambda_O &= 0\\
CO&: & -47{,}940 + RT\ln y_{CO} + \lambda_C + \lambda_O &= 0\\
CO_2&: & -94{,}610 + RT\ln y_{CO_2} + \lambda_C + 2\lambda_O &= 0\\
H_2&: & RT\ln y_{H_2} + 2\lambda_H &= 0
\end{aligned}$$

From (*7.165*) there are three material balances:

$$\begin{aligned}
C&: & y_{CH_4} + y_{CO} + y_{CO_2} &= 2/\Sigma n_i\\
H&: & 4y_{CH_4} + 2y_{H_2O} + 2y_{H_2} &= 14/\Sigma n_i\\
O&: & y_{H_2O} + y_{CO} + 2y_{CO_2} &= 3/\Sigma n_i
\end{aligned}$$

In addition (*7.166*) requires

$$y_{CH_4} + y_{H_2O} + y_{CO} + y_{CO_2} + y_{H_2} = 1$$

Simultaneous solution of these nine equations provides the following results:

$$\begin{aligned}
y_{CH_4} &= 0.0199 & & \\
y_{H_2O} &= 0.0995 & \Sigma n_i &= 8.656\\
y_{CO} &= 0.1753 & \lambda_C/RT &= 0.797\\
y_{CO_2} &= 0.0359 & \lambda_O/RT &= 25.1\\
y_{H_2} &= 0.6694 & \lambda_H/RT &= 0.201\\
\hline
\Sigma y_i &= 1.0000 & &
\end{aligned}$$

The λ_i are of no real interest, but are included to make the results complete.

Solved Problems

SOLUTION THERMODYNAMICS (Secs. 7.1 through 7.5)

7.1. Derive (*7.2*).

The functions M^t discussed in connection with (*7.2*) represent extensive properties. This means that they are homogeneous in the first degree; i.e.

$$M^t(\alpha n_1, \alpha n_2, \ldots) = \alpha M^t(n_1, n_2, \ldots) \tag{1}$$

where T and P are presumed constant. For differential changes in $M^t(\alpha n_1, \alpha n_2, \ldots)$ resulting from differential variations in α at constant $n_1, n_2, \ldots$, we have from (*1*):

$$dM^t(\alpha n_1, \alpha n_2, \ldots) = M^t(n_1, n_2, \ldots)\, d\alpha$$

or

$$\frac{\partial M^t(\alpha n_1, \alpha n_2, \ldots)}{\partial \alpha} = M^t(n_1, n_2, \ldots) \tag{2}$$

However, by application of the chain rule we can also write

$$\frac{\partial M^t(\alpha n_1, \alpha n_2, \ldots)}{\partial \alpha} = \sum \left[\frac{\partial M^t(\alpha n_1, \alpha n_2, \ldots)}{\partial(\alpha n_i)}\right]_{\alpha n_j} \left[\frac{\partial(\alpha n_i)}{\partial \alpha}\right]_{n_i}$$

or

$$\frac{\partial M^t(\alpha n_1, \alpha n_2, \ldots)}{\partial \alpha} = \sum n_i \left[\frac{\partial M^t(\alpha n_1, \alpha n_2, \ldots)}{\partial(\alpha n_i)}\right]_{\alpha n_j} \tag{3}$$

where the summation is over all i.

Equating the right-hand sides of (*2*) and (*3*), we get

$$M^t(n_1, n_2, \ldots) = \sum n_i \left[\frac{\partial M^t(\alpha n_1, \alpha n_2, \ldots)}{\partial(\alpha n_i)}\right]_{\alpha n_j} \tag{4}$$

Since (*4*) is general, it must hold for the particular case of $\alpha = 1$, for which it becomes

$$M^t(n_1, n_2, \ldots) = \sum n_i \left[\frac{\partial M^t(n_1, n_2, \ldots)}{\partial n_i}\right]_{n_j} \tag{5}$$

Noting that $M^t(n_1, n_2, \ldots) \equiv nM$ and with the implicit condition of constant T and P that

$$\left[\frac{\partial M^t(n_1, n_2, \ldots)}{\partial n_i}\right]_{n_j} \equiv \left[\frac{\partial(nM)}{\partial n_i}\right]_{n_j} = \bar{M}_i$$

we see that (*5*) reduces to

$$nM = \sum n_i \bar{M}_i$$

which is (*7.2*). This result is a special case of *Euler's theorem on homogeneous functions.*

7.2. (*a*) Derive the *Gibbs-Duhem equation* for partial molar properties $\bar{M}_i$:

$$\sum x_i\, d\bar{M}_i = \left(\frac{\partial M}{\partial T}\right)_{P,x} dT + \left(\frac{\partial M}{\partial P}\right)_{T,x} dP \tag{1}$$

(*b*) Apply this equation to the two components of a binary solution at constant T and P, and give graphical interpretations of the result.

(*a*) According to (*7.2*), the total differential $d(nM)$ of the total property nM ($\equiv M^t$) is

$$d(nM) = \sum n_i\, d\bar{M}_i + \sum \bar{M}_i\, dn_i \tag{2}$$

However, nM ($\equiv M^t$) can be considered a function of T, P, and the n_i, and we can therefore write as an alternative expression

$$d(nM) = \left[\frac{\partial(nM)}{\partial T}\right]_{P,n} dT + \left[\frac{\partial(nM)}{\partial P}\right]_{T,n} dP + \sum \bar{M}_i\, dn_i \tag{3}$$

where the coefficient of dn_i in the summation has been abbreviated through use of the definition (*7.1*). Both (*2*) and (*3*) are valid for differential changes in nM resulting from differential variations in T, P, or the n_i. Therefore they may be combined to give

$$\sum n_i \, d\bar{M}_i \quad = \quad \left[\frac{\partial(nM)}{\partial T}\right]_{P,n} dT + \left[\frac{\partial(nM)}{\partial P}\right]_{T,n} dP \tag{4}$$

which is one form of the Gibbs-Duhem equation. The desired equation (*1*) is obtained by division of (*4*) by n, since constant n implies constant x.

(*b*) For a binary system at constant T and P, (*1*) becomes

$$x_1 \, d\bar{M}_1 + x_2 \, d\bar{M}_2 \quad = \quad 0 \tag{5}$$

However, the pure-component properties M_i are functions of T and P only, so that (*5*) may be rewritten as

$$x_1 \, d(\bar{M}_1 - M_1) + x_2 \, d(\bar{M}_2 - M_2) \quad = \quad 0 \tag{6}$$

Dividing (*5*) by dx_1 and noting that $x_2 = 1 - x_1$, we obtain

$$x_1 \frac{d\bar{M}_1}{dx_1} \quad = \quad -(1 - x_1) \frac{d\bar{M}_2}{dx_1} \tag{7}$$

Alternatively, we may divide (*6*) by dx_1, giving

$$x_1 \frac{d(\bar{M}_1 - M_1)}{dx_1} \quad = \quad -(1 - x_1) \frac{d(\bar{M}_2 - M_2)}{dx_1} \tag{8}$$

Equations (*7*) and (*8*) assert that the composition derivatives of the $\bar{M}_i$ (or the $\bar{M}_i - M_i$) at constant T and P are not independent, but are related as shown. For example, the slopes of the two representative $\bar{M}_i - M_i$ curves shown in Fig. 7-24 are of opposite sign at every value of x_1, as required by (*8*).

The composition extremes at $x_1 = 0$ and $x_1 = 1$ are of special interest. If $d\bar{M}_1/dx_1$ is finite at $x_1 = 0$, then from (*7*) and (*8*),

$$\frac{d\bar{M}_2}{dx_1} \quad = \quad \frac{d(\bar{M}_2 - M_2)}{dx_1} \quad = \quad 0 \qquad \text{at } x_1 = 0$$

Similarly, if $d\bar{M}_2/dx_2$ is finite at $x_2 = 0$, i.e. if $d\bar{M}_2/dx_1$ is finite at $x_1 = 1$, then

$$\frac{d\bar{M}_1}{dx_1} \quad = \quad \frac{d(\bar{M}_1 - M_1)}{dx_1} \quad = \quad 0 \qquad \text{at } x_1 = 1$$

The limiting derivatives $d\bar{M}_i/dx_i$ at $x_i = 0$ *are* finite for nonelectrolyte solutions, and the last two equations require that the $\bar{M}_i - M_i$ curves become tangent to the composition axis at $x_i = 1$. The two curves in Fig. 7-24 conform to this requirement.

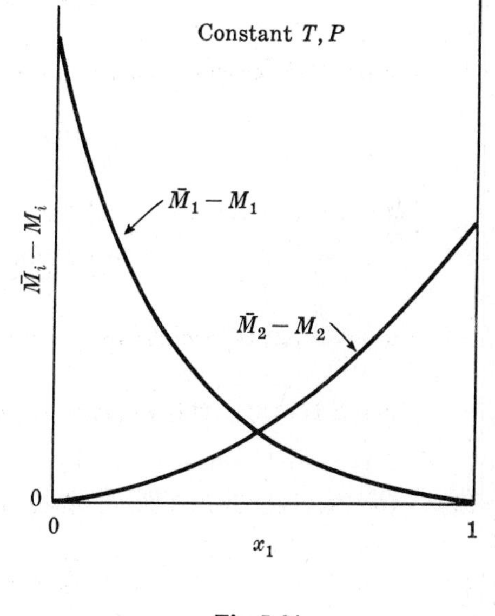

Fig. 7-24

A further important property of the $\bar{M}_i - M_i$ plot follows from (*6*). Consider the property change of mixing ΔM referred to standard states based on the Lewis and Randall rule:

$$\Delta M \quad = \quad M - \sum x_i M_i \quad = \quad x_1(\bar{M}_1 - M_1) + x_2(\bar{M}_2 - M_2)$$

The total differential of ΔM is

$$d\,\Delta M = x_1\,d(\bar{M}_1 - M_1) + (\bar{M}_1 - M_1)\,dx_1 + x_2\,d(\bar{M}_2 - M_2) + (\bar{M}_2 - M_2)\,dx_2$$

Since $x_1 + x_2 = 1$, $dx_2 = -dx_1$, and the first and third terms on the right in the last equation are related according to (*6*). The expression for $d\,\Delta M$ thus becomes

$$d\,\Delta M = [(\bar{M}_1 - M_1) - (\bar{M}_2 - M_2)]\,dx_1$$

Now $\Delta M = 0$ at $x_1 = 0$ and at $x_1 = 1$; as a result we obtain on integrating the last equation from $x_1 = 0$ to $x_1 = 1$

$$\int_0^1 [(\bar{M}_1 - M_1) - (\bar{M}_2 - M_2)]\,dx_1 = 0 \tag{9}$$

or, equivalently,

$$\int_0^1 (\bar{M}_1 - M_1)\,dx_1 = \int_0^1 (\bar{M}_2 - M_2)\,dx_1 \tag{10}$$

Equation (*10*) requires that the areas under the two $\bar{M}_i - M_i$ versus x_1 curves be equal, while (*9*) requires that the net area under a plot of $[(\bar{M}_1 - M_1) - (\bar{M}_2 - M_2)]$ versus x_1 be zero.

Equations (*7*) through (*10*) are examples of *thermodynamic consistency* requirements for the partial molar properties $\bar{M}_i$. Such requirements constitute necessary, but not sufficient, conditions for the validity of thermodynamic data. Thus, if $\bar{M}_1$ and $\bar{M}_2$ are determined for a binary solution at constant T and P, they may be subjected to a "slope test" [(*7*) or (*8*)] or to an "area test" [(*9*) or (*10*)]. If data do not satisfy any one of these tests they are inconsistent and therefore incorrect. Even if data prove consistent, however, they are not necessarily valid, for purely thermodynamic considerations can do no more than provide a rational basis for the rejection of data.

7.3. Prove that for an ideal solution of real gases $\delta_{jk} = 0$ for all j and k, where

$$\delta_{jk} = 2B_{jk} - B_{jj} - B_{kk}$$

Substitution of coefficients from (*5.10*) allows the virial equation (*5.8*) to be written in the form

$$Z = 1 + \frac{BP}{RT} + \frac{(C - B^2)}{(RT)^2}P^2 + \cdots$$

An expression for $\ln\phi$ then follows from application of (*7.20*):

$$\ln\phi = \int_0^P (Z-1)\frac{dP}{P} \qquad \text{(const. } T, x)$$

$$= \frac{BP}{RT} + (\quad)P^2 + (\quad)P^3 + \cdots \tag{1}$$

where the coefficients represented by parentheses are functions of temperature and composition only. For a pure component i, (*1*) gives

$$\ln\phi_i = \frac{B_{ii}P}{RT} + (\quad)P^2 + (\quad)P^3 + \cdots \tag{2}$$

An expression for the fugacity coefficient of component i in solution is obtained from (*1*) as indicated in Example 7.4. The extension of (*7.41*) of that example is

$$\ln\hat{\phi}_i = \frac{B_{ii}P}{RT} + \left[\frac{1}{2RT}\sum_j\sum_k y_j y_k(2\delta_{ji} - \delta_{jk})\right]P + \{\quad\}P^2 + \{\quad\}P^3 + \cdots \tag{3}$$

where the coefficients represented by braces are functions of temperature and composition only.

For standard states based on the Lewis and Randall rule, the activity coefficient of component i in a gas mixture is

$$\gamma_i = \frac{\hat{f}_i}{y_i f_i} = \frac{\hat{f}_i/y_i P}{f_i/P} = \frac{\hat{\phi}_i}{\phi_i}$$

from which we obtain

$$\ln \gamma_i = \ln \hat{\phi}_i - \ln \phi_i \tag{4}$$

Substitution of (2) and (3) into (4) yields

$$\ln \gamma_i = \left[\frac{1}{2RT} \sum_j \sum_k y_j y_k (2\delta_{ji} - \delta_{jk})\right] P + [\quad] P^2 + [\quad] P^3 + \cdots \tag{5}$$

where again the coefficients represented by brackets are functions of temperature and composition only.

For an ideal solution, $\ln \gamma_i = \ln(1) = 0$ for every component at all temperatures, pressures, and compositions. Thus the coefficient in brackets for each term of the expansion given by (5) must be identically zero. In particular, from the coefficient of P we obtain

$$\sum_j \sum_k y_j y_k (2\delta_{ji} - \delta_{jk}) = 0$$

The only way this equation can be valid for arbitrary compositions and temperatures is for δ_{jk} (and hence δ_{ji}) to be zero for all possible pairs j, k. Similar but much more complex relationships can be shown to hold for the third and higher virial coefficients of ideal solutions of gases; these relationships are obtained by equating the coefficients of the higher-order terms in (5) to zero.

7.4. A cleaning solution is to be manufactured from equal masses of acetone(a) and dichloromethane(d), both at 25(°C). If these components are mixed adiabatically at a pressure of 1(atm), with the addition of negligible stirring work, what is the temperature of the cleaning solution formed?

Data: Heat capacities of the pure components at 25(°C) and 1(atm):

$$C_{P_a} = 0.519(\text{cal})/(\text{g})(\text{K}) \qquad C_{P_d} = 0.285(\text{cal})/(\text{g})(\text{K})$$

Heats of mixing for the equal-mass acetone-dichloromethane solution at 1(atm):

T(°C)	ΔH(cal)/(g)
20	−2.978
25	−2.957
30	−2.936

For purposes of calculation we consider the process to consist of two steps, as indicated in Fig. 7-25. Since the process may be considered mechanically reversible, application of the first

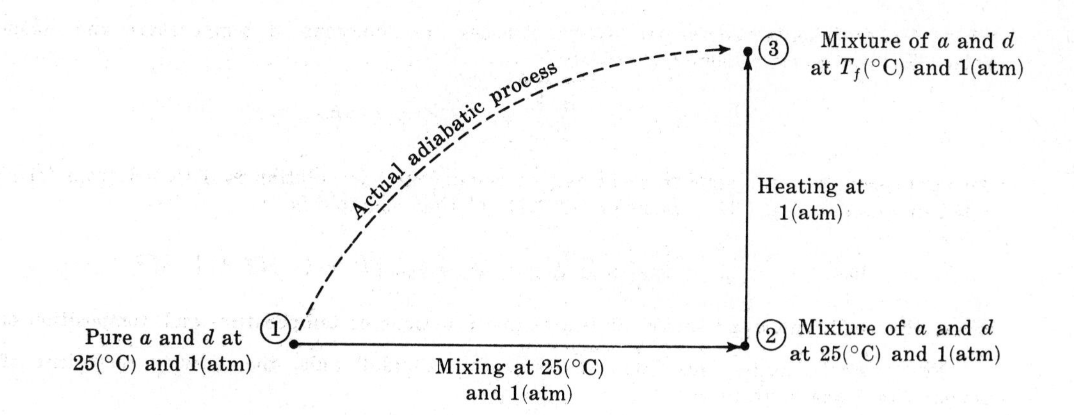

Fig. 7-25

law to the overall process gives simply (see Example 1.7):

$$\Delta H_{13}^{t} = \Delta H_{12}^{t} + \Delta H_{23}^{t} = 0 \tag{1}$$

An expression for the heat of mixing at constant T and P is given by (*7.59*), and we apply it to step 12:

$$\Delta H_{12}^{t} = (m_a + m_d)\,\Delta H \tag{2}$$

where ΔH is the heat of mixing per unit mass of solution.

The enthalpy change for step 23 is given by

$$\Delta H_{23}^{t} = (m_a + m_d)\int_{25}^{T_f} C_P\,dT = (m_a + m_d)C_P(T_f - 25)$$

where C_P is the specific heat capacity of the mixture, taken to be independent of T for the small temperature range considered here. This heat capacity is found by application of (*7.58*):

$$C_P = x_a C_{P_a} + x_d C_{P_d} + \Delta C_P$$

The heat capacity change of mixing ΔC_P is related to ΔH in the same way that C_P is related to H:

$$\Delta C_P = \left(\frac{\partial\,\Delta H}{\partial T}\right)_{P,x}$$

Combination of the last three equations yields the required expression for ΔH_{23}^{t}:

$$\Delta H_{23}^{t} = \left[m_a C_{P_a} + m_d C_{P_d} + (m_a + m_d)\left(\frac{\partial\,\Delta H}{\partial T}\right)_{P,x}\right](T_f - 25) \tag{3}$$

Substitution of (*2*) and (*3*) into (*1*) and solution for T_f gives

$$T_f = 25 - \frac{\Delta H}{x_a C_{P_a} + x_d C_{P_d} + (\partial\,\Delta H/\partial T)_{P,x}} \tag{4}$$

All the quantities in (*4*), except for the derivative, are given with the problem statement. The derivative can be approximated from the values of ΔH given for 20 and 30(°C):

$$\left(\frac{\partial\,\Delta H}{\partial T}\right)_{P,x} \cong \frac{-2.936 - (-2.978)}{30 - 20} = 0.0042(\text{cal})/(\text{g})(\text{K})$$

Substituting numerical values into (*4*), we get the required answer:

$$T_f = 25 - \frac{-2.957}{(0.50)(0.519) + (0.50)(0.285) + 0.0042} = 32.3(°\text{C})$$

The adiabatic mixing process considered here is seen to cause a temperature rise of 7.3(°C). We note that the sign of the temperature change is determined by the sign of ΔH. In the present case ΔH is negative (an *exothermic* heat of mixing). Had ΔH been positive (an *endothermic* heat of mixing), the temperature would have decreased. If ΔH were zero, e.g. if acetone-dichloromethane were an ideal solution, the temperature would not have changed upon mixing. We note further that the temperature change is independent of the *amount* of 50% solution formed. A temperature rise of 7.3(°C) occurs whether 1(lb$_m$) or 1(ton) of cleaning solution is produced.

7.5. Show how V_i°, H_i°, G_i°, and S_i° are related to f_i° and its derivatives.

The standard states considered here are regarded as real or fictitious states of pure materials, and equations valid for pure materials may therefore be used to find the necessary relationships. The equation for V_i° follows from (*7.38*) applied to the standard state:

$$V_i^\circ = RT\left[\frac{\partial(\ln f_i^\circ)}{\partial P}\right]_T \tag{1}$$

Similarly, H_i° is found from (*7.37*):

$$H_i^\circ = H_i' - RT^2\left[\frac{\partial(\ln f_i^\circ)}{\partial T}\right]_P \tag{2}$$

The expression for G_i° is obtained from (7.30):

$$G_i^\circ = G_i' + RT \ln (f_i^\circ/P) \qquad (3)$$

Finally, we find S_i° by differentiation of (3) as indicated by (3.39):

$$S_i^\circ = -\left(\frac{\partial G_i^\circ}{\partial T}\right)_P = -\left(\frac{\partial G_i'}{\partial T}\right)_P - RT\left[\frac{\partial(\ln f_i^\circ)}{\partial T}\right]_P + R \ln (f_i^\circ/P)$$

Again from (3.39), $S_i' = -(\partial G_i'/\partial T)_P$, and therefore

$$S_i^\circ = S_i' - RT\left[\frac{\partial(\ln f_i^\circ)}{\partial T}\right]_P - R \ln (f_i^\circ/P) \qquad (4)$$

7.6. Derive (7.65), (7.66), and (7.67).

Equation (7.62) provides a general relation between property changes of mixing and partial molar properties. Thus

$$\frac{P\,\Delta V}{RT} = \frac{P}{RT}\sum x_i(\bar{V}_i - V_i^\circ)$$

$$\frac{\Delta H}{RT} = \frac{1}{RT}\sum x_i(\bar{H}_i - H_i^\circ) \qquad (1)$$

$$\frac{\Delta S}{R} = \frac{1}{R}\sum x_i(\bar{S}_i - S_i^\circ)$$

The standard-state properties are interpreted as properties of the pure components, and all may be related to the single property G_i° by equations developed in Chapter 3. Thus from (3.38) and (3.39):

$$V_i^\circ = \left(\frac{\partial G_i^\circ}{\partial P}\right)_T \quad \text{and} \quad S_i^\circ = -\left(\frac{\partial G_i^\circ}{\partial T}\right)_P \qquad (2)$$

Similarly, by the Gibbs-Helmholtz equation (Problem 3.9),

$$H_i^\circ = -RT^2\left[\frac{\partial(G_i^\circ/RT)}{\partial T}\right]_P \qquad (3)$$

Analogous relationships hold for the corresponding partial molar properties. The basis for their derivation is provided by Problem 3.12, in which was developed the equation

$$d\mu_i = d\bar{G}_i = -\bar{S}_i\,dT + \bar{V}_i\,dP \qquad \text{(constant } x\text{)}$$

Interpretation of $-\bar{S}_i$ and $\bar{V}_i$ as partial differential coefficients of dT and dP yields by inspection [see (3.2) and (3.3)]:

$$\bar{S}_i = -\left(\frac{\partial \bar{G}_i}{\partial T}\right)_{P,x} \quad \text{and} \quad \bar{V}_i = \left(\frac{\partial \bar{G}_i}{\partial P}\right)_{T,x} \qquad (4)$$

Since, by (3.30) and (3.32), $nH = nG + T(nS)$, differentiation with respect to n_i at constant T, P, and n_j gives

$$\bar{H}_i = \bar{G}_i + T\bar{S}_i$$

Substitution for $\bar{S}_i$ yields the analogue of the Gibbs-Helmholtz equation:

$$\bar{H}_i = \bar{G}_i - T\left(\frac{\partial \bar{G}_i}{\partial T}\right)_{P,x} = -RT^2\left[\frac{\partial(\bar{G}_i/RT)}{\partial T}\right]_{P,x} \qquad (5)$$

Combination of (2), (3), (4), and (5) with (1) gives, on rearrangement:

$$\frac{P\,\Delta V}{RT} = \frac{1}{RT}\sum x_i\left[\frac{\partial(\bar{G}_i - G_i^\circ)}{\partial(\ln P)}\right]_{T,x}$$

$$\frac{\Delta H}{RT} = -\sum x_i\left\{\frac{\partial[(\bar{G}_i - G_i^\circ)/RT]}{\partial(\ln T)}\right\}_{P,x} \qquad (6)$$

$$\frac{\Delta S}{R} = -\frac{1}{RT}\sum x_i\left[\frac{\partial(\bar{G}_i - G_i^\circ)}{\partial(\ln T)}\right]_{P,x}$$

However, by (*7.63*), $\bar{G}_i - G_i^\circ = RT \ln(\hat{f}_i/f_i^\circ)$. Substitution of this expression into (*6*) gives (*7.65*), (*7.66*), and (*7.67*), as required.

7.7. (*a*) Using the results of Example 7.8, calculate and plot as a function of composition the fugacities of benzene and cyclohexane in liquid mixtures of these components at 40(°C) and 1(atm). The fugacities of pure liquid benzene and cyclohexane at these conditions are 0.240(atm) and 0.244(atm), respectively.

(*b*) Calculate the Henry's law constants for benzene and cyclohexane at 40(°C) and 1(atm).

(*a*) From the definition (*7.85*), the fugacity $\hat{f}_i$ is related to γ_i and f_i° by $\hat{f}_i = x_i f_i^\circ \gamma_i$. The activity-coefficient expressions of Example 7.8 are for standard states based on the Lewis and Randall rule, for which $f_i^\circ = f_i$. Thus combination of the defining equation with (*7.96a*) and (*7.96b*) yields for the two components benzene(1) and cyclohexane(2):

$$\hat{f}_1 = x_1 f_1 e^{B(1-x_1)^2}$$

$$\hat{f}_2 = (1-x_1) f_2 e^{Bx_1^2}$$

By the statement of the problem, $f_1 = 0.240$(atm) and $f_2 = 0.244$(atm); from Example 7.8, $B = 0.458$. Substitution of these values in the last two equations gives

$$\hat{f}_1 = 0.240\, x_1 e^{0.458(1-x_1)^2} \qquad (1)$$

$$\hat{f}_2 = 0.244(1-x_1) e^{0.458 x_1^2} \qquad (2)$$

These equations are represented by solid lines in Fig. 7-26.

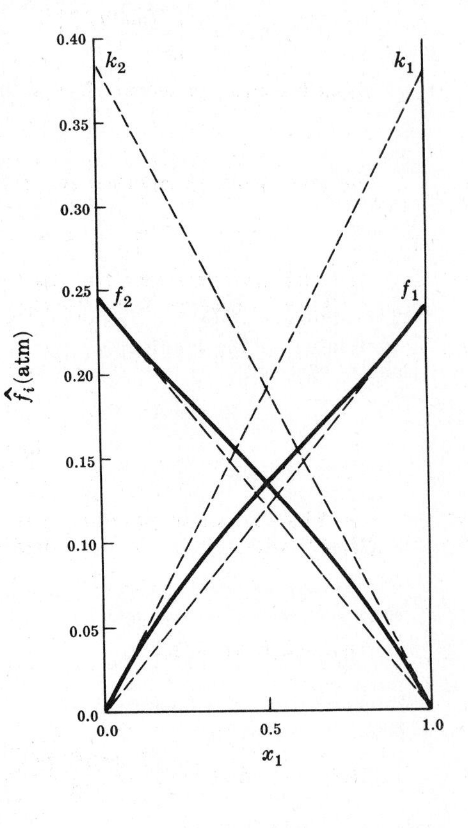

Fig. 7-26

(*b*) The Henry's law constants are calculated from (*1*) and (*2*) according to the definition (*7.56*):

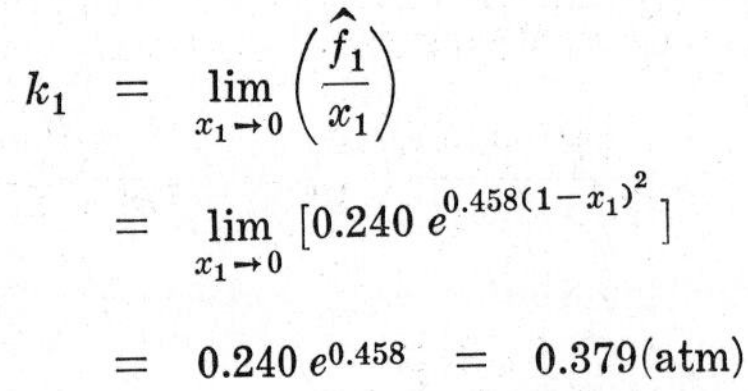

$$k_1 = \lim_{x_1 \to 0} \left(\frac{\hat{f}_1}{x_1}\right)$$

$$= \lim_{x_1 \to 0} [0.240\, e^{0.458(1-x_1)^2}]$$

$$= 0.240\, e^{0.458} = 0.379(\text{atm})$$

Similarly,
$$k_2 = \lim_{x_2 \to 0} \left(\frac{\hat{f}_2}{x_2}\right) = 0.244\, e^{0.458} = 0.386(\text{atm})$$

The upper pair of dashed lines in Fig. 7-26 represents the approximation to the actual $\hat{f}_i$ versus x_1 behavior provided by Henry's law, whereas the lower pair of dashed lines represents the approximation based on the Lewis and Randall rule. Henry's law clearly provides an excellent approximation for a component at high dilution, whereas the Lewis and Randall rule serves best for a component at high concentration.

7.8. The composition dependence of activity coefficients in real binary systems is only infrequently described by the symmetrical equations (*7.96a*) and (*7.96b*). One can, however, generate more flexible expressions for $\ln \gamma_i$ by retaining more terms in the dimensionless G^E equations (*7.91*) and (*7.92*). Derive the two-constant *Margules* and *van Laar* expressions for $\ln \gamma_i$ by letting $D = \cdots = 0$ and $D' = \cdots = 0$ in (*7.91*) and (*7.92*), respectively. Give graphical comparisons of the equations for representative values of the constants.

Solution of (*7.91*) for G^E/RT, with $D = \cdots = 0$, yields

$$G^E/RT = Bx_1x_2 + Cx_1x_2(x_1 - x_2)$$

from which the expression for nG^E/RT is found to be

$$\frac{nG^E}{RT} = \frac{Bn_1n_2}{n} + \frac{Cn_1n_2(n_1 - n_2)}{n^2}$$

Application of (*7.90*) to the last equation gives

$$\ln \gamma_1 = \left[\frac{\partial(nG^E/RT)}{\partial n_1}\right]_{T,P,n_2} = \frac{Bn_2}{n} - \frac{Bn_1n_2}{n^2} + \frac{C(2n_1 - n_2)n_2}{n^2} - \frac{2C(n_1 - n_2)n_1n_2}{n^3}$$

which becomes, in terms of mole fractions,

$$\ln \gamma_1 = x_2^2[B - C(4x_2 - 3)]$$

The corresponding equation for $\ln \gamma_2$ is found analogously, and is

$$\ln \gamma_2 = x_1^2[B + C(4x_1 - 3)]$$

The last two equations are the two-constant Margules equations for a binary mixture. They take a slightly different form through use of the substitutions $A_{12} = B - C$, $A_{21} = B + C$:

$$\ln \gamma_1 = A_{12}x_2^2\left[1 + 2x_1\left(\frac{A_{21}}{A_{12}} - 1\right)\right]$$
$$\ln \gamma_2 = A_{21}x_1^2\left[1 + 2x_2\left(\frac{A_{12}}{A_{21}} - 1\right)\right] \qquad (1)$$

The van Laar equations are found from (*7.92*) in a similar manner. Solution of (*7.92*) for G^E/RT, with $D' = \cdots = 0$, yields

$$\frac{G^E}{RT} = \frac{x_1x_2}{B' + C'(x_1 - x_2)}$$

from which we obtain

$$\frac{nG^E}{RT} = \frac{n_1n_2}{nB' + C'(n_1 - n_2)}$$

Then

$$\ln \gamma_1 = \left[\frac{\partial(nG^E/RT)}{\partial n_1}\right]_{T,P,n_2} = \frac{(B' - C')n_2^2}{[nB' + C'(n_1 - n_2)]^2} = \frac{(B' - C')x_2^2}{[B' - C'(2x_2 - 1)]^2}$$

Similarly, we find

$$\ln \gamma_2 = \frac{(B' + C')x_1^2}{[B' + C'(2x_1 - 1)]^2}$$

These are the van Laar equations. Making the substitutions

$$A'_{12} = \frac{1}{B' - C'} \qquad A'_{21} = \frac{1}{B' + C'}$$

we can write them as

$$\ln \gamma_1 = \frac{A'_{12}x_2^2}{\left[1 + \left(\frac{A'_{12}}{A'_{21}} - 1\right)x_1\right]^2}$$
$$\ln \gamma_2 = \frac{A'_{21}x_1^2}{\left[1 + \left(\frac{A'_{21}}{A'_{12}} - 1\right)x_2\right]^2} \qquad (2)$$

The constants A_{12} and A_{21} (or A'_{12} and A'_{21}) are respectively equal to $\ln \gamma_1^\infty$ and $\ln \gamma_2^\infty$, the logarithms of the infinite-dilution values of the activity coefficients. Thus one basis for comparison of (*1*) and (*2*) is provided if we set

$$A_{12} = A'_{12} \quad \text{and} \quad A_{21} = A'_{21}$$

Equivalently, we can set

$$A_{12} = A'_{12} \quad \text{and} \quad A_{21}/A_{12} = A'_{21}/A'_{12}$$

Using the second pair of equations, we will illustrate the behavior of the Margules and van Laar equations by specification of the following values:

$$A_{12} = A'_{12} = 0.5 \quad \text{(fixed)}$$

$$A_{21}/A_{12} = A'_{21}/A'_{12} = 1.0, 1.5, 2.0, 2.5 \quad \text{(four cases)}$$

In each case the Margules and van Laar equations will yield the same limiting values of the activity coefficients at $x_1 = 0$ and $x_1 = 1$, but intermediate values will depend on the form of the equation. Results for the Margules equations (*1*) are shown by Fig. 7-27 and for the van Laar equations (*2*) by Fig. 7-28. When $A_{21}/A_{12} = A'_{21}/A'_{12} = 1$, the two sets of equations (*1*) and (*2*)

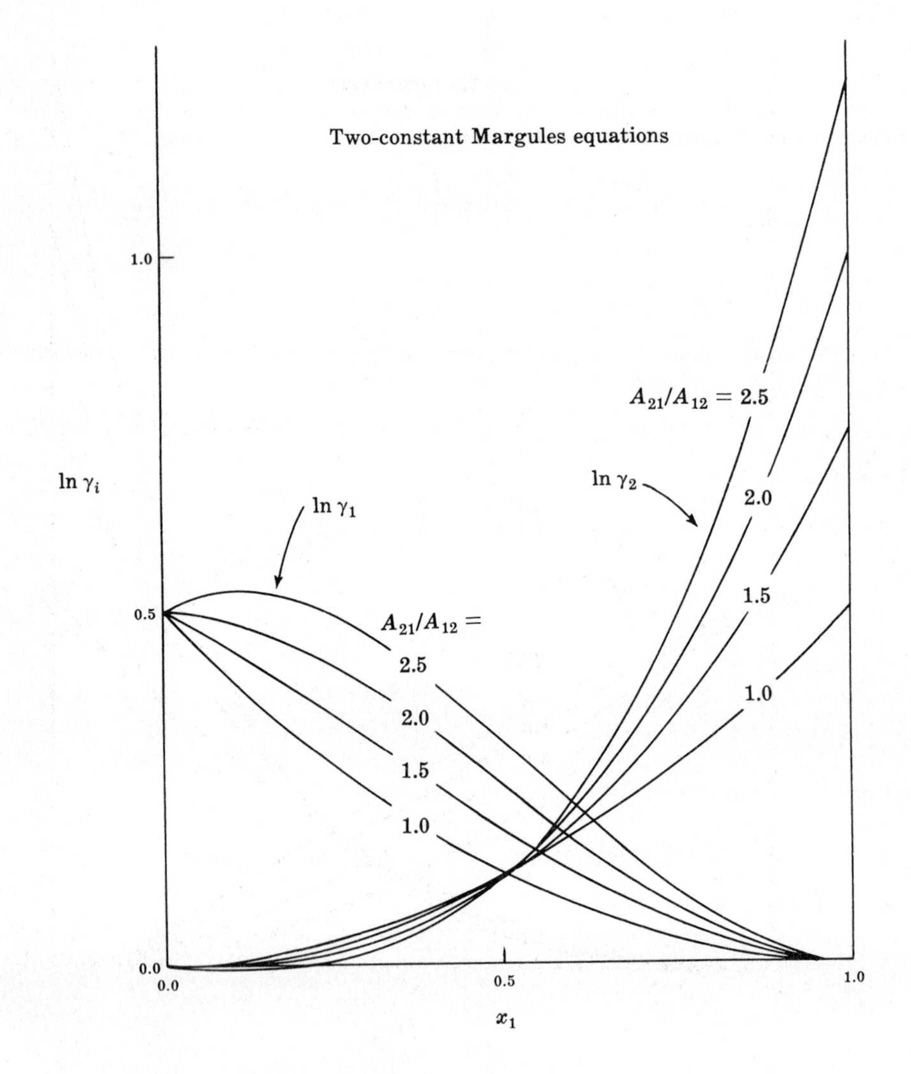

Fig. 7-27

become identical, and in fact reduce to the symmetrical equations (*7.96a*) and (*7.96b*) with $B = 0.5$. However, as the ratio $A_{21}/A_{12} = A'_{21}/A'_{12}$ increases, differences between the sets of equations become apparent. In particular, the $\ln \gamma_2$ curves given by the van Laar equation approach their limiting values at $x_1 = 1$ with steeper slopes than do the corresponding curves given by the Margules equation. On the other hand, the $\ln \gamma_1$ curves given by the Margules equation show a much stronger dependence on the value of $A_{21}/A_{12} = A'_{21}/A'_{12}$ than do the corresponding curves determined from the van Laar equation. In fact, as seen from Fig. 7-27 the $\ln \gamma_1$ curves given by the Margules equation exhibit a maximum (with a corresponding minimum in the $\ln \gamma_2$ curve) for values of $A_{21}/A_{12} > 2$.

The use of the Margules and van Laar equations is in the fitting of experimental data for activity coefficients, and the different characteristics of the two equations provide for the accommodation of a considerable range of behavior.

7.9. Find expressions for the temperature dependence of the coefficient B in (*7.93*) for solutions for which (*a*) $H^E = 0$, and (*b*) $S^E = 0$.

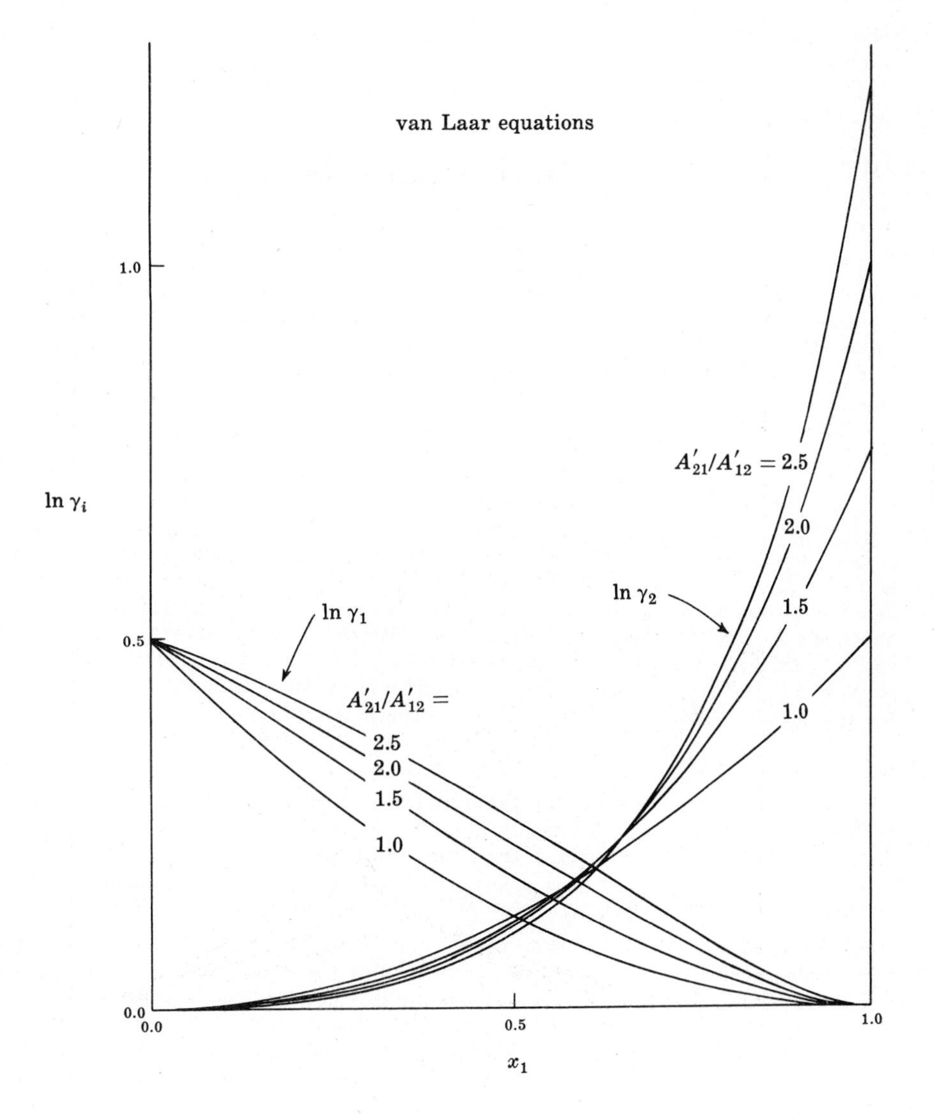

Fig. 7-28

(*a*) By (*7.94*) and the problem statement,

$$-x_1x_2T\left(\frac{\partial B}{\partial T}\right)_P = \frac{H^E}{RT} = 0 \quad \text{or} \quad \left(\frac{\partial B}{\partial T}\right)_P = 0$$

Thus, B is *independent* of temperature. Solutions for which $H^E = 0$ are called *athermal.*

(*b*) By (*7.95*) and the problem statement,

$$-x_1x_2\left\{B + \left[\frac{\partial B}{\partial(\ln T)}\right]_P\right\} = \frac{S^E}{R} = 0 \quad \text{or} \quad \left[\frac{\partial(\ln B)}{\partial(\ln T)}\right]_P = -1$$

Integration of this equation gives

$$B = \frac{\beta}{RT} \tag{1}$$

where the coefficient β is a function of P only, and has been defined so as to have units of (energy)/(mole). According to (*1*), B varies inversely with T; the corresponding expression for G^E, (*7.93*), then becomes

$$G^E = \beta x_1x_2 \tag{2}$$

Solutions for which S^E *and* V^E are zero are called *regular*; for such solutions β is a constant, independent of T or P, and G^E becomes a function of composition only.

PHASE EQUILIBRIUM CALCULATIONS (Secs. 7.6 and 7.7)

7.10. Discuss the liquid-solid Tx diagrams shown in Fig. 7-29(*a*), (*b*), and (*c*).

(*a*) This diagram is for a system in which the components form a continuous series of homogeneous *solid solutions.* The properties of such diagrams are analogous to those of vapor-liquid Tx diagrams of the type shown in Fig. 7-12. Thus, mixtures with Tx coordinates falling above the freezing curve ACD are homogeneous liquid solutions; similarly, states below the melting curve ABD are solid solutions. States of coexisting liquid and solid solutions are connected as shown by horizontal tie lines. Examples of pairs of compounds having this type of liquid-solid phase diagram are nitrogen-carbon monoxide, silver chloride-sodium chloride, and copper-nickel.

(*b*) Figure 7-29(*b*) represents a system which exhibits a eutectic, but in which the pure components are completely insoluble in the solid phase. This type of diagram is a limiting case of the kind shown in Fig. 7-18, for which partial solubility obtains in the solid phase. Thus, in Fig. 7.29(*b*), the regions $AECA$ and $BEDB$ are two-phase liquid-solid regions. In $AECA$, liquid mixtures with compositions given by the curve AE are in equilibrium with crystals of pure solid 2, while in $BEDB$ liquid mixtures with compositions given by BE are in equilibrium with crystals of pure solid 1. Below the melting line CED, pure solid 1 is in equilibrium with pure solid 2. *Complete* solid-solid immiscibility is an idealization which never strictly obtains for real systems; however, some systems approach this behavior.

(*c*) Figure 7-29(*c*) depicts a system for which a third chemical species, represented by the vertical line $DFEK$, is formed in the solid phase and for which all species are completely insoluble as solids. If the composition coordinate is mole fraction, then the chemical formula of the compound can be directly determined from the value of x_1 for the line $DFEK$. In the present case the compound is found at a mole fraction of 0.5, so that the chemical formula is MN (or a multiple of MN, say M_lN_l), where M and N represent species 1 and 2.

Figure 7-29(*c*) can be considered a combination of two diagrams of the type of Fig. 7-29(*b*) placed side by side, with the properties of the two halves of the diagram (to the left and to the right of $DFEK$) separately similar to those of Fig. 7-29(*b*). Thus, points C and G represent eutectic mixtures of N (component 2) with MN, and M (component 1) with MN, respectively. Because of the formation of the intermediate compound, pure solid M and N cannot coexist as equilibrium phases; within the area $BCEKLB$, pure solid N is in equilibrium with solid MN, while pure solid M coexists with solid MN in the area $EFGHJKE$. As with Fig. 7-29(*b*), Fig. 7-29(*c*) represents a type of limiting behavior which is only approximated by real systems.

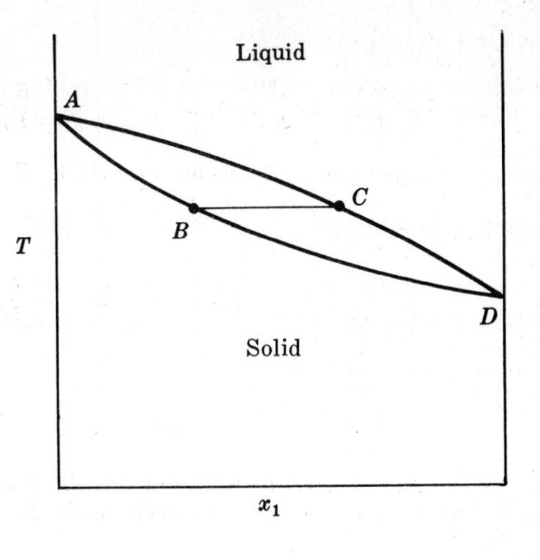

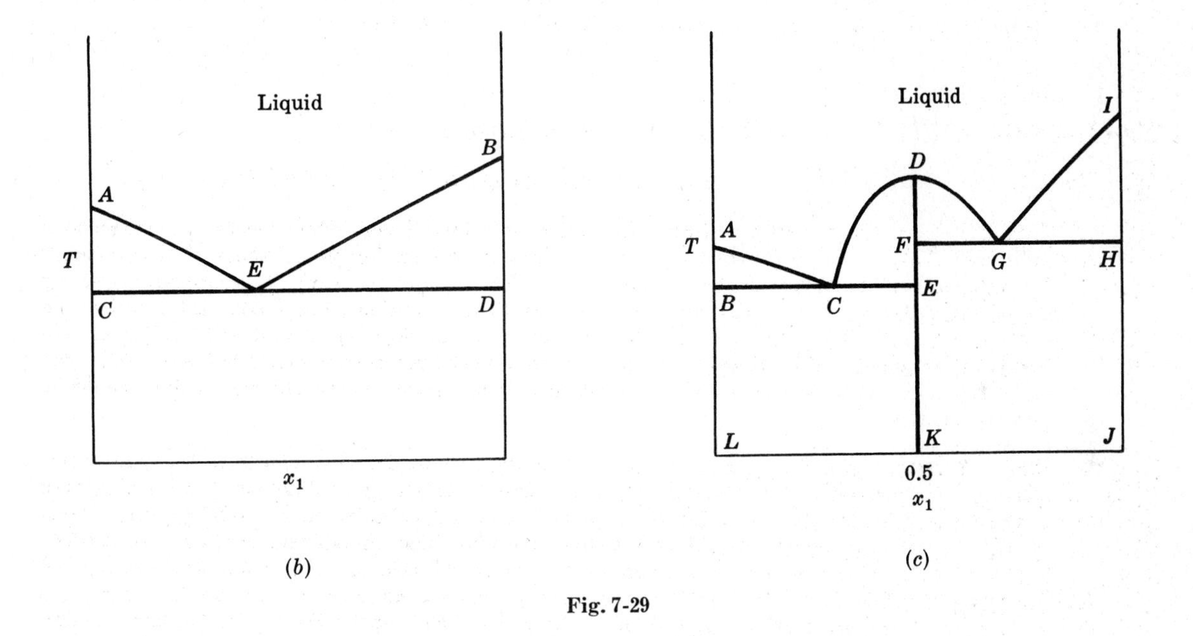

Fig. 7-29

7.11. Experimental VLE data for the isopropanol(1)-benzene(2) system at 45(°C) are [Brown, I., Aust. J. Chem., *9*, 364 (1956)]:

x_1	y_1	P(mm Hg)	x_1	y_1	P(mm Hg)
0.0000	0.0000	223.74	0.5504	0.3692	264.92
0.0472	0.1467	252.50	0.6198	0.3951	259.35
0.0980	0.2066	264.13	0.7096	0.4378	247.70
0.2047	0.2663	272.06	0.8073	0.5107	227.14
0.2960	0.2953	273.40	0.9120	0.6658	189.28
0.3862	0.3211	272.22	0.9655	0.8252	159.80
0.4753	0.3463	269.49	1.0000	1.0000	136.05

(a) Plot the dew- and bubble-point curves versus mole fraction of isopropanol. Calculate and plot versus x_1 on the same diagram the vapor-phase partial pressures P_1 and P_2. Compare these curves with the bubble-point and partial-pressure curves given by Raoult's law.

(b) Derive values of $\ln\gamma_1$ and $\ln\gamma_2$ from the data and plot them against x_1. Plot on the same graph the G^E/x_1x_2RT curve, and show for comparison the G^E/x_1x_2RT curve one would obtain from the two-constant Margules equation (see Problem 7.8) if the constants were determined from experimental activity coefficients at infinite dilution.

(a) The dew-point $(P - y_1)$ and bubble-point $(P - x_1)$ curves are shown on Fig. 7-30; the corresponding bubble-point curve for Raoult's law is given as the straight line $P - x_1$ (RL). This system exhibits sufficiently large positive deviations from Raoult's law to give a maximum-pressure azeotrope at about $x_1 = 0.29$ and $P = 273.5$(mm Hg).

Values of the partial pressures are calculated from the given data by the definition $P_i = y_iP$, and are shown as the curves P_1 and P_2. The corresponding curves for Raoult's law are the straight lines P_1 (RL) and P_2 (RL).

(b) For these low pressures, the approximate equation (*7.101*) is entirely adequate for description of the VLE of the system. Solving for γ_i and taking logarithms, we obtain

$$\ln\gamma_i = \ln\left(\frac{y_iP}{x_iP_i^{sat}}\right) \tag{1}$$

Equation (*1*) is used to reduce the given data to $\ln\gamma_1$ and $\ln\gamma_2$ values for the liquid phase. The value of G^E/x_1x_2RT corresponding to each of these pairs of $\ln\gamma_i$ values is calculated from

$$\frac{G^E}{x_1x_2RT} = \frac{\ln\gamma_1}{x_2} + \frac{\ln\gamma_2}{x_1} \tag{2}$$

where (*2*) is obtained by writing (*7.86*) for a binary system and dividing the result by x_1x_2.

We apply these equations to the data for $x_1 = 0.4753$. The pure-component vapor pressures are equal to the total pressures at $x_1 = 0.0$ and $x_1 = 1.0$; thus $P_1^{sat} = 136.05$(mm Hg) and $P_2^{sat} = 223.74$(mm Hg). Equation (*1*) then gives

$$\ln\gamma_1 = \ln\left(\frac{y_1P}{x_1P_1^{sat}}\right) = \ln\left(\frac{0.3463\times 269.49}{0.4753\times 136.05}\right) = 0.3669$$

Similarly $$\ln\gamma_2 = 0.4059$$

Substitution of these values into (*2*) then gives the corresponding value of G^E/x_1x_2RT:

$$\frac{G^E}{x_1x_2RT} = \frac{0.3669}{1-0.4753} + \frac{0.4059}{0.4753} = 1.553$$

The complete curves for $\ln\gamma_1$, $\ln\gamma_2$, and G^E/x_1x_2RT are plotted on Fig. 7-31. It can be shown that the limiting values of G^E/x_1x_2RT as $x_1 \to 0$ and $x_1 \to 1$ are equal to $\ln\gamma_1^\infty$ and $\ln\gamma_2^\infty$, respectively. Thus we find by extrapolation that $\ln\gamma_1^\infty = 2.18$ and $\ln\gamma_2^\infty = 1.44$. The $\ln\gamma_i$ curves, which are more difficult to extrapolate, are drawn so as to yield these intercepts.

According to the results of Problem 7.8, the Margules equation gives

$$G^E/x_1x_2RT = A_{12} + (A_{21} - A_{12})x_1$$

where A_{12} and A_{21} are the intercepts with the G^E/x_1x_2RT axes at $x_1 = 0$ and $x_1 = 1$. If these intercepts are made equal to the experimental values of $\ln\gamma_1^\infty$ and $\ln\gamma_2^\infty$, respectively, then the dashed straight line shown on Fig. 7-31 results. Clearly, the two-constant Margules equation is incapable of accurately representing the liquid-phase activity coefficients for this system; for even if other values were chosen for A_{12} and A_{21}, a straight line would always result for the G^E/x_1x_2RT curve.

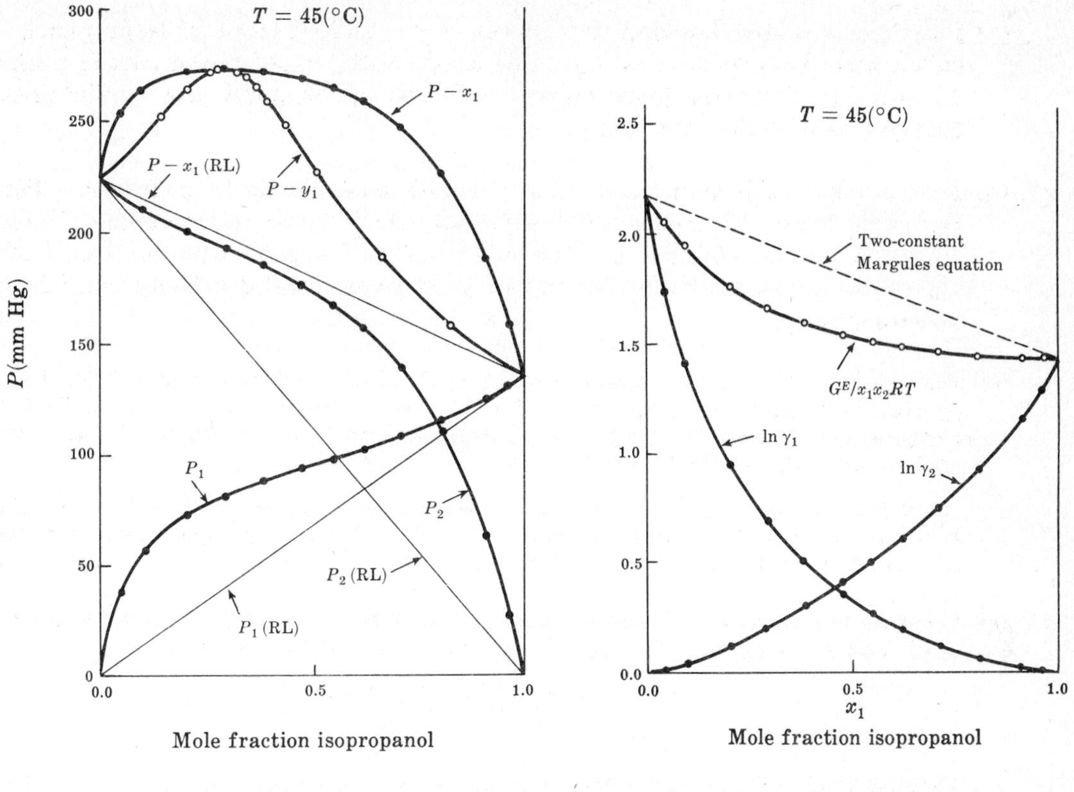

Fig. 7-30

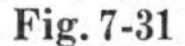

Fig. 7-31

7.12. Derive equations for description of the VLE behavior of a binary system in which the components are completely immiscible in the liquid phase. Base the derivation on the approximate VLE relationship (*7.101*).

Solution of the problem is expedited by consideration of the Tx diagram of the system, shown in Fig. 7-32. Comparison with Fig. 7-29(*b*) shows that the present system is the vapor-liquid analogue of liquid-solid systems in which the solid phases are immiscible, and for which eutectic behavior is observed. Temperatures T_2 and T_1 are the boiling points at pressure P of pure 2 and 1, respectively. There are two distinct VLE regions. Within the area $AECA$ (region I) vapor mixtures with compositions given by the curve AE are in equilibrium with pure liquid 2; within the area $BEDB$ (region II) vapor mixtures with compositions given by BE are in equilibrium with pure liquid 1. At the temperature T^{az}, vapor of composition y_1^{az} is in equilibrium with *two* liquid phases containing pure 2 and pure 1, respectively. The exceptional point E represents a *heterogeneous azeotrope,* the vapor-liquid analogue of a liquid-solid eutectic.

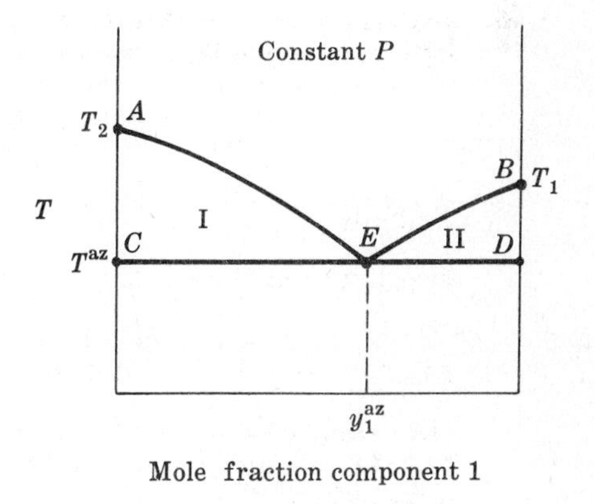

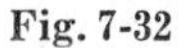

Fig. 7-32

For VLE in region I, (*7.101*) written for component 2 is $y_2P = x_2\gamma_2P_2^{sat}$. But $x_2 = 1.0$ and therefore $\gamma_2 = 1.0$ in region I (we assume the use of standard states based on the Lewis and Randall rule). Thus we obtain for region I

$$y_2(\text{I}) = \frac{P_2^{sat}}{P} \qquad y_1(\text{I}) = 1 - \frac{P_2^{sat}}{P} \tag{1}$$

For VLE in region II, (*7.101*) written for component 1 is $y_1 P = x_1 \gamma_1 P_1^{\text{sat}}$. But $x_1 = 1.0$ and $\gamma_1 = 1.0$, and we obtain

$$y_1(\text{II}) = \frac{P_1^{\text{sat}}}{P} \qquad y_2(\text{II}) = 1 - \frac{P_1^{\text{sat}}}{P} \tag{2}$$

Either of the equations (*1*) represents the dew-point curve AE; similarly, either of the equations (*2*) represents the dew-point curve BE. These equations are the basis for the familiar statement that the liquid components of such systems "exert their own vapor pressures." Thus, the partial pressure P_2 ($\equiv y_2 P$) of 2 in region I is given by (*1*) as $P_2(\text{I}) = P_2^{\text{sat}}$. Similarly, the partial pressure P_1 ($\equiv y_1 P$) of 1 in region II is given by (*2*) as $P_1(\text{II}) = P_1^{\text{sat}}$.

The two dew-point curves converge at E, and therefore at this point $y_1(\text{I}) = y_1(\text{II})$, or $y_2(\text{I}) = y_2(\text{II})$. Combination of (*1*) and (*2*) thus gives, on rearrangement, the following implicit equation for T^{az}:

$$P = P_1^{\text{sat}} + P_2^{\text{sat}} \tag{3}$$

The azeotropic composition y_1^{az} is found by substitution of (*3*) into (*1*) or (*2*):

$$y_1^{\text{az}} = \frac{P_1^{\text{sat}}}{P_1^{\text{sat}} + P_2^{\text{sat}}} \tag{4}$$

7.13. (*a*) Derive equations for determination of equilibrium conditions in two-component, liquid-liquid systems. (*b*) Determine the equation of the solubility curve of a binary liquid system for which the G^E expression is given by (*7.93*).

(*a*) We designate the two liquid phases by the superscripts α and β, Equation (*7.46*), written for the two components 1 and 2, gives $\hat{f}_1^\alpha = \hat{f}_1^\beta$ and $\hat{f}_2^\alpha = \hat{f}_2^\beta$. Elimination of the $\hat{f}_i$ in favor of activity coefficients by use of (*7.85*) gives

$$x_1^\alpha \gamma_1^\alpha = x_1^\beta \gamma_1^\beta \tag{1}$$

$$x_2^\alpha \gamma_2^\alpha = x_2^\beta \gamma_2^\beta \tag{2}$$

where it has been assumed that $(f_i^\circ)^\alpha = (f_i^\circ)^\beta$, i.e. that γ_i^α and γ_i^β for a given component are referred to the same standard state.

The γ_i are functions of T, P, and composition; thus, for fixed T and P, (*1*) and (*2*) constitute two equations in the two unknowns x_1^α and x_1^β (or x_2^α and x_2^β). The solubility curve is the curve determined by the set of x_1^α, x_1^β, and T values (or x_2^α, x_2^β, and T values) which satisfy both (*1*) and (*2*) for a given pressure P.

(*b*) Equations (*1*) and (*2*) may be written in alternate forms which are more convenient for application. Thus we obtain, on taking logarithms and rearranging,

$$\ln\left(\frac{x_1^\alpha}{x_1^\beta}\right) = \ln \gamma_1^\beta - \ln \gamma_1^\alpha \tag{3}$$

$$\ln\left(\frac{1 - x_1^\alpha}{1 - x_1^\beta}\right) = \ln \gamma_2^\beta - \ln \gamma_2^\alpha \tag{4}$$

The expressions for $\ln \gamma_i$ consistent with (*7.93*) are given by (*7.96a*) and (*7.96b*) of Example 7.8. When applied to components 1 and 2 in the α and β phases, they yield the following expressions:

$$\ln \gamma_1^\beta - \ln \gamma_1^\alpha = B(x_1^\beta - x_1^\alpha)[(x_1^\beta - 1) + (x_1^\alpha - 1)] \tag{5}$$

$$\ln \gamma_2^\beta - \ln \gamma_2^\alpha = B(x_1^\beta - x_1^\alpha)(x_1^\alpha + x_1^\beta) \tag{6}$$

Substitution of (*5*) into (*3*), and (*6*) into (*4*), and elimination of B from the resulting equations, gives

$$(x_1^\alpha + x_1^\beta) \ln\left(\frac{x_1^\alpha}{x_1^\beta}\right) = [(1 - x_1^\beta) + (1 - x_1^\alpha)] \ln\left(\frac{1 - x_1^\beta}{1 - x_1^\alpha}\right) \tag{7}$$

Equation (*7*) is satisfied by all pairs (x_1^α, x_1^β) for which

$$x_1^\beta = 1 - x_1^\alpha \tag{8}$$

This requires that the solubility curve be symmetrical about $x_1 = 1/2$. The equation of the curve is found by combination of (3), (5), and (8); the result is

$$\ln\left(\frac{x_1^\alpha}{1 - x_1^\alpha}\right) = B(2x_1^\alpha - 1) \tag{9}$$

Numerical values of the solubilities x_1^α and x_1^β for specified T and P can be found from (8) and (9) if B is given as a function of T and P.

7.14. For the completely miscible binary system benzene(1)-chloroform(2) the VLE K-values at 70(°C) and 1(atm) are $K_1 = 0.719$, $K_2 = 1.31$. Calculate the equilibrium compositions of the liquid and vapor phases and the fraction of the total system which is liquid for a total system composition of (*a*) 40 mole % benzene, and (*b*) 50 mole % benzene.

The mole fractions of components 1 and 2 must sum to unity in each phase:

$$x_1 + x_2 = 1 \qquad y_1 + y_2 = 1$$

Formal elimination of y_1 and y_2 in favor of x_1 and x_2 by the definition (7.112) reduces the second equation to $K_1x_1 + K_2x_2 = 1$, which combines with the first equation to give

$$x_1 = \frac{K_2 - 1}{K_2 - K_1} \tag{1}$$

or, since $y_1 = K_1x_1$,

$$y_1 = K_1\frac{K_2 - 1}{K_2 - K_1} \tag{2}$$

For specified T and P, there is only one possible set of equilibrium compositions, and hence one set of K-values, in a two-phase, two-component system. Thus the equilibrium compositions (*if* two phases actually exist) are independent of the overall system composition for given values of T and P. This is reflected by the fact that (1) and (2) do not contain the overall compositions z_1 and z_2. Thus we obtain both for part (*a*) and for part (*b*)

$$x_1 = \frac{1.31 - 1}{1.31 - 0.719} = 0.525 \qquad x_2 = 1 - x_1 = 0.475$$

$$y_1 = (0.719)(0.525) = 0.377 \qquad y_2 = 1 - y_1 = 0.623$$

Although L (the fraction of the system that is liquid) could be determined by application of the general equation (7.114), it is more simply calculated for a binary system by direct material balance. Thus, on the basis of a total system of one mole, we can write $x_1L + y_1(1 - L) = z_1$. Substitution of expressions (1) and (2) for x_1 and y_1 and solution for L then gives

$$L = \frac{z_1}{K_2 - 1} + \frac{z_1 - K_1}{1 - K_1} \tag{3}$$

Using the data we obtain for (*a*):

$$L = \frac{0.40}{1.31 - 1} + \frac{0.40 - 0.719}{1 - 0.719} = 0.155$$

and for (*b*):

$$L = 0.834$$

Thus, although the equilibrium compositions resulting from a flash calculation for a binary system are independent of the feed composition, the relative amounts of liquid and vapor formed are not.

CHEMICAL-REACTION CALCULATIONS (Secs. 7.8 and 7.9)

7.15. The single independent chemical reaction $0 = \sum_i \nu_i X_i(p)$ is to be carried out in a steady-state-flow reactor. All chemical species present in either the feed or the product streams are in their standard states at the same temperature T. Derive an expression for the rate of heat transfer $\dot{Q}$ to the chemical reactor.

Equation (*6.9*) is applicable. Neglecting potential- and kinetic-energy terms, and assuming that no shaft work is done, we obtain

$$\dot{Q} = \dot{n}_P H_P - \dot{n}_F H_F \tag{1}$$

where $\dot{n}_P$ is the total molar product rate, $\dot{n}_F$ is the total molar feed rate, H_P is the molar enthalpy of the products, and H_F is the molar enthalpy of the feed.

All of the chemical species are in their standard states, and therefore

$$H_P = \sum_i (z_i)_P H_i^\circ \tag{2}$$

$$H_F = \sum_i (z_i)_F H_i^\circ \tag{3}$$

where $(z_i)_P$ is the overall mole fraction of species i in the product, $(z_i)_F$ is the overall mole fraction of i in the feed, and H_i° is the standard-state enthalpy of i. Equations (*2*) and (*3*) are valid when each species i enters and leaves the reactor in its standard state as a *pure component*. However, they also hold for ideal-gas mixtures, because the pure ideal-gas state at 1(atm) *is* a standard state, and because there is neither a pressure effect on the enthalpy nor an enthalpy change of mixing for ideal gases.

Combination of (*1*), (*2*), and (*3*) gives

$$\dot{Q} = \sum_i [(\dot{n}_i)_P - (\dot{n}_i)_F] H_i^\circ \tag{4}$$

where $(\dot{n}_i)_P$ and $(\dot{n}_i)_F$ are the molar product and feed rates, respectively, of component i. These rates are not independent, however, but are related through the reaction stoichiometry to the rate of conversion, as measured by the time derivative $\dot{\epsilon}$ of the reaction coordinate ϵ. The required relationship is obtained from (*7.125*) on formal substitution of $(\dot{n}_i)_P$ for n_i, $(\dot{n}_i)_F$ for n_{i_0}, and $\dot{\epsilon}$ for ϵ:

$$(\dot{n}_i)_P = (\dot{n}_i)_F + \nu_i \dot{\epsilon} \tag{5}$$

Substitution of (*5*) into (*4*) gives

$$\dot{Q} = \sum_i \nu_i \dot{\epsilon} H_i^\circ = \left(\sum_i \nu_i H_i^\circ\right) \dot{\epsilon}$$

or

$$\dot{Q} = \Delta H^\circ \dot{\epsilon} \tag{6}$$

where, by (*7.135*), $\Delta H^\circ = \sum_i \nu_i H_i^\circ$. Equation (*6*) asserts that $\dot{Q}$ for the prescribed process is proportional to the standard enthalpy change of reaction. This expression is the basis for many isothermal "heat of reaction" calculations in chemical reaction engineering.

7.16. One (g mole)/(min) of hydrogen gas is to be completely oxidized to water vapor by burning with the theoretical amount of air in an adiabatic reactor. The hydrogen and air enter the reactor as ideal gases at 1(atm) and 300(K), and the products leave the reactor as an ideal-gas mixture at 1(atm) and at some higher temperature. Estimate this product temperature. For the reaction

$$H_2(g) + \tfrac{1}{2}O_2(g) \longrightarrow H_2O(g)$$

$\Delta H^\circ = -57{,}804$(cal)/(g mole) at 300(K). Equations for C_P° of $H_2O(g)$ and $N_2(g)$ [C_P° in (cal)/(g mole)(K), T in (K)] are

$$C_P^\circ(H_2O) = 7.17 + 2.56 \times 10^{-3}T \qquad C_P^\circ(N_2) = 6.66 + 1.02 \times 10^{-3}T$$

Take air to contain 21 mole % O_2 and 79 mole % N_2.

Neglecting potential- and kinetic-energy changes, assuming that $\dot{W}_s = 0$, and noting that $\dot{Q} = 0$ by the problem statement, we may write (*6.9*) on a molar basis as

$$\dot{n}_P H_P(T) - \dot{n}_F H_F(300) = 0$$

where $\dot{n}_F$ and $\dot{n}_P$ are molar feed and product rates, $H_F(300)$ is the molar enthalpy of the feed at 300(K), and $H_P(T)$ is the molar enthalpy of the products at the unknown temperature T. Addition and subtraction of the term $\dot{n}_P H_P(300)$ from the left-hand side of this equation gives

$$\dot{n}_P[H_P(T) - H_P(300)] + [\dot{n}_P H_P(300) - \dot{n}_F H_F(300)] = 0$$

But, by Problem 7.15, the second term in brackets is just equal to $\dot{\epsilon}\Delta H^\circ(300)$. Morever, the first term is given by

$$\dot{n}_P[H_P(T) - H_P(300)] = \sum (\dot{n}_i)_P \int_{300}^{T} C_{P_i}^\circ \, dT$$

$$= \int_{300}^{T} \sum (\dot{n}_i C_{P_i}^\circ)_P \, dT$$

Thus the equation to be solved for the final temperature T is

$$\int_{300}^{T} \sum (\dot{n}_i C_{P_i})_P \, dT + \dot{\epsilon}\Delta H^\circ(300) = 0 \tag{1}$$

The quantities $(\dot{n}_i)_P$ and $\dot{\epsilon}$ must now be found. Letting $H_2 \equiv (1)$, $O_2 \equiv (2)$, $H_2O \equiv (3)$, and $N_2 \equiv (4)$, we have from the statement of the problem

$$(\dot{n}_1)_F = 1(\text{g mole})/(\text{min}) \qquad (\dot{n}_2)_F = 0.5(\text{g mole})/(\text{min})$$

$$(\dot{n}_3)_F = 0(\text{g mole})/(\text{min}) \qquad (\dot{n}_4)_F = 0.5 \times \frac{79}{21} = 1.881(\text{g mole})/(\text{min})$$

The stoichiometric coefficients of the species are $\nu_1 = -1$, $\nu_2 = -\frac{1}{2}$, $\nu_3 = 1$, and $\nu_4 = 0$. Since the hydrogen is completely oxidized, $(\dot{n}_1)_P = 0$ and $\dot{\epsilon}$ is

$$\dot{\epsilon} = \frac{(\dot{n}_1)_P - (\dot{n}_1)_F}{\nu_1} = \frac{0 - 1}{-1} = 1(\text{g mole})/(\text{min})$$

This last equation follows from the extension of (*7.125*) to a steady-flow system [see (*5*) of Problem 7.15]. The molar product rates of the other species are then found to be

$$(\dot{n}_2)_P = 0(\text{g mole})/(\text{min}) \qquad (\dot{n}_3)_P = 1(\text{g mole})/(\text{min}) \qquad (\dot{n}_4)_P = 1.881(\text{g mole})/(\text{min})$$

Substitution of the given heat-capacity equations and the numerical values of $\Delta H^\circ(300)$, $(\dot{n}_3)_P$, and $(\dot{n}_4)_P$ into (*1*) then gives

$$\int_{300}^{T} [(1)(7.17 + 2.56 \times 10^{-3}T) + (1.881)(6.66 + 1.02 \times 10^{-3}T)] \, dT + (1)(-57{,}804) = 0$$

which becomes, on integration and rearrangement,

$$19.70(T - 300) + 2.24 \times 10^{-3}[T^2 - (300)^2] - 57{,}804 = 0$$

Solution of this equation for T gives the required answer: $T = 2522(\text{K})$.

7.17. Estimate the equilibrium composition at 400(K) and 1(atm) of a gaseous mixture containing the three isomers n-pentane(1), isopentane(2), and neopentane(3). Standard formation data at 400(K) are: $\Delta G_f^\circ(1) = 9600(\text{cal})/(\text{g mole})$, $\Delta G_f^\circ(2) = 8220(\text{cal})/(\text{g mole})$, and $\Delta G_f^\circ(3) = 8990(\text{cal})/(\text{g mole})$.

We must first determine the independent chemical reactions. The three formation reactions are

$$5C(s) + 6H_2(g) \longrightarrow n\text{-}C_5H_{12}(g) \tag{1}$$

$$5C(s) + 6H_2(g) \longrightarrow \text{iso-}C_5H_{12}(g) \tag{2}$$

$$5C(s) + 6H_2(g) \longrightarrow \text{neo-}C_5H_{12}(g) \tag{3}$$

Elimination of $C(s)$ and $H_2(g)$ by subtraction of equation (*1*) from (*2*) and from (*3*) gives *two* independent reactions:

$$n\text{-}C_5H_{12}(g) \longrightarrow \text{iso-}C_5H_{12}(g) \qquad (4)$$

$$n\text{-}C_5H_{12}(g) \longrightarrow \text{neo-}C_5H_{12}(g) \qquad (5)$$

Reaction equilibrium calculations for the case of more than one independent chemical reaction can occasionally be done quite simply by straightforward extensions of the method based on (*7.152*). At 1(atm) pressure, we assume ideal-gas behavior, and the generalization of the ideal-gas equilibrium equation (*7.156*) becomes

$$\prod_i y_i^{\nu_{i,j}} = P^{-(\Sigma\nu)_j} K_j \qquad (j = 1, 2, \ldots, r) \qquad (6)$$

where r is the number of independent reactions, $\nu_{i,j}$ is the stoichiometric coefficient of i in the jth reaction, and $\left(\sum \nu\right)_j = \sum_i \nu_{i,j}$. Designating reaction (*4*) by the subscript 1 and reaction (*5*) by the subscript 2, the stoichiometric coefficients for the present problem are $\nu_{1,1} = -1$, $\nu_{2,1} = 1$, $\nu_{1,2} = -1$, and $\nu_{3,2} = 1$. Thus $\left(\sum\nu\right)_1 = 0$ and $\left(\sum\nu\right)_2 = 0$, and (*6*) becomes

$$\begin{aligned} y_2 &= y_1 K_1 \\ y_3 &= y_1 K_2 \end{aligned} \qquad (7)$$

where

$$\begin{aligned} K_1 &= \exp(-\Delta G_1^\circ/RT) \\ K_2 &= \exp(-\Delta G_2^\circ/RT) \end{aligned} \qquad (8)$$

There are only two independent mole fractions; therefore, for specified T, (*7*) constitutes two equations in two unknowns. The equilibrium mole fractions are found by combination of (*7*) with the equation $y_1 + y_2 + y_3 = 1$:

$$\begin{aligned} y_1 &= 1/(1 + K_1 + K_2) \\ y_2 &= K_1/(1 + K_1 + K_2) \\ y_3 &= K_2/(1 + K_1 + K_2) \end{aligned} \qquad (9)$$

Numerical values of ΔG_1° and ΔG_2° are found from the given standard formation values. Thus

$$\Delta G_1^\circ = 8220 - 9600 = -1380\text{(cal)/(g mole)}$$

$$\Delta G_2^\circ = 8990 - 9600 = -\ 610\text{(cal)/(g mole)}$$

and, from (*8*),

$$K_1 = \exp[-(-1380)/(1.987 \times 400)] = 5.676$$

$$K_2 = \exp[-(-610)/(1.987 \times 400)] = 2.154$$

The equilibrium mole fractions calculated from (*9*) with these values for K_1 and K_2 are

$$y_1 = 0.113 \qquad y_2 = 0.643 \qquad y_3 = 0.244$$

7.18. One method for the production of hydrogen cyanide (HCN) is the gas-phase nitrogenation of acetylene (C_2H_2) according to the reaction

$$N_2(g) + C_2H_2(g) \longrightarrow 2HCN(g) \qquad (1)$$

The feed to a reactor in which the above reaction is to take place contains gaseous N_2 and C_2H_2 in their stoichiometric proportions. The reaction temperature is controlled at 300(°C). Estimate the maximum mole fraction of HCN in the product stream if the reactor pressure is (*a*) 1(atm), and (*b*) 200(atm). At 300(°C), ΔG° for the reaction is 7190(cal)/(g mole). Physical properties for HCN are: $T_c = 456.7$(K), $P_c = 48.9$(atm), and acentric factor $\omega = 0.4$.

(*a*) The maximum composition of HCN will be that corresponding to equilibrium conversion. At 1(atm) the gas mixtures will be assumed ideal, and (*7.156*) therefore applies:

$$\prod_i y_i^{\nu_i} = P^{-\nu} K \qquad (7.156)$$

Letting $N_2 \equiv (1)$, $C_2H_2 \equiv (2)$, and $HCN \equiv (3)$, we see from (*1*) that $\nu_1 = -1$, $\nu_2 = -1$, and $\nu_3 = 2$. Thus $\nu = \sum \nu_i = 0$. The equilibrium constant K is found from (*7.153*) and the given value for ΔG°:

$$K = \exp\left[\frac{-7190}{1.987 \times (273+300)}\right] = 1.809 \times 10^{-3}$$

By the statement of the problem, the initial mole fractions y_{i_0} are $y_{1_0} = 0.5$, $y_{2_0} = 0.5$, and $y_{3_0} = 0.0$. Thus the material balance equation (*7.155*) gives

$$y_1 = 0.5 - \xi$$
$$y_2 = 0.5 - \xi \qquad (2)$$
$$y_3 = 2\xi$$

Substitution of these equations and numerical values into (*7.156*) yields

$$\frac{\xi^2}{(0.5-\xi)^2} = 4.523 \times 10^{-4}$$

from which we obtain $\xi = 0.0104$. The equilibrium mole fractions of the product stream are then found from (*2*):

$$y_1 = 0.4896 \qquad y_2 = 0.4896 \qquad y_3 = 0.0208$$

The maximum mole fraction of HCN at 300(°C) and 1(atm) is thus 0.0208.

(*b*) At 200(atm), the assumption of ideal-gas behavior is certain to be invalid. However, we shall assume that the gas mixtures are *ideal solutions*, for which $\hat{\phi}_i = \phi_i$. The general equilibrium equation (*7.154*) for gas-phase reactions then simplifies to

$$\prod y_i^{\nu_i} = \left(\prod \phi_i^{-\nu_i}\right) P^{-\nu} K \qquad (3)$$

Calculation of the equilibrium composition therefore requires numerical values for the pure-component fugacity coefficients ϕ_i. These may be estimated from the generalized correlation of Sec. 5.5 by use of (*7.29*):

$$\ln \phi = \frac{P_r}{T_r}(B^0 + \omega B^1) \qquad (7.29)$$

The results of the calculations are summarized below in tabular form. The physical constants for N_2 and C_2H_2 are from Appendix 3; those for HCN were given in the problem statement.

Compound	T_c(K)	P_c(atm)	ω	T_r	P_r	B^0	B^1	ϕ
N_2	126.2	33.5	0.040	4.54	5.97	+0.065	+0.15	1.10
C_2H_2	309.5	61.6	0.190	1.85	3.25	−0.075	+0.17	0.928
HCN	456.7	48.9	0.4	1.26	4.09	−0.21	+0.05	0.540

Thus, $\phi_1 = 1.10$, $\phi_2 = 0.928$, and $\phi_3 = 0.540$. Substitution of (*2*) and the numerical values of ν, the ν_i, and the ϕ_i into (*3*) gives, on rearrangement,

$$\frac{\xi^2}{(0.5-\xi)^2} = 1.583 \times 10^{-3}$$

from which we obtain $\xi = 0.0192$. The equilibrium mole fractions are found from (*2*):

$$y_1 = 0.4808 \qquad y_2 = 0.4808 \qquad y_3 = 0.0384$$

The effect of increasing the reaction pressure from 1(atm) to 200(atm) is to increase the equilibrium composition of HCN by nearly a factor of two. This effect results solely from the vapor-phase nonidealities, and not from the reaction stoichiometry, because $\nu = 0$ and by (*7.161*) *no* pressure effect on conversion would obtain for the ideal-gas reaction.

7.19. Derive an expression for determination of the equilibrium states of a system in which there occurs the single independent reaction

$$0 = \nu_1 X_1(s1) + \nu_2 X_2(s2) + \nu_3 X_3(g)$$

The designations s1 and s2 indicate that the compounds X_1 and X_2 are present as two distinct pure solid phases.

All three species are present as pure components in their phases, and therefore (*7.152*) becomes

$$(f_1/f_1^\circ)^{\nu_1}(f_2/f_2^\circ)^{\nu_2}(\phi_3 P)^{\nu_3} \;=\; K \qquad (1)$$

where we have used the relationships $f_3^\circ = 1(\text{atm})$ and $f_3 = \phi_3 P$ for gaseous species 3. The standard states for the solids are *not* states of unit fugacity, and hence $f_i^\circ \neq 1(\text{atm})$ for $i = 1, 2$. However, expressions for the fugacity ratios of the solids may be obtained by integration of (*7.17*) from the standard-state pressure $[P^\circ = 1(\text{atm})]$ to the system pressure P:

$$f_i/f_i^\circ \;=\; \exp\left[\frac{1}{RT}\int_1^P V_i\,dP\right] \qquad (i = 1, 2) \qquad (2)$$

where V_i is the molar volume of the pure solid, and the integration is carried out at the system temperature. Combination of (*1*) with (*2*) then gives

$$(\phi_3 P)^{\nu_3}\exp\left[\frac{1}{RT}\int_1^P (\nu_1 V_1 + \nu_2 V_2)\,dP\right] \;=\; K \qquad (3)$$

For the type of system under consideration, there are three phases ($\pi = 3$), three components ($m = 3$), and one independent chemical reaction ($r = 1$). The extended phase rule (*7.151*) then gives $F = 1$. Thus there is only a single degree of freedom for the system. This is in accord with (*3*), which shows that the equilibrium pressure is determined on specification of the equilibrium temperature. The compound present in the gas phase (species 3) is invariably a *product* species ($\nu_3 > 0$) for this type of reaction, and the equilibrium pressures defined by (*3*) are called *decomposition pressures* (see Problem 7.35).

Supplementary Problems

SOLUTION THERMODYNAMICS (Secs. 7.1 through 7.5)

7.20. (*a*) Show that the "partial molar mass" of a component in solution is equal to the molecular weight of the component. (*b*) Write (*7.7*) for the components in a ternary solution.

Ans. (*b*)

$$\bar{M}_1 \;=\; M - x_2(\partial M/\partial x_2)_{T,P,x_3} - x_3(\partial M/\partial x_3)_{T,P,x_2}$$

$$\bar{M}_2 \;=\; M - x_1(\partial M/\partial x_1)_{T,P,x_3} - x_3(\partial M/\partial x_3)_{T,P,x_1}$$

$$\bar{M}_3 \;=\; M - x_1(\partial M/\partial x_1)_{T,P,x_2} - x_2(\partial M/\partial x_2)_{T,P,x_1}$$

7.21. At 30(°C) and 1(atm), the volumetric data for liquid mixtures of benzene(*b*) and cyclohexane(*c*) are represented by the simple quadratic expression $V = 109.4 - 16.8\,x_b - 2.64\,x_b^2$, where x_b is the mole fraction of benzene and V has units of $(\text{cm})^3/(\text{g mole})$. Find expressions for $\bar{V}_b$, $\bar{V}_c$, and ΔV (for standard states based on the Lewis and Randall rule) at 30(°C) and 1(atm).

Ans. $\bar{V}_b = 92.6 - 5.28\,x_b + 2.64\,x_b^2$, $\bar{V}_c = 109.4 + 2.64\,x_b^2$, $\Delta V = 2.64\,x_b x_c$

7.22. The following pair of equations has been suggested for representation of partial-molar-volume data for simple binary systems at constant T and P:

$$\bar{V}_1 - V_1 \;=\; a + (b-a)x_1 - bx_1^2$$

$$\bar{V}_2 - V_2 \;=\; a + (b-a)x_2 - bx_2^2$$

where a and b are functions of T and P only, and V_1 and V_2 are the molar volumes of the pure components. Are these equations thermodynamically sound? (*Hint:* See Problem 7.2.)

Ans. No. Although the equations give $\bar{V}_1 = V_1$ and $\bar{V}_2 = V_2$ for $x_1 = 1$ and $x_2 = 1$, and in addition satisfy the area test for thermodynamic consistency [(*9*) or (*10*) of Problem 7.2], they do *not* satisfy the slope test [(*7*) or (*8*) of Problem 7.2]. Values generated from such equations would be thermodynamically inconsistent, and therefore incorrect.

7.23. Data for properties of components in solution are sometimes reported as "apparent molar properties." For a binary system of constituents 1 and 2, the apparent molar property $\mathcal{M}_1$ of component 1 is defined by

$$\mathcal{M}_1 = \frac{M - x_2 M_2}{x_1}$$

where x is mole fraction, M is the molar property of the mixture, and M_2 is the molar property of pure 2 at the solution T and P.

(*a*) Derive equations for determination of the partial molar properties $\bar{M}_1$ and $\bar{M}_2$ from knowledge of $\mathcal{M}_1$ as a function of x_1 at constant T and P. The equations should include only the quantities x_1, M_2, $\mathcal{M}_1$, and $d\mathcal{M}_1/dx_1$.

(*b*) Obtain expressions for the limiting cases $x_1 = 0$ and $x_1 = 1$.

Ans. (*a*) $\bar{M}_1 = \mathcal{M}_1 + x_1(1-x_1)\dfrac{d\mathcal{M}_1}{dx_1}, \quad \bar{M}_2 = M_2 - x_1^2\dfrac{d\mathcal{M}_1}{dx_1}$

(*b*) For $x_1 = 0$, $\bar{M}_1^\infty = \mathcal{M}_1^\infty$; for $x_1 = 1$, $\bar{M}_2^\infty = M_2 - \left(\dfrac{d\mathcal{M}_1}{dx_1}\right)_{x_1=1}$

7.24. A ternary gas mixture contains 20 mole % A, 35 mole % B, and 45 mole % C. At a pressure of 60(atm) and a temperature of 75(°C), the fugacity coefficients of components A, B, and C in this mixture are 0.7, 0.6, and 0.9, respectively. What is the fugacity of the mixture? *Ans.* 44.6(atm)

7.25. Determine a good estimate for the fugacity of liquid $CHClF_2$ at 0(°F) and 2000(psia).

Data: (*a*) Molecular weight = 86.5

(*b*) P^{sat} = 38.78(psia) at 0(°F)

(*c*) Z = 0.932 for the saturated vapor at 0(°F)

(*d*) Volumetric data at 0(°F) are

P(psia)	V(ft)^3/(lb$_m$)	P(psia)	V(ft)^3/(lb$_m$)
10	5.573	1000	0.0077
40	0.0119	1500	0.0056
500	0.0098	2000	0.0035

[*Hint:* See (*7.28*) and Example 7.5.]

Ans. $f = 47.2$(psia)

PHASE EQUILIBRIUM CALCULATIONS (Secs. 7.6 and 7.7)

7.26. Derive the following alternative expressions to (*7.98*):

$$y_i\hat{\phi}_i^v = x_i\hat{\phi}_i^l, \qquad y_i\gamma_i^v f_i^{\circ v} = x_i\gamma_i^l f_i^{\circ l}, \qquad y_i\hat{\phi}_i^v = x_i\gamma_i^l\phi_i^{\circ l}$$

7.27. The equilibrium total pressure P for VLE in a binary system described by (*7.101*) is

$$P = x_1\gamma_1 P_1^{\text{sat}} + x_2\gamma_2 P_2^{\text{sat}}$$

If some or all of the assumptions upon which (*7.101*) is based are waived, then a similar but more rigorous equation may be written for P:

$$P = x_1\gamma_1P_1^* + x_2\gamma_2P_2^*$$

but now P_1^* and P_2^* are in general functions of T, P, and the vapor compositions. If the vapor phase is described by the truncated virial equation (*5.22*), and if the molar volumes of the pure liquids are independent of pressure, show that

$$P_1^* = P_1^{sat} \exp\left[\frac{(V_1^{sat} - B_{11})(P - P_1^{sat}) - y_2^2\delta_{12}P}{RT}\right]$$

$$P_2^* = P_2^{sat} \exp\left[\frac{(V_2^{sat} - B_{22})(P - P_2^{sat}) - y_1^2\delta_{12}P}{RT}\right]$$

[*Hint:* See (*7.43*), (*7.99*), and (*7.100a*).]

7.28. Components 1 and 2 are essentially insoluble in the liquid phase. Estimate the dew-point temperatures and the compositions of the first drops of liquid formed when vapor mixtures of 1 and 2 containing (*a*) 75 mole % component 1, and (*b*) 25 mole % component 1, are cooled at 1(atm) total pressure. (*Hint:* See Problem 7.12.) Vapor-pressure data for the pure components are

T(°C)	P_1^{sat}(atm)	P_2^{sat}(atm)
85	0.2322	0.5706
90	0.3312	0.6921
95	0.3906	0.8342
100	0.4598	1.000
105	0.5390	1.192
110	0.6313	1.414
115	0.7334	1.668
120	0.8521	1.959
125.6	1.000	2.364

Ans. (*a*) 115.7(°C), liquid is pure 1; (*b*) 92.1(°C), liquid is pure 2

7.29. (*a*) Derive expressions for the bubble-point pressure P_b and the dew-point pressure P_d of a system which obeys Raoult's law.

(*b*) Assuming that Raoult's law applies, determine P_b and P_d at 110(°C) for an overall system composition of 45 mole % n-octane(1), 10 mole % 2,5-dimethylhexane(2), and 45 mole % 2,2,4-trimethylpentane(3). The pure-component vapor pressures at 110(°C) are:

$$P_1^{sat} = 0.634(\text{atm}), \quad P_2^{sat} = 1.027(\text{atm}), \quad \text{and} \quad P_3^{sat} = 1.36(\text{atm})$$

Ans. (*a*) $P_b = \Sigma z_i P_i^{sat}$, $P_d = [\Sigma (z_i/P_i^{sat})]^{-1}$

(*b*) $P_b = 1.00$(atm), $P_d = 0.879$(atm)

7.30. An auxiliary function frequently employed in VLE computations is the *relative volatility* α_{ij}, defined as $\alpha_{ij} = K_i/K_j$, where K_i and K_j are the VLE K-values of components i and j, respectively.

(*a*) Show that $\alpha_{ij} = 1$ for all component pairs (i, j) at an azeotropic state.

(*b*) Derive an expression for α_{ij} of the components in a system described by (*7.101*).

(*c*) The liquid-phase activity coefficients at infinite dilution (γ_i^∞) are approximately 2.3 and 7.0, respectively, for chloroform(1) and methanol(2) in the chloroform-methanol system at 50(°C). Pure-component vapor pressures at 50(°C) are $P_1^{sat} = 0.667$(atm) and $P_2^{sat} = 0.174$(atm). Show that the VLE of this system exhibits a maximum-pressure azeotrope at 50(°C).

Ans. (*b*) $\alpha_{ij} = \gamma_i P_i^{sat}/\gamma_j P_j^{sat}$

CHEMICAL-REACTION CALCULATIONS (Secs. 7.8 and 7.9)

7.31. In compilations of standard formation data, one also often finds entries for *standard heats of combustion* ΔH_c° for standard combustion reactions. A standard combustion reaction for a given chemical species is the reaction between one mole of that species and oxygen to form specified products, with all the reactant and product species present in their standard states. For compounds containing only carbon, hydrogen, and oxygen, the products are customarily taken as $CO_2(g)$ and $H_2O(l)$. Thus, the standard combustion reaction for gaseous n-butane $[n\text{-}C_4H_{10}(g)]$ is

$$n\text{-}C_4H_{10}(g) + \tfrac{13}{2}O_2(g) \longrightarrow 4CO_2(g) + 5H_2O(l)$$

where all the gaseous species are present as pure ideal gases at 1(atm), and water is present as a pure liquid at 1(atm).

The standard heats of combustion for benzene $[C_6H_6(g)]$, hydrogen, and cyclohexane $[C_6H_{12}(g)]$ at 25(°C) are −789,080(cal)/(g mole), −68,317(cal)/(g mole), and −944,790(cal)/(g mole), respectively. Using these data, calculate ΔH° at 25(°C) for the reaction

$$C_6H_6(g) + 3H_2(g) \longrightarrow C_6H_{12}(g)$$

Ans. $\Delta H^\circ = -49{,}241$(cal)/(g mole)

7.32. The reaction of Example 7.16 is carried out at a constant pressure of 1(atm) and a constant temperature of 150(°C). How much heat is transferred from the gases to the surroundings? Assume ideal-gas mixtures. At 150(°C), for units of (cal)/(g mole), $\Delta H_f^\circ[C_3H_8(g)] = -22{,}229$, $\Delta H_f^\circ[CO_2(g)] = -92{,}865$, and $\Delta H_f^\circ[H_2O(g)] = -56{,}783$. *Ans.* 9670(cal)

7.33. Determine the number of degrees of freedom of the following systems:

(*a*) A chemically reactive mixture of $ZnO(s)$, $ZnSO_4(s)$, $SO_2(g)$, $SO_3(g)$, and $O_2(g)$. The two solid phases are completely immiscible.

(*b*) A reactive gas mixture containing k isomeric hydrocarbons.

(*c*) A system in which the reaction $H_2(g) + \frac{1}{2}O_2(g) \longrightarrow H_2O(l)$ takes place.

(*d*) Same as part (*c*), except that air (21 mole % O_2, 79 mole % N_2) is the oxidant.

(*e*) Same as part (*c*), only "real" air (21.0 mole % O_2, 78.1 mole % N_2, 0.9 mole % argon) is the oxidant.

Ans. (*a*) 2, (*b*) 2, (*c*) 2, (*d*) 3, (*e*) 3 [The argon-to-N_2 ratio is constant.]

7.34. (*a*) A compound X is known to polymerize to the compound X_l in the gas phase according to the reaction $lX(g) \rightarrow X_l(g)$, where l is the number of X-units in the polymer ($l > 1$). For constant l, and assuming ideal-gas behavior, show that the extent of equilibrium polymerization increases with increasing pressure at constant T.

(*b*) The following experimental data were recorded for the equilibrium mole fractions of monomer X in two gas-phase monomer-polymer mixtures:

T(°C)	P(atm)	y_X
100	1.0	0.807
100	1.5	0.750

What is the value of l for the polymerization reaction of part (*a*)?

Ans. (*b*) $l = 2$

7.35. Limestone ($CaCO_3$) decomposes upon heating to yield quicklime (CaO) and carbon dioxide (CO_2). Determine the temperature at which limestone exerts a decomposition pressure of 1(atm) [see

Problem 7.19]. Standard formation data at 25(°C) [in (cal)/(g mole)] and constants for the heat-capacity equation (*7.144*) [C_P° in (cal)/(g mole)(K) and T in (K)] are given below.

Compound	a	$10^3 b$	$10^{-5} d$	ΔH_f°	ΔG_f°
$CaCO_3$(s)	24.98	5.24	−6.20	−288,450	−269,780
CaO(s)	11.67	1.08	−1.56	−151,900	−144,400
CO_2(g)	10.55	2.16	−2.04	−94,052	−94,260

Ans. 890(°C)

Chapter 8

Thermodynamic Analysis of Processes

Many of the calculations of thermodynamics depend on the assumption of reversibility; real processes, carried out with real machines and equipment, are never reversible. Nevertheless, it is entirely possible to analyze real processes by the methods of thermodynamics. Such an analysis requires no more than application of the first and second laws of thermodynamics to a process considered in relation to its surroundings. For this purpose it is convenient to develop a few simple equations which provide a particularly useful formulation of these two laws. Following this brief development, the remainder of the chapter is devoted to applications of these equations to a variety of practical processes. This material not only demonstrates the practical utility of thermodynamics, but it also affords a review of all the preceding chapters.

Thermodynamic analysis of a real process has as its purpose the determination of the efficiency of the process with respect to energy utilization. In addition it shows the influence that each irreversibility in the process exerts on this efficiency. Since the processes to be considered all involve steady-state flow, we limit the theoretical treatment to this case.

8.1 THE WORK OF A *COMPLETELY* REVERSIBLE PROCESS

For simplicity we consider steady-state flow processes for which the energy equation is given by (*6.11*):

$$\Delta H + \frac{\Delta u^2}{2g_c} + \Delta z\left(\frac{g}{g_c}\right) = \sum Q - W_s \qquad (6.11)$$

Figure 8-1 represents the essential features of any such process. We presume that the process exists in surroundings which constitute a heat reservoir at the constant tempera-

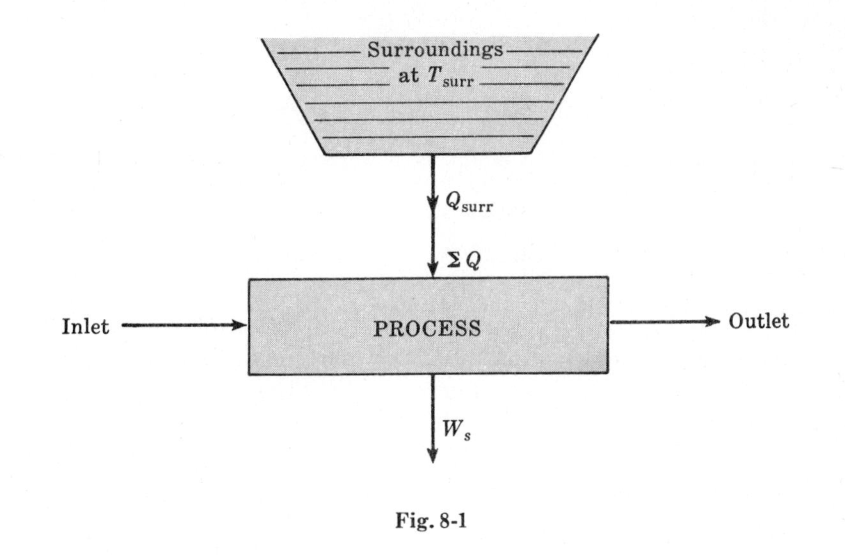

Fig. 8-1

ture T_{surr}. Heat exchange between the process and its surroundings causes an entropy change in the surroundings given by $\Delta S_{surr} = Q_{surr}/T_{surr}$. Since $Q_{surr} = -\Sigma Q$, this may also be written $\Sigma Q = -T_{surr}\Delta S_{surr}$. Combination of this last equation with (*6.11*) gives

$$W_s \quad = \quad -T_{surr}\Delta S_{surr} \;-\; \Delta H \;-\; \frac{\Delta u^2}{2g_c} \;-\; \Delta z\left(\frac{g}{g_c}\right) \tag{8.1}$$

Since the entropy change of the surroundings is rarely known, this equation is of little use. However, for the special case of a *completely reversible* process, we have from the second law that $\Delta S + \Delta S_{surr} = 0$ or $\Delta S_{surr} = -\Delta S$, where ΔS is the entropy change of the system. Since the process considered is one of steady-state flow, the entropy of the control volume is constant, and ΔS must therefore represent the difference in entropies between the outlet stream and the inlet stream. Substitution into (*8.1*) then gives

$$W_{rev} \quad = \quad T_{surr}\Delta S \;-\; \Delta H \;-\; \frac{\Delta u^2}{2g_c} \;-\; \Delta z\left(\frac{g}{g_c}\right)$$

In this equation W_{rev} is the work associated with a *completely reversible* process that causes the change of state implied by the changes ΔS, ΔH, Δu, and Δz. When these property changes are the result of a real (irreversible) process, then this equation gives the work that would be associated with a *completely reversible* process that accomplished exactly the *same change of state.* This stipulation of *complete* reversibility implies not only reversibility within the process, but also reversibility of heat transfer between the system and its surroundings. Such a completely reversible process is taken as the *ideal* against which to measure the efficiencies of real processes that accomplish the same change of state. In order to indicate this explicitly we rewrite this fundamental equation as

$$\boxed{W_{ideal} \quad = \quad T_0\Delta S \;-\; \Delta H \;-\; \frac{\Delta u^2}{2g_c} \;-\; \Delta z\left(\frac{g}{g_c}\right)} \tag{8.2}$$

In addition to replacing W_{rev} by W_{ideal} we have for the sake of brevity also substituted T_0 for T_{surr}.

Equation (*8.2*) gives the *minimum* work requirement for a given change of state when the process *requires* work. It gives the *maximum* work obtainable for a given change of state when the process *produces* work. Thus we define the *thermodynamic efficiency* η_t differently for the two cases:

$$\eta_t(\text{work produced}) \;\equiv\; W_s/W_{ideal} \tag{8.3a}$$

$$\eta_t(\text{work required}) \;\equiv\; W_{ideal}/W_s \tag{8.3b}$$

Example 8.1. Problem 6.18 asks for the minimum power requirement to cool an air stream from 38(°C) to 15(°C) when the surroundings are at 38(°C). If we take the kinetic- and potential-energy terms to be negligible, then (*8.2*) applied to the air stream becomes

$$W_{ideal} \quad = \quad T_0\Delta S \;-\; \Delta H$$

Since the process requires work W_{ideal} is the minimum work. If ΔS and ΔH represent the entropy and enthalpy changes for each mole of air passing through the process, then to get the *rate* at which work is required, i.e. the power, we need only multiply by $\dot{n}$, the molar flow rate of air:

$$\dot{W}_{ideal} \quad = \quad (T_0\Delta S \;-\; \Delta H)\dot{n}$$

This is, in fact, the equation derived in Problem 6.18, though with a slightly different notation, and it was used directly to provide the answer to the problem.

Example 8.2. In Example 6.9 saturated steam at 250(psia) enters a process and serves as the source of energy for making heat available at a temperature level of 500(°F). The steam is assumed to leave the process as liquid at 70(°F), the temperature of the surroundings (cooling water). The question is:

How much heat can be made available at a temperature level of 500(°F) for every pound of steam passing through the process?

The maximum work obtainable from each pound of steam is given by (*8.2*). Neglecting the kinetic- and potential-energy terms, we get

$$W_{\text{ideal}} = T_0 \Delta S - \Delta H = (70+460)(0.0746-1.5274) - (38.1-1202.1) = 394.0(\text{Btu})/(\text{lb}_\text{m})$$

This amount of work may be used to operate a reversible heat pump between the surroundings (cooling water) at a temperature level of 70(°F) and a heat reservoir at a temperature level of 500(°F). For such a device we have the relation (*2.4a*), which may be rearranged to give:

$$|Q_H| = |W|\frac{T_H}{T_H - T_C} = (394.0)\frac{500+460}{500-70} = 879.6(\text{Btu})$$

This is the heat made available at a temperature level of 500(°F) for each pound of steam passing through the system. The value calculated here agrees exactly with the value determined in Example 6.9.

Example 8.3. What is the thermodynamic efficiency η_t of the turbine described in Problem 6.12 if the temperature of the surroundings is $T_0 = 70(°\text{F})$ or 530(R)?

For this work-producing process, η_t is given by (*8.3a*). We need to know both W_s, the actual shaft work of the turbine, and W_{ideal} as given by (*8.2*) for the actual change of state. Since the turbine of Problem 6.12 is adiabatic and the kinetic- and potential-energy terms are considered negligible, the energy equation (*6.11*) gives

$$W_s = -\Delta H = 268.0(\text{Btu})/(\text{lb}_\text{m})$$

where we have used the value for ΔH determined in Problem 6.12.

The assumption of negligible kinetic- and potential-energy changes reduces (*8.2*) to

$$W_{\text{ideal}} = T_0 \Delta S - \Delta H$$

In this equation $\Delta H = -268.0$, as before, and ΔS is given by

$$\Delta S = S_2 - S_1 = 1.9779 - 1.7234 = 0.2545(\text{Btu})/(\text{lb}_\text{m})(\text{R})$$

and we therefore have

$$W_{\text{ideal}} = (530)(0.2545) - (-268.0) = 402.9(\text{Btu})/(\text{lb}_\text{m})$$

From (*8.3a*) we get

$$\eta_t = \frac{W_s}{W_{\text{ideal}}} = \frac{268.0}{402.9} = 0.665$$

In comparison with a turbine efficiency arbitrarily based on isentropic expansion of the steam to P_2 and calculated in Problem 6.12 to be $\eta = 0.652$, the thermodynamic efficiency η_t is seen to be slightly higher. The two efficiencies are not directly related, being based on two different standards of performance.

It is instructive to devise a completely reversible process by which the actual change of state experienced by the steam could be accomplished. One such process would consist of two steps: (1) A reversible adiabatic expansion of the steam from its initial state to the final pressure of 1(psia). As shown in Problem 6.12 this produces wet steam. (2) Reversible transfer of heat from the surroundings at 530(R) to the wet steam so as to evaporate the moisture and produce saturated steam at 1(psia). These two steps are shown on Fig. 8-2.

As shown by Problem 6.12, step (1) produces work in the amount

$$W_1 = -(\Delta H)_S = 411.1(\text{Btu})/(\text{lb}_\text{m})$$

The resulting wet steam has an enthalpy at point 2′ of $H_2' = 962.7(\text{Btu})/(\text{lb}_\text{m})$.

Step (2) is a constant-pressure, constant-temperature heating process that vaporizes the moisture content of the steam, producing saturated steam at $P_2 = 1(\text{psia})$ and $T_2 = 101.7(°\text{F})$ with an enthalpy $H_2 = 1105.8(\text{Btu})/(\text{lb}_\text{m})$. The heat required is given by the energy equation $Q_2 = H_2 - H_2'$. Since this heat transfer is to be accomplished reversibly, we must imagine a reversible heat pump operating on the Carnot cycle to take heat from the surroundings at 70(°F) and to release heat to the steam at 101.7(°F). The amount of work required by the heat pump is given by (*2.4a*):

$$|W| = |Q_H|\left(1 - \frac{T_C}{T_H}\right)$$

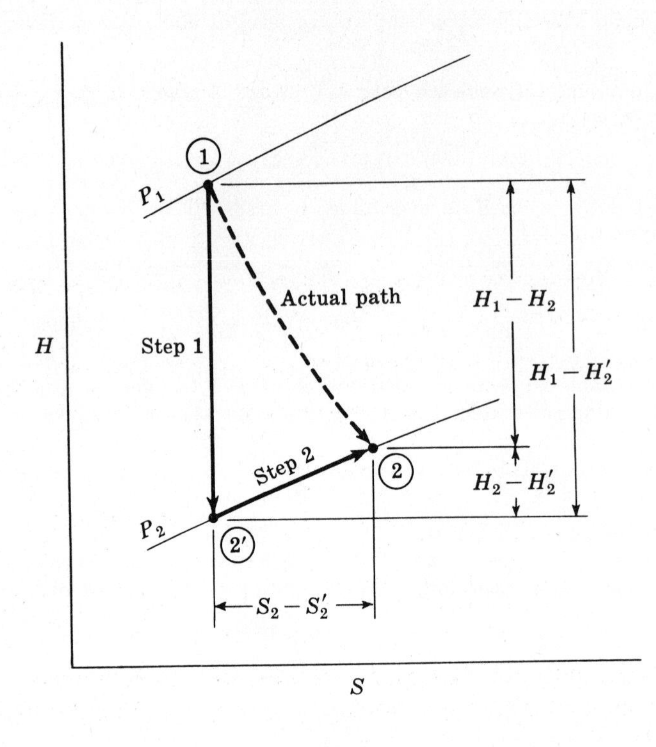

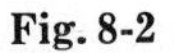
Fig. 8-2

We identify $|W|$ with the work of step (2), and since work is required, we have $|W| = -W_2$. We take $|Q_H|$ to be the heat added to the steam; T_C is T_0, and T_H is T_2. Thus

$$W_2 = -Q_2\left(1 - \frac{T_0}{T_2}\right) = -Q_2 + T_0\frac{Q_2}{T_2}$$

However,
$$Q_2 = H_2 - H_2' = 1105.8 - 962.7 = 143.1(\text{Btu})/(\text{lb}_\text{m})$$

$$\frac{Q_2}{T_2} = S_2 - S_2' = 1.9779 - 1.7234 = 0.2545(\text{Btu})/(\text{lb}_\text{m})(\text{R})$$

Therefore
$$W_2 = -143.1 + (530)(0.2545) = -8.2(\text{Btu})/(\text{lb}_\text{m})$$

For the two steps combined, we have

$$W_{\text{ideal}} = W_1 + W_2 = 411.1 - 8.2 = 402.9(\text{Btu})/(\text{lb}_\text{m})$$

This is the same value we determined earlier and much more simply by direct application of (*8.2*). This calculation does make clear the difference between the two efficiencies η and η_t. The former is given by

$$\eta = \frac{W_s}{W_1}$$

whereas the thermodynamic efficiency is

$$\eta_t = \frac{W_s}{W_{\text{ideal}}} = \frac{W_s}{W_1 + W_2}$$

8.2 ENERGY THAT BECOMES UNAVAILABLE FOR WORK

The difference between the ideal work for a given change of state and the actual work of an irreversible process that results in the same change of state is clearly the energy that becomes unavailable for work as a result of the irreversibility of the actual process. For want of a better term, we will call this quantity the *lost work*. It is given by

$$W_{lost} = W_{ideal} - W_s \tag{8.4}$$

Substituting for W_{ideal} by (*8.2*) and for W_s by (*6.11*), where both equations are written for the same change of state, we get:

$$\boxed{W_{lost} = T_0 \Delta S - \sum Q} \tag{8.5}$$

Thus the lost work for any process is simply related to the entropy change of the system ΔS and the heat transferred to the system ΣQ.

Alternative forms of (*8.5*) may be developed by noting that $\Sigma Q = -Q_{surr}$. Then

$$W_{lost} = T_0 \Delta S + Q_{surr}$$

However, $Q_{surr} = T_{surr} \Delta S_{surr} = T_0 \Delta S_{surr}$. Therefore

$$W_{lost} = T_0 \Delta S + T_0 \Delta S_{surr} = T_0 (\Delta S_{sys} + \Delta S_{surr})$$

or

$$\boxed{W_{lost} = T_0 \Delta S_{total}} \tag{8.6}$$

where ΔS_{total} is the total entropy change in system *and* surroundings as a result of the process. The second law of thermodynamics as given by (*2.2*) requires that $\Delta S_{total} \geqq 0$. Combining this with (*8.6*), we see that

$$W_{lost} \geqq 0$$

When a process is completely reversible, the equality holds, and the lost work is zero. For irreversible processes the inequality holds, and we see that the lost work or energy that becomes unavailable for work is positive. The significance of this result is clear: The greater the irreversibility of a process, the greater the increase in total entropy, and the greater the amount of energy that becomes unavailable for work. Thus every irreversibility carries with it a price.

For processes of more than one step it is advantageous to calculate W_{lost} for each step separately. Then (*8.4*) becomes

$$\sum W_{lost} = W_{ideal} - W_s \tag{8.7}$$

For work-requiring processes this equation is best expressed as

$$W_s = W_{ideal} - \sum W_{lost} \tag{8.7a}$$

The terms on the right-hand side of (*8.7a*) represent an *analysis* of the components of the actual work, showing what part of it is ideally required to bring about the change of state (W_{ideal}) and the parts which result from irreversibilities in the various steps of the process (ΣW_{lost}).

For work-producing processes (*8.7*) is best written

$$W_{ideal} = W_s + \sum W_{lost} \tag{8.7b}$$

Here the terms on the right-hand side represent an *analysis* of the ideal work, showing the part actually produced (W_s) and the parts which become unavailable because of irreversibilities in the various steps of the process (ΣW_{lost}).

Example 8.4. Problem 6.23 describes a steam turbine which operates adiabatically and produces 4000(HP). Superheated steam at 300(psia) and 900(°F) enters the turbine; the discharge is saturated vapor at 1.5(psia), which is condensed and cooled to a final state of liquid at 90(°F). This second step is accomplished by the transfer of heat to cooling water, taken from the surroundings and presumably subsequentlv discarded to the surroundings. Since the entering cooling water is at 65(°F), we can take this to be the temperature of the surroundings. We have here a two-step process — expansion of steam through the turbine, and condensation of the exhaust steam in a condenser — and we wish to analyze it.

We denote the initial state of the steam by subscript 1, the state of the exhaust steam by subscript 2, and the final state of liquid water by subscript 3. The following data are then taken from the steam tables:

$$H_1 = 1473.6(\text{Btu})/(\text{lb}_\text{m}) \qquad S_1 = 1.7589(\text{Btu})/(\text{lb}_\text{m})(\text{R})$$

$$H_2 = 1111.7 \qquad S_2 = 1.9438$$

$$H_3 = 58.1 \qquad S_3 = 0.1116$$

The overall change of state resulting from the process is from state 1 to state 3, and for this change of state (*8.2*) gives W_{ideal}. Neglecting kinetic- and potential-energy terms we have

$$W_{\text{ideal}} = T_0\,\Delta S - \Delta H = (65+460)(0.1116-1.7589) - (58.1-1473.6)$$

$$= 550.7(\text{Btu})/(\text{lb}_\text{m})$$

The lost work quantities are given by (*8.5*). For the turbine:

$$W_{\text{lost}} = T_0(S_2-S_1) - Q = (525)(1.9438-1.7589) - 0 = 97.1(\text{Btu})/(\text{lb}_\text{m})$$

For the condenser:

$$W_{\text{lost}} = T_0(S_3-S_2) - Q$$

However, the energy equation for the condenser is $Q = \Delta H = H_3 - H_2$; therefore

$$W_{\text{lost}} = T_0(S_3-S_2) - (H_3-H_2) = 525(0.1116-1.9438) - (58.1-1111.7)$$

$$= 91.7(\text{Btu})/(\text{lb}_\text{m})$$

The actual work of the process on a pound-mass basis is given by the energy equation (*6.11*):

$$W_s = -\Delta H = -(H_2-H_1) = -(1111.7-1473.6) = 361.9(\text{Btu})/(\text{lb}_\text{m})$$

Since this is a work-producing process, our analysis is done with respect to (*8.7b*),

$$W_{\text{ideal}} = W_s + \sum W_{\text{lost}}$$

and is shown in the following table.

	$(\text{Btu})/(\text{lb}_\text{m})$	% of W_{ideal}
W_s, the actual work	361.9	65.72
W_{lost} in the turbine	97.1	17.63
W_{lost} in the condenser	91.7	16.65
Total or W_{ideal}	550.7	100.00

These results show that of the work which could theoretically or ideally be obtained for the given change of state 17.63% was unrealized because of irreversibilities in the turbine and 16.65% was unrealized because of irreversibilities in the condenser. The actual work represents 65.72% of the ideal, and this is η_t, the thermodynamic efficiency of the process. This efficiency is quite different from the turbine efficiency $\eta = 77.26\%$ determined in Problem 6.32.

Solved Problems

8.1. Compare (*8.2*) with the mechanical-energy balance (*6.17*), and combine the result with (*8.4*) to show the relation of ΣF in (*6.17*) to the lost work.

The two equations to be compared are

$$W_{\text{ideal}} = T_0\,\Delta S - \Delta H - \frac{\Delta u^2}{2g_c} - \Delta z\left(\frac{g}{g_c}\right) \tag{8.2}$$

$$-W_s = \int_1^2 V\,dP + \frac{\Delta u^2}{2g_c} + \Delta z\left(\frac{g}{g_c}\right) + \sum F \tag{6.17}$$

Addition gives, in view of (*8.4*),

$$W_{\text{lost}} = T_0\,\Delta S - \Delta H + \int_1^2 V\,dP + \sum F \tag{1}$$

However, the property relation (*3.48*) requires that $dH = T\,dS + V\,dP$ and integration yields

$$\Delta H = \int_1^2 T\,dS + \int_1^2 V\,dP$$

Combining this with (*1*) we have

$$W_{\text{lost}} = T_0\,\Delta S - \int_1^2 T\,dS + \sum F \tag{2}$$

In our formal treatment of ideal work and lost work, we assumed that T_0 was fixed at a constant value in any particular application. However, the selection of a surroundings temperature is inevitably a matter of judgment and we could have considered the more general case of arbitrary "surroundings" temperatures. In particular we could consider an imaginary surroundings made up of an infinite number of heat reservoirs, one for every temperature traversed by the system. Heat transfer between system and "surroundings" could then always be reversible, and T_0 would be variable and equal to T. The $T_0\,\Delta S$ terms in our equations would become $\int_1^2 T\,dS$, and (*2*) would reduce to

$$W_{\text{lost}} = \sum F$$

This provides an interpretation of the fluid-friction term as the lost work with respect to the *system* temperature, rather than with respect to a real surroundings temperature. What this does in effect is to leave out of the lost-work concept all work lost as the result of direct heat transfer between system and surroundings. It is a lost-work concept that takes into account only irreversibilities internal to the system.

8.2. If $T_0 = 20(°\text{C}) = 293(\text{K})$, what is W_{lost} for the process described in Problem 6.13? In addition, determine η_t.

By (*8.5*), since the process is adiabatic

$$W_{\text{lost}} = T_0\,\Delta S$$

The entropy change of an ideal gas is given by (*2.10*), which with C_P constant becomes

$$\Delta S = C_P \ln\frac{T_2}{T_1} - R\ln\frac{P_2}{P_1}$$

Substitution of the numerical values given in Problem 6.13 gives

$$\begin{aligned}\Delta S &= 5(\text{cal})/(\text{g mole})(\text{K}) \times \ln\frac{646}{293} - 1.987(\text{cal})/(\text{g mole})(\text{K}) \times \ln\frac{5}{1} \\ &= 0.7552(\text{cal})/(\text{g mole})(\text{K})\end{aligned}$$

Therefore $\quad W_{\text{lost}} = 293(\text{K}) \times 0.7552(\text{cal})/(\text{g mole})(\text{K}) = 221.3(\text{cal})/(\text{g mole})$

The actual work is

$$W_s = -C_P \Delta T = -5(\text{cal})/(\text{g mole})(\text{K}) \times 353(\text{K}) = -1765(\text{cal})/(\text{g mole})$$

By (8.4), $W_{\text{ideal}} = W_s + W_{\text{lost}}$. Therefore

$$W_{\text{ideal}} = -1765 + 221.3 = -1543.7(\text{cal})/(\text{g mole})$$

and

$$\eta_t = \frac{W_{\text{ideal}}}{W_s} = \frac{-1543.7}{-1765} = 0.875$$

8.3. The process described in Problem 6.19 was shown to satisfy the laws of thermodynamics and therefore to be possible. We can get an idea of the efficiency of such a device by comparing the lost work of the process with the work required to produce the compressed nitrogen used to operate it. The minimum work required to generate nitrogen at 6(atm) and 21(°C) is the work of reversible, isothermal compression of nitrogen from 1(atm) to 6(atm) at 21(°C). Compare this minimum work requirement with the lost work of the device, taking $T_0 = 21(°\text{C}) = 294(\text{K})$.

The work of isothermal, reversible compression in a flow process is given by (6.17). Neglecting the potential- and kinetic-energy terms and setting $\Sigma F = 0$ because the process is reversible, we get simply

$$-W_s = \int_1^2 V\,dP$$

As in Problem 6.19 we assume nitrogen behaves as an ideal gas, and in this case

$$W_s = -\int_1^2 \frac{RT}{P}\,dP = -RT \ln\frac{P_2}{P_1} = -1.987 \times 294 \times \ln 6 = -1046.7(\text{cal})/(\text{g mole})$$

The lost work of the device is most easily calculated from (8.6), $W_{\text{lost}} = T_0\,\Delta S_{\text{total}}$. Since there is no heat exchange with the surroundings, ΔS_{total} is just the entropy change of the nitrogen flowing through the device, and this was determined in Problem 6.19 to be 6.81(cal)/(K) on the basis of one gram mole of *each* exit steam. If we change the basis to one gram mole of *entering* nitrogen then $\Delta S_{\text{total}} = 3.405(\text{cal})/(\text{g mole})(\text{K})$ and

$$W_{\text{lost}} = (294)(3.405) = 1001.1(\text{cal})/(\text{g mole})$$

Thus, of the 1046.7(cal)/(g mole) of work required as a minimum to compress the nitrogen, 1001.1(cal)/(g mole) is lost because of the irreversibilities of the device, and only 45.6(cal)/(g mole) or 4.36% of the work is effective in bringing about the separation into two streams at different temperatures. The device is clearly highly inefficient. (Such devices do exist, and are known as Hilsch tubes or Ranque-Hilsch tubes.)

8.4. Make a thermodynamic analysis of the process described in Problem 6.14. Take $T_0 = 80(°\text{F}) = 540(\text{R})$.

The overall process results in the compression of methane from the initial conditions $T_1 = 80(°\text{F})$ and $P_1 = 100(\text{psia})$ to the final conditions $T_3 = 100(°\text{F})$ and $P_3 = 500(\text{psia})$. For this change of state the data of Problem 6.14 give

$$\Delta H = 408.0 - 407.0 = 1.0(\text{Btu})/(\text{lb}_\text{m})$$

$$\Delta S = 1.257 - 1.450 = -0.193(\text{Btu})/(\text{lb}_\text{m})(\text{R})$$

By (8.2)

$$W_{\text{ideal}} = T_0\,\Delta S - \Delta H = 540(-0.193) - 1.0 = -105.2(\text{Btu})/(\text{lb}_\text{m})$$

The lost-work terms are calculated from (8.5), $W_{\text{lost}} = T_0\,\Delta S - \Sigma Q$. For the compressor

$$W_{\text{lost}} = 540(1.490 - 1.450) - 0 = 21.6(\text{Btu})/(\text{lb}_\text{m})$$

For the cooler, assuming the heat removed is discarded to the surroundings, we have

$$Q = H_3 - H_2 = 408.0 - 569.4 = -161.4(\text{Btu})/(\text{lb}_\text{m})$$

and $$W_{\text{lost}} = 540(1.257 - 1.490) - (-161.4) = 35.6(\text{Btu})/(\text{lb}_{\text{m}})$$

The thermodynamic analysis follows from *(8.7a)* and is given by the following table.

	$(\text{Btu})/(\text{lb}_{\text{m}})$	%
W_{ideal}	−105.2	64.8 $(=\eta_t)$
$-W_{\text{lost}}$, compressor	− 21.6	13.3
cooler	− 35.6	21.9
W_s	−162.4	100.0

The value of W_s agrees exactly with the value calculated in Problem 6.14.

8.5. Make a thermodynamic analysis of the process described in Problem 6.24. Take $T_0 = 27(°\text{C}) = 300(\text{K})$.

The ideal work of the process is found from *(8.2)* which in this case becomes

$$W_{\text{ideal}} = T_0\,\Delta S - \Delta H - \frac{\Delta u^2}{2g_c}$$

Although the potential-energy term has been taken to be negligible, the kinetic-energy term must be retained. Since the initial and final states are both saturated steam at 1(bar), both ΔS and ΔH are zero. Thus

$$W_{\text{ideal}} = -\frac{\Delta u^2}{2g_c}$$

$$\dot{W}_{\text{ideal}} = -\dot{m}\frac{\Delta u^2}{2g_c} = -\frac{2.5(\text{kg})/(\text{s}) \times 600^2(\text{m})^2/(\text{s})^2}{2 \times 1(\text{kg})(\text{m})/(\text{N})(\text{s})^2 \times 1000(\text{N-m})/(\text{s})(\text{kW})} = -450(\text{kW})$$

The lost-work quantities are found from *(8.5)*,

$$W_{\text{lost}} = T_0\,\Delta S - \sum Q$$

and we wish to apply this equation to the compressor and to the nozzle. We must therefore determine the intermediate state of the steam at the compressor discharge or nozzle inlet. The energy equation for the compressor is *(6.10)*, which reduces to

$$\dot{m}\,\Delta H = \dot{Q} - \dot{W}_s$$

The rate of heat transfer is given in the statement of Problem 6.24 as $\dot{Q} = -150(\text{kJ})/(\text{s})$ or $-150(\text{kW})$; the work $\dot{W}_s$ is given by the answer to Problem 6.24 as $\dot{W}_s = -600(\text{kW})$. Thus

$$2.5(\text{kg})/(\text{s}) \times \Delta H(\text{kJ})/(\text{kg}) = -150 + 600 = 450(\text{kW})$$

and $$\Delta H = H_2 - H_1 = 180.0(\text{kJ})/(\text{kg}) = 180.0(\text{J})/(\text{g})$$

In addition we have from the steam tables for saturated steam at 1(bar) that

$$H_1 = H_3 = 2675.5(\text{J})/(\text{g}) \qquad S_1 = S_3 = 7.3594(\text{J})/(\text{g})(\text{K})$$

Therefore $$H_2 - 2675.5 = 180.0 \quad \text{or} \quad H_2 = 2855.5(\text{J})/(\text{g})$$

We find from the vapor tables that steam has this enthalpy at 3(bar) at a temperature of 195.12(°C), and that $S_2 = 7.2900(\text{J})/(\text{g})(\text{K})$. The lost work for the compressor is therefore

$$\dot{W}_{\text{lost}} = 300(\text{K}) \times 2.5(\text{kg})/(\text{s}) \times [7.2900 - 7.3594](\text{kJ})/(\text{kg})(\text{K}) + 150(\text{kJ})/(\text{s})$$
$$= 98.0(\text{kJ})/(\text{s}) = 98.0(\text{kW})$$

For the nozzle

$$\dot{W}_{\text{lost}} = 300 \times 2.5 \times [7.3594 - 7.2900] = 52.0(\text{kW})$$

The thermodynamic analysis is made with reference to (*8.7a*): $\dot{W}_s = \dot{W}_{ideal} - \Sigma \dot{W}_{lost}$.

	(kW)	%
$\dot{W}_{ideal}$	−450	75.0 ($=\eta_t$)
$-\dot{W}_{lost}$, compressor	−98	16.3
nozzle	−52	8.7
$\dot{W}_s$	−600	100.0

8.6. A standard refrigeration cycle is to be used in the production of a continuous supply of chilled water at a temperature of 50(°F) and at a rate of 3000(lb_m)/(min). The process is shown in Fig. 8-3. The refrigerant circulating in the cycle is H_2O. Saturated steam at 0.12(psia) at point 1 enters an adiabatic compressor of 80% efficiency based on isentropic operation, and is compressed to 1(psia) at point 2, from which point it enters a condenser. It emerges at point 3 as a saturated liquid at 1(psia), heat Q having been discharged to the surroundings at 80(°F). The saturated liquid flashes through a valve, which acts as a throttle to reduce the pressure to 0.12(psia) at point 4. The remaining liquid is vaporized in the evaporator, producing saturated vapor at point 1. The water to be chilled also passes through the evaporator and by heat exchange with the evaporating refrigerant is cooled from its feed temperature of 80(°F) to 50(°F).

Determine the power requirement of the compressor, and the rate of heat rejection to the surroundings. Make a thermodynamic analysis of the process.

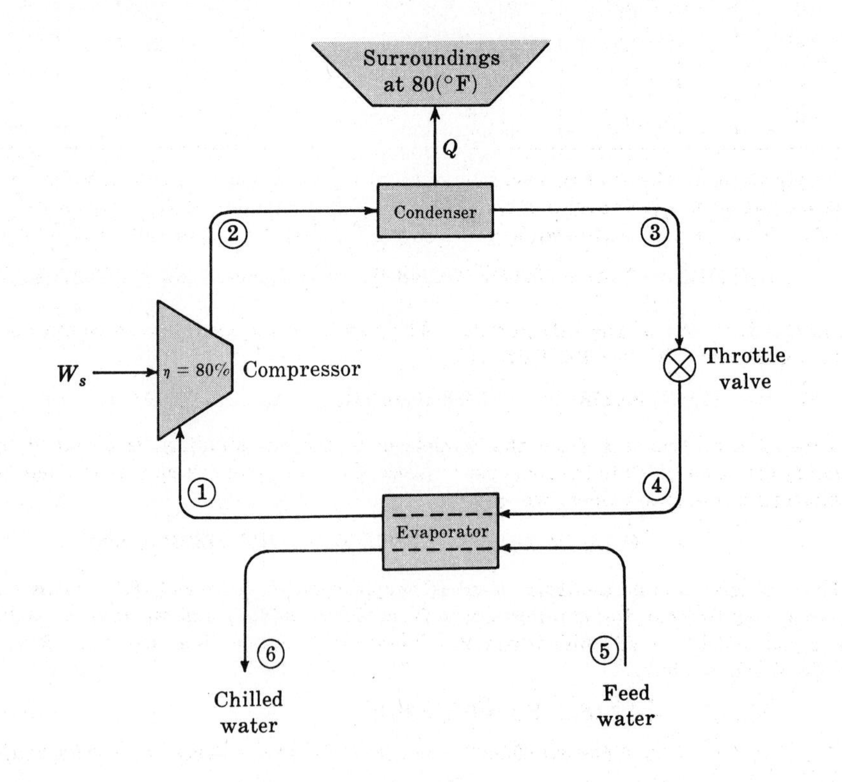

Fig. 8-3

First we find the thermodynamic properties at the various points indicated on the diagram. Data for points 1, 3, 5 and 6 can be found directly from the steam tables, and are listed in the table below. Calculations are required only for points 2 and 4. Consider first the compressor. The work W_s is equal to the isentropic work divided by the efficiency $\eta = 0.8$ (see Fig. 6-10). From the steam tables we find

$$W_s = \frac{W_s(\text{rev})}{\eta} = \frac{-(\Delta H)_S}{\eta} = \frac{-(H_2' - H_1)}{\eta} = \frac{-(1231.9 - 1078.8)}{0.8}$$

$$= -191.4(\text{Btu})/(\text{lb}_\text{m})$$

Since $W_s = -\Delta H = -(H_2 - H_1) = -(H_2 - 1078.8) = -191.4$, we have

$$H_2 = 1078.8 + 191.4 = 1270.2(\text{Btu})/(\text{lb}_\text{m})$$

and from the steam tables for vapor at 1(psia)

$$T_2 = 461.06(°\text{F}) \quad \text{and} \quad S_2 = 2.2040(\text{Btu})/(\text{lb}_\text{m})(\text{R})$$

For the throttle valve, expansion from 1.00 to 0.12(psia) occurs at constant enthalpy, and

$$H_4 = H_3 = 69.74 = H_{f_4} + x_4 H_{fg_4} = 7.67 + x_4(1071.1)$$

Solving for x_4, we find the quality (fraction vapor) to be $x_4 = 0.05795$. Finally,

$$S_4 = S_{f_4} + x_4 S_{fg_4} = 0.01547 + 0.05795(2.1449) = 0.13977(\text{Btu})/(\text{lb}_\text{m})(\text{R})$$

All computed property values are listed in the following table.

Point	State	T (°F)	P (psia)	H (Btu)/(lb$_\text{m}$)	S (Btu)/(lb$_\text{m}$)(R)
1	sat. vapor	39.65	0.12	1078.8	2.1604
2	vapor	461.06	1.00	1270.2	2.2040
3	sat. liquid	101.70	1.00	69.74	0.13266
4	wet vapor	39.65	0.12	69.74	0.13977
5	liquid	80.00	atmospheric	48.09	0.09332
6	liquid	50.00	atmospheric	18.06	0.03607

Application of the energy equation (*6.9*) to the evaporator allows calculation of the flow rate of the refrigerant. This equation reduces to $(\dot{m}_1 H_1 - \dot{m}_4 H_4) + (\dot{m}_6 H_6 - \dot{m}_5 H_5) = 0$. Since $\dot{m}_4 = \dot{m}_1$ and $\dot{m}_5 = \dot{m}_6$, we can write $\dot{m}_1(H_1 - H_4) + \dot{m}_6(H_6 - H_5) = 0$. Substitution of numerical values gives

$$\dot{m}_1(1078.8 - 69.74) + 3000(18.06 - 48.09) = 0 \quad \text{or} \quad \dot{m}_1 = 89.278(\text{lb}_\text{m})/(\text{min})$$

This is the flow rate of the refrigerant. The power requirement of the compressor is the product of this figure and $W_s = -191.4(\text{Btu})/(\text{lb}_\text{m})$:

$$\dot{W}_s = -(191.4)(89.278) = -17{,}088(\text{Btu})/(\text{min}) \quad \text{or} \quad 300.5(\text{kW}) \quad \text{or} \quad 402.9(\text{HP})$$

The rate of heat transfer from the condenser to the surroundings is obtained by application of the energy equation (*6.10*) to the condenser. Since $\dot{m}_2 = \dot{m}_3 = \dot{m}_1$, it may be written $(H_3 - H_2)\dot{m}_1 = \dot{Q}$. Substituting numerical values, we get

$$\dot{Q} = (69.74 - 1270.2)(89.278) = -107{,}178(\text{Btu})/(\text{min})$$

The thermodynamic analysis requires application of (*8.2*) and (*8.5*). From (*8.2*) we get the ideal-work requirement for chilling water from 80 to 50(°F), and we take as a basis for making the analysis 3000(lb$_\text{m}$) of chilled water. Neglecting kinetic and potential energies, and taking $T_0 = 539.67(\text{R})$, we have:

$$\dot{W}_{\text{ideal}} = \dot{m}_6[T_0(S_6 - S_5) - (H_6 - H_5)]$$

$$= 3000[539.67(0.03607 - 0.09332) - (18.06 - 48.09)] = -2598(\text{Btu})/(\text{min})$$

The lost work for each part of the process is given by (*8.5*):

$$W_{lost} = T_0 \Delta S - \sum Q$$

For the compressor, since $Q = 0$:

$$\dot{W}_{lost} = \dot{m}_1[T_0(S_2 - S_1)]$$

$$= 89.278[539.67(2.2040 - 2.1604)] = 2101(\text{Btu})/(\text{min})$$

For the condenser:

$$\dot{W}_{lost} = \dot{m}_2[T_0(S_3 - S_2)] - \dot{Q}$$

$$= 89.278[539.67(0.13266 - 2.2040)] - (-107{,}178) = 7380(\text{Btu})/(\text{min})$$

For the throttle valve, with $Q = 0$,

$$\dot{W}_{lost} = \dot{m}_3[T_0(S_4 - S_3)]$$

$$= 89.278[539.67(0.13977 - 0.13266)] = 342(\text{Btu})/(\text{min})$$

For the evaporator, Q is again zero, since Q always represents heat transfer to the surroundings, and

$$\dot{W}_{lost} = T_0[\dot{m}_4(S_1 - S_4) + \dot{m}_6(S_6 - S_5)]$$

$$= 539.67[89.278(2.1604 - 0.13977) + 3000(0.03607 - 0.09332)]$$

$$= 4667(\text{Btu})/(\text{min})$$

Since this is a work-requiring process, the thermodynamic analysis is based on (*8.7a*),

$$\dot{W}_s = \dot{W}_{ideal} - \sum \dot{W}_{lost}$$

and is represented by the following table.

	(Btu)/(min)	%
$\dot{W}_{ideal}$	−2598	15.2 ($= \eta_t$)
$-\dot{W}_{lost}$, compressor	−2101	12.3
condenser	−7380	43.2
throttle valve	−342	2.0
evaporator	−4667	27.3
$\dot{W}_s$	−17,088	100.0

The thermodynamic efficiency η_t of this process is seen to be 15.2%. The mechanical irreversibilities in the compressor result in the loss (or waste) of 12.3% of the actual work of the process. The irreversibilities in both the condenser and the evaporator resulting from heat transfer across finite temperature differences cause large lost-work terms. In fact these lost-work terms together amount to over 70% of the actual work. Surprisingly, the mechanical irreversibilities of the throttling process produce a very small lost-work term.

A thermodynamic analysis allocates the losses in a process according to the irreversibilities of the process, but it tells nothing about how to eliminate them. This is a matter which requires creative engineering.

8.7. The basic stationary-power-plant cycle is shown in Fig. 8-4. The source of energy is a nuclear reactor, and the diagram therefore depicts what is called a nuclear power plant. It does not differ basically from a fossil-fuel power plant, except in the source of energy. Heat q flows from the nuclear reactor to the boiler to produce saturated steam at point 1 at a pressure of 1050(psia). This steam flows to an adiabatic turbine which exhausts at 1(psia) at point 2. The turbine efficiency is 70%

compared with isentropic operation. The exhaust steam enters a condenser, from which it emerges at point 3 as liquid water, slightly subcooled to 100(°F). Heat Q is discarded from the condenser to the surroundings at 70(°F). The liquid from the condenser is fed to an adiabatic pump of 80% efficiency, which raises its pressure at point 4 to that of the boiler, 1050(psia). The plant has a rated capacity of 750,000(kW).

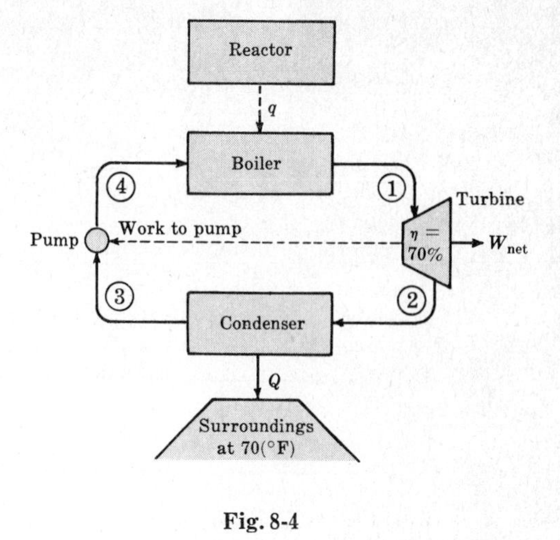

Fig. 8-4

Determine $\dot{Q}$ and the steam rate when the plant is operating at rated capacity, and make a thermodynamic analysis of the process. Treat the nuclear reactor as a heat reservoir at 600(°F).

We first determine the property values of the circulating steam at the various points of the cycle. At point 1 there is saturated steam at 1050(psia), and its properties can be read directly from the steam tables. These are entered in the table below. The steam at point 2 will be wet, and to find its properties we first assume an isentropic expansion of steam from point 1 to point 2 (see Fig. 6-8). In this case

$$S_2' = S_1 = 1.3844(\text{Btu})/(\text{lb}_\text{m})(\text{R})$$

Since the steam is wet,

$$S_2' = S_{f_2} + x_2' S_{fg_2} = 0.1327 + x_2'(1.8453) = 1.3844$$

from which the quality is $x_2' = 0.6783$. The enthalpy is then

$$H_2' = H_{f_2} + x_2' H_{fg_2} = 69.7 + (0.6783)(1036.0) = 772.4(\text{Btu})/(\text{lb}_\text{m})$$

Thus $$(\Delta H)_S = 772.4 - 1190.4 = -418.0(\text{Btu})/(\text{lb}_\text{m})$$

The work of the turbine is given by

$$W_s = \eta W_s(\text{rev}) = -\eta(\Delta H)_S = -0.70(-418.0) = 292.6(\text{Btu})/(\text{lb}_\text{m})$$

Since $$W_s = -\Delta H = -(H_2 - H_1) = -(H_2 - 1190.4) = 292.6$$

we find $H_2 = 897.8(\text{Btu})/(\text{lb}_\text{m})$. The quality is given by

$$H_2 = 897.8 = 69.7 + x_2(1036.0) \qquad \text{or} \qquad x_2 = 0.7993$$

and the entropy is calculated from this quality as

$$S_2 = 0.1327 + (0.7993)(1.8453) = 1.6076(\text{Btu})/(\text{lb}_\text{m})(\text{R})$$

These computed properties are also listed in the table below.

At point 3 we have a slightly subcooled liquid at 100(°F) and 1(psia). At 100(°F) the pressure of *saturated* liquid is about 0.95(psia). Since the effect of pressure on liquid properties is small, liquid at 100(°F) has almost exactly the same properties at 1(psia) as at the saturation pressure of 0.95(psia), and we therefore take the properties at point 3 to be those of saturated liquid at 100(°F).

The liquid at point 4 is compressed (or subcooled) liquid at 1050(psia). Data for such liquids are given in the steam tables under the heading "liquid." If we initially assume the pump to compress the liquid isentropically from point 3 to point 4, then

$$S_4' = S_3 = 0.1296(\text{Btu})/(\text{lb}_\text{m})(\text{R})$$

and we must find the point in the tables for liquid where the entropy has this value at 1050(psia). (See Problem 6.22.) After considerable interpolation, we find the enthalpy at this point to be

$H_4' = 71.2(\text{Btu})/(\text{lb}_m)$. Thus $(\Delta H)_S = 71.2 - 68.1 = 3.1(\text{Btu})/(\text{lb}_m)$ and

$$W_{\text{pump}} = \frac{W_{\text{pump}}(\text{rev})}{\eta} = \frac{-(\Delta H)_S}{\eta} = \frac{-3.1}{0.8} = -3.9(\text{Btu})/(\text{lb}_m)$$

Since $W_{\text{pump}} = -\Delta H = -(H_4 - H_3)$,

$$H_4 = H_3 - W_{\text{pump}} = 68.1 - (-3.9) = 72.0(\text{Btu})/(\text{lb}_m)$$

Further interpolation in the tables for liquid at 1050(psia) gives

$$T_4 = 101.2(°\text{F}) \quad \text{and} \quad S_4 = 0.1310(\text{Btu})/(\text{lb}_m)(\text{R})$$

All property values are summarized in the following table.

Point	State	T (°F)	P (psia)	H (Btu)/(lb$_m$)	S (Btu)/(lb$_m$)(R)
1	sat. vapor	550.7	1050	1190.4	1.3844
2	wet vapor	101.7	1	897.8	1.6076
3	liquid	100.0	1	68.1	0.1296
4	liquid	101.2	1050	72.0	0.1310

At the rated capacity of 750,000(kW), the net work rate is

$$\dot{W}_{\text{net}} = 750{,}000(\text{kW}) \times 56.8699(\text{Btu})/(\text{min})(\text{kW}) = 42{,}652{,}400(\text{Btu})/(\text{min})$$

Since $W_{\text{net}} = W_s + W_{\text{pump}} = 292.6 - 3.9 = 288.7(\text{Btu})/(\text{lb}_m)$, the steam rate is given by

$$\dot{m} = \frac{\dot{W}_{\text{net}}}{W_{\text{net}}} = \frac{42{,}652{,}400(\text{Btu})/(\text{min})}{288.7(\text{Btu})/(\text{lb}_m)} = 147{,}740(\text{lb}_m)/(\text{min})$$

or about 74(ton) per minute.

Energy equations written for the boiler and for the condenser allow determination of $\dot{q}$ and $\dot{Q}$:

$$\dot{q} = \dot{m}(H_1 - H_4) = (147{,}740)(1190.4 - 72.0) = 165{,}232{,}000(\text{Btu})/(\text{min})$$

$$\dot{Q} = \dot{m}(H_3 - H_2) = (147{,}740)(68.1 - 897.8) = -122{,}579{,}600(\text{Btu})/(\text{min})$$

Of course,

$$\dot{W}_{\text{net}} = \dot{q} + \dot{Q} = 165{,}232{,}000 - 122{,}579{,}600 = 42{,}652{,}400(\text{Btu})/(\text{min})$$

The thermodynamic analysis of the plant follows from (*8.7b*): $\dot{W}_{\text{ideal}} = \dot{W}_{\text{net}} + \Sigma \dot{W}_{\text{lost}}$. The simplest means for calculation of $\dot{W}_{\text{ideal}}$ for this case is the Carnot equation (*2.4a*), which becomes in the present notation

$$\dot{W}_{\text{ideal}} = \dot{q}\left(1 - \frac{T_0}{T}\right)$$

where T is the reactor temperature. With

$$T = 600 + 459.67 = 1059.67(\text{R}) \quad \text{and} \quad T_0 = 70 + 459.67 = 529.67(\text{R})$$

we get

$$\dot{W}_{\text{ideal}} = (165{,}232{,}000)\left(1 - \frac{529.67}{1059.67}\right) = 82{,}641{,}700(\text{Btu})/(\text{min})$$

The lost-work terms are given by (*8.5*): $W_{\text{lost}} = T_0 \Delta S - \Sigma Q$. For the turbine, $Q = 0$, and

$$\dot{W}_{\text{lost}} = \dot{m}T_0(S_2 - S_1) = (147{,}740)(529.67)(1.6076 - 1.3844) = 17{,}466{,}200(\text{Btu})/(\text{min})$$

For the condenser

$$\dot{W}_{\text{lost}} = \dot{m}T_0(S_3 - S_2) - \dot{Q}$$

$$= (147{,}740)(529.67)(0.1296 - 1.6076) - (-122{,}579{,}600) = 6{,}921{,}000(\text{Btu})/(\text{min})$$

For the pump, with $Q = 0$,

$$\dot{W}_{\text{lost}} = \dot{m}T_0(S_4 - S_3) = (147{,}740)(529.67)(0.1310 - 0.1296) = 109{,}500(\text{Btu})/(\text{min})$$

For the boiler and reactor considered together, Q is also zero. In no event can q be treated as any part of ΣQ in (*8.5*). In this equation ΣQ is *always* heat transfer between the system *and the surroundings.* Since the reactor is not part of the surroundings, it must be part of the system, and heat transfer between parts of the system is not included in ΣQ. It is for this reason that the symbol q is used for the internal transfer of heat from the reactor to the boiler. The lost work associated with this process can be attributed to neither the reactor nor boiler alone, but is shared between them. Thus the boiler and reactor are treated as a single part of the process, and for it we have

$$\dot{W}_{\text{lost}} = \dot{m}T_0(S_1 - S_4) + T_0\left(\frac{-\dot{q}}{T}\right)$$

where $-\dot{q}/T$ is the rate of entropy change (decrease) of the reactor, treated as a heat reservoir at T. Thus

$$\begin{aligned}\dot{W}_{\text{lost}} &= (147{,}740)(529.67)(1.3844 - 0.1310) - (529.67)\left(\frac{165{,}232{,}000}{1059.67}\right) \\ &= 15{,}492{,}600(\text{Btu})/(\text{min})\end{aligned}$$

Analysis according to (*8.7b*) is represented by the following table.

	(Btu)/(min)	%
$\dot{W}_{\text{net}}$	42,652,400	51.6 ($=\eta_t$)
$\dot{W}_{\text{lost}}$, turbine	17,466,200	21.1
condenser	6,921,000	8.4
pump	109,500	0.1
reactor-boiler	15,492,600	18.8
$\dot{W}_{\text{ideal}}$	82,641,700	100.0

The thermodynamic efficiency of this plant η_t is 51.6%, which is very high in comparison with most commercial processes. It is quite different from the so-called thermal efficiency η of the process:

$$\eta = \dot{W}_{\text{net}}/\dot{q} = 42{,}652{,}400/165{,}232{,}000 = 0.258$$

This figure means that of the energy supplied by the reactor only 25.8% appears as work, and 74.2% is discarded as heat to the surroundings. However, even if the plant were perfect ($\Sigma\, W_{\text{lost}} = 0$), the thermal efficiency would be

$$\eta' = \dot{W}_{\text{ideal}}/\dot{q} = 82{,}641{,}700/165{,}232{,}000 = 0.500$$

and half of $\dot{q}$ would still appear as heat discarded to the surroundings.

8.8. Fresh water is produced from sea water by the process of Fig. 8-5, page 328. Sea water at 25(°C) and containing 3.45 weight % dissolved salts enters at point 1, and first flows through a heat exchanger, where its temperature is raised. It then goes to an evaporator where 50% of its water content is vaporized at a pressure of 1(bar), producing a more concentrated brine of (it will be shown) 6.67 weight % dissolved salts, which flows out of the evaporator at point 3. If the liquid in the evaporator is well mixed, its concentration of salts must be 6.67 weight % (except immediately adjacent to the sea-water inlet). Because of the boiling-point elevation this solution boils at 100.7(°C) at 1(bar), producing a slightly superheated vapor, which leaves the evaporator at point 5 and flows to a compressor, which raises the pressure to 1.6(bar). The compressor operates adiabatically and is 75% efficient compared with an isentropic process. The compressed vapor enters coils in the evaporator where it condenses at 1.6(bar) [saturation temperature, 113.32(°C)], providing heat for the evaporation process. Design conditions call for saturated liquid water to leave the coils at point 7. This condensate and the hot brine solution from the evaporator are

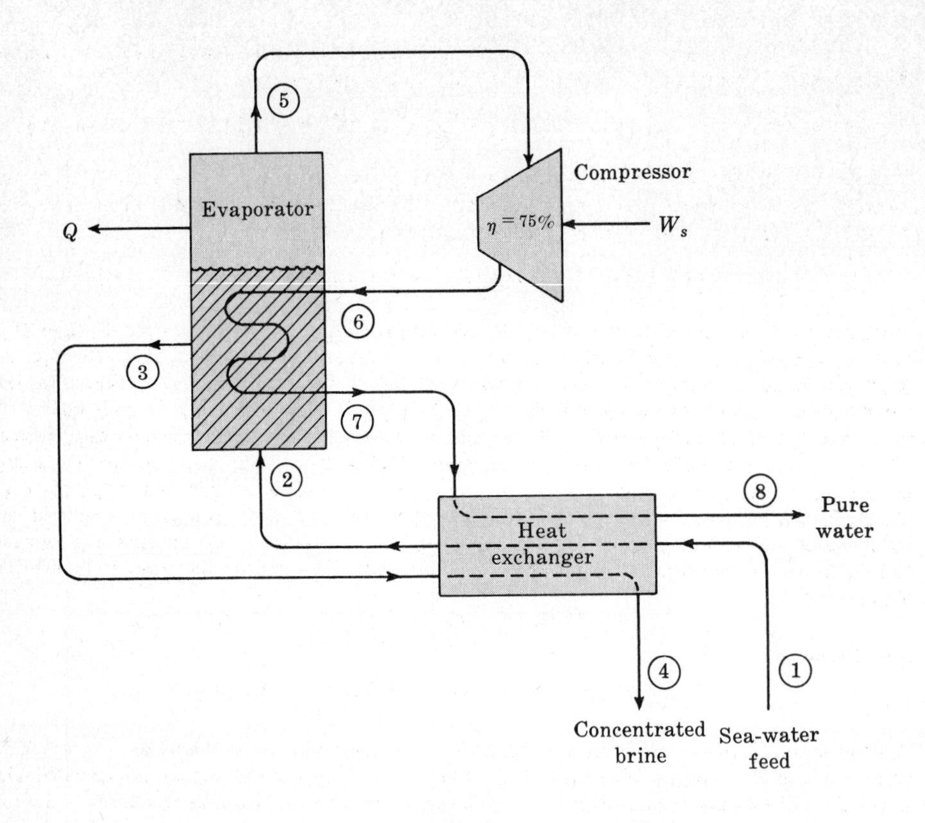

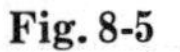
Fig. 8-5

used in the heat exchanger to preheat the incoming sea water. Heat losses to the surroundings are considered negligible from the compressor and heat exchanger, but a heat loss of 1.7(J) per gram of pure water produced is estimated for the evaporator.

Make a thermodynamic analysis of the process.

We take as a *basis* for working the problem 1(g) of pure water produced at point 8. Since this represents half of the water in the entering sea water, the mass of sea water entering is

$$m_1 = \frac{\text{mass of water entering}}{\text{fraction water in sea water}} = \frac{2(\text{g})}{1.0000 - 0.0345} = 2.0715(\text{g})$$

The mass of salt entering is therefore $2.0715 - 2.0000 = 0.0715(\text{g})$, and the amount of concentrated brine produced at point 4 is $2.0715 - 1.0000 = 1.0715(\text{g})$, with a concentration of

$$\frac{0.0715}{1.0715} \times 100 = 6.67\% \text{ dissolved salts}$$

We first calculate W_s for the compressor, since all required information is available. The actual work of the compressor is equal to the isentropic work divided by the efficiency (see Fig. 6-10). Making use of the steam tables we find

$$W_s = \frac{W_s(\text{rev})}{\eta} = \frac{-(\Delta H)_S}{\eta} = \frac{-(H_6' - H_5)}{\eta} = \frac{-(2762.2 - 2677.6)}{0.75} = -112.8(\text{J})/(\text{g})$$

Since $W_s = -\Delta H = -(H_6 - H_5) = -(H_6 - 2677.6) = -112.8(\text{J})/(\text{g})$, we get

$$H_6 = 2677.6 + 112.8 = 2790.4(\text{J})/(\text{g})$$

and from the steam tables

$$T_6 = 159.16(^\circ\text{C}) \quad \text{and} \quad S_6 = 7.4314(\text{J})/(\text{g})(\text{K})$$

These values for point 6 are tabulated along with data for the other points on the next page.

The next step is to apply the energy equation (*6.9*) to the *overall* process. Neglecting kinetic- and potential-energy terms, (*6.9*) written for unit time becomes

$$m_4H_4 + m_8H_8 - m_1H_1 = Q - W_s \tag{1}$$

Values for Q, W_s, and all masses are known. In addition, the properties at point 1 are fixed, because the conditions are fully specified. Available data for sea water give its enthalpy as $H_1 = 99.65$(J)/(g). Substitution of known values into (*1*) gives

$$(1.0715)(H_4) + (1.0000)(H_8) - (2.0715)(99.65) = -1.7 - (-112.8) \tag{2}$$

where each term has units of (J). This equation contains two unknowns, H_4 and H_8, and we must make an additional specification, if we are to determine both of them. To this end, we assume that the heat exchanger is so designed that $T_4 = T_8$. In this case, the selection of a single temperature allows values for both H_4 and H_8 to be found from available data. The procedure is to select a trial value of $T = T_4 = T_8$ and to substitute the corresponding enthalpies into (*2*). If the equation is satisfied, then the correct T was used. The temperature so found by this trial procedure is 38.4(°C), and properties for points 4 and 8 at this temperature are listed in the accompanying table. Data for all points except point 2 have now been specified either by given information or calculation. The enthalpy at point 2 is found by application of the energy equation (*6.9*) to an appropriate portion of the process. We choose to apply it to the heat exchanger. Since

$$m_4 = m_3, \qquad m_8 = m_7, \qquad \text{and} \qquad m_2 = m_1$$

(*6.9*) becomes

$$m_4(H_4 - H_3) + m_1(H_2 - H_1) + m_8(H_8 - H_7) = 0$$

The only unknown is H_2, and substitution of numerical values leads to $H_2 = 376.59$(J)/(g). Sea water has this enthalpy at 93.86(°C). The table shows these values along with the entropy for point 2. The table is now complete, and the thermodynamic analysis can be made.

Point	State	wt. % salts	T(°C)	P(bar)	m(g) per unit time	H(J)/(g)	S(J)/(g)(K)
1	liquid	3.45	25	1	2.0715	99.65	0.3342
2	liquid	3.45	93.86	1	2.0715	376.59	1.1689
3	sat. liq.	6.67	100.7	1	1.0715	388.11	1.2010
4	liquid	6.67	38.4	1	1.0715	146.21	0.4618
5	vapor	0.0	100.7	1	1.0000	2677.6	7.3651
6	vapor	0.0	159.16	1.6	1.0000	2790.4	7.4314
7	sat. liq.	0.0	113.32	1.6	1.0000	475.36	1.4550
8	liquid	0.0	38.4	1.6	1.0000	160.88	0.5512

The ideal work is given by (*8.2*). Neglecting kinetic- and potential-energy terms, and applying it to the process considered, we get $W_{\text{ideal}} = T_0(m_4S_4 + m_8S_8 - m_1S_1) - (m_4H_4 + m_8H_8 - m_1H_1)$. If T_0 is arbitrarily taken to be 298.15(K), i.e. 25(°C), this becomes

$$\begin{aligned} W_{\text{ideal}} &= 298.15(1.0715 \times 0.4618 + 0.5512 - 2.0715 \times 0.3342) \\ &\qquad - (1.0715 \times 146.21 + 160.88 - 2.0715 \times 99.65) \\ &= -5.6(\text{J}) \end{aligned}$$

The lost work is given by (*8.5*) as $W_{\text{lost}} = T_0 \Delta S - Q$. Application of this equation to the three parts of the process, the heat exchanger, the evaporator, and the compressor, gives the following results.

Heat exchanger, $Q = 0$:

$$W_{\text{lost}} = T_0[m_4(S_4 - S_3) + m_1(S_2 - S_1) + m_8(S_8 - S_7)]$$

$$= 298.15[1.0715(0.4618 - 1.2010) + 2.0715(1.1689 - 0.3342) + (0.5512 - 1.4550)]$$

$$= 9.9(J)$$

Evaporator, $Q = -1.7(J)$:

$$W_{lost} = T_0[m_5S_5 + m_3S_3 - m_2S_2 + m_6(S_7 - S_6)] - (-1.7)$$

$$= 298.15[7.3651 + 1.0715 \times 1.2010 - 2.0715 \times 1.1689 + (1.4550 - 7.4314)] + 1.7$$

$$= 77.5(J)$$

Compressor, $Q = 0$:

$$W_{lost} = T_0[m_6(S_6 - S_5)]$$

$$= 298.15[7.4314 - 7.3651] = 19.8(J)$$

Since the process is one that requires work, the thermodynamic analysis is based on (*8.7a*),

$$W_s = W_{ideal} - \sum W_{lost}$$

and is given by

	(J)/(g) of pure H_2O	%
W_{ideal}	−5.6	5.0 ($= \eta_t$)
$-W_{lost}$, heat exchanger	−9.9	8.8
evaporator	−77.5	68.7
compressor	−19.8	17.5
W_s	−112.8	100.0

The value of W_s is in agreement with that calculated earlier. If electric power costs 0.01($)/(kW-hr), then the power cost comes to just over 0.001($)/(gal) of pure water.

8.9. We wish to consider the work that can be obtained by carrying out the following reactions in steady-flow processes:

$$H_2(g) + \tfrac{1}{2}O_2(g) \rightarrow H_2O(l) \qquad (1)$$

$$H_2(g) + \tfrac{1}{2}O_2(g) \rightarrow H_2O(g) \qquad (2)$$

(*a*) What is the maximum work that can be obtained by carrying out reaction (*1*) at 1(atm) and 300(K)?

(*b*) What is the maximum work that can be obtained by carrying out reaction (*2*) with pure H_2 and *air* (21 mole % O_2 and 79 mole % N_2) at 1(atm) and 300(K) as reactants and a mixture of H_2O and N_2 gases at 1(atm) and 500(K) as products?

(*c*) If H_2 is burned completely with the theoretical amount of air, both initially at 1(atm) and 300(K), in an adiabatic reactor at constant pressure, what is the maximum work that can be obtained from the flue gases if they are brought to 500(K) at a constant pressure of 1(atm)? Where is the irreversibility in this process? What has increased in entropy and by how much? The temperature of the surroundings is 300(K).

(*d*) If the flue gases of part (*c*) are cooled to 500(K) at a constant pressure of

1(atm) by the transfer of heat to a boiler used to generate superheated steam at 500(psia) and 1000(°F), what is the work that will be obtained from a simple power-plant cycle operating as follows? A turbine ($\eta = 80\%$) expands the steam from the boiler to 2(psia). A condenser at 2(psia) delivers saturated liquid water to a pump ($\eta = 80\%$) that returns the water to the boiler. Make a thermodynamic analysis of the process. Take $T_0 = 300$(K).

Data. For reaction (*1*) at 300(K):

$$\Delta H^\circ = -68{,}323\text{(cal)/(g mole)} \qquad \Delta G^\circ = -56{,}671\text{(cal)/(g mole)}$$

For reaction (*2*) at 300(K):

$$\Delta H^\circ = -57{,}804\text{(cal)/(g mole)} \qquad \Delta G^\circ = -54{,}615\text{(cal)/(g mole)}$$

For N_2 the molar heat capacity, (cal)/(g mole)(K), is:

$$C_P^\circ = 6.66 + 1.02 \times 10^{-3}\, T\text{(K)}$$

For H_2O(g) the molar heat capacity, (cal)/(g mole)(K), is:

$$C_P^\circ = 7.17 + 2.56 \times 10^{-3}\, T\text{(K)}$$

We will assume throughout that all gases at $P = 1$(atm) are ideal. For our calculations we will need the standard entropy change of reaction, given by (*7.138*): $\Delta S^\circ = (\Delta H^\circ - \Delta G^\circ)/T$. Application of this equation gives the following results.

For reaction (*1*) at 300(K):

$$\Delta S^\circ = (-68{,}323 + 56{,}671)/300 = -38.84\text{(cal)/(g mole)(K)}$$

For reaction (*2*) at 300(K):

$$\Delta S^\circ = (-57{,}804 + 54{,}615)/300 = -10.63\text{(cal)/(g mole)(K)}$$

(*a*) Since the gases are at 1(atm) and are assumed to be ideal, the reactants and products are in their standard states at 300(K). The maximum work is the ideal work, given by (*8.2*):

$$W_{\text{ideal}} = T_0\,\Delta S - \Delta H$$

where ΔS and ΔH are the property changes when 1 mole of H_2O(l) is produced from the pure reactants of (*1*). In this case these property changes are the standard property changes of reaction. Thus

$$W_{\text{ideal}} = T_0\,\Delta S^\circ - \Delta H^\circ = (300)(-38.84) - (-68{,}323) = 56{,}671\text{(cal)}$$

This value is the negative of ΔG° because T_0 has been taken equal to T, and is the work that could be obtained if the reaction were carried out reversibly at 300(K). This would indeed be the electrical work of a reversible H_2/O_2 fuel cell operating at 300(K). Note that such a fuel cell would discharge heat to the surroundings in an amount determined by the energy equation $\Delta H = Q - W$. For the present application this becomes

$$Q = \Delta H^\circ + W_{\text{ideal}} = -68{,}323 + 56{,}671 = -11{,}652\text{(cal)}$$

(*b*) The reaction of a mole of H_2 with half a mole of O_2 in air requires

$$0.5/0.21 = 2.381\text{(mole) of air}$$

made up of 0.5(mole) of O_2 and 1.881(mole) of N_2. The reaction described is not the *standard* reaction, and the property changes ΔH and ΔS for the process are calculated from the scheme of Fig. 8-6, page 332. All states are considered ideal-gas states at 1(atm). We now determine ΔH and ΔS for the various steps of this scheme.

For the unmixing of air: Since an ideal-gas mixture is a special case of an ideal solution, the applicable equations are (*7.75*) and (*7.76*) written in terms of y_i for gas-phase mole fraction:

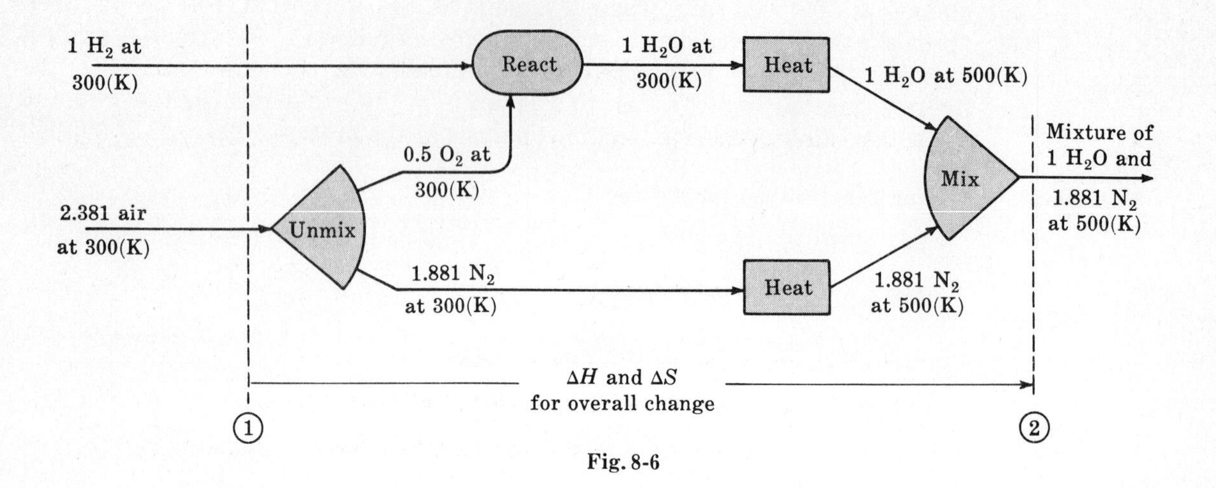

Fig. 8-6

$$\Delta H^{\text{id}} = 0 \qquad \Delta S^{\text{id}} = -R \sum y_i \ln y_i$$

These equations are for a *mixing* process. For the reverse process we merely change sign. Thus

$$\Delta H^{\text{unmix}} = 0 \qquad \Delta S^{\text{unmix}} = R \sum y_i \ln y_i$$

Substitution of numerical values in the last equation gives

$$\Delta S^{\text{unmix}} = 1.987(\text{cal})/(\text{g mole})(\text{K})\left[\frac{0.5}{2.381}\ln\left(\frac{0.5}{2.381}\right) + \frac{1.881}{2.381}\ln\left(\frac{1.881}{2.381}\right)\right]$$

$$= -1.021(\text{cal})/(\text{g mole})(\text{K})$$

This is the value for 1(g mole) of mixture; the value for 2.381(g mole) of air is

$$\Delta S^{\text{unmix}} = (2.381)(-1.021) = -2.432(\text{cal})/(\text{K})$$

and, of course, $\Delta H^{\text{unmix}} = 0$.

For the reaction: Since the reactants and products are in their standard states at 300(K), we have simply

$$\Delta H^{\text{react}} = \Delta H^\circ = -57{,}804(\text{cal}) \qquad \Delta S^{\text{react}} = \Delta S^\circ = -10.63(\text{cal})/(\text{K})$$

For the heating steps: The property changes per mole are

$$\Delta H^{\text{heat}} = \int_{300}^{500} C_P\,dT = \int_{300}^{500} (a + bT)\,dT$$

$$= a(500 - 300) + \frac{b(500^2 - 300^2)}{2}$$

$$\Delta S^{\text{heat}} = \int_{300}^{500} \frac{C_P}{T}\,dT = \int_{300}^{500} \frac{a + bT}{T}\,dT$$

$$= a \ln\left(\frac{500}{300}\right) + b(500 - 300)$$

Thus for 1(g mole) of $H_2O(g)$

$$\Delta H^{\text{heat}} = (7.17)(200) + \frac{0.00256}{2}(160{,}000) = 1639(\text{cal})$$

$$\Delta S^{\text{heat}} = 7.17 \ln\left(\frac{500}{300}\right) + (0.00256)(200) = 4.175(\text{cal})/(\text{K})$$

and for 1.881(g mole) N_2

$$\Delta H^{\text{heat}} = 1.881\left[(6.66)(200) + \frac{0.00102}{2}(160{,}000)\right] = 2659(\text{cal})$$

$$\Delta S^{\text{heat}} = 1.881\left[6.66 \ln\left(\frac{500}{300}\right) + (0.00102)(200)\right] = 6.783(\text{cal})/(\text{K})$$

For mixing the product streams: Again we are mixing ideal gases and therefore

$$\Delta H^{\text{mix}} = 0 \qquad \Delta S^{\text{mix}} = -nR \sum y_i \ln y_i$$

Thus

$$\Delta S^{\text{mix}} = -(2.881)(1.987)\left[\frac{1}{2.881}\ln\left(\frac{1}{2.881}\right) + \frac{1.881}{2.881}\ln\left(\frac{1.881}{2.881}\right)\right] = 3.696(\text{cal})/(\text{K})$$

For the entire process we sum the above values:

$$\Delta H = 0 - 57{,}804 + 1639 + 2659 + 0 = -53{,}506(\text{cal})$$

$$\Delta S = -2.432 - 10.63 + 4.175 + 6.783 + 3.696 = 1.59(\text{cal})/(\text{K})$$

Thus for the process described

$$W_{\text{ideal}} = T_0\,\Delta S - \Delta H = (300)(1.59) + 53{,}506 = 53{,}980(\text{cal})$$

(*c*) The process described involves two steps: first, adiabatic combustion to produce flue gases at a high temperature T, and second, reversible cooling to produce the maximum work from the hot flue gases as they are cooled to 500(K). These two steps can be depicted as in Fig. 8-7. The pressure is everywhere 1(atm).

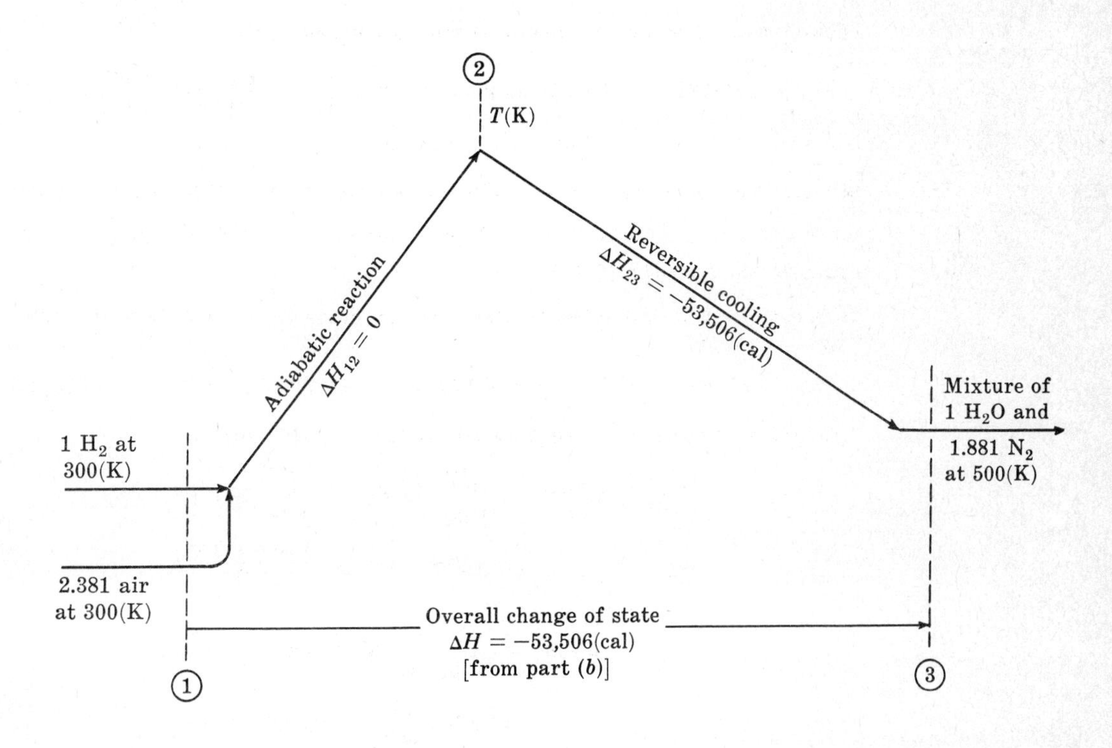

Fig. 8-7

Our first task is to determine T, the temperature resulting from complete combustion of the reactants in an adiabatic reactor. Exactly this problem was solved in Problem 7.16, and we simply make use of the answer: $T = 2522(\text{K})$. This value is needed to allow calculation of the entropy change of step $2 \rightarrow 3$. Since this step is a simple cooling process, we have

$$\Delta S_{23} = \int_{2522}^{500} \left(\sum n_i C^\circ_{P_i} \right) \frac{dT}{T}$$

Use of the data provides:

$$\sum n_i C^\circ_{P_i} = 19.70 + 4.48 \times 10^{-3} T$$

and

$$\Delta S_{23} = 19.70 \ln \left(\frac{500}{2522} \right) + (4.48 \times 10^{-3})(500 - 2522) = -40.94(\text{cal})/(\text{K})$$

Thus for the cooling process, application of (*8.2*) gives

$$W_{\text{ideal}} = 300(-40.94) - (-53{,}506) = 41{,}220(\text{cal})$$

The difference between the process considered here and that of part (*b*), for which $W_{\text{ideal}} = 53{,}980(\text{cal})$, is that the reaction itself was taken to be reversible in part (*b*). Here an adiabatic combustion process takes place, and such processes are inherently irreversible. In Fig. 8-8 we examine the entropy changes resulting from steps $1 \to 2$ and $2 \to 3$. The value of $\Delta S_{13} = 1.59(\text{cal})/(\text{K})$ comes from part (*b*), where exactly the same overall change of state was considered. Since $\Delta S_{12} + \Delta S_{23} = \Delta S_{13}$, the value for ΔS_{12} is given by

$$\Delta S_{12} = \Delta S_{13} - \Delta S_{23} = 1.59 - (-40.94) = 42.53(\text{cal})/(\text{K})$$

Since the cooling process is reversible, $\Delta S_{23} + \Delta S_{\text{surr}} = 0$ and therefore, as shown,

$$\Delta S_{\text{surr}} = 40.94(\text{cal})/(\text{K})$$

Since $\Delta S_{\text{surr}} = Q_{\text{surr}}/T_0$,

$$Q_{\text{surr}} = T_0 \Delta S_{\text{surr}} = (300)(40.94) = 12{,}280(\text{cal})$$

Thus the process results in the transfer of this quantity of heat to the surroundings, and the overall result of the process is an entropy increase in the surroundings of 40.94(cal)/(K) and an entropy increase in the system of 1.59 (cal)/(K). The total entropy increase is

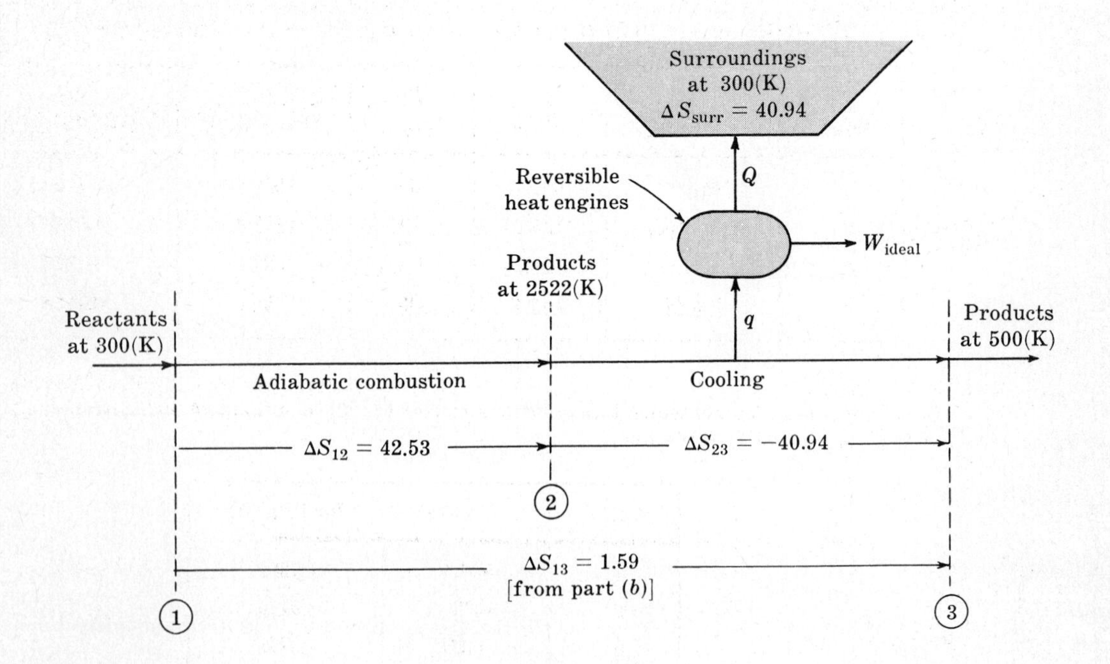

Fig. 8-8

$$\Delta S_{total} = 40.94 + 1.59 = 42.53\text{(cal)/(K)}$$

exactly the increase caused by the combustion process. Since by (8.6), $W_{lost} = T_0\,\Delta S_{total}$, the lost work of the process is

$$W_{lost} = 300(42.53) = 12{,}760\text{(cal)}$$

This is exactly the difference between the values of W_{ideal} for the process of part (b) and the present process.

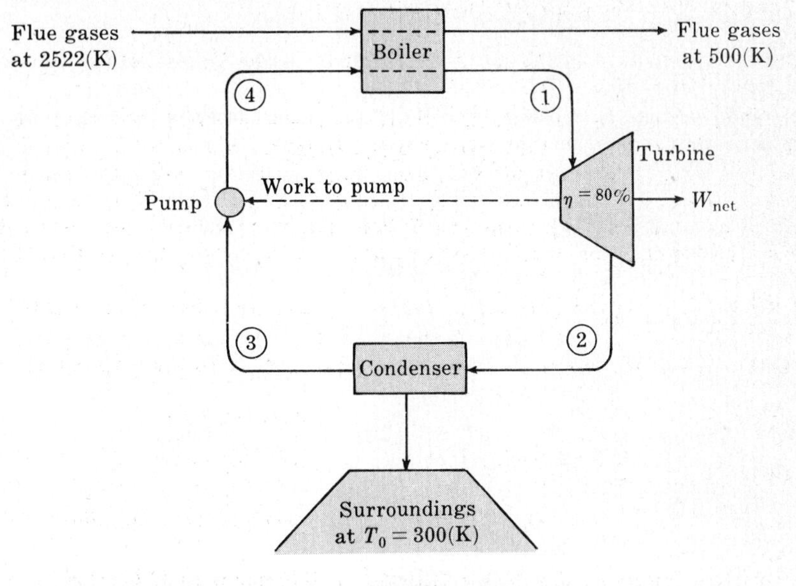

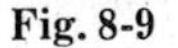

Fig. 8-9

(d) The process described is shown schematically in Fig. 8-9. It is a power-plant cycle little different from that of Problem 8.7, and property values at the various points are determined just as in that problem. Present results are summarized in the following table.

Point	State	T (°F)	P (psia)	H (Btu)/(lb$_m$)	S (Btu)/(lb$_m$)(R)
1	vapor	1000	500	1520.7	1.7371
2	wet vapor	126.0	2	1111.4	1.9118
3	sat. vapor	126.0	2	94.0	0.1750
4	liquid	127.3	500	96.5	0.1758

It is convenient for subsequent calculations to convert these enthalpies and entropies into units of (cal), (g), and (K). We find:

Point	H(cal)/(g)	S(cal)/(g)(K)
1	845.4	1.7382
2	617.9	1.9130
3	52.3	0.1751
4	53.7	0.1759

From these figures we get the total shaft work of the turbine:

$$W_s = -\Delta H = -(H_2 - H_1) = -(617.9 - 845.4) = 227.5(\text{cal})/(\text{g})$$

The pump work is

$$W_{\text{pump}} = -\Delta H = -(H_4 - H_3) = -(53.7 - 52.3) = -1.4(\text{cal})/(\text{g})$$

The net work of the cycle is then

$$W_{\text{net}} = 227.5 - 1.4 = 226.1(\text{cal})/(\text{g})$$

Finally, the heat transferred to the surroundings is

$$Q = \Delta H = H_3 - H_2 = 52.3 - 617.9 = -565.6(\text{cal})/(\text{g})$$

If we take as a basis 1(g mole) of H_2 burned to form the flue gases, then the energy equation for the boiler is

$$\Delta H_{\text{flue gases}}(\text{cal}) + m_{\text{steam}}(\text{g}) \times [H_1 - H_4](\text{cal})/(\text{g}) = 0$$

The value of $\Delta H_{\text{flue gases}}$ is just ΔH_{23} of part (*c*). Thus

$$-53{,}506(\text{cal}) + m_{\text{steam}}(\text{g}) \times [845.4 - 53.7](\text{cal})/(\text{g}) = 0 \qquad \text{or} \qquad m_{\text{steam}} = 67.58(\text{g})$$

On this basis

$$W_{\text{net}} = 226.1(\text{cal})/(\text{g}) \times 67.58(\text{g}) = 15{,}280(\text{cal})$$

$$Q = -565.6 \times 67.58 = -38{,}220(\text{cal})$$

The lost work, $W_{\text{lost}} = T_0 \Delta S - \Sigma Q$, has the following values.

Boiler: The entropy change of the flue-gas stream is ΔS_{23} of part (*c*). Thus

$$W_{\text{lost}} = 300[-40.94 + (1.7382 - 0.1759)(67.58)] + 0 = 19{,}390(\text{cal})$$

Turbine: $$W_{\text{lost}} = 300(1.9130 - 1.7382)(67.58) + 0 = 3540(\text{cal})$$

Condenser: $$W_{\text{lost}} = 300(0.1751 - 1.9130)(67.58) - (-38{,}220) = 2990(\text{cal})$$

Pump: $$W_{\text{lost}} = 300(0.1759 - 0.1751)(67.58) + 0 = 20(\text{cal})$$

The thermodynamic analysis of the process follows from (*8.7b*), which for this process is written:

$$W_{\text{ideal}} = W_{\text{net}} + \sum W_{\text{lost}}$$

The results are given by the following table.

	$\frac{(\text{cal})}{(\text{g mole } H_2 \text{ burned})}$	%
W_{net}	15,280	37.07 (= η_t)
W_{lost}, boiler	19,390	47.04
turbine	3,540	8.59
condenser	2,990	7.25
pump	20	0.05
W_{ideal} [see part (*c*)]	41,220	100.00

In summary, the reaction $H_2 + \frac{1}{2}O_2 \rightarrow H_2O$ is theoretically capable of providing work in the amount determined in part (*a*): 56,671(cal)/(g mole of H_2). However, if oxygen is supplied by air and the products of the reaction are discarded at 500(K), then the maximum possible work is reduced to that of part (*b*): 53,980(cal)/(g mole of H_2). If in addition an actual burning or combustion process is used to produce the reaction, then the maximum work is that of part (*c*): 41,220(cal)/(g mole of H_2). If beyond that one superimposes the irreversibilities of a practical power plant, then the work actually obtained is approximately the net work of part (*d*): 15,280(cal)/(g mole of H_2). Thus a conventional power plant would produce only

$$\frac{15{,}280}{56{,}671} \times 100$$

or about 27% of the work theoretically available from the reaction.

Supplementary Problems

8.10. Determine η_t for the turbine of Problem 6.23. Take $T_0 = 65(°\text{F}) = 525(\text{R})$. *Ans.* 0.7885

8.11. Determine η_t for the compressor of Problem 6.37. Take $T_0 = 70(°\text{F}) = 530(\text{R})$. *Ans.* 0.8698

8.12. Find η_t for the process of Problem 6.38. Take $T_0 = 65(°\text{F}) = 525(\text{R})$. *Ans.* 0.8484

8.13. Rework Problem 6.40 making use of (*8.2*).

8.14. What is the lost work for the process described in Example 6.4? Take $T_0 = 60(°\text{F}) = 520(\text{R})$.

Ans. $\dot{W}_{\text{lost}} = 10{,}450(\text{Btu})/(\text{min})$

8.15. Refrigeration at a temperature level of 150(R) is required for a certain process. A cycle using helium gas has been proposed to operate as follows. Helium at 1(atm) is compressed adiabatically to 5(atm), water-cooled to 60(°F), and sent to a heat exchanger where it is cooled by returning helium. From there it goes to an adiabatic expander which delivers work to be used to help drive the compressor. The helium then enters the refrigerator, where it absorbs enough heat to raise its temperature to 140(R). It returns to the compressor by way of the heat exchanger.

Helium may be considered an ideal gas with a constant molar heat capacity at constant pressure of 5(Btu)/(lb mole)(R). If the efficiencies of the compressor and expander are 80 percent and if the minimum temperature difference in the exchanger is 10(°F), at what rate must the helium be circulated to provide refrigeration at a rate of 100(Btu)/(min)? What is the net power requirement of the process? Sketch the cycle, and show the temperatures at the various points. What is the coefficient of performance of the cycle? How does it compare with the Carnot coefficient of performance? Make a thermodynamic analysis of the process. Take $T_0 = 60(°\text{F})$.

Ans. $\dot{W}_{\text{net}} = 26.0(\text{HP})$, $\omega = 0.0906$, $\omega_{\text{carnot}} = 0.405$

Thermodynamic analysis:

W_{ideal}	22.2%
W_{lost}, compressor	11.2%
cooler	35.4%
exchanger	4.9%
expander	16.6%
refrigerator	9.7%
W_{net}	100.0%

Review Questions for Chapters 5 through 8

For each of the following statements indicate whether it is true or false.

______1. The compressibility factor Z is always less than or equal to unity.

______2. For any real gas at constant temperature, as the pressure approaches zero the residual volume $\Delta V'$ approaches zero.

______3. The virial coefficients B, C, etc., of a gaseous mixture are functions of temperature and composition only.

______4. The residual enthalpy and residual entropy of a real gas approach zero as the pressure approaches zero.

______5. Three-parameter corresponding-states correlations are more useful than two-parameter correlations because they work for any compound whatever.

______6. The inversion curve of a real fluid defines the states for which the Joule-Thomson coefficient is zero.

______7. The second virial coefficient B of a binary gas mixture is in general calculable from values of B for the pure gases.

______8. The critical properties T_c, P_c, and Z_c are constants for a given compound.

______9. All real fluids become simple fluids in the limit as the pressure approaches zero.

______10. The Redlich-Kwong equation is superior to the van der Waals equation because its mixing rules are exact.

______11. A closed system is one of constant volume.

______12. A steady-state flow process is one for which the velocities of all streams may be assumed negligible.

______13. Gravitational potential-energy terms may be ignored in the steady-state energy equation if all streams entering and leaving the control volume are at the same elevation.

______14. Frictional effects are difficult to incorporate explicitly in the energy equations because such effects constitute violations of the second law of thermodynamics.

______15. In an adiabatic flow process, the entropy of the fluid must increase as the result of any irreversibilities within the system.

______16. The temperature of a gas undergoing a continuous throttling process may either increase or decrease across the throttling device, depending on conditions.

______17. The Mach number $\mathbf{M}$ is negative for a subsonic flow.

______18. When an ideal gas is compressed adiabatically in a flow process and is then cooled to the initial temperature, the heat removed in the cooler is equal to the work done by the compressor. (Assume potental- and kinetic-energy effects are negligible.)

______19. A total property M^t of a homogeneous mixture is always equal to $\Sigma n_i M_i$, where n_i is the number of moles of species i and M_i is the corresponding molar property of pure i.

______20. As $x_i \to 1$ the partial molar volume $\bar{V}_i$ of a component in solution becomes equal to V_i, the molar volume of pure i at the T and P of the solution.

______21. In the limit as $P \to 0$, the ratio f/P for a gas goes to infinity, where f is the fugacity.

______22. The fugacity coefficient ϕ has units of pressure.

______23. The residual Gibbs function $\Delta G'$ is related to ϕ by $\Delta G' = -RT \ln \phi$.

______24. For equilibrium among contacting phases, the fugacity of a given component must be the same in all phases.

______25. For an ideal solution at constant T and P, the fugacity of a component in solution is proportional to its mole fraction.

______26. The numerical value of a property change of mixing depends on the standard state chosen for each component in the solution.

______27. The heat of mixing to form a given binary solution at constant T and P increases with increasing temperature if the total heat capacity of the solution formed is greater than the total heat capacity of the pure constituents that are mixed.

______28. The entropy change of mixing at constant T and P to form a binary solution from pure constituents is equal to the heat of mixing at the same conditions divided by the absolute temperature.

______29. The use of standard states based on the Lewis and Randall rule is necessarily more realistic than the use of standard states based on Henry's law.

______30. A mixture of ideal gases is an ideal solution.

______31. All property changes of mixing are zero for an ideal solution.

______32. All excess properties are zero for an ideal solution.

______33. The activity coefficient is zero for a component in an ideal solution.

______34. The bubble-point curve of a binary VLE system represents the states of saturated vapor mixtures.

______35. At a binary azeotrope the dew-point and bubble-point curves become tangent to one another.

______36. The number of degrees of freedom for an azeotropic state in a two-component VLE system is 1.

______37. The freezing curves and melting curves are in general different in a binary liquid-solid system.

______38. Liquid-phase activity coefficients are generally less than zero for systems which exhibit negative deviations from Raoult's law.

______39. The VLE K-values for systems described by Raoult's law are true constants, independent of T, P, and composition.

______40. There are two independent reactions for a chemically reactive system containing air (21 mole % O_2, 79 mole % N_2), S(s), SO_2(g), and SO_3(g).

______41. The reaction coordinate ϵ is a quantity which characterizes only the *equilibrium* conversion of a given chemical reaction.

______42. $\Delta G_{T,P}$ for a reaction is zero if all substances participating in the reaction are in their standard states.

______43. ΔS° for a reaction is zero at a temperature for which $\Delta H^\circ = \Delta G^\circ$.

______44. The equilibrium constant K for a chemical reaction is independent of pressure.

______45. If the standard Gibbs function change of a reaction is zero, the reaction is thermodynamically impossible.

______46. The chemical equilibrium constant K increases with increasing T, provided that the standard enthalpy change of reaction ΔH° is positive.

______47. There are two degrees of freedom in a chemically reactive system containing the gaseous species N_2, H_2, and NH_3.

______48. At constant temperature, an increase in pressure will cause an increase in the yield of methanol (CH_3OH) from the ideal-gas reaction

$$CO(g) + 2H_2(g) \longrightarrow CH_3OH(g)$$

______49. W_{ideal} is the same for all steady-flow processes that produce the same change in state, provided that the temperature of the surroundings is the same.

______50. Lost work is a quantity devised to account for exceptions to the first law of thermodynamics.

Ans.

1	2	3	4	5	6	7	8	9	10	11	12	13	14	15
F	F	T	T	F	T	F	T	F	F	F	F	T	F	T

16	17	18	19	20	21	22	23	24	25	26	27	28	29	30
T	F	T	F	T	F	F	T	T	T	T	T	F	F	T

31	32	33	34	35	36	37	38	39	40	41	42	43	44	45
F	T	F	F	T	T	T	F	F	T	F	F	T	T	F

46	47	48	49	50
T	F	T	T	F

Appendix 1

Conversion Factors

For conciseness, the conversion factors given below for each quantity are referred to a single basic or derived SI or cgs unit. Conversions between other pairs of units for a given quantity are made by employing the usual rules for manipulation of units.

Example. Find the factor for converting (ft)^3 to (gal).

From the volume entries we find

$$1(\mathrm{m})^3 = 35.3147(\mathrm{ft})^3 = 264.172(\mathrm{gal})$$

from which

$$1(\mathrm{ft})^3 = \frac{264.172}{35.3147} = 7.48051(\mathrm{gal})$$

Quantity	Conversion	Quantity	Conversion
Length	1(m) = 100(cm)	Energy	1(J) = 1(kg)(m)^2/(s)^2
	= 3.28084(ft)		= 1(N-m)
	= 39.3701(in)		= 1(W-s)
Mass	1(kg) = 10^3(g)		= 10^7(dyne-cm)
	= 2.20462(lb$_m$)		= 10^7(erg)
Force	1(N) = 1(kg)(m)/(s)^2		= 10(cm^3-bar)
	= 10^5(dyne)		= 0.239006(cal)
	= 0.224809(lb$_f$)		= 9.86923(cm^3-atm)
Pressure	1(bar) = 10^5(kg)/(m)(s)^2		= 5.12197 $\times$ 10^{-3}(psia)(ft)^3
	= 10^5(N)/(m)^2		= 0.737562(ft-lb$_f$)
	= 10^6(dyne)/(cm)^2		= 9.47831 $\times$ 10^{-4}(Btu)
	= 0.986923(atm)	Power	1(kW) = 10^3(kg)(m)^2/(s)^3
	= 14.5038(psia)		= 10^3(W)
	= 750.061(mm Hg)		= 10^3(J)/(s)
Volume	1(m)^3 = 10^6(cm)^3		= 10^3(V)(A)
	= 10^3(liter)		= 239.006(cal)/(s)
	= 35.3147(ft)^3		= 737.562(ft-lb$_f$)/(s)
	= 264.172(gal)		= 56.8699(Btu)/(min)
Density	1(g)/(cm)^3 = 10^3(kg)/(m)^3		= 1.34102(HP)
	= 10^3(g)/(liter)		
	= 62.4278(lb$_m$)/(ft)^3		
	= 8.34540(lb$_m$)/(gal)		

Note: atm ≡ standard atmosphere
cal ≡ thermochemical calorie
Btu ≡ International Steam Table Btu

Appendix 2

Values of the Universal Gas Constant

$$\begin{aligned}
R &= 8.314(\mathrm{J})/(\mathrm{g\ mole})(\mathrm{K}) = 1.987(\mathrm{cal})/(\mathrm{g\ mole})(\mathrm{K}) \\
&= 83.14(\mathrm{cm^3\text{-}bar})/(\mathrm{g\ mole})(\mathrm{K}) = 82.05(\mathrm{cm^3\text{-}atm})/(\mathrm{g\ mole})(\mathrm{K}) \\
&= 0.7302(\mathrm{atm})(\mathrm{ft})^3/(\mathrm{lb\ mole})(\mathrm{R}) = 10.73(\mathrm{psia})(\mathrm{ft})^3/(\mathrm{lb\ mole})(\mathrm{R}) \\
&= 1545(\mathrm{ft\text{-}lb_f})/(\mathrm{lb\ mole})(\mathrm{R}) = 1.986(\mathrm{Btu})/(\mathrm{lb\ mole})(\mathrm{R})
\end{aligned}$$

Appendix 3

Critical Constants and Acentric Factor

Compound	T_c(K)	P_c(atm)	V_c(cm)^3/(g mole)	Z_c	ω
Argon	151	48.0	75.2	0.290	−0.002
Xenon	289.8	58.0	118.8	0.290	+0.002
Methane	190.7	45.8	99.4	0.290	0.013
Oxygen	154.8	50.1	74.4	0.293	0.021
Nitrogen	126.2	33.5	90.1	0.291	0.040
Carbon monoxide	133	34.5	93.1	0.294	0.049
Ethylene	283.1	50.5	124	0.270	0.085
Hydrogen sulfide	373.6	88.9	98	0.284	0.100
Propane	369.9	42.0	200	0.277	0.152
Acetylene	309.5	61.6	113	0.274	0.190
Cyclohexane	553.2	40	308	0.271	0.209
Benzene	562.1	48.6	260	0.274	0.211
Carbon dioxide	304.2	72.9	94.0	0.274	0.225
Ammonia	405.6	112.5	72.5	0.245	0.250
n-Pentane	469.5	33.3	311	0.269	0.252
n-Hexane	507.3	29.9	368	0.264	0.298
Acetone	509.1	47	211	0.237	0.318
Water	647.3	218.0	56.8	0.233	0.344
n-Heptane	540.3	27.0	426	0.259	0.349
n-Octane	568.6	24.6	486	0.256	0.398
Methanol	513.2	78.5	118	0.220	0.556
Ethanol	516.3	63.0	167	0.248	0.635

INDEX

The letter *e* following a page number indicates that the entry refers to an Example, and similarly the letter *p* refers to a Problem.

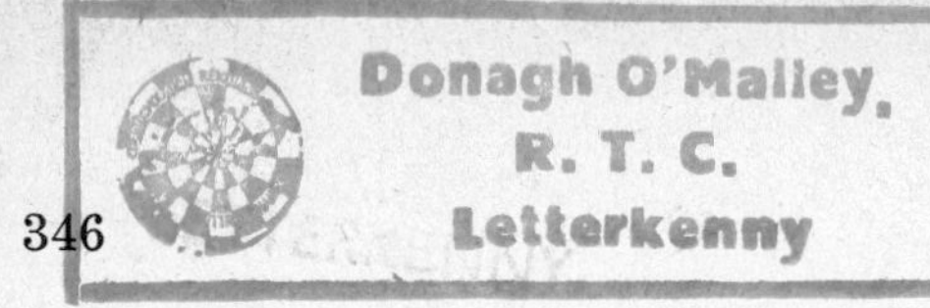